Fascinating Particle Physics

Philip Bechtle · Florian Bernlochner · Herbi Dreiner ·
Christoph Hanhart · Josef Jochum · Jörg Pretz ·
Kristin Riebe

Fascinating Particle Physics

From Quarks, Neutrinos and Higgs to the Mysteries of the Universe

Springer

Contact
fascinating_particle_physics@listen.uni-bonn.de

ISBN 978-3-662-72043-1 ISBN 978-3-662-72044-8 (eBook)
https://doi.org/10.1007/978-3-662-72044-8

The book was initially translated by an artificial intelligence machine translation tool. It was subsequently substantially revised and extended by the authors in both content and form.

Translation from the German language edition: "Faszinierende Teilchenphysik" by Philip Bechtle, Florian Bernlochner, Herbi Dreiner, Christoph Hanhart, Josef Jochum, Jörg Pretz, Kristin Riebe, © Der/die Herausgeber bzw. der/die Autor(en), exklusiv lizenziert an Springer-Verlag GmbH, DE, ein Teil von Springer Nature 2023. Published by Springer Berlin Heidelberg. All Rights Reserved.

Planing/Editorial: Lisa Edelhäuser and Caroline Strunz
Project Management: Bianca Alton
Illustrations: Authors
Typesetting and layout by the authors, based on a template by Glaeser/Polthier
Cover Illustration:
Event display (bottom): CERN, ATLAS experiment; Mexican hat potential (top): Kristin Riebe
Cover Design: deblik, Berlin

This Springer imprint is published by the registered company Springer-Verlag GmbH, DE, part of Springer Nature.
The registered company address is: Heidelberger Platz 3, 14187 Berlin, Germany

If disposing of this product, please recycle the paper.

To our families,
to our students who again and again provide inspiration for our curiosity, and to all the dedicated teachers who inspire their students for the wonders of Nature.

Preface

Where do we come from? How did everything begin, and how did it all evolve? How did the Universe come into being? What is matter made of? These are questions that have preoccupied humankind since time immemorial and that we have even partially answered, thanks to research.

People have always been fascinated by the observation of the stars and the phenomena at play in the Universe in general. Equally fascinating is "the invisible", the tiniest constituents of matter and how they interact, because it is only by investigating these building blocks of matter that we can understand the astrophysical processes. Astronomy, astrophysics and particle physics are all intertwined, and the models of particle physics are needed to explain what matter was like at the beginning of our Universe.

Particle physics deals with the elementary constituents of matter and the fundamental forces that operate between them. The goal is to further our knowledge, in particular to allow us to answer the questions mentioned above. Particle physics is fundamental research, and fundamental research is a precious cultural commodity, with an intrinsic value for society. The knowledge gained from it is key to the achievements of civilization, to technological progress and change. Fundamental research promotes critical thinking, broadens our horizons and encourages creativity and innovation. It lies at the heart of every "innovation", but here's the rub: real breakthroughs, "disruptive innovation", can't be planned, because technological challenges first have to be overcome before the research can go ahead. After the Brout-Englert-Higgs field was first predicted by theory, almost 50 years went by before it was experimentally proven and the Higgs boson was discovered. How were Brout, Englert and Higgs able to make their predictions back in 1964? Obviously, the concepts of our theories involve a deep understanding of Nature and not just a simple description of phenomena.

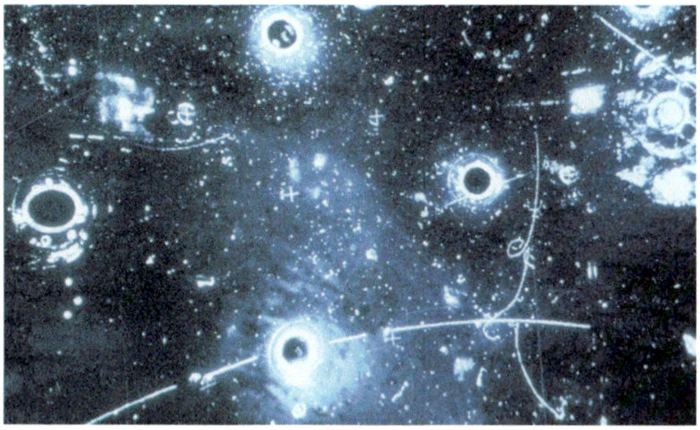

The discovery of the Higgs boson not only excited the scientific world but also captured the imagination of many non-scientists and continues to do so today, more than ten years later. Why? Because it's extremely interesting to explore, through research, the fields that form the basis of our knowledge and our existence. In so doing, we see the many connections that exist between knowledge and belief, science and philosophy. This is really exciting. A couple of years ago, when students in Great Britain were asked what had attracted them to physics, it was precisely these big, to some extent esoteric questions that they spoke of.

Fundamental research is not only a matter of "gaining knowledge" but also leads to considerable technological development. This is the principle of the virtuous circle, the positive, self-reinforcing cycle: fundamental research—driver of innovation—applied research—driver of innovation—further fundamental research. Without this virtuous circle, new approaches and solutions in fields like medicine, information technology and sustainable energy would be impossible. But all this takes time, and in some cases a great deal of time.

Back to the Higgs boson. Even after the technological challenges had been overcome, it still took several years of data taking and analyses before a signal proving the existence of the Higgs boson slowly emerged from the copious amounts of data produced by the LHC experiments. When our analysis of this data had exceeded the discovery threshold, it was clear that we could announce that we had uncovered the last building block of the Standard Model of particle physics. This success would not have been possible without years of cooperation between technicians, engineers and physicists from all over the world, from different cultures and many other disciplines. International cooperation is an important key to progress, not only in science.

But was that it? Can particle physics be considered complete? No. Theories can only ever be verified to a certain extent. Only the future will tell us whether the Standard Model is also applicable at much higher energies than we can contemplate today. As Alexander von Humboldt already knew, "Every law of Nature that reveals itself to the observer suggests a higher, as yet unrecognized one". In other words, every step forward in knowledge brings its own new puzzle for us to solve. And this is precisely what I find so fascinating about research, and about particle physics in particular. Yes, we have experimental confirmation of the Standard Model, which represents a resounding success. But many questions remain that it cannot answer. The Model does not describe even 5 % of the matter in the Universe. It's high time that we finally get to grips with the remaining 95 % and delve deeper in our exploration of the invisible. The search for answers to fundamental questions about our Universe, its origins and its evolution, is bringing scientists from different branches of research together.

Of course, we need to master the basics of physics before we attempt to tackle these big questions. This book bridges the gap between those basics and the many questions that remain open. Even if you don't have much of a maths background but would like to delve deeper into the subject, this is the book for you. It's worth getting to know more about particle physics and so I encourage you to dive in.

Particle physics *is* fascinating.

by Professor Rolf-Dieter Heuer,
CERN Director General 2009–2015

Preface to the First English Edition

This book is a translation of the original book written in German: Faszinierende Teilchenphysik (Springer 2023) by the same authors. We have translated the book ourselves with the help of Christa Börsch. It is fair to say that it is *not* a direct translation. We have taken the opportunity while going through the book to correct minor mistakes, to improve explanations, or otherwise make the book clearer and easier to read. Furthermore, in the intermediate 1.5 years there have been new developments which we have here taken into account. In the process, we have expanded some double pages to two double pages, and we have also incorporated a few new articles.

The new articles include in Chapter 1 a new path from the visible to the invisible and a new introduction to composite particles, in Chapter 5 several large images of different types of experiments, in Chapter 6 a new graphical explanation of the analysis of particle reactions, in Chapter 8 an article on the Quark-Gluon Plasma, in Chapter 11 two articles on planned future accelerators, for example to study the Higgs boson in detail, as well as on the new FASER experiment at the LHC, which has for the first time observed man-made very high energy neutrinos from collider collisions (2024). Furthermore, the original article in Chapter 10 on Grand Unified Theories (GUTs), has here been split into two articles: one just on coupling constant unification, the main motivation for GUTs, and another giving more details on the structure of GUTs.

We hope this English edition finds as much interest with readers as the German edition has.

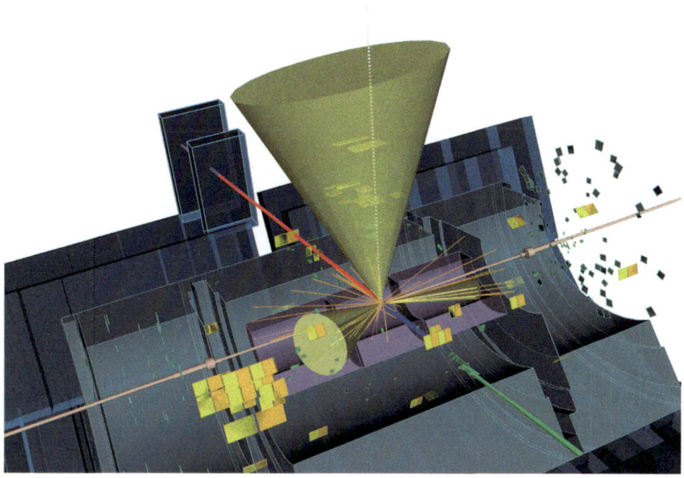

About This Book

The book you are holding in your hands presents nuclear and particle physics on about 150 self-contained double pages. It exhibits the results of our field of science and also shows the methods, possibilities, and limits of modern research.

Particle physics is characterized by a diverse and highly specialized research landscape. This book will introduce you to the subareas of physics that have emerged and are dedicated to various tasks connected to particle physics. For example, many experimental investigations need particle accelerators, that is, facilities that accelerate particles to high energy. These are developed and operated by researchers in accelerator physics in collaboration with engineers. However, an accelerator in itself does not yet provide data usable for particle physics—for this, dedicated measuring instruments, the detectors, are necessary. These are researched, built, and operated by experimental physicists. Finally, theoretical physicists are dedicated to the development of mathematical theories with which we describe and explain Nature. They attempt, on the one hand, to *capture* the information contained in the data obtained from various measurements in a mathematical expression and, on the other hand, to derive predictions and proposals for future experiments.

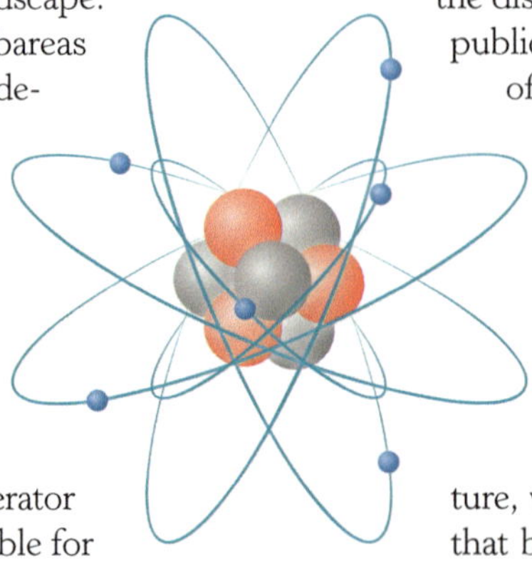

Reflecting this complexity, the team of authors consists of physicists with very different scientific backgrounds. Philip Bechtle and Florian Bernlochner work as experimental physicists in high-energy physics. Jörg Pretz has focused his career on precision measurements at low energies and is also linked to accelerator physics. Josef Jochum's research centers on (non-accelerator) experimental astroparticle physics with a main focus on neutrinos. Herbi Dreiner and Christoph Hanhart are theoretical physicists, the former focusing on high-energy physics and physics beyond the Standard Model, and the latter on physics at low and medium energies with a focus on the strong interaction. Kristin Riebe is an astrophysicist and is dedicated to the dissemination of knowledge in press and public relations work, as well as the design of graphics, illustrations, and visualizations, and sometimes books.

When we drafted this book, we deliberately decided against following the historical sequence of scientific discoveries. Instead, we primarily present the current state of knowledge in its logical structure, while not forgetting the developments that brought us to where we are. Most importantly, the book is modular in structure: Each article is a double-page spread, where the topic mentioned in the title and subtitle is discussed as comprehensively as possible. It is supplemented by images as well as cross references to other double pages that either lay the foundations or delve deeper into certain aspects. This also means that individual articles can be read on their own and that the same topic can appear several times in the book from different perspectives and at different levels of depth. In order to provide a comprehensive overview of nuclear and particle physics, we have included also more complex topics. Some of these more advanced articles might not be easy to read at first. We specifically draw attention to this in the corresponding headers. We encourage you to continue reading anyway, while possibly making use of the cross references to other double pages, as well as the glossary for assistance. All of this provides a path to deeper understanding.

In this way, our goal is to provide unique insight into our extremely exciting field of research. We took a lot of care to present the facts in an accessible, yet correct way, while largely leaving out a deeper mathematical description. Whether we have succeeded in this balancing act can only be judged by you; we would be very pleased to receive corresponding feedback and constructive criticism.[1]

The outline of this book is as follows: Chapter 1 provides a general overview and describes what this book is about. Building on this, Chapter 2 presents the general conceptual foundations of experimental and theoretical particle physics, which are deepened in Chapters 3 and 4. In Chapters 5 and 6, we describe the specific apparatuses and methods that can make seemingly invisible objects visible. In Chapters 7 and 8, we introduce the elements of the "Standard Model", the best theory of fundamental building blocks and interactions that we have. Chapter 9 shows how incredibly well this theory has already been experimentally confirmed, and where ever more precise measurements aim to find its possible imperfections. However, more understanding leads to more and more precisely formulated open questions! These are discussed in Chapter 10. The searches for possible new and yet unknown physics beyond the Standard Model are presented in Chapter 11. From the related fields of astronomy and cosmology we also learn a lot about particle physics. With this exciting research and a description of the connection of the largest object known—the Universe—to the smallest—the elementary particles, we conclude the book in Chapter 12.

In reading the book, it makes sense to follow the above structure, but thanks to the modularity of the independent articles on each double-page spread it is not necessary to do so. Instead, you can also move through the book along different content lines according to your interests following the cross links from double page to double page. The comprehensive glossary at the end of the book can provide you with assistance, in which frequently appearing terms and concepts are briefly described.

It remains for us to wish you a lot of fun reading this book. Hopefully, we will succeed in satisfying your curiosity in many aspects and draw forth your interest in what is yet to be discovered.

Philip Bechtle,
Florian Bernlochner,
Herbi Dreiner,
Christoph Hanhart,
Josef Jochum,
Jörg Pretz,
and Kristin Riebe

[1] fascinating_particle_physics@listen.uni-bonn.de

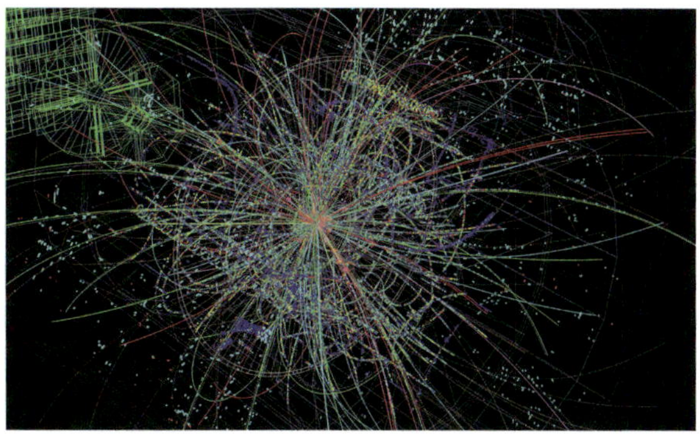

Content

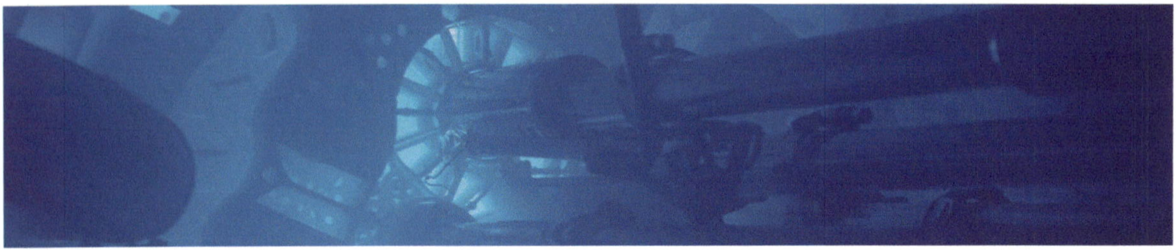

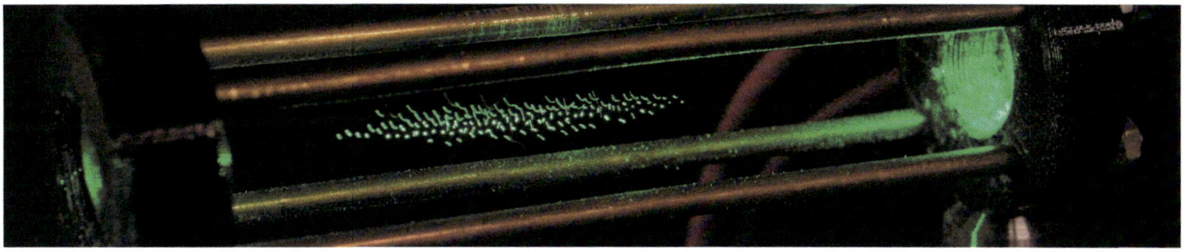

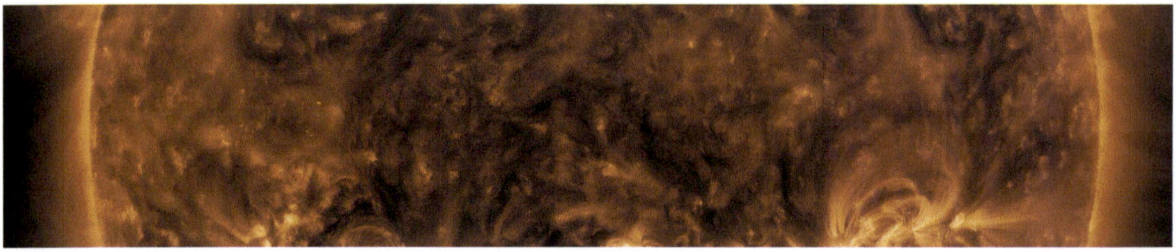

The flip-book starting on this page shows how two hydrogen nuclei (protons) collide with each other. In addition to many other particles, a Higgs boson is produced, which decays into two photons after a short time. You can find out more about the discovery of the Higgs boson on page 246.

1 What Is Particle Physics?

This question can be explored from several perspectives—for instance, by examining the objects of study, the fundamental questions being asked, or the instruments used in research. Particle physics can also be understood by contrast, through comparison with other natural sciences. These various approaches are presented in this chapter. Of course, the wide range of topics in particle physics can only be briefly introduced here, but they will be revisited and explored in greater depth throughout the book. We hope this chapter serves as an invitation to embark on an exciting journey—one that leads not only to profound insights but also to the many open questions at the heart of modern nuclear and particle physics.

Particle Physics as a Global Endeavor

A Pictorial Journey Around the World

Particle physics is a global endeavor with experiments spanning the entire globe: scientists from many nations unite to further our understanding of the Universe and its building blocks. The content of this book focuses on how such research is conducted, aiming to provide the reader with an idea of the questions pursued and their significance. Before delving in deeper, the images on this page are intended to convey an impression of the diversity of research in particle physics.

The depicted experiments take us on a journey from west to east around the globe: starting with the BaBar experiment at the Stanford Linear Accelerator Center near San Francisco (1), which explored the asymmetry between matter and antimatter (*CP* violation ↓), next to an accelerator cavity at Fermilab near Chicago (2), where the heaviest elementary particle to date was discovered (top quark ↓). The journey continues to the remote parts of Argentina, where the Pierre Auger Observatory (3) investigates high-energy particles from outer space (↓) in the midst of grazing animals. From there, we travel to the South Pole, where the IceCube experiment (4) (Detectors Deep in a Mountain ↓) studies cosmic neutrinos. From the H.E.S.S. experiment (Multimessenger Astronomy ↓) in Namibia in Southern Africa (5), which investigates the properties of cosmic rays, we return to the Northern Hemisphere to Geneva (6), where the Higgs boson (↓) was discovered at CERN in 2012. Subsequently, we travel to Karlsruhe (7), Germany, where the KATRIN experiment investigates neutrinos (↓) and their mass. From Europe our journey takes us to China (8), where the Daya Bay experiment (Neutrino oscillations ↓) investigates the mysterious world of weakly interacting neutrinos. These and many other topics, such as the possible decay of the hydrogen nucleus (GUT ↓), are also explored in Japan at the Super-Kamiokande experiment (9). Finally, we travel south to Australia (10), where the SABRE detector is located, which will search for dark matter (↓) in a gold mine. Just as these showcased experiments circumnavigate the globe, the following chapters invite you on a journey through the wonderful and mysterious world of particle physics. Enjoy!

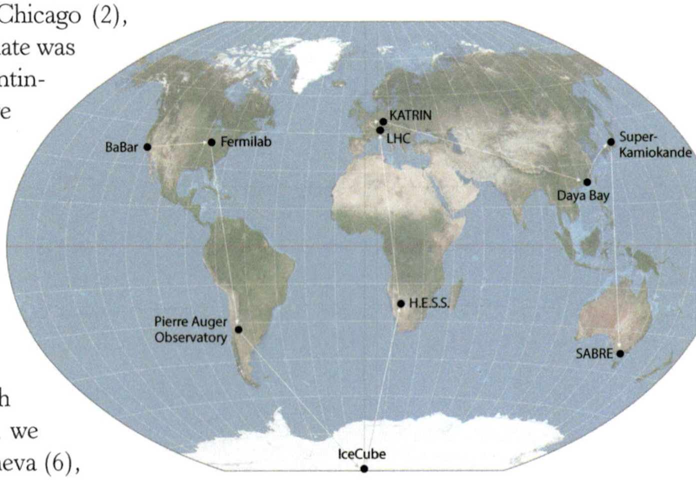

Particle physics is a global endeavor: The map shows the location of the depicted experiments.

CP Violation in Beauty → p. 258
Discovery and Properties of the Top Quark → p. 242
Ultra-High-Energy Cosmic Rays → p. 338
Large Detectors Deep in a Mountain → p. 132
Multimessenger Astronomy → p. 340

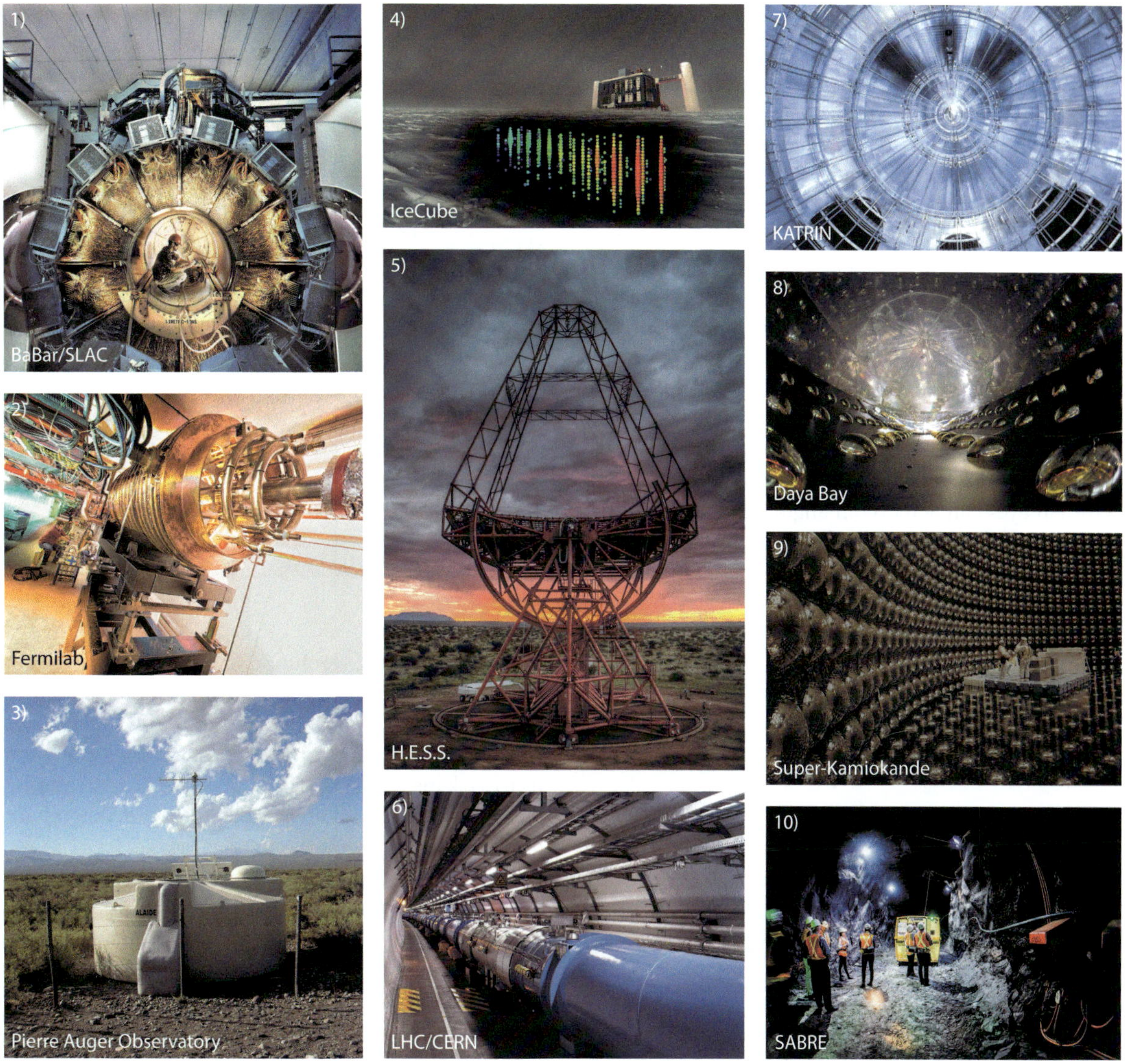

Ten particle physics experiments that span the world, from California (top left) to Geneva in Switzerland (bottom right)

The Discovery of the Higgs Boson → p. 246
Measurement of the Neutrino Masses → p. 256
Neutrino Oscillations → p. 198
GUT → p. 276
Dark Matter → p. 282

The Crystal
Where Does Its Beauty Come From?

Our senses enable us to experience ourselves and the world around us. Each sense employs a specific physical or chemical principle: Scent and taste are based on chemical reactions between molecules; tactile senses feel the pressure on and below the surface of our skin. Hearing uses pressure waves in a medium: air. Last but not least, our eyes (↓) employ visible light as a probe.

An example of an amethyst crystal with its vibrant purple color

Our journey to the world within requires the recognition and observation of smaller and smaller objects. Therefore, it seems appropriate to focus on the sense with the highest spatial resolution. This is our eyesight. Human eyes can achieve an angular resolution of 0.02 degrees. At a comfortable viewing distance of twenty centimeters (cm) or about eight inches, this corresponds to a spatial resolution of about one fifteenth of a milimeter—or about one four-hundredth of an inch. That is sufficient to see microbiotic life, such as an amoeba, without magnification, but it does not allow us to look inside the amoeba. However, we can employ technical devices to increase the resolution: magnifying glasses and microscopes.

We can try to come as close to a crystal as we want—we still cannot see its constituents using visible light as a probe between the crystal and our eye. However, we can see the regular shapes, odd angles and straight lines that this quartz crystal naturally possesses.

Although microscopes enable us to look inside the amoeba, they fail in our quest to see the internal structure of matter when it comes to objects like crystals. Crystals are fascinating: These natural objects "grow" from a liquid. Their wealth and depth of colors, and the breaking of light within them, have enchanted humans since the beginning of time.

Detectors in Everyday Life → p. 24

However, for the sake of our journey towards the fundamental structure of matter, their most interesting property is that they grow in regular shapes, just by themselves. You can test that out yourself: Heat up some water, pour kitchen salt and stir until the newly added salt does not dissolve anymore. Then, let the remaining salt crystals settle, and pour the water into a glass. Hang a hair in your salt water, put the glass on the window sill and wait. The water will start to evaporate, and by themselves salt crystals will start to grow along the hair and elsewhere in the glass. Let them grow and then look at them closely. You will find many perfect cubes and right angles. Some internal order of Nature seems to know about geometry! Wouldn't it be fascinating to uncover the microscopic building principles and the inner order which creates this visible order?

Crystals grow in regular shapes. The orderly cubes of this salt crystal were created by Nature, not by man. What underlying symmetry might have caused these beautiful cubes to emerge all by themselves?

Unfortunately, no microscope and no other tool that uses visible light can have enough resolution to show us the objects within the crystal, and the internal guiding principles which create the orderly shapes. Might there be a fundamental physical challenge which prevents visible light from uncovering these miracles? Let us study and overcome that in the following pages.

Maybe a symmetry of a microscopic "building block" can create a symmetry of a macroscopic object?

Molecules and Atoms

Order Creates Order

Why can't we see the internal structure of a crystal using visible light? The answer to this question lies deep in the nature of light: At the macroscopic level, we can understand it as a *wave* of electric and magnetic fields. At the scale of particle physics, electromagnetic waves themselves are nothing but collections of particles—the *photons* (↓)—whose movement and appearance however is governed by wave formulae.

For now, it suffices to consider light as an electromagnetic wave. We see an object using light because light is reflected or scattered from the object. The first picture on the left exemplifies this process using water waves as an example: Plane water waves come from the left and hit a big round object—exemplified randomly by a rubber plug. There is a shadow and waves are reflected from the object. We "see" the object because the eye (↓) observes that reflected wave.

In this example, the object is larger than the distance from one peak of the plane wave to the next: the *wavelength*. Let us now study what happens if the object is significantly smaller than the wavelength. In the second image, a small screw is standing upright in the water basin. The diameter of the screw—and hence the cross-section it presents towards the incoming surface waves—is significantly smaller than the wavelength. Looking at the wave from the top shows a peculiar result: The wave passes the object as if it were not there.

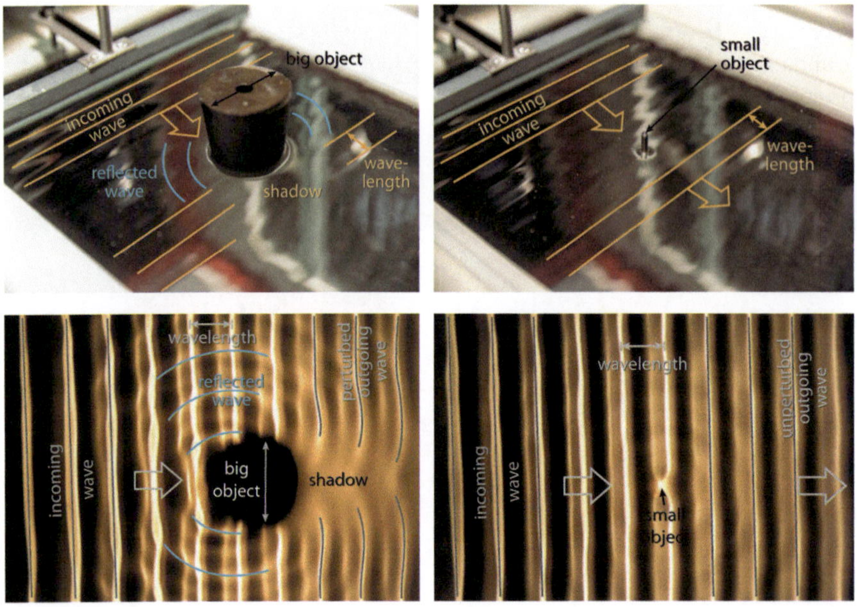

A water wave can be used as a model for the wave structure of light. A wave can "see" an object that is bigger than its wavelength (left). The wave is deflected, and we can observe the deflection: That is what seeing means. If an object is smaller than the wavelength (right), the object becomes invisible: The wave passes the object without noticeable effect, and it is as if the object were not there.

Photons: Electromagnetism → p. 176
Detectors in Everyday Life → p. 24

The plane wave looks the same before and after the screw. Observing the wave, our eye (or a more fancy detector for visible light) cannot see any scatter or reflection from the screw. The screw is invisible to the water wave.

In order to use a wave as a probe to detect an object, the wavelength needs to be of the same size or shorter than the scale of the object. To see the constituents of a crystal, we thus need light of a wavelength shorter than the crystal structures. This is not possible with visible light. The wavelength of visible light is connected to its color. The shortest wavelength in the visible spectrum is that of violet light, at about 400 nanometers (nm). One nanometer is one billionth of a meter, or 10^{-9} m. The constituents of most crystals are atoms or ions (electrically charged atoms), which are about 0.1 to 1 nm in size, about 1/4000th to 1/400th of the wavelength of violet light. Much too small for seeing them using visible light.

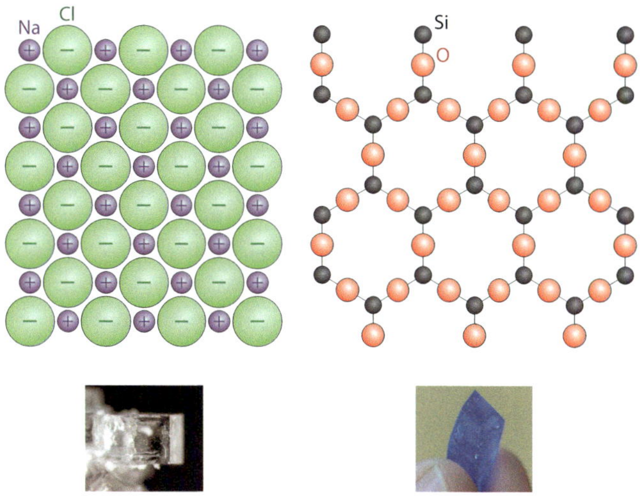

The interaction between ions (left, a rectangular salt crystal) and atoms (right, a hexagonal quartz crystal) on the microscopic level creates regular structures. The angular structure found when looking at the crystals using visible light (see previous article) coincides with the structure of the binding between ions, atoms, and molecules.

Luckily, our quest to find the source of the regular structure of crystals does not end here. In 1901, Wilhelm Conrad Röntgen was the first recipient of the newly created Nobel Prize in Physics, honored for the discovery of X-rays (↓) in 1895. X-rays are electromagnetic waves, just like visible light. However, their wavelengths are significantly shorter. To achieve such short wavelengths, more energy per "quantum" of light is necessary. This connection of higher energy with higher resolution will accompany us throughout the whole book, and will be further explored towards the end of this section. By using X-rays of a wavelength less than 0.1 nm, the inner structure of crystals can be made visible. The result is striking: The geometry of atomic bindings, and the most efficient spacing of ions, follows exactly the same geometric structure as the visible growth of the salt crystal as a whole! The interaction between atoms of 0.1 nm size governs the geometry of crystals that are 100 million times larger!

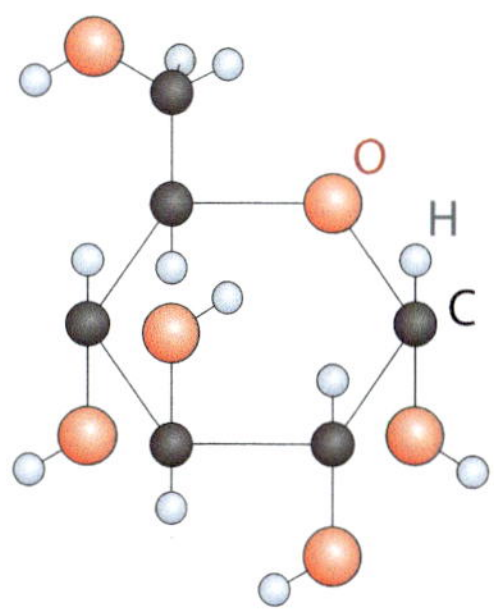

Complex structures can emerge through molecular bindings between atoms, as we can see using this example of a glucose molecule.

X-Rays → p. 124

What Is Particle Physics?

A Fundamental Science Never Walks Alone

In a nutshell, the goal of particle physics is to find the fundamental building blocks (↓) of Nature and the laws that govern their interactions (↓). This search is not new. Our current knowledge has been acquired over many generations with increasingly refined experimental and theoretical methods. We have delved deeper and deeper into the underlying structure of matter, extracting ever more fundamental secrets from Nature. We present the current state of this exciting endeavor in this book.

All natural sciences do this, but it is still possible to clearly distinguish the different scientific fields. Let us take a look at the image below. It shows objects of very different sizes. As a consequence, they are investigated within very different areas of physics. On the far left, for example, crystals are studied in solid state physics. Molecules and atoms are studied in molecular and atomic physics. Their reactions with each other are investigated in detail in chemistry. If we build more complex structures out of the atoms and molecules we enter the field of biology.

The interaction of light with matter, which takes place at the atomic level, is investigated within (quantum) optics. If one deals with even smaller objects, the atomic nuclei or even the building blocks of their building blocks, the quarks, one has entered the field of nuclear or particle physics. Since the last two disciplines have a considerable overlap, we typically do not distinguish them in this book. What is exciting is that the physics of the largest objects, studied in astrophysics, astronomy, and even cosmology, and the physics of the smallest objects, which are the focus of particle physics, impinge on each other. Only together can they explore the history and composition of the Universe (↓), as

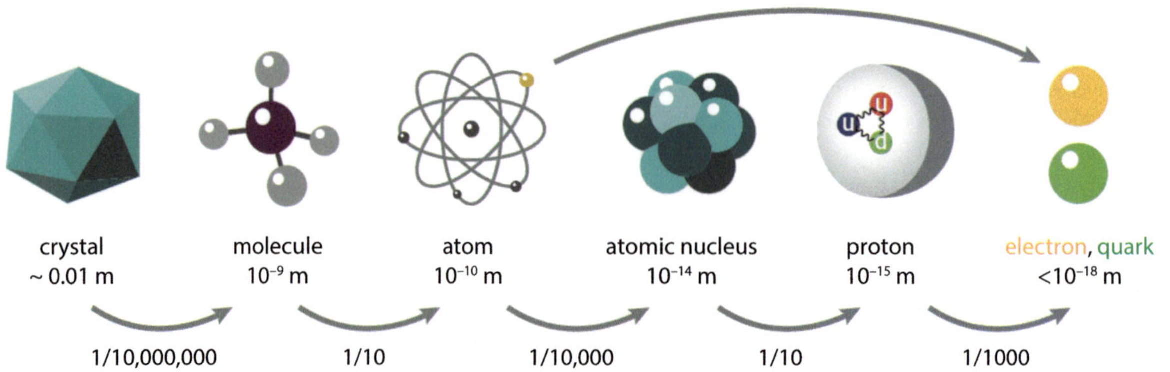

What do you find when you break a crystal down into its building blocks?

The Elementary Particles → p. 14
The Four Fundamental Forces of Nature → p. 16

well as the elementary structure of matter. The decisive element that distinguishes particle physics from other areas of physics is therefore its search for the fundamental building blocks of matter and the forces that act between them—and of course also how those form more complex structures like nuclei.

A helpful analogy to understand the methodology of particle physics is the circumstantial evidence acquired in criminology. Also in physics we start from some initial evidence. For example, today there is almost no antimatter (↓) in the Universe, although in the Big Bang initially matter and antimatter should have been created in equal numbers. We then try to figure out why or how this came about. The particle physics investigation team is made up of different groups: experimental physicists develop, build and operate detectors, they collect additional evidence in form of data, which allow them to test hypotheses. Theoretical physicists try to find a more precise or even new descriptive mathematical model to match the observations or predict possible new discoveries. The accelerator physicists make all the investigations possible by providing the necessary particle beams of the appropriate energy. Furthermore, essential contributions are made by many dedicated and highly qualified technicians.

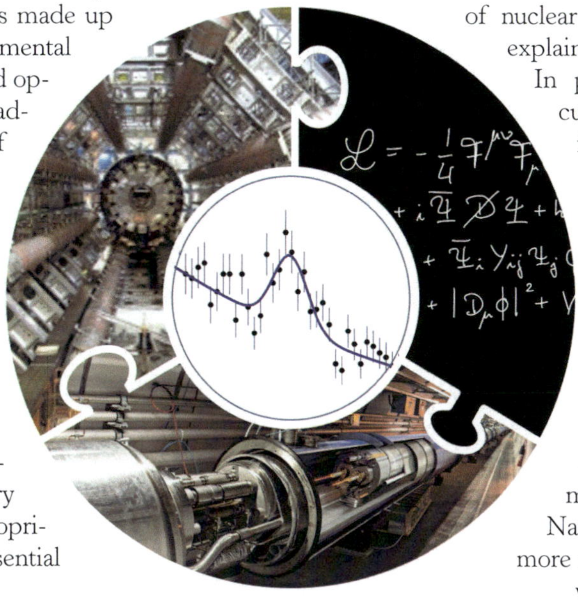
The discovery of the Higgs boson—a close cooperation of theoretical, detector, accelerator and experimental physics

In particle physics, the special moments arise when a piece of the puzzle falls into place and suddenly one gets an idea of the larger picture. This can happen for example with more precise and detailed data, measured under conditions never studied before, or by discovering a new mathematical connection within the theory. Similarly a detective might suddenly realize how all the clues point to the solution of the crime. It is particularly exciting if, based on some freshly gained understanding, one formulates a prediction, for example a new particle, which is then experimentally *confirmed*! It is these moments that make a career in particle physics so special, so exciting.

In this book, we introduce various topics of nuclear and particle physics and explain how they are connected. In particular, we present the current state of affairs in the field. We show in detail what a wonderful treasure, what a fantastic cultural achievement the *Standard Model of elementary particle physics* (↓) represents. We wish to demonstrate how it came about, how successful it is, where it reaches its limits, and with which methods we try to coax from Nature, what a possible, even more fundamental, theory of the world might look like.

Evolution of the Universe → p. 346
What Is Antimatter? → p. 22
Chapter 7 The Electroweak Standard Model of Particle Physics → p. 175

Origins of Particle Physics

How and Who?

The beginning of particle physics is marked by the discovery of the electron by Joseph J. Thompson in 1897. He performed experiments with a cathode ray tube, as shown in the figures. This is a small accelerator, similar to what used to be included in TV sets.

We can use the drawing below to explain how it works. At point C a wire is placed at a negative potential and heated, thus electrons are emitted. They are accelerated to the anode A (at a positive potential). They can fly through a hole in the anode, and are focused by the aperture B, which is grounded. They are then deflected by the electric field between the plates (D, E) and can be observed on the screen on the far right. It is possible to determine the ratio of the electric charge to the mass of the particles (q/m) by measuring the deflection on the screen as a function of the strength of the electric field between D and E.—In the experiment, the ratio was always the same, which is indicative of an elementary particle.

In 1896, Eugen Goldstein had already investigated ionized gas atoms, including hydrogen (H^+), in similar fashion in an anode ray tube. He observed many different values of q/m. Much later, in 1898, Wilhelm Wien was able to identify the particle with the largest value of q/m as the proton.

In 1911 the atomic nucleus was discovered by Ernest Rutherford, Hans Geiger and Ernest Marsden with the help of a scattering experiment. In similar fashion in 1932 James Chadwick discovered the neutron. Thus the building blocks of the atoms were determined.

In 1896 Henri Becquerel and Marie Curie discovered radioactivity: α (*alpha*), β (*beta*) and γ (*gamma*) rays. The α-rays were found to be helium nuclei. Their emission from a nucleus is governed by the strong interaction. β^--rays consist of energetic electrons and β decay is governed by the weak interaction. γ-rays consist of highly energetic photons, the quanta of the electromagnetic force, introduced in 1905 by Albert Einstein.

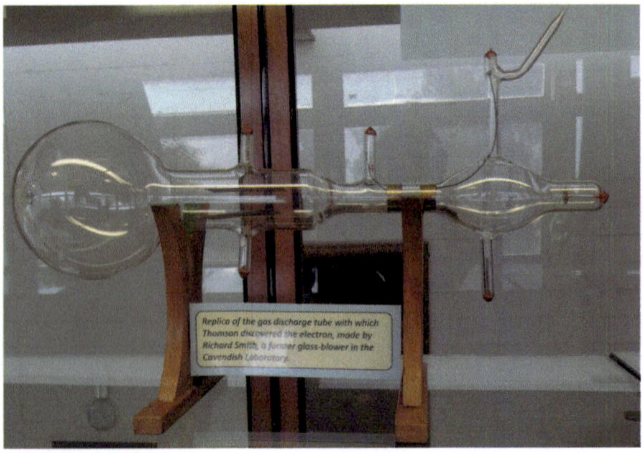

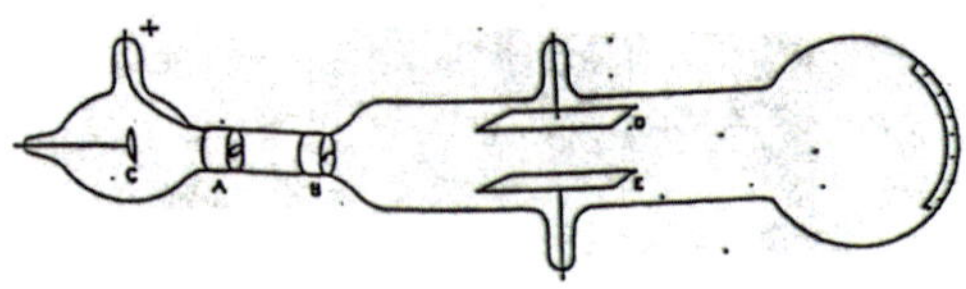

Reconstruction of J. J. Thompson's cathode ray tube (top) and original drawing of J. J. Thompson's cathode ray tube (bottom). The electrons fly from right to left in the glass tube as shown, and reversed in the drawing..

[1] For a very good history of elementary particle physics up until 1950 see the book by A. Pais, *Inward Bound,* Oxford University Press.

Nuclear isotopes: From Nucleus to Nuclei → p. 226
Noether's Theorem → p. 100, Symmetries → p. 60
Discrete Symmetries: Violating P and CP → p. 190

Thus already in 1896 there were first hints to the three fundamental forces of the Standard Model of particle physics, and since 1905 also hints that these forces must be quantized, i.e., the elctromagnetic field consists of many quanta.[1] In 1936 the first particle beyond our everyday matter was discovered in cosmic rays: the muon. The physicist Isidor Rabi's response was: "Who ordered that?" We still don't know the answer.

Many of the physicists who contributed to the development of particle physics are well-known, such as Marie Curie and her co-discovery of radioactivity, Albert Einstein and the development of relativity, Niels Bohr and his atomic model, as well as Erwin Schrödinger and his cat. Many Nobel Prizes have been awarded in recognition of the development of the Standard Model of particle physics. However recognition for the work performed by female scientists in these developments has often been lacking. Therefore we here draw particular attention to important historical and current contributions made by female scientists in the last 125 years. They have managed to make extraordinary contributions despite adverse circumstances[2], which have often not been recognized or publicly acknowledged. By way of example we briefly discuss the work of nine women whose contributions are also directly relevant to this book:

Marie Curie (1867–1934) was a co-discoverer of radioactivity and she made fundamental contributions to its understanding. She was the first person to receive two science Nobel Prizes (physics: 1903, chemistry: 1911).

Lise Meitner (1878–1968) discovered several nuclear isotopes (↓). She is the co-discoverer of nuclear fission and presented the first physical explanation.

Emmy Noether (1882–1935) was a mathematical physicists who discovered the fundamental connection between symmetries and conserved quantities (↓).

Maria Goeppert Mayer (1906–1972) made decisive contributions to our understanding of nuclear structure and received the Nobel Prize for physics for this work, in 1963.

Chien-Shiung Wu (1912–1997) experimentally discovered parity violation (↓) in 1956.

Vera Rubin (1928–2016) measured galactic rotation curves which was a decisive contribution to the discovery of dark matter (↓).

Helen Edwards (1936–2016) made decisive contributions to superconducting accelerator technology, and thus made accelerators such as the Tevatron and the LHC possible.

Helen Quinn (1943–) made decisive suggestions for the understanding of *CP* violation in the strong interactions (↓) and contributed significantly to the development of Grand Unified Theories (↓).

Fabiola Gianotti (1960–) was the spokesperson of the ATLAS experiment when the Higgs boson (↓) was discovered in 2012. From 2016—2025 she was the first woman as Director General of CERN.

Fabiola Gianotti at CERN

[2] In 1899, Elsa Neumann was the first woman in Germany to gain a doctorate in physics (Berlin University). In 1988(!) Annette Zippelius became the first female full professor of physics in Germany (University of Göttingen).

Dark Matter → p. 282
The Strong *CP* Problem → p. 290 and GUT → p. 276
The Discovery of the Higgs Boson → p. 246

Length Scales and Units

From Vast to Miniscule—It's All Physics

In this book, a multitude of different length scales play a role: On the one hand, there is the vast extent of the Universe, the galaxies, and the planetary systems, which we can hardly grasp by looking at the night sky. On the other hand, there are the objects of our everyday life: from kilometer-high mountains to, e.g., biological systems or a grain of sand. Then there are chemical processes characterized by the typical size of atoms as well as subatomic processes inside the atomic nucleus.

As indicated in the figure, different forces dominate at different scales, leading to correspondingly different phenomena. *Gravity* is responsible for the structure formation in planetary systems, galaxies, and the Universe. It is by far the weakest of all forces. However, it has an infinite range (why this is the case is described in the article On the Ranges of Interactions ↓). The visible processes around us are predominantly determined by electromagnetism. This is stronger than gravity, but is mostly shielded by opposite charges, and thus usually short-ranged, except, e.g., in hot plasmas in galaxies. *Electromagnetism* is responsible for the structure of atomic shells and molecular bonds, as well as the structure of solid bodies and pretty much all pro-

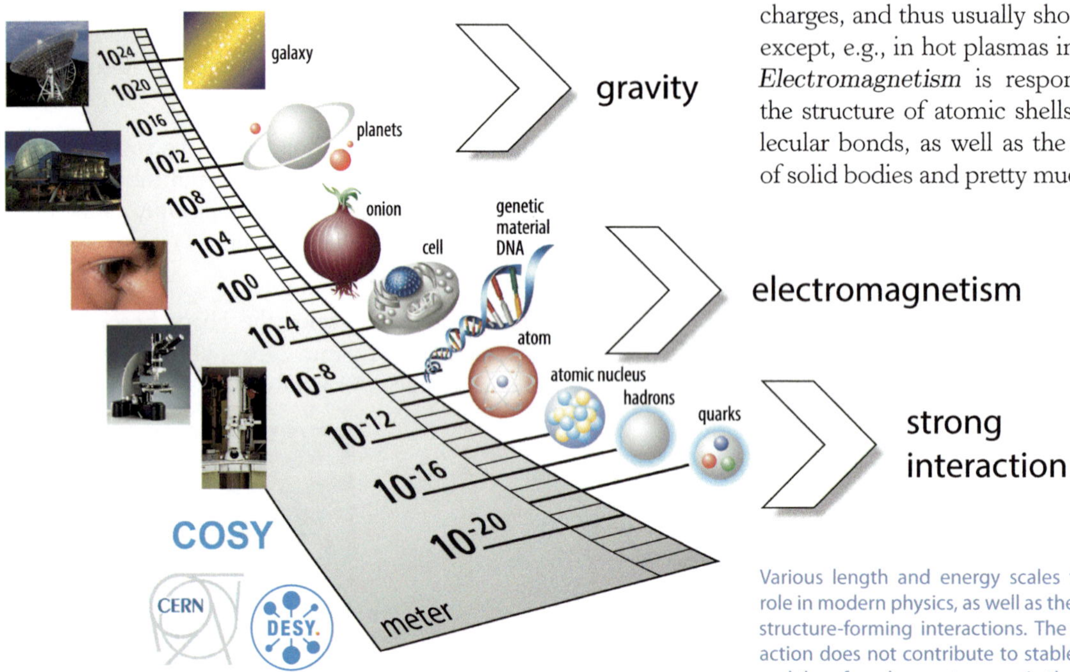

Various length and energy scales that play a role in modern physics, as well as the respective structure-forming interactions. The weak interaction does not contribute to stable structures and therefore does not appear in the picture.

On the Ranges of Interactions → p. 58

cesses that we can perceive with our senses and that are not brought about by gravity. The *strong interaction* is even stronger and holds together not only protons and neutrons inside atomic nuclei, but also their building blocks, the quarks, inside protons and neutrons. The *weak interaction*, on the other hand, is very weak and is mainly manifested in radioactive decays such as the beta decay of the neutron, or the conversion of protons to neutrons in the Sun.

The strength of an interaction and its range determine the size of the structures it can generate. Since gravity has an infinite range and its strength increases with the mass of the involved objects, it creates structures that are gigantic in the truest sense of the word. From our own experience with electromagnetism, we know that, for example, commercially available magnets act on each other at a distance of a few centimeters. A magnet that is significantly larger in extent is the Earth, whose magnetic field reaches over ten thousand kilometers. It moves compass needles and protects us from the shower of charged particles from space. Directly relevant for this book is the effect of electromagnetism at the atomic level with typical extensions of 10^{-10} m. The strong interaction is relevant at the subatomic level, i.e., within atomic nuclei at 10^{-15} m. The weak interaction acts on even smaller scales: With a range of 10^{-18} m, it is so weak and short-ranged that it does not contribute to structure formation.

As shown in the figure, modern physics deals with length scales that span over 30 orders of magnitude. The further one moves away from everyday objects in the illustration, the larger the devices one needs for their observation. The reason for this is that the resolution needs to be increased. Astronomy uses huge telescopes, or even links many telescopes together to be able to look deep into space. On the other hand, if one wants to investigate very small objects, an optical microscope quickly becomes insufficient. However, one can use the relationship between the quantum mechanical wavelength and the momentum of a probe (Quantum Mechanics ↓): The greater the energy and thus the momentum, the smaller the wavelength and thus also the object that one can still "see" (Molecules and Atoms ↓). An electron microscope allows insights at the atomic level in this way. However, if one wants to look even more closely, the wavelength of the probe must be further reduced. For this, accelerators (↓) with ever greater energy are needed.

For convenience, researchers dealing with very different systems use different units. Within our galaxy, for example, the light year, the distance that light travels in one year, is a useful length scale. This is about 10^{16} m. In everyday life, we use kilometers, meters and centimeters, depending on what we are talking about. For the distance Hamburg-Munich we use kilometers, for the thickness of a screw millimeters. Atoms have a size of 10^{-10} m. This length scale is called *angstrom*, denoted Å. Atomic nuclei have a size of 10^{-15} m. The length scale is called *fermi* or *femtometer*, with the symbol "fm".

The same applies to masses (M) and energies (E). Since according to special relativity (↓) these are connected via $E = Mc^2$, in particle physics typically no distinction is made between these two quantities. For example, a proton has a mass of $1.67 \cdot 10^{-27}$ kg $=$ 938 MeV$/c^2 = 938 \cdot 10^6$ eV$/c^2$, where the electron volt (eV) is the energy (↓) that an electron gains after passing through an electric potential of one volt.

Molecules and Atoms → p. 6
Concepts of Quantum Mechanics → p. 50
How Does a Particle Accelerator Work? → p. 66
$E = mc^2$: Special Relativity → p. 38
Energy and Power → p. 36

The Elementary Particles

The Building Blocks of the Universe

Elementary particles (↓) are thought to be point-like and not composite. According to our current understanding they comprise the fundamental building blocks of Nature. Although everything around us including ourselves shows a fascinating variety, three building blocks are sufficient for this high level of complexity: the up quark (↓), the down quark and the electron. Three quarks form the proton and the neutron in the combinations up-up-down and up-down-down, respectively. Atomic nuclei (↓) are composed of protons and neutrons, and are surrounded by electrons to form atoms.

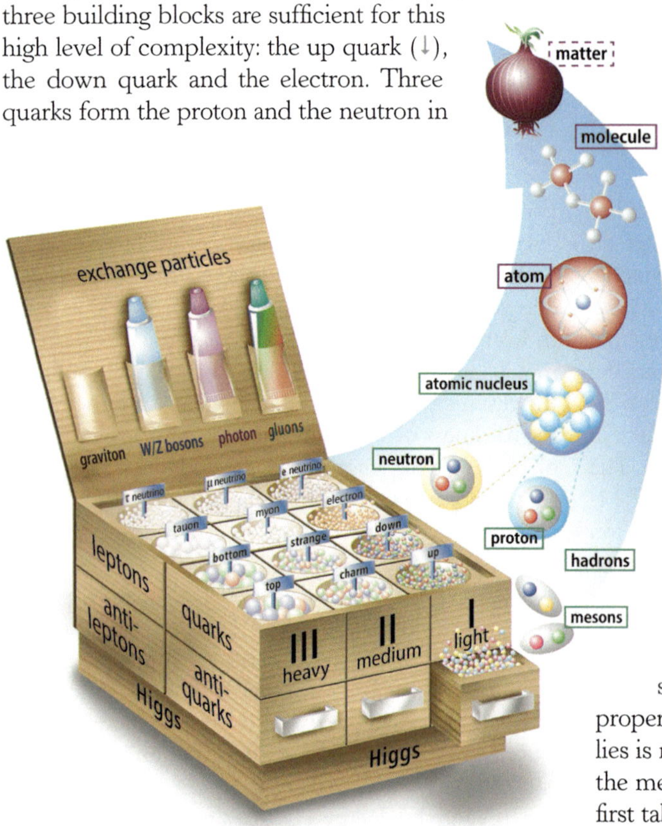

The building blocks of matter and their interactions

However this is not all. In decays, such as neutron beta decay, electrons and electron antineutrinos are produced, and there are also atomic nuclei in which bound protons decay into bound neutrons, producing the antiparticles of the electron and electron antineutrino, the positron and the electron neutrino. The protons and neutrons also have antiparticles, and accordingly there are antiquarks, anti-up and anti-down. Antinucleons do not occur significantly in Nature, but can be produced in accelerators.

The elementary particles named so far in this article build the so-called *first family*, see the drawer "I—light" in the figure. It has been shown that all members of the family, electron, electron neutrino, up quark and down quark, are necessary for the theory to be mathematically consistent. There are more families in the Standard Model—they have the same number of particles with basically the same properties, just different masses. The number of families is not determined by the theory. The properties of the members of the different families are listed in the first table (↓) at the end of this book and in the figure on the right page.

What Makes a Particle Elementary? → p. 18
The Quark Model: A First Look → p. 20 and The Quark Model Revisited → p. 202
From Nucleus to Nuclei → p. 226
Tables: Properties of Elementary Particles → p. 353

Accelerator experiments have shown that in addition to the first family just described, there are two more: The second consists of the strange and the charm quark as well as the muon and its neutrino, the third consists of the bottom and the top quark and the tau lepton and its neutrino. Today we know that there is no further family. Why there are three, we do not know.

The masses of the elementary particles are shown in the picture at the bottom right. They are not predicted by the Standard Model, but are parameters, which must be determined in experiments. It is important to note that quarks do not exist freely, but only bound together according to strict rules (Confinement ↓). Moreover, their masses only make up a part of the masses of these bound systems (↓), collectively called hadrons. Therefore measuring the quark masses is complicated.

The particle masses pose a puzzle. It is not understood why they span so many orders of magnitude: the top quark is about as many times heavier than an up quark as an elephant is heavier than a mouse. It is also not understood why in the first family the positively charged quark is the lighter one, whereas in the other two families it is the heavier one. However, it is important that the up quark is lighter than the down quark, for otherwise there would be no life on Earth as we know it (Quark Mass Effect ↓).

In addition to the building blocks of matter just described, the Standard Model includes the field particles of the interactions, which act as transmitters of the forces. These are the photon (γ) for electromagnetism, the electrically neutral Z boson, as well as the two charged W bosons (W^+ and W^-) for the weak interaction, and the eight gluons for the strong interaction. While photons interact with each other only indirectly, Z and W bosons as well as gluons show self-interactions. Especially the latter feature has dramatic phenomenological implications (QCD ↓). The properties of the force particles are summarized in the second table of the appendix. Gravity, not part of the Standard Model, is also attributed an exchange particle, the graviton, even though it has not yet been discovered. Finally, there is the Higgs boson (↓). Its role in the Standard Model is to break the electroweak symmetry and give the elementary particles their masses. It is the only spinless elementary particle and was the last of the building blocks of the Standard Model to be discovered in 2012.

The profile of the elementary particles is of course only complete once their other properties, such as spin, charge, baryon number and lepton number are assigned—these are also given in tables at the end of this book.

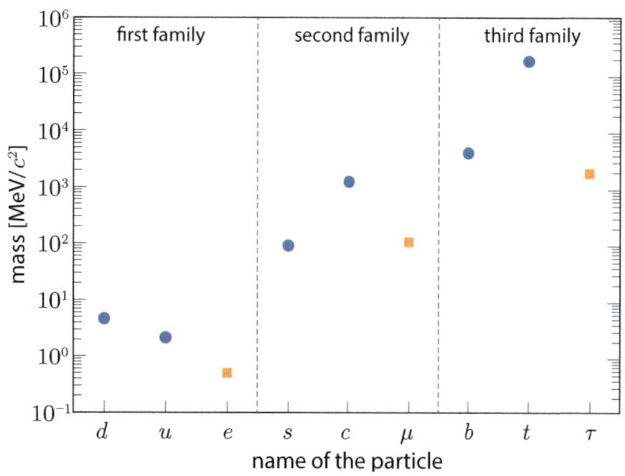

Comparison of particle masses. The black circles represent the quark masses, the orange squares the charged leptons. The masses of the neutrinos are very small and therefore not shown.

Confinement → p. 206
Where Does Our Mass Come From? → p. 220
The Quark Mass Effect → p. 224
QCD: Quantum Chromodynamics → p. 204
The Discovery of the Higgs Boson → p. 246

The Four Fundamental Forces of Nature

Four Very Unequal Partners

We know of four fundamental forces: gravitation, electromagnetism, the weak interaction, and the strong interaction. The last three are combined in the Standard Model of elementary particle physics. The interactions of the Standard Model are all built according to the same principle: the forces are mediated by the exchange of particles specific to the force. We expect that this also applies to gravity. The four fundamental forces differ drastically in range (↓) and strength and are therefore relevant for very different systems, which consequently exhibit very different properties (Scales ↓).

To illustrate the above, we invite you on a journey, a journey within your minds. Imagine we are traveling on a proton which is approaching a hydrogen atom—that is, another proton with an electron in its surrounding shell. Let us start our journey at least at a distance of 10^{-8} m—that is about a factor 1000 less than the thickness of a hair—away from the electrically neutral hydrogen atom, which is only about 10^{-10} m in size. From this "large" distance, the effects of the electric charges of the electron and proton in the hydrogen atom cancel each other. This shields the electromagnetism from us and our proton and the very weak attraction of gravitation is the only force that significantly affects our proton, pulling it weakly towards the atom.

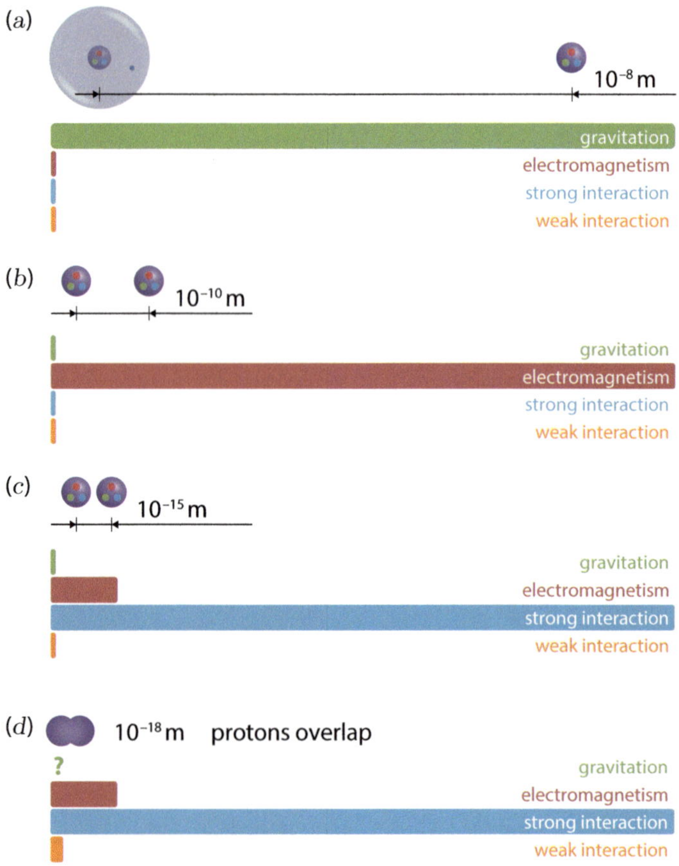

The relative importance of the fundamental forces at different distances —the bar length indicates the relative strength of the forces.

On the Ranges of Interactions → p. 58
Length Scales and Units → p. 12

However, this changes dramatically as soon as we approach to within atomic distances of the hydrogen nucleus, i.e., penetrate into the electron cloud. Now our positively charged proton "sees" the also positively charged nucleus. Although the gravitational force is now 10 000 times stronger than before due to the reduced distance, this is negligible compared to the electromagnetic repulsion that our proton now feels—gravitation is over 30 orders of magnitude weaker than electromagnetism at these distances, mediated by the exchange of photons. On the other hand, we are still way too far from the nucleus for the strong interaction to play a role. Out here, it is even negligible compared to gravity. The diversity of Nature we experience in our everyday lives is thus primarily determined by electromagnetism. For example, it makes objects shine in color and holds together all the many substances around us, such as wood, stones, and metal. The effects of the strong interaction are indirect, as it determines (in interplay with electromagnetism), which atomic nuclei (↓) are stable with which nuclear charges.

To continue the journey, let's imagine that we are energetic enough to overcome the electric repulsion. Then again something dramatic happens as soon as our proton has come closer to the nuclear proton to within 10^{-15} m—another factor of 100 000. Now the strong force becomes noticeable, initially through the exchange of light quark-antiquark systems—this completely overwhelms electromagnetism and our proton is attracted to the atomic nucleus.

At even smaller distances ($< 10^{-15}$ m) the protons begin to overlap. Now the strong force is directly mediated by the exchange of gluons, which are the fundamental force carriers of the strong interaction between the quarks, the building blocks of protons. However, this does not change the hierarchy between the forces: The strong force remains by far the strongest, followed by electromagnetism. However, at these small distances, the weak interaction, mediated by the exchange of W and Z bosons, becomes relevant. But at a distance of 10^{-18} m it is still a factor of 10 000 weaker than the strong force and a factor of 100 weaker than electromagnetism. At such small distances, gravity can no longer be assigned a readily calculable force because within a proton there is a high energy density, which also gravitates. Here classical gravity is no longer applicable.

Thus, which of the four forces dominates depends on which particles are involved and also on how closely they approach each other. To illustrate the former point further we may repeat the above journey, however, now not travelling on a proton but on an electron. Then the relative importance of the forces shown in images (*a*) and (*b*) of the figure remain unchanged—however, the blue bars need to be removed since electrons do not feel the strong interaction. Accordingly, image (*c*) will be the same as image (*b*). In image (*d*) the onset of the weak interaction remains correct.

Among the forces of the Standard Model, the weak interaction plays a special role: It is so weak and short-ranged that it cannot form stable structures. However, it is the only interaction that changes particle types—it can, for example, convert a down quark into an up quark. This happens, e.g., in the beta decay of a neutron (Particles' Births and Deaths ↓). Without the weak interaction, neutrons would be stable. Moreover, all the rich physics of neutrinos needs the weak interaction, but this story is told in other articles.

From Nucleus to Nuclei → p. 226
Particles' Births and Deaths → p. 54

What Makes a Particle Elementary?

Are There Always Smaller Building Blocks?

An elementary particle is an entity in Nature which has no measurable extension. It is not divisible and has no substructure. In the special theory of relativity (↓) elementary particles are defined as *point-like* and are characterized by a fixed mass and a fixed spin (↓), i.e., a fixed magnetic moment. Furthermore, they can carry three kinds of charges: electric, weak and color (strong). Through these they experience the corresponding forces.

Based on our current understanding, the electron is such a point-like particle and thus elementary. It has mass $m_e = 9.1 \cdot 10^{-31}$ kg $= 511$ keV$/c^2$ (Length Scales and Units ↓) and spin 1/2. It has electric charge $-e$, weak charge (isospin) $-1/2$, and has no color or strong charge (see tables at page 353). This holds for *all* electrons; it is the definition of an electron.

How do we decide if a particle is elementary? If it is divisible it has smaller building blocks. These could in turn be elementary or again divisible. Even if we can't divide an object into independent parts because of technical limitations (e.g. accelerator energy) or on principle (e.g. the proton), there can still be observable effects of the substructure. This could be like shaking a closed box to decide if there are loose balls inside. What happens then, if we look ever closer into the building blocks of Nature?

Consider halving a 20 cm long carrot again and again. The relevant piece keeps getting smaller, but it is initially still a carrot. It has some substructre with the stem and the fibers inside. After halving about 12 times we reach the biological cells. These are much smaller building blocks of the carrot. Under the microscope we can see that they are in the physical sense not elementary. If we could continue to halve the bits with a knife, after about 30 halvings, i.e., at about 10^{-10} m, we would reach the molecules and atoms of the carrot cells. This is not possible with a knife, since even the sharpest blade is about 2 500 iron atoms wide, i.e., about $5 \cdot 10^{-6}$ m.

Do atoms have a substructure? In 1909, the physicist Ernest Rutherford and his collaborators bombarded a thin gold foil with energetic alpha particles, i.e., helium nuclei. Almost all of the alpha particles flew straight through the foil (Scattering Experiments ↓). Only about one in every 100 000 particles was scattered by 90 degrees or more (red arrow in the figure). The larger the deflection angle the rarer the occurrence. The angu-

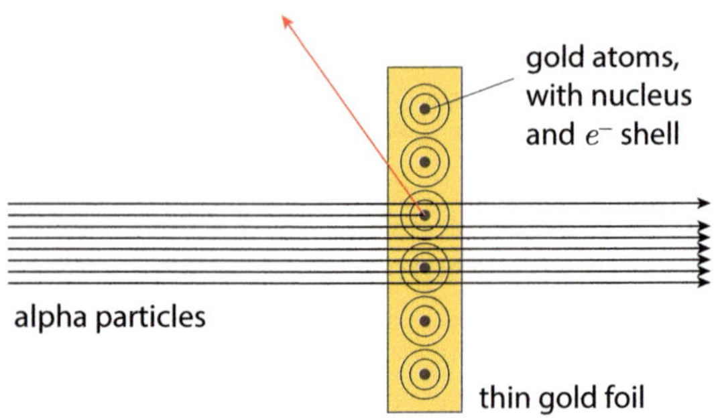

Schematic depiction of the Rutherford scattering experiment.

Special Relativity → p. 38, Special Theory of Relativity in Particle Physics → p. 92
Spin → p. 44
Length Scales and Units → p. 12
Tables: Properties of Elementary Particles → p. 353
Scattering Experiments → p. 46

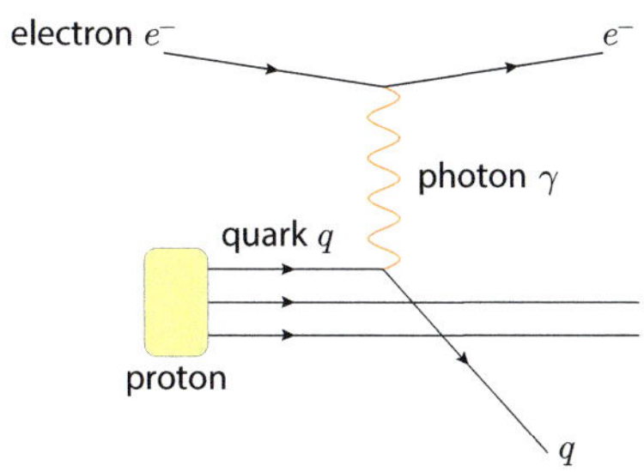

Scattering of an electron at a quark inside the proton by exchanging a photon

lar distribution corresponded exactly to that expected from electrodynamics for the scattering of two point-like electrically charged particles. Rutherford could conclude that the atom must have a positively electrically charged point-like structure at its center, the nucleus, which also carries almost all of the mass (99.9 %) of the atom. Based on this distribution he could set an upper bound of about 10^{-14} m for the diameter of the nucleus. This is very small compared to the shell containing the electrons of the gold atoms with a diameter of about 10^{-10} m.

As the energy of the alpha particle is increased, it can approach the gold nucleus ever closer, despite the electrical repulsion (both are positively charged). You can then distinguish in the scattering distribution that the target nucleus has a finite charge distribution spread over a sphere of about 10^{-14} m diameter. This is no longer the distribution of the scattering on a point-like particle. Without breaking it apart you can "see" that the nucleus has a substructure. With even higher energy it was possible to break off parts of the nucleus. This is a form of fission and is how the neutron was discovered as a new particle.

Thus the nucleus consists of protons and neutrons, each about 10^{-15} m in diameter. Are *they* elementary?

In 1967 at the SLAC laboratory in California, electrons were scattered off protons at even higher energies (21 GeV) (Structure of the Proton ↓, Quark Distributions in the proton ↓). The proton was broken apart, but the scattering distribution was again that of point-like particles scattering. This was the discovery of ***quarks***, the constituents of the proton. The electrons and the quarks deep *inside* the proton scattered as point-like particles on each other. The quark wasn't liberated but could be seen!

In the 1990s at the LEP accelerator at CERN, electrons and positrons were scattered off each other with even higher energies (see Feynman graphs ↓). They still scattered as point-like particles. They approached to within about 10^{-18} m. From this we conclude that electrons are point-like down to at least 10^{-18} m. At the LHC we have been able to reach a similar conclusion for quarks down to about the same length scale.

Whether we call a particle elementary thus depends on the experimental possibilities. Today we consider electrons and quarks as elementary. This view could be altered by future experiments.

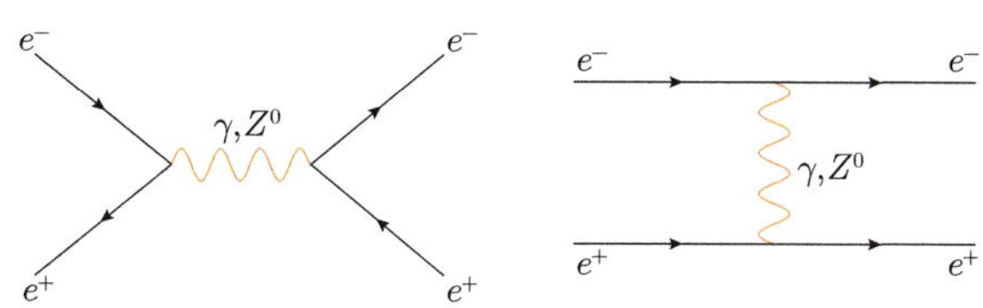

Feynman diagrams for electron-positron scattering by the exchange of a photon or a Z^0 boson

The Structure of the Proton → p. 212
Quark Distributions Inside the Proton → p. 214
Feynman Diagrams → p. 48 and Feynman Diagrams II → p. 102

The Quark Model: A First Look

Assembling the LEGO Blocks

Soon after the proton and the neutron were established as building blocks of the nucleus of the atom, many other particles were discovered. This raised the question whether they can all be called *elementary*. Or are all these particles composed of more elementary entities, much like all atoms in the periodic system are made of three components: protons, neutrons and electrons? Looking at the proton and neutron gives some hints. On the one hand they are very different. The proton has a positive electric charge, the neutron, as the name indicates, is electrically neutral. On the other hand they are very similar. Their masses only differ by 0.1 %. That cannot be a coincidence.

In the previous article we learned that bombarding protons with electrons revealed that the proton is not elementary but rather consists of more fundamental particles called quarks. They play an essential role throughout this book and are discussed in many articles.

But how can quarks explain the similarities and differences between the proton and neutron? To describe the proton and neutron, we introduce two types of quarks called up and down. The up quarks carry an electric charge of +2/3 of the elementary charge e, the down quarks −1/3. Now it is just like putting LEGO blocks together, while respecting a few simple rules: Combine any three quarks, up or down, with the constraint that the electric charge of the object is a multiple of the elementary charge e. In this quark model the proton consists of two up and one down quark:

$$\begin{array}{llcccccc} \text{proton:} & p = & u & + & u & + & d \\ \text{charge:} & 1 = & +2/3 & & +2/3 & & -1/3 \end{array}$$

Thus the charges add up to one elementary charge as it should be. The neutron consists of two down and one up quark:

$$\begin{array}{llcccccc} \text{neutron:} & n = & u & + & d & + & d \\ \text{charge:} & 0 = & +2/3 & & -1/3 & & -1/3 \end{array}$$

This makes the neutron electrically neutral. Assigning approximately the same masses to up and down quarks explains why neutron and proton have a similar mass.

One could argue that the quark model does not really simplify things. We replaced two objects (proton and neutron) by two other objects (up and down quarks). But, following the rules established above, the quark model also predicts particles consisting of three up quarks or three down quarks. These particles are also observed. With the rule that the object has to have an electric charge being a multiple of the elementary charge, it is not allowed to create objects with two or four quarks. Indeed, such particles have never been observed.

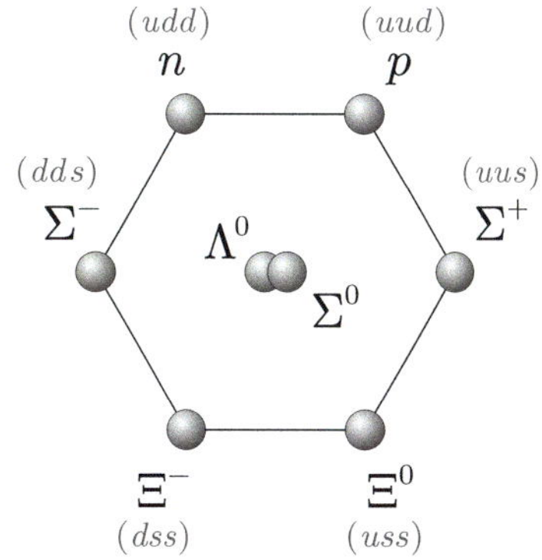

The so-called baryon octett consisting of particles composed of three quarks with the flavors u, d and s.

The story does not end here. Meanwhile six quarks types, also called *flavors* in particle physics, have been discovered. In addition to up and down, we have strange, charm, bottom and top quarks, as indicated in the table. More than 100 particles can be described with these six buildings blocks. The figure shows particles made up of three quarks containing the flavors up, down and strange. In the top line we find the well known proton and neutron. The other particles are labeled with greek letters. Why the particles are arranged in such an hexagonal structure and why two particles are present in the center has a fascinating mathematical background and is explained in dedicated articles (see below).

Furthermore, for every quark there is a corresponding antiquark carrying the opposite electric charge. This allows us now to build objects out of one quark and one antiquark like $u\bar{d}$ and $d\bar{u}$, leading to integer charges, which are also realized in Nature. For example, particles known as pions (π^+ and π^-) are described by this quark content.

Quarks play such an essential role in particle physics that they are taken up in many articles in this book. Almost the entire chapter on strong interaction (↓) is dedicated to quarks and their interactions. A more detailed discussion of the development of the quark model is given in The Quark Model Revisted (↓). Failed attempts to isolate them and to detect a single quark are described in The Search for Free Quarks (↓). The theory which describes how quarks interact with each other can be found in the articles on QCD (↓) and Confinement (↓).

quark	**charge/e**	**antiquark**	**charge/e**
up (u)	2/3	anti-up ($\bar{u}$)	−2/3
down (d)	−1/3	anti-down ($\bar{d}$)	1/3
strange (s)	−1/3	anti-strange ($\bar{s}$)	1/3
charm (c)	2/3	anti-charm ($\bar{c}$)	−2/3
bottom (b)	−1/3	anti-bottom ($\bar{b}$)	1/3
top (t)	2/3	anti-top ($\bar{t}$)	−2/3

The six quark flavors sorted by increasing mass. Antiquarks are denoted with a horizontal bar over the corresponding symbol, e.g., $\bar{u}$.

Chapter 8 The Peculiar Strong Interaction → p. 201
The Quark Model Revisited → p. 202
The Search for Free Quarks → p. 208
QCD: Quantum Chromodynamics → p. 204
Confinement → p. 206

What Is Antimatter?

Science or Fiction?

We are made of atoms, and these in turn consist of a nucleus made of protons and neutrons, as well as an atomic shell made of electrons. The proton carries positive electric charge, $q = e$, the electron has negative charge, $q = -e$ and the neutron has no electric charge $q = 0$. These are the building blocks of matter in our world. Since about 100 years, we know that the interactions of these subatomic particles are governed by quantum mechanics (↓). For example, we compute the atomic energy levels via the Schrödinger equation. The underlying kinetic energy of a particle employed in the Schrödinger equation has the non-relativistic form, i.e., for velocities much slower than the speed of light, $E_{\text{kin}} = \frac{1}{2} m v^2$, where m is the mass of the particle (electron or nucleus) and v is its velocity. This does not agree with special relativity. There the total energy of a free particle is

$$E = \sqrt{m^2c^4 + p^2c^2} \approx mc^2 + \frac{1}{2}mv^2 - \frac{mv^4}{8c^2} + \dots ,$$

where p is the momentum of the particle and c is the speed of light. The non-relativistic Schrödinger equation is insufficient to describe the atomic spectra to a very high precision. There are small but measurable discrepancies, the so-called *fine structure* of atomic spectra. This problem can be resolved in *ad hoc* fashion through a first relativistic correction to the kinetic energy ($-\frac{mv^4}{8c^2}$), which modifies the Schrödinger equation accordingly, as shown above. This *ad hoc* solution is unsatisfactory for a fundamental theory. Furthermore, not all relativistic corrections are taken into account by this one term.

In 1928, Paul Dirac succeeded in unifying quantum mechanics and special relativity into a single relativistic wave equation, the Dirac equation. This equation holds for all energies, respectively all velocities arbitrarily close to the speed of light. In 1931, Dirac furthermore showed that this equation made an astonishing prediction: for each particle known at the time (electron, proton) there should be a partner particle with the identical mass and spin but with equal and opposite electric charge. Specifically, the partner of the electron should be the positron, with identical mass but with positive electric charge $q = +e$. Similarly, the antiproton should have the same mass as the proton, but the electric charge $q = -e$.

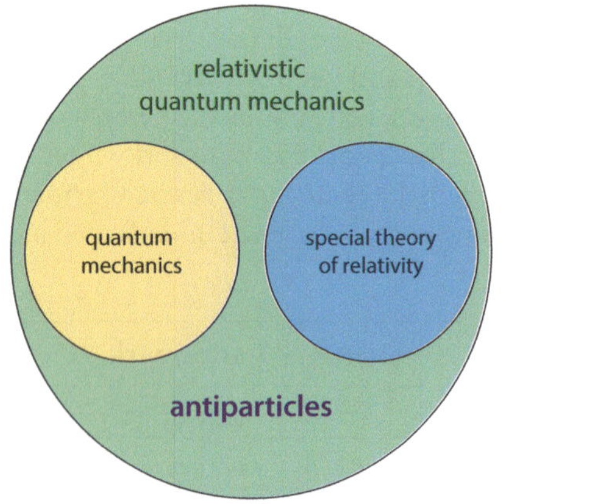

These so-called *antiparticles* are matter particles and have a mass. Thus we call them *antimatter*. They are a necessary consequence of the unification of quantum mechanics and special relativity—and they actually

Concepts of Quantum Mechanics → p. 50
Atomic energy levels: Chance, Interference and Predictions → p. 52
Neutrino Masses → p. 272

exist. In 1932, Carl Anderson discovered the positron in reactions of cosmic rays traversing a detector. The antiproton was discovered in accelerator experiments in Berkeley by Emilio Segrè and Owen Chamberlain in 1955. The antineutron (discovered in 1956, also in Berkeley) exists even though like the neutron it is electrically neutral. This is because, as we know today, the neutron is made of three quarks: two down quarks and one up quark. These in turn are charged and have matching antiparticles, which make up the antineutron.

Meanwhile the antiparticles of the other matter particles in the Standard Model have been discovered: the anti-muon, anti-tau, three antineutrinos and six anti-quarks. The neutrinos are electrically neutral and, as far as we know, they have no inner structure. In principle they could each be their own antiparticle. The neutrino has spin 1/2 and the observed antineutrinos would simply be neutrinos with opposite helicity(↓). Such neutrinos, which are their own antiparticle are called *Majorana neutrinos* (Neutrino Masses ↓). Experimentally they could contribute to neutrinoless double beta decay (↓), which is being searched for intensively today.

The photon has no electric charge and no other inner quantum number. Therefore photons are their own antiparticle. They are not subject to charge conservation and they can be produced and destroyed unrestrictedly. This can be seen, for example, in the effect of a normal light switch. On the other hand, due to electric charge conservation, electrons can only be produced or destroyed together with positrons. In order to describe these processes for electrons, the relativistic quantum mechanics must be extended to relativistic quantum field theory (↓). This theory for example explains the annihilation of an electron and a positron to two or

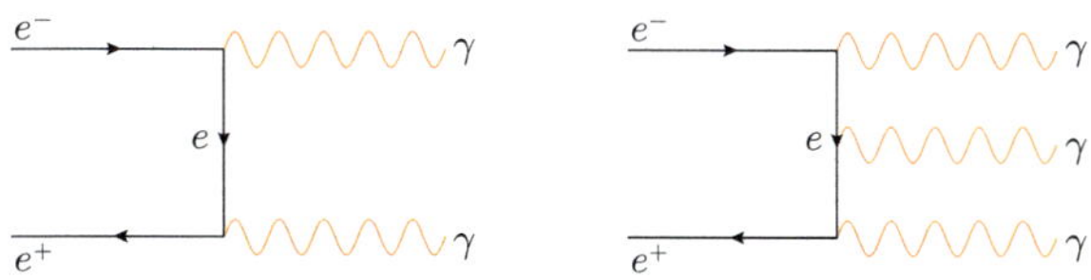

Feynman graphs for the annihilation of electron and positron into two or three photons

three photons. Two photons each with energy greater than $m_e c^2$, the minimal energy of an electron or positron, can also collide to produce an electron and a positron (pair production).

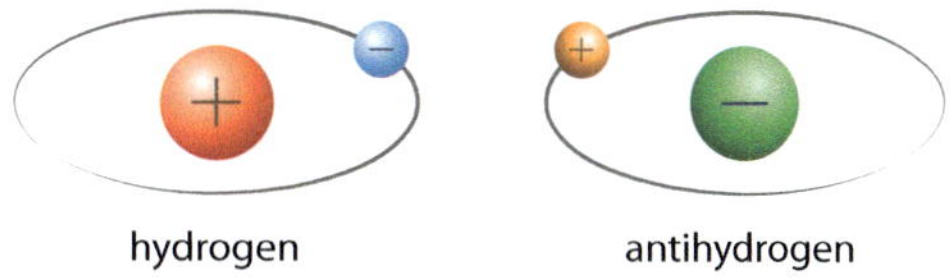

An antiproton can even capture a positron to form an antihydrogen atom. In 1995 this was first observed with the PS210-experiment at CERN. Quantum field theory predicts that the anti-atomic spectra are identical to the corresponding atomic spectra. This is a consequence of the so-called *CPT theorem* (↓). In 2002, the ALPHA collaboration at CERN has shown that this is indeed the case, within the experimental errors. Furthermore in 2023 ALPHA has shown that antimatter behaves identically to matter under gravity.

In the beginning we mentioned that we are made of matter. Now we can also say: we are *not* made of antimatter. Looking out into the Universe we seem to see only galaxies made of matter and none of antimatter. How this imbalance might have come about in the early Universe is discussed in the article about baryogenesis (↓).

Helicity: Up, Down, Left, Right → p. 180
Neutrinoless double beta decay: Searching for a New Experiment → p. 296
Quantum Field Theory → p. 104
CPT theorem: Discrete Symmetries: Violating *P* and *CP* → p. 190
Baryogenesis → p. 328

Detectors in Everyday Life

A Particle Detector in Our Head

In our everyday lives there are surprisingly many particle detectors. For example, in a light barrier at an elevator door, a diode measures particles of light, the photons, which cross the door entrance. The same is true for the sensor of a camera or, in the past, for a light-sensitive film in an analog camera. An electricity meter indirectly measures the number of electrons that flow through the power lines to the house. We even carry a very special particle detector with us at all times: our eyes.

The eye is an amazing instrument: It performs an optical imaging process in which all the light rays coming from a specific point in the field of view, and which fall anywhere onto the surface of the pupil are collected onto a single point on the back side of the eye. Here is our own detection device for the light particles: the retina. It is only 0.2 mm thick, very similar to some detectors in particle physics (Semiconductor Detectors ↓).

The retina contains two different "subdetectors". On the one hand are the cones. They can distinguish between light of different colors and thus different frequency, or correspondingly photons of different energy. This happens because the cones are filled with different pigments, which allow light particles with a certain energy to penetrate into deeper layers. Light particles corresponding to other colors are intercepted. Thus, the eye contains some kind of photon energy detector (↓).

On the other hand are the rods. They are sensitive to all energies of the visible-light particles and are built much narrower than the cones. With their help, healthy human eyes can separately see structures of about 1 mm size at a distance of 5 m. Thus, the eye also contains a detector capable of providing directional information (↓), as in particle physics experiments. The similarities do not end there. Like modern particle physics detec-

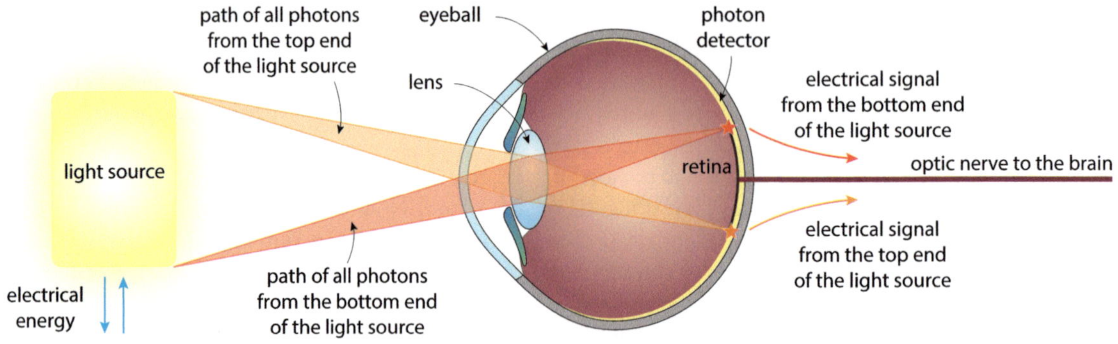

The path of a light signal from the source to the brain

Semiconductor Detectors → p. 70
Energy Measurement → p. 78
Position and Momentum Measurement → p. 76

tors, the eye does not simply send all data to the brain but processes it in the retina before the resulting "reconstructed" data (reconstruction of mass and more ↓) is forwarded—this saves data volume in the optic nerve and accelerates visual perception, as in particle physics experiments.

Another fascinating feature of the eye is its sensitivity. In order to generate a signal, a rod cell needs to be hit by only five light particles (photons) with energy in the range of visible light within about 1/30 s. Particle detectors are usually designed to be very sensitive and, if possible, to detect every particle with a minimum amount of energy. Even more fascinating is the dynamic range: Between the light intensity at which one can just barely see something in the dark and the greatest brightness in which one can still open one's eyes without pain, there is a ratio of about 1/1 000 000 in the density of detected photons.

The dynamic range is also important for particle physics experiments. However, in most experiments, the ratio between the smallest energy transfer that barely generates a signal and the largest signal that can still be distinguished from "even larger" signals is only around a factor of 1 000, rather than 1 000 000, as for the eye.

The eye can thus be regarded as a particle detector. However, despite all the similarities, particle physics detectors are usually not built exactly like the eye. The comparison of the schematic illustrations of the eye and a tracking detector shows the differences. In particle experiments, the particles often move through several layers of individual detectors. The particles are much higher in energy than the photons of visible light and can thus penetrate more than one layer. In some experiments, such as at the Large Hadron Collider (LHC), a single particle must even possess about 200 million times the energy of a typical light particle in order to be identified as a specific particle within the zoo of particles. Therefore, it cannot be stopped in a thin layer or focused with a lens. Instead, a sequence of detectors is used so that the path of the particle can be traced even without focusing. From the curvature of the track of the charged particles in a magnetic field, its momentum can be calculated—just as in the eye the energy of the photons corresponds to the perceived color. Interestingly, in early variants of particle detectors scientists used their own eyes to identify and classify detector signals.

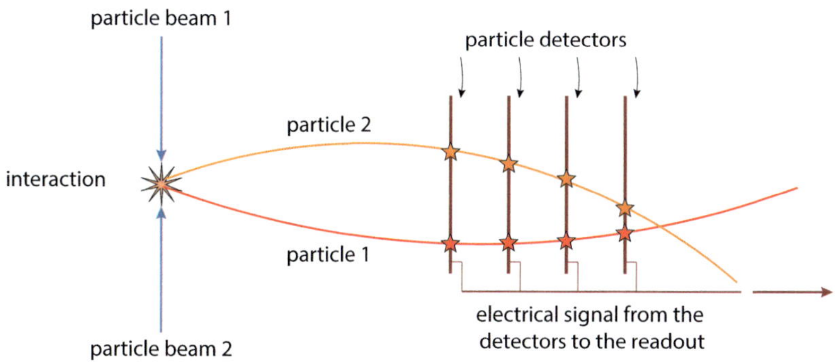

The path of a particle from its creation to the read out of the detector signals

Identifying the Particle Type via the Mass → p. 156

How Does Physics Make Progress?

A Complicated Search For Simple Answers

Scientists have successfully explained many of the mysteries that the Universe and the world around us present. But when can we claim success in our quest to understand Nature on a deeper level? We attempt to illustrate this by recounting the development of our knowledge about gravity. Let's go back to the late seventeenth century. There had been many quantitative observations of the movement of the planets, for example by Tycho Brahe. The geocentric worldview, which places the Earth at the center of the Universe, was overturned and Kepler's three laws of planetary motion formulated. It was Isaac Newton who, in the late seventeenth century, succeeded in formulating a comprehensive, mathematical description of all these observations. This great achievement is a blueprint for the scientific method, even today.

Why was Newton's law of gravitation recognized as a success? One could, for example, have claimed that planets are held on their orbits by angels just so that the orbits seem to follow fixed laws. The difference between this approach and Newton's laws, however, is that the former is not testable and also not suitable for making predictions for other phenomena. In contrast, Newton's theory not only provided a mathematical explanation for Kepler's laws, but also for the laws of free fall, empirically found by Galileo. So, the guiding principles of simplicity, universality, and predictive power have convinced us of the correctness of Newton's approach. And even today these principles play a key role in all sciences.

Newton's theory employs the same mathematical formalism to explain gravity on Earth, as well as the observation that planets move on fixed elliptical orbits (Kepler's 1s law). It matches *almost all* observations excellently. In fact, there is a tiny deviation from Newton's original prediction for Mercury's orbit. It is the closest planet to the Sun, and thus feels its attraction the strongest: The elliptical orbit is *not* fixed in space, but rotates by 0.16 degrees per 100 years. About 90 % of this is due to the influence of the other planets, understood within Newton's theory. Kepler's 1st law only holds for a single planet rotating the Sun. The remaining 0.01 degrees per 100 years *cannot* be explained within Newton's theory. The quantitative description of this deviation is one of the successes of the general theory of relativity by Albert Einstein (1915), which reproduces Newton's theory in the limit of weak attraction.

Does this make Newton's theory "wrong"? As a *universal* exact description of gravity it is indeed refuted. However, it is appropriate to use Newton's theory in almost all everyday and technical applications, as it provides a sufficiently accurate description of Nature. In this respect, the development of Newton's theory was a great success, but in general relativity we have an even more fundamental theory, required, e.g., for GPS navigation. We do not know where this development will end.

To make progress in particle physics, we basically proceed analogously. In the beginning we might have an experimental observation that demands an explanation. Once this is found in the form of a suitable mathematical description, testable predictions can be derived, which in turn may initiate new experiments. Ideally, a consistent picture with greater predictive power gradually emerges.

Similar to the case of gravity, this process has lead to a very successful theory: the Standard Model of particle physics. It allows us to understand the structure of matter and the interactions of its components very well, so that from a few basic principles we can derive the outcome of many experiments with extremely high accuracy. In this book we present many examples (↓). All the same, we already know that the Standard Model is not the long-sought truly fundamental theory, because there are phenomena that it cannot explain. These include the fact that our Universe is dominated by matter, very little antimatter can be found (↓), and we understand neither dark matter (↓) nor dark energy (↓). Therefore, a lot of effort is being put into the search for a more fundamental theory today.

The development of the Standard Model of elementary particle physics, as well as our current understanding of its limits, is the subject of this book. An accurate description of atoms, atomic nuclei and their substructure, as well as the zoo of new particles discovered requires tremendous effort and constant progress in the development of measurement methods, instruments, and accelerator facilities. All this is only possible with the help of suitable specialization among the participating researchers. Parallel to this, the theoretical description of the observed phenomena must be advanced. Therefore, in modern particle physics one finds scientists working in accelerator physics, experimental physics, and theoretical physics. Their research is carried out in a symbiotic relationship. In this book we present the research fields of the three groups. You will gain insight into their work, but also into the employed methods. We hope to show where this exciting journey could go.

Newton realized that the fall of the apple to the Earth has the same reason as the "fall" of the Earth and Moon around their common center of gravity.

The Hurdles of Success → p. 266
Matter and Antimatter → p. 314
Dark Matter → p. 282
Dark Energy → p. 284

Experimental Research in Particle Physics

Diverse Experiments Addressing Fundamental Questions

In order to investigate the smallest building blocks of matter, we must construct the largest and most complex scientific machines ever built by humankind. This seemingly paradoxical connection is related to the quantum mechanical (↓) nature of these smallest building blocks. In quantum mechanics large energy corresponds to small wavelengths, which are needed to *resolve* the small length scales. This can be achieved by accelerating particles to very high energies within our large facilities. The Universe itself is also a source of very high-energy particles, in the form of cosmic rays (↓). However, investigating cosmic rays and their reactions also requires large and sophisticated detectors.

This explains the scale, but not the complexity. Why do we need such sophisticated detectors to make sense of the natural world? The reason is unexpectedly simple: Particle physics is already well-acquainted with nearly all common processes. To deepen our understanding, we must study rare processes. However, to observe such processes without waiting for several human lifetimes, we have only one option: generate an extremely large number of particle interactions in a short time and attempt to find the proverbial needle in the haystack, a special event within billions of particle collisions (Trigger and Data Processing ↓). This can only be achieved with ever more complex experimental equipment. The required detectors and accelerator facilities act as technological drivers with often unanticipated benefits.

Present-day experiments have reached such a level of sophistication that their realization is only possible through intense international collaborations. Physicists and other experts in electronics, accelerator technologies, semiconductor manufacturing and high-performance computing from all over the world are required to jointly build them. If we trace the origin of particles in a typical experiment, their journey often begins with the work of accelerator physicists: they develop, construct, and operate the facilities (Principle of an Accelerator ↓), which accelerate the initial particles to high energies. For instance, at the world's largest accelera-

A small fraction of the members of the ATLAS Collaboration

An underground experimental cavern under construction at CERN

Concepts of Quantum Mechanics → p. 50
Cosmic Rays → p. 336
Trigger and Data Processing → p. 150
How Does a Particle Accelerator Work? → p. 66
How Does Physics Make Progress? → p. 26 and A Deep Dive Into the Scientific Method → p. 30

tor, the Large Hadron Collider (LHC) at the European Organization for Nuclear Research, CERN, in Geneva, thousands of physicists from around the world collaborate on the development of the necessary technology, and hundreds work at CERN on its operation. These accelerators are typically housed within large particle physics laboratories such as CERN.

When a new accelerator complex is planned and developed, another group also forms in order to measure the results of particle collisions created by the accelerator: international collaborations of experimenters plan, construct, operate, and later analyze the data obtained from detectors to investigate the reactions of high-energy particles. For instance, at the LHC, four major experiments were built, with each collaboration consisting of several thousand members, who self-organize. The funding required to operate and continuously refine their detectors comes from many countries across the globe.

Experimental particle physicists measure the fundamental properties of Nature and the forces that act between them. Another group, theoretical particle physicists, develop mathematical theories of Nature and then compute predictions for experimental outcomes based on them. They work individually or in groups and regularly meet all over the world to discuss their ideas (Progress, Scientific Method ↓).

Besides the LHC, there are many smaller accelerators and experiments. Despite their smaller size and lower energy, they can also contribute important results to particle physics. Additionally, various types of experiments do not require accelerators at all, as Nature or other technical facilities provide particles in alternative ways. These include experiments conducted on satellites in space (↓), deep within mountains (↓), neutrino detectors adjacent to nuclear reactors, or a cubic-kilometer-scale detector embedded in the ice cap at the South Pole (↓).

Collaboration often fosters innovation. Particle physics experiments have given us profound insights into the inner workings of the Universe, but in addition, several intriguing byproducts relevant to everyday life have been developed. An example for this is the World Wide Web (WWW), whose invention marked the beginning of the modern information age. The Web was created by Tim Berners-Lee at CERN to facilitate easier information exchange across international collaborations of particle physicists. At its core, research in particle physics is a global collaborative effort with open exchange of ideas and information. The knowledge particle physics creates is pieced together by physicists to form a coherent picture of the fundamental building blocks of Nature, their interactions, derivable properties, and the history of the Universe. This is the essence of this book.

The magnetic system of the ATLAS detector at the Large Hadron Collider during construction

Measurement Uncertainties → p. 86 and Statistics → p. 84
Particle Physics in Space → p. 316
Direct Search for Dark Matter → p. 310
IceCube: Large Detectors Deep in a Mountain → p. 132
The Invention of the World Wide Web → p. 172

A Deep Dive Into the Scientific Method

Establishing the Difference Between "Truth" And "Knowledge"

The scientific method has been the catalyst for unprecedented progress in our understanding of the natural world in the last centuries. On these pages, we want to outline some of its interesting features and investigate what makes it so successful. From the start it is important to note that *the scientific method* as one specific, well defined procedure does not exist. The method has been adapted by different research areas to meet their respective challenges and answer their research questions, and its implementation has changed over time. Furthermore, not all scientists agree on the details of the scientific method. One key feature most adaptions possess, however, is that progress depends on critically questioning what is known, and frequently revising past findings and methods. There are further common features among all quantitatively oriented fields in the natural sciences on how they use the scientific method.

An essential aspect of the scientific method is the comparison of theory and experiment. Here, however, it is important to realize the method's inability to assert the 'truthfulness' of explanations for observed phenomena. Instead, quantitative science constructs mathematical models that delineate predictable observable phenomena. Or in different words, we cannot ascertain the absolute truth of any explanation in all their statements and under every condition, as conceptually there could be an infinite number of possible different explanations for the same given observation. However, we possess a precise understanding of the measurable properties inherent of an observable process, and which known sets of theories correctly or incorrectly describe these observations. In other domains of mathematically-driven natural sciences, this process is similar but not always immediately evident. For instance, a statement about the melting of a glacier might initially seem more tangible to many observers than a statement about the properties of the Higgs boson (↓). However, upon closer examination, the distinctions fade: While one can physically touch the glacier, making it more directly perceptible, all predictions regarding its past or future behavior rely on mathematical models. The effectiveness of its sliding on the subsoil, the evolution of heat distribution within, or the uncertainties in measurements—all are based on cautious extrapolation from one measurement to another and from laboratory insights to real-world scenarios. The quality of such a sophisticated model for the glacier can only be assessed when predictions are experimentally tested.

The situation in particle physics is very similar: a model is considered adequate if, with the simplest assumptions, it can precisely describe all phenomena whose existence was used to construct it. It rises from adequacy to greatness if it makes new predictions that have never been experimentally explored and are confirmed only after the creation of the model! Hence, there is a dedicated branch of research: theoretical physics, which, in the pursuit of an increasingly sophisticated theory of Nature, derives predictions from fundamental assumptions. These predictions, in turn, can be tested in experiments. If a prediction agrees with an observation, within the uncertainties (↓), then the model is considered successful and will be further scrutinized.

Higgs mechanism: Spontaneous Symmetry Breaking III → p. 188
Determining Measurement Uncertainties → p. 158

If a prediction does not agree with the measurement, even when considering all possible uncertainties in the numerical calculation or the measurement, then the model, along with the assumptions underlying its predictions, is refuted. A measurement not agreeing with a well-founded prediction is not a failure of science, but opens the door for alternative explanations. Often, these situations are the basis for profound and deeper understanding, resulting in a state that is the most crucial and creative science ever can be in.

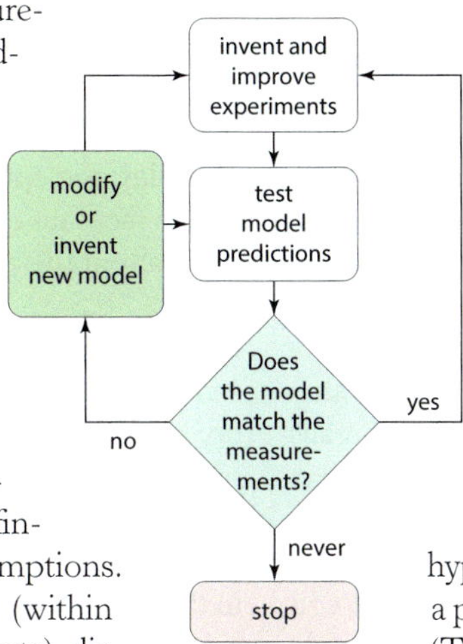

It is an intrinsic property of the natural sciences that they can never definitively prove their theories or assumptions. They can, however, with certainty (within the framework of statistical statements) disprove them. A currently successful prediction could always be contradicted by a more precise experiment in the future. If that happens, however, it is unclear *what* exactly has been disproven: the fundamental idea of the theory, the numerical calculation of its predictions, the translation of the prediction to a measuring instrument, the analysis of the signals from the measuring instrument, and so on. The entirety of this process of challenging our understanding of Nature is incredibly fascinating. Unraveling the reasons for each deviation between theory and experiment is the creative art of the natural sciences. Although scientific theories can never ultimately be considered "true", they are far from arbitrary. The successful theories, models, and methods are the few survivors of a ruthless selection process, sieving out all disproven models and uncertain measurement methods. Thus the natural sciences have gained immense, objective *knowledge* about Nature, without claiming to know the absolute *truth* of the explanations.

In this context, particle physics must continually grapple with situations where there is no single model that truly explains *everything*, as it has not yet been possible to construct a single model that accurately describes all measurements. However, this is neither a crisis nor a downside of the natural sciences; rather, it is their normal state. There are often various ways out of such a situation. For example, one might create a new additional hypothesis, which might later pave the way to a possible successful expansion of the theory. (This is not an easy task—this new component must not make a single incorrect prediction for *all* previous experiments.) Alternatively, one can embark on the more challenging but sometimes more promising path: attempting to restructure the entire theoretical framework on a new foundation, which is simpler, more comprehensive, and often aesthetically perceived as more *beautiful*.

Such revolutions in our conceptual understanding of the Universe—born out of pure curiosity—generate immense leaps in knowledge about the natural world, laying the indispensable foundation for our modern, highly technological knowledge-based society, often centuries later. Whether such further revolutions will succeed in the future cannot be guaranteed, but scientific pursuit does not waver.

Imagining Reality

How Different Descriptions of Reality Can Coexist

Most people have a natural sense of what "exists": something you can see, touch, hear, taste, or smell. An example of this could be a beach goer witnessing a distant ship sail over the horizon. The ship evidently exists because it's visible. However, this is not as straightforward as it seems: our brain registers signals sent by our eyes. These signals are generated by the eye from photons hitting the retina and translated into signals in a complex process (Detectors in everyday Life ↓). So, we perceive processed signals from the ship, not the ship itself. Every movie is proof that a deceptively realistic-looking ship may not be a real ship at all. Moreover, our eyes are not always truthful, as optical illusions demonstrate—or did you notice that the circles in the left illustration are not interwoven, but arranged concentrically? Or that the lines running from left to right in the right illustration are not slanted, but parallel?

Optical Illusions: Do all circles on the left have a common center? Are all horizontal lines on the right parallel with respect to each other?

Observations in particle physics also rely on complex measurement processes, and their interpretation is further complicated by the fact that the objects being investigated are so small that they completely elude our everyday perception. The situation is not made simpler by the fact that quantum mechanics (↓) compels us to think about these objects in a completely different way.

Particle physics creates mathematical models that describe observable and predictable phenomena in a theory of the most fundamental particles and forces. The process of how these concepts or theories are created, developed, and altered can be well illustrated by the progress of our knowledge about the atom. Let's start in the year 1908. It was known that chemical and physical processes could be well described by assuming that matter is composed of atoms as immutable building blocks (Elementary Particles ↓). However, the structure of these atoms was unclear. This is where the experiments of Hans Geiger and Ernest Marsden, known as the *Rutherford Experiment* (Scattering Experiments ↓), come into play. Geiger and Marsden observed that alpha particles (From Nucleons to Nuclei ↓) shot at a thin gold foil typically passed through the foil without deflection, but in rare instances were scattered almost back in the direction of the source. What was observed was the result of a scattering experiment, not the structure of the atom directly. The atom was probed with energetic alpha

Detectors in Everyday Life → p. 24
Concepts of Quantum Mechanics → p. 50
What Makes a Particle Elementary? → p. 18
Scattering Experiments → p. 46
From Nucleus to Nuclei → p. 226

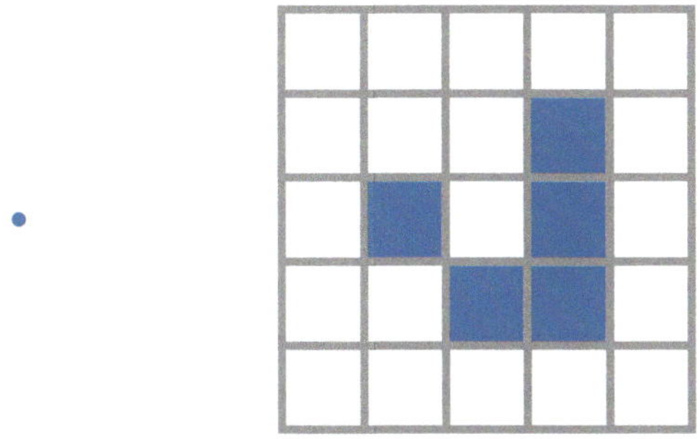

"Is" the electron a point-like object without any intrinsic volume (left) or a hypothetical "glider" in an extreme hypothetical simulation[1] (right)? As long as both descriptions match the experimental data, we cannot distinguish them from each other. Both would be equally grounded and justified in science.

particles, not with our eyes. Ernest Rutherford interpreted these results as follows: the atom does not have a continuous mass distribution, but is mostly empty. Most of the mass is concentrated in a very small and very heavy nucleus that is rarely hit by the alpha particles. But if the alpha particle does encounter the nucleus, it undergoes a large deflection, sometimes even right back to its source with a deflection angle of 180°.

The atomic nucleus, "seen" as a heavy point-like object in the atom for the first time in the imagination of Ernest Rutherford, is "real" because physics can also precisely explain other experiments using it, not because it is directly "visible" to us humans. Nowadays, physicists know that the atomic nucleus is not point-like, but is extended to about 10^{-15} m, and consists of neutrons and protons. These, in turn, are made up of quarks and gluons. And is it possible that these are not point-like and elementary constituents, but consist of yet unknown particles themselves?

The conclusion drawn from this development cannot be that Rutherford's abstract notion of the atomic nucleus as a heavy point-like object is wrong or that the atomic nucleus has no correspondence in reality. Although physics changes and improves its precise ideas and content of theories about Nature, the experimental results once observed and described do not disappear retroactively, even when an improved description emerges. Hence, the nucleus has survived the progress of science as a useful concept, even as over the course of 100 years our knowledge of it evolved from a positively charged point particle to an extended object governed by the strong force and made of protons and neutrons.

Hence, the concepts of modern physics are not merely formulas without any underlying significance for reality. Crucial for the advancement of understanding fundamental building blocks is the notion of *representation*: Progress in understanding fundamental physics often occurs when new phenomena are not simply described by an empirically derived formula but by an *idea* represented by the formula. To stick with Rutherford's example: One could describe the frequency of backscattering of alpha particles with a spontaneously formulated probability and draw no further conclusions from Geiger's and Marsden's experiment. Alternatively, one can investigate whether the idea of the atomic nucleus, explaining the observed probability of backscattering, has far-reaching implications in many other areas. This is precisely what physics has done. Scattering experiments have since helped to unravel a much deeper substructure of matter, and are accompanying us up to the Large Hadron Collider of our time. The latter approach, deliberately using theories *representative* of ideas and exploring all their predicted consequences, is the path that runs throughout this entire book.

[1] For the interested reader: This pun is grounded in "Conway's Game of Life", https://playgameoflife.com/

γ
e⁻
N
S

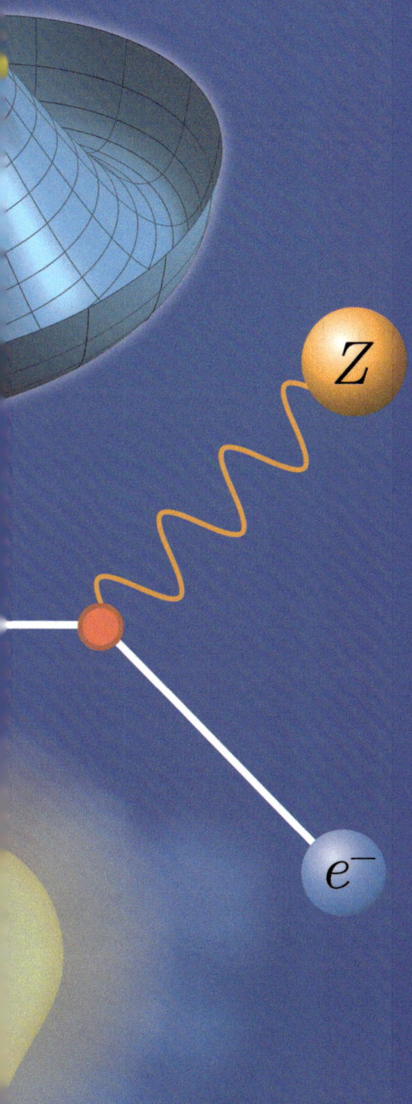

2 Fundamental Principles

Before delving into the main content of this book, let us first review some basic concepts that will be essential throughout. At the Large Hadron Collider (LHC), protons in the accelerator and the newly produced particles typically travel at speeds very close to the speed of light. To understand their behavior, we must draw on the principles of special relativity.

In particle physics, we work at the subatomic level, so it is crucial to understand key concepts of quantum mechanics. This includes how we can make accurate predictions despite the inherent randomness of individual events, such as in radioactive decay. We also explore the concept of spin, a fundamental property of particles that describes their intrinsic angular momentum. Additionally, we discuss how particles can be created, destroyed, and even decay spontaneously.

In order to learn something about these unimaginably small particles, we carry out huge scattering experiments. What are these particles and why are we performing these experiments? And how can we use relatively simple graphical means, the so-called Feynman graphs, to illustrate and explain the reactions of elementary particles?

Finally, we examine the crucial role that symmetries play in understanding particle physics—even when they are only approximate or have been broken!

Energy and Power

Kilowatt Hours, Electron Volts—How Does It All Fit Together?

The article about plasma acceleration (↓) mentions an 850 terawatt ($= 850 \cdot 10^{12}$ watt) laser. However, a power plant typically only delivers one gigawatt (10^9 watt) of electrical power, almost a million times smaller. How does this fit together? This book repeatedly states the energy in electron volts. What does this have to do with the unit kilowatt hours, which you find on your electricity bill? This article addresses these questions in detail.

The key to understanding these questions is the distinction between energy and power. Consider two Tour-de-France riders, who climb a mountain pass. They both have done the same amount of work and thus gained the same amount of energy, if we, as so often in physics, make some approximations. Here we neglect friction losses and assume that both riders plus bikes have the same mass m. The work done is then simply calculated as

A pumped storage power plant stores energy by raising water from the lake below to the lake above.

$$W = m \cdot g \cdot h \ ,$$

where h is the height gained and $g = 9.81\ \mathrm{m/s^2}$ is the acceleration due to gravity. This work has increased the energy stored in the cyclists, here the potential energy. This is exactly what is used in a pumped storage power plant. Water is pumped up, e.g., with excess energy from wind turbines. This increases its potential energy. Later, when needed, the potential energy of the water can be converted into electrical energy by releasing the water to power a turbine.

Now let's move on to power. The cyclist who climbed the mountain faster has exerted the greater amount of power because power is work divided by the time it took to perform the work. Let us take a closer look at the units. According to the previous formula, work has the unit $\mathrm{kg\,m^2/s^2}$. This is also known as one joule (J). Therefore, power (work per time) has the unit $\mathrm{J/s = kg\,m^2/s^3}$, for which the unit watt is also common. More details can be found in the information box on the right.

On your electricity bill you will find the unit kilowatt hour or short kWh. This corresponds to an energy, as a power ($\mathrm{kW} = 10^3\,\mathrm{W}$) is multiplied by a time (h = hour). On a light bulb, the power rating is given in watt. What is this all about? You pay the electricity company for work or energy. So it doesn't matter whether you let a 100 watt bulb burn for one hour or a 50 watt bulb for two hours. In both cases, the electricity company has to provide the same amount of energy.

Plasma Acceleration → p. 120

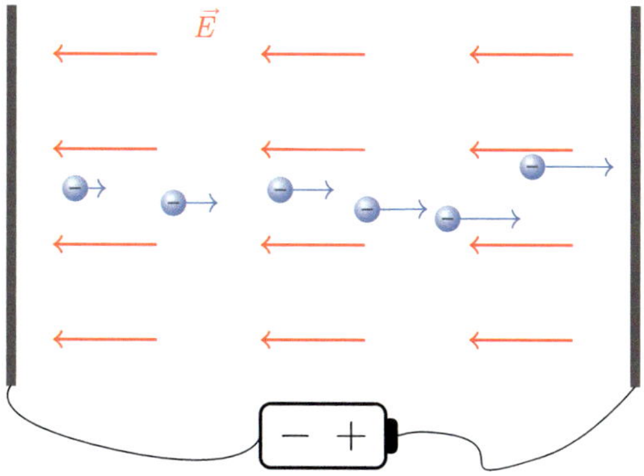

Electrons are accelerated between two oppositely charged plates. In the process, they gain kinetic energy.

Let's return to the example with the laser. The enormous power of the laser only needs to be provided for a fraction of a second, typically 10^{-14} s. This means that the laser only requires an energy of 8.5 joules for such a short pulse. This can be calculated by multiplying $850 \cdot 10^{12}\,\mathrm{W} \cdot 10^{-14}\,\mathrm{s}$. As shown in the information box, this energy could be provided by a simple battery.

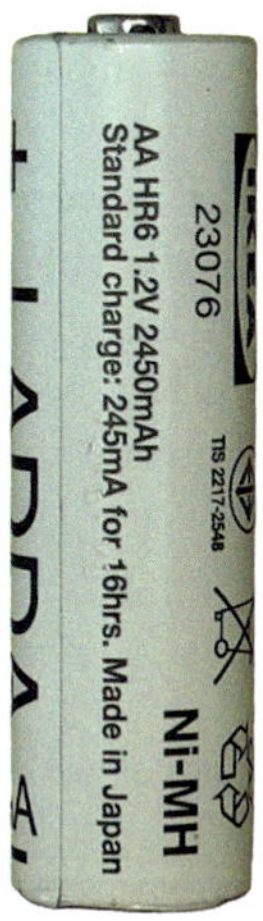

So, what does the unit electron volt, which is used so often in this book, mean? An electron that passes through a voltage of one volt gains one electron volt (eV) of energy. As shown in the lower part of the information box, an electron volt corresponds to $1.6 \cdot 10^{-19}$ J. Expressed in joules, the numerical value is very small. Therefore, in particle physics, calculations are preferably performed in electron volts.

Conversion of Energy Units

Expressed in the basic units of meter, kilogram, and second, energy has the unit

$$1\,\mathrm{kg\,m^2/s^2} = 1\,\mathrm{J}\,.$$

To simplify notations, the unit joule (J) was introduced. Since power is energy per time, the unit is

$$1\,\mathrm{kg\,m^2/s^3} = 1\,\mathrm{J/s} = 1\,\mathrm{W}\,,$$

which is also referred to as watt (W). Of course, we also have $1\,\mathrm{J} = 1\,\mathrm{Ws}$. Since an hour (unit: h) has 3600 s and $1\,\mathrm{W} = 1/1000\,\mathrm{kW}$, one can write

$$\begin{aligned} 1\,\mathrm{J} &= \mathrm{W} \cdot 1/3600 \cdot \mathrm{h} = 1/3\,600\,000\,\mathrm{kWh} \\ &= 2.78 \cdot 10^{-7}\,\mathrm{kWh}\,. \end{aligned}$$

An electron passing through a voltage of one volt gains one electron volt (eV) of energy.

The elementary charge is $e = 1.6 \cdot 10^{-19}$ coulomb (C). In electricity, one learns that 1 coulomb × volt equals 1 joule, so

$$1\,\mathrm{eV} = 1.6 \cdot 10^{-19}\,\mathrm{J}\,.$$

A coulomb is the same as an ampere × second. Looking at a standard 1.2 V battery with the specification 2450 mAh (milliampere hours), you can calculate the energy stored in it:

$$E = 1.2\,\mathrm{V} \cdot 2450\,\mathrm{mAh} = 10\,584\,\mathrm{J}\,.$$

Exactly this energy is gained by a mass of $m = 1$ kg, when it is raised in the gravitational field of the Earth by $h = 1079$ m, because

$$E = m\,g\,h = 10\,584\,\mathrm{J}\,.$$

Here, $g = 9.81\,\mathrm{m/s^2}$ is the acceleration due to gravity.

Special Relativity

Is There a Speed Limit on the Autobahn?

Light literally moves as quick as a flash, with the speed $c = 2.998 \cdot 10^8$ m/s. In our daily lives this is unimaginably fast. But a light or radio signal takes about 1.25 seconds from the Earth to the Moon and it takes on average 12 minutes for a signal to reach a Mars rover. This has a strong impact on communication. In particle physics the speed of light is also of immediate relevance. At the LHC the protons fly through the accelerator at 99.9999989 % of the speed of light.

Motion close to the speed of light is described by the special theory of relativity. It is based on two principles, both of which are experimentally confirmed

1. The relativity principle: Let there be two observers who move relative to each other at a constant linear velocity, e.g., one person on a uniformly passing train and the other on the platform. The principle of relativity states that the laws of Nature are the same for these two observers. Such stationary or uniformly moving observers, or reference systems, are also called *inertial systems*.

2. Constant speed of light: The speed of light is the same in all inertial frames of reference even if the light was emitted in a different system. Thus, a beam of light emitted by a moving train also has the speed c for the observer on the platform.

We can immediately understand the principle of relativity: A game of table tennis or billiards on a smoothly moving train is no different from one in the station concourse. The second postulate, however, is strange: a ball thrown from a moving train in the direction of motion is of course faster for the observer on the platform than for the one on the train. However, surprisingly, this is *not* the case with light.

As we shall see the resulting effects are very important for particle physics. To understand this better, let's first consider as a simpler example the measurement of the transit time of a light signal in a train, compared with the measurement of the transit time of the *same* light signal viewed from the platform. The extension to particle physics is discussed in a separate article (↓).

Let's assume a woman is standing in a train compartment, where there are two horizontal mirrors separated by a vertical distance ℓ_A. One mirror is on the ceiling and one on the floor (figure on the left). Between the mirrors, the woman can send light signals back and

Since the speed of light is constant, but the distance ℓ_B is greater than ℓ_A, the transit times of light signals in the two reference systems on the left and right are different.

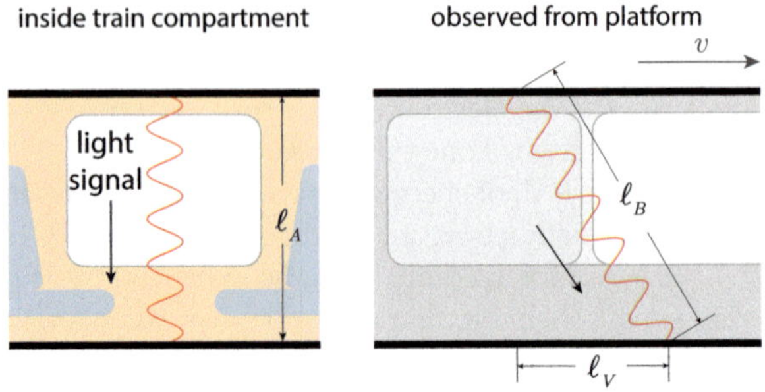

Special Theory of Relativity in Particle Physics → p. 92

forth. By counting the laps traveled by the light she has a clock with a unit of time $t_A = \ell_A / c$, instead of, for example, seconds. The length, i.e., the distance between the mirrors as a function of time, is: $\ell_A = c\, t_A$.

Now the train passes a man on the platform at speed v. For him, the two mirrors on the train are also a clock, however, as we will see, with a different unit of time t_B. In the time in which the light signal flies once from the top mirror to the bottom one, the train moves forward a distance $\ell_V = v\, t_B$. Here we use t_B as the man on the platform measures this distance and we must employ his measure of time. Due to this movement, he observes the light to follow a longer, a diagonal path. He sees the distance covered in the time $\ell_B = c\, t_B$ (see figure on the right). According to the second principle, we must use *the same* speed of light c, even though the light was emitted by the moving train. As you can see from the figure, the three lines form a right-angled triangle. The Pythagorean theorem therefore results in $\ell_V^2 + \ell_A^2 = \ell_B^2$. If we replace the distances by the corresponding products of velocity and time, we get

$$v^2 t_B^2 + c^2 t_A^2 = c^2 t_B^2 \,.$$

We can solve this equation for t_B and obtain

$$t_B = \frac{t_A}{\sqrt{1 - \frac{v^2}{c^2}}} = \gamma\, t_A \,,$$

where we have introduced the so-called *boost factor* "gamma" (γ).

$$\gamma = \frac{1}{\sqrt{1 - \frac{v^2}{c^2}}} \,.$$

We have found: In order to guarantee that both observers measure the same speed of light, they must measure different time intervals! This is due to the longer distance in the second case. The difference is called *time dilation*. Since v is always smaller than c, we must have $\gamma \geq 1$ and therefore $t_B \geq t_A$. Thus the clock on the platform runs *slower* than the clock on the train.

In the special theory of relativity, a *length contraction* can be derived in a similar way for distances in the direction of the velocity v. In our case, the woman, who is at rest in relation to the train carriage, measures its length as L_W. For the man on the platform, the train is passing by at velocity v. He measures the length L_M. The special theory of relativity says that $L_M = L_W / \gamma$. The man sees the carriage shortened by a factor $1/\gamma$.

With related arguments (not shown here) and by comparing the energies of an object seen from the train and from the platform, one can derive Einstein's famous formula

$$E_0 = mc^2 \,.$$

Thus the mass of a particle is a form of energy, here more precisely the *rest energy*, which is indicated by the index 0. The mass (in kg) is multiplied by the very large factor of the speed of light squared. This formula applies in the particle's rest system. If, for example, a proton with mass m is at rest in a laboratory experiment, it has the energy $E_0 = mc^2$. However, if a proton moves with the speed v, e.g., at the LHC, it has the total energy in the laboratory system

$$E = \gamma m c^2.$$

Here γ is the boost factor again. When v approaches the speed of light, γ becomes larger and larger. You would need an infinite amount of energy to bring the proton to the speed of light. Therefore the speed of light is an absolute upper limit, even on German highways.

Kinematics

Momentum and Angular Momentum

Kinematics is the mathematical description and measurement of motion. In everyday life, we describe movement in a simple way: an object is located at a given time t at the location x, y and z in the three spatial dimensions. Here, e.g., x represents the location in the left/right, y in the up/down and z in the forward/backward direction. The three components x, y, and z together form a three-dimensional *vector*, $\vec{r} = (x, y, z)$. An object moves with speed v_x, v_y and v_z to the left or right, up or down, and forward or backward. The three components v_x, v_y and v_z also form a vector, the velocity $\vec{v} = (v_x, v_y, v_z)$, which describes the rate of change of location in the three spatial dimensions. The size of the mass m describes how much matter is moving. Only these four quantities (m, v_x, v_y, v_z) are required to quantify movement. The speed v is defined as the magnitude of the velocity vector $v = |\vec{v}| = \sqrt{v_x^2 + v_y^2 + v_z^2}$.

What about relativistic speeds in particle physics? The principle is exactly the same, with the restriction that the speed is necessarily smaller than the speed of light $v < c$, due to the special theory of relativity (↓). It is important to note that the speed is limited, but *not the energy*. In relativity the energy is $E = \gamma m c^2$, where $\gamma = 1/\sqrt{1 - v^2/c^2}$ is the Lorentz boost factor. The closer v gets to c, the greater γ and thus the energy. However, the speed itself always remains below c for particles with $m > 0$.

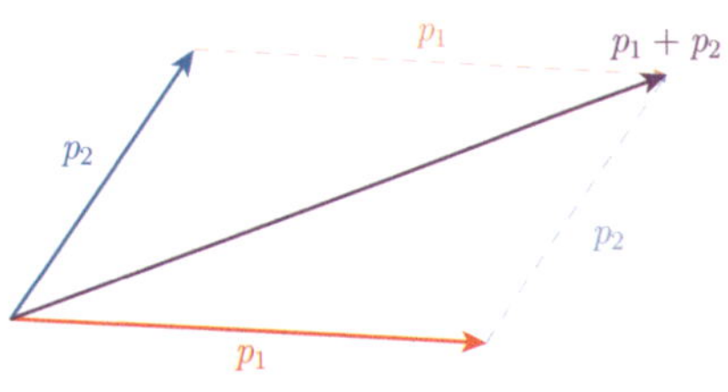

The addition of two momenta $\vec{p}_1$ and $\vec{p}_2$

In the relativistic case, employing the speed is not so useful, e.g., at the LHC almost all values are very close to c. A proton with energy $E = 20$ GeV has a speed of 99.89 % of c. At 50 GeV it is 99.98 % of c. The speed changes by less than one per mille, whereas the energy has more than doubled. It turns out that the momentum is a more suitable quantity instead. It is also a vector and has three components for the three spatial directions: $\vec{p} = (p_x, p_y, p_z)$.

In classical mechanics, the momentum is defined as $\vec{p} = m\vec{v}$, (i.e., $p_x = m v_x$, $p_y = m v_y$, $p_z = m v_z$), i.e., as mass times velocity for small velocities (γ very close to 1). This must be generalized in the special theory of relativity to $\vec{p} = \gamma m \vec{v}$, where γ is again the Lorentz boost factor. Thus the momentum, like the energy, is not limited. Overall we have the relationship between energy and momentum

$$E^2 = (m c^2)^2 + (\vec{p} c)^2 .$$

Unlike momentum, energy is not a vector, it is just a number (with units). It contains contributions from the motion, $\vec{p} c$, as well as from the rest energy $m c^2$.

The momentum of electrically charged particles can be measured in the LHC detectors. The detectors have a constant magnetic field B. In this field, the trajectory of the charged particles is curved, with a radius of cur-

Special Relativity → p. 38
Position and Momentum Measurement → p. 76
Concepts of Quantum Mechanics → p. 50

vature R. Then the momentum is $p = qBR$, where q is the electric charge of the particle. Since q and B are fixed, the momentum is directly proportional to the measured radius of curvature. Therefore the trajectory must be determined as accurately as possible.

Momentum is a particularly important quantity because it is conserved in all particle physics collisions. The sum of the momenta of all particles before the collision is equal to the sum of all momenta after the collision. Here, the momenta must be added as vectors, such as the momenta $\vec{p}_1$ and $\vec{p}_2$ in the figure. If, for example, a neutrino is produced in an event at the LHC, which leaves the detector unobserved, the observed momenta perpendicular to the incoming proton beam will not add up to zero, although the initial state has no momentum in the perpendicular direction.[1] This is referred to as ***missing transverse momentum***, to indicate that not all final state particles in the reaction were observed and thus reconstructed.

A photon can collide with an electron at rest, after which the electron moves, i.e., has momentum. Accordingly, we must also attribute momentum to the photon, even though it has zero rest mass. The momentum p_γ of a photon of energy E is $p_\gamma = E/c = hf/c$, where f is the frequency of the photon (see Quantum Mechanics ↓).

We now consider another variable, the ***angular momentum***. It is a measure of the rotational movement around a point. For a circular motion of a particle with momentum p, the angular momentum L is $L = rp$, where r is the distance of the particle from the center of the circle. The angular momentum is also a vector whose direction is given by the "right-hand rule", see figure. It points vertically out of the plane of rotation.

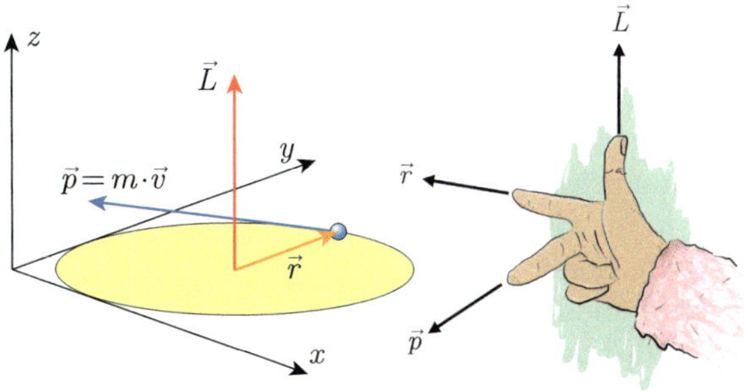

The right-hand rule for the direction of the angular momentum.

Angular momentum is also a conserved quantity, like momentum and energy. Skaters, for example, make use of this: When they pull their arms inwards during a pirouette, they reduce r. Since L is conserved, p must be increased and they turn spectacularly faster (see also the illustration in the article on P and CP violation ↓).

In quantum mechanics, angular momentum is quantized, i.e., only fixed values are permitted: 0, 1/2, 1, 3/2, ... The orbital angular momentum of moving particles discussed above can only take on integer values. However, in quantum mechanics all particles have an ***intrinsic*** angular momentum, called ***spin*** (↓). This can also take on half-integer values. The electron and the proton each have spin 1/2. They always carry this angular momentum with them even if they are not moving ($v = 0$). Half-integer spins, as with the electron and proton, are a purely quantum mechanical effect. The photon has spin 1, *i.e.*, it has twice as much angular momentum. In the interactions in particle physics, the angular momentum is conserved. Often only the spin of the particles contributes, such as in the decay of the pion to a muon and a neutrino (↓).

[1] Here we only look at the perpendicular direction, since (charged) particles in the extreme forward direction can escape undetected within the beam tube.

Discrete Symmetries: Violating P and CP → p. 190
Spin → p. 44
The Weak Decay of Pions → p. 192

Acceleration and Radiation

If You Accelerate, You Radiate

When an electric charge is accelerated, it radiates energy in the form of electromagnetic waves. There are many technical applications for this, e.g., radios, cell phones, and X-rays. In Nature, sunlight or the infrared heat radiation from embers in a fire are just such waves. In circular particle accelerators with electrons this radiation is very important because it leads to a large loss of energy, which must be continuously resupplied. However, this *synchrotron* radiation is also extracted from the accelerator and employed to analyze molecular structures (Synchrotron Radiation ↓).

In daily life, "acceleration" denotes an *increase* in speed. If you pedal harder while cycling, you speed up—you accelerate. In physics, acceleration means that the velocity *changes*. Braking to slow down is therefore included; we speak of negative acceleration.

If you whirl a ball on a string around your head, the speed can always be the same, but the direction is constantly changing. In physics, velocity is a vector (Kinematics ↓): it has a magnitude *and* a direction. Even if only the direction changes this is an acceleration, and for a charged particle, energy is radiated.

Let's take a closer look at an accelerated charge. In the metal of a vertical radio antenna, the electrons oscillate up and down. They first move upwards, are slowed down, turn around and accelerate downwards. Then they slow down again, turn around and fly upwards. They constantly change the magnitude and direction of their velocity, and emit radio waves.

In a cell phone, the antenna is smaller and the frequency is higher, but the same thing happens. In an X-ray tube, very fast electrons are shot at a piece of metal. There they are stopped over a very short distance. The electrons are therefore negatively accelerated and emit X-rays. In a hot piece of metal, the electrons wobble back and forth strongly and emit red or white light, depending on the temperature of the metal. In a circular accelerator, the electrons or protons constantly change direction because they fly in a circle: they radiate energy. At the LHC, the protons lose around 7 keV or one billionth of their energy per revolution.

But how exactly is this radiation generated? In the first figure we see a representation of the electric field lines of a stationary electron, the blue dot in the center. They extend radially outwards to infinity and do not change.

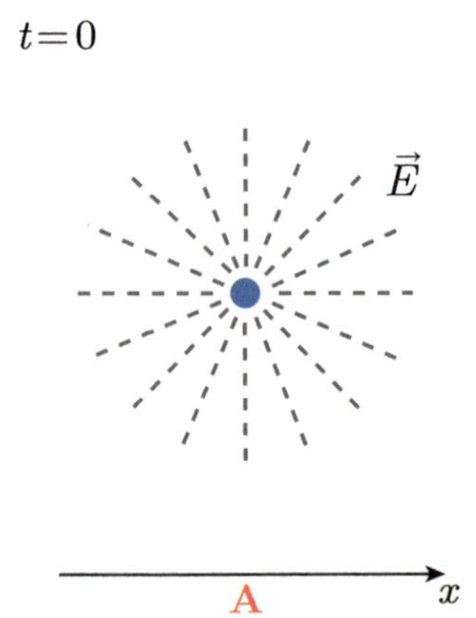

Electric field lines of a stationary electron

Synchrotron Radiation → p. 126
Kinematics → p. 40

The electron does not radiate, the field does not transmit energy. If the electron now moves along a straight-line with uniform velocity, it can also not radiate. The special theory of relativity (↓) states that the same physics applies in uniformly moving reference system as in stationary ones. But why do *accelerated* charged particles radiate? We will try to clarify this using the other two pictures.

Now, as shown in the second figure, within a short time we move the electron from point A to the right to point B. Then the electric field shifts, but only in a small range. In this short time, a light signal could only cover the distance to the light blue ring. This means that the electric field could only change within the light blue ring. An observer outside the light blue ring cannot even know that the electron has moved. Accordingly, the field has not yet changed there. This is what special relativity tells us: The speed of light is high, but finite. Furthermore, the electric field must be continuous, i.e., the lines must remain connected. This results in a kink in the field line at the transition along the light blue circle, as can be seen in the figure. The kink itself is where only part of the movement was observable.

If we accelerate the charge back to point A, a second light blue ring is created with a new inverted kink (third image). In addition, the first ring has traveled further away at the speed of light. If the charge is regularly moved back and forth, a series of kinks is created, which look exactly like a wave. This wave propagates at the speed of light and carries energy. This is electromagnetic radiation. In a receiving antenna the wave can accelerate an electron and produce music in the radio.

If you dip a stick periodically in a lake, a beautiful surface wave spreads out. In electromagnetism, there is no medium like water to transport the wave. But if you wiggle an electric charge, the electric field *must* wobble. These disturbances then propagate as waves. This happens in almost empty space and even in a vacuum. This is how we can communicate with the Mars rover.

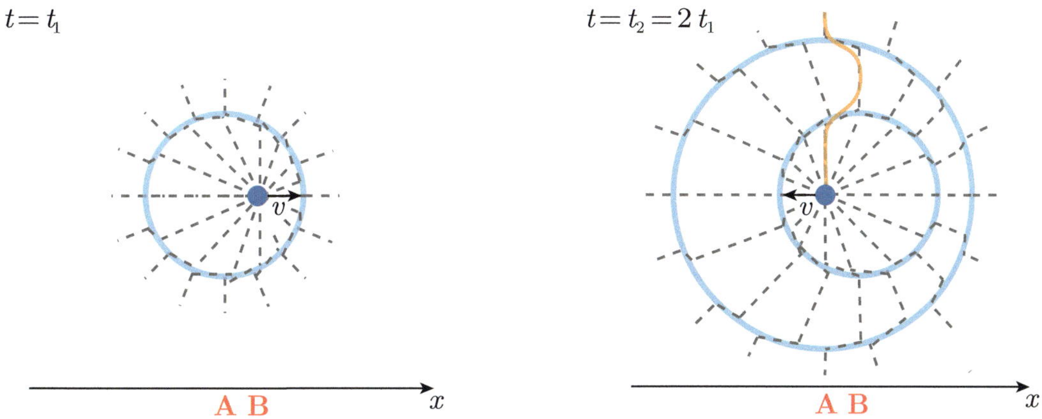

Electromagnetic radiation is generated when a charge is accelerated.

Special Relativity → p. 38

Spin

A Peculiar Property of Quantum Mechanics

Why does a magnet stick to the refrigerator? This is due to a peculiar quantity in quantum mechanics, called *spin*, for which there is no analogue in classical physics.

When electrons move through a metal wire, an electric current flows. From the theory of electricity, it is known that this generates a magnetic field. On an atomic level, an electron orbiting the atomic nucleus represents an electric current and thus also generates a magnetic field. In this context, one speaks of a magnetic moment, which is closely related to the angular momentum (↓) of the rotational motion. However, electrons and other elementary particles also generate a magnetic field without moving. This is due to the intrinsic angular momentum, called spin, which contributes to the magnetic moment. Spin is a fundamental property of every elementary particle (↓), like mass and charge.

But how did one even come up with the idea of attributing such an additional property to particles? If you shoot an electrically neutral particle that has a magnetic moment into a spatially variable magnetic field, it experiences a force and is deflected. The magnetic moment behaves like a little magnet and interacts with the applied magnetic field. Due to the quantization of angular momentum in quantum mechanics (see information box), the magnetic moments experience deflections only in certain discrete directions. For example, if the magnetic moment in an atom is caused solely by the circular motion of an electron, one would expect deflections into an odd number of fixed directions according to the laws of quantum mechanics.

An electron on a circular path around an axis has an angular momentum $L = rv$, which is the product of the distance r and the velocityt v. This is associated with a magnetic moment, in the opposite direction, denoted: $\vec{\mu}$.

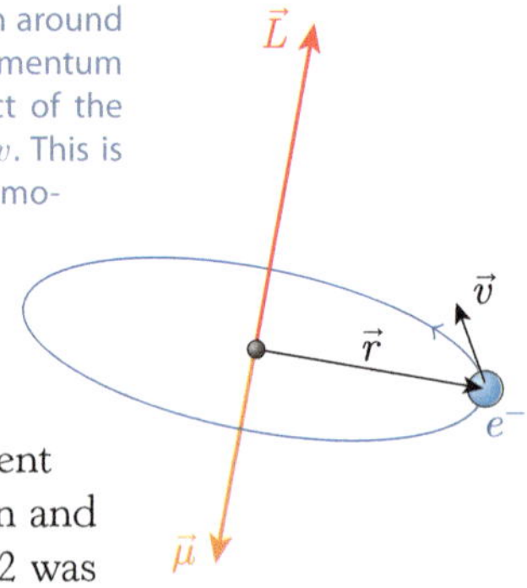

The aim of an experiment conducted by Otto Stern and Walther Gerlach[1] in 1922 was to demonstrate this effect of angular momentum quantization. They sent electrically neutral silver atoms through a spatially variable magnetic field. In silver atoms, the magnetic moment is determined by one single electron in the shell. The result of the experiment was indeed a deflection in very specific discrete directions, however not as predicted in an odd number. The silver atoms surprisingly settled in exactly two separate regions.

This puzzle was solved by Samuel Goudsmith and George Uhlenbeck. They proposed in 1925 that the electron has an additional intrinsic angular momentum, the spin, with the angular momentum quantum number 1/2, which according to the rules of quantum mechanics explains the observed deflections in two directions (see information box).

This initially peculiar quantum mechanical property has proven to be fundamental. The magnetism of permanent magnets, like those on our refrigerators, ulti-

[1] W. Gerlach, O. Stern 1924, *Über die Richtungsquantelung im Magnetfeld*, Annalen der Physik, 74, 673, doi:10.1002/andp.19243791602
W. Gerlach, O. Stern 1922, *The experimental proof of direction quantization in the magnetic field*, Zeitschrift der Physik, 9, 349, doi:10.1007/BF01326983

Angular momentum: Kinematics → p. 40

Angular Momentum in Quantum Mechanics

In quantum mechanics angular momentum plays a central role. It can only take on integer multiples of Planck's constant $h/(2\pi) = \hbar$. The corresponding angular momentum quantum number can therefore only take on the values $L = 0, 1, 2, ...$

The spin of a particle, which is also an angular momentum, can take on half-integer values. The electron is a particle with the spin quantum number 1/2. The deflections mentioned in the text can be understood, such that for an angular momentum L

$$2\,L + 1$$

different substates are possible. These are the possible projections of the angular momentum vector $\vec{L}$ onto a fixed axis for measurement. This results in three states in the case $L = 1$. If one replaces L with the spin quantum number 1/2, one obtains exactly the splitting into two levels observed in the Stern-Gerlach experiment.

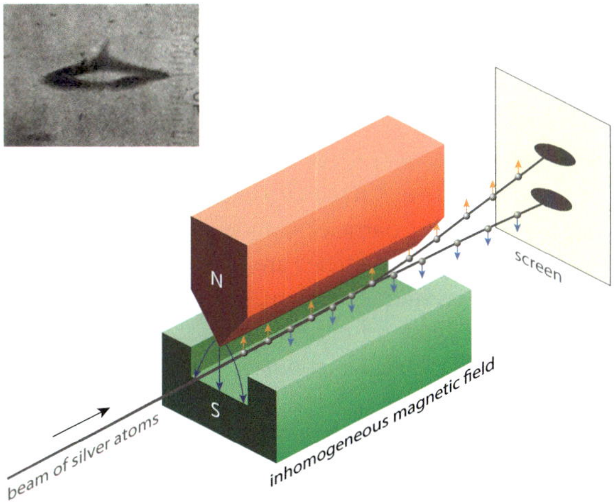

Principle of the Stern-Gerlach experiment with a black-and-white image from the original publication by Stern and Gerlach. The silver atoms (dark areas) are found in two regions. These are only schematically indicated as ellipses in the image of the experimental setup.

mately is due to the spins of the electrons in the magnet. Also, the splitting of atomic spectra in magnetic fields can only be understood with the help of spin.

On the theoretical side, spin is needed to reconcile quantum mechanics with special relativity. The so-called *spin-statistics theorem* provides a theoretical justification for the empirical finding that particles with half-integer spin may not coincide in all their quantum numbers. This has far-reaching consequences for atomic structure. In the Standard Model of particle physics, particles are characterized according to their spin. Particles with half-integer spin (1/2, 3/2, 5/2,...) are called *fermions* (↓). Examples include the electron, the quarks, the muon, the neutrinos, which are all spin-1/2 particles. Particles with integer spin (0, 1, 2,...) are referred to as *bosons* (↓). Examples include the photon with spin 1 and the Higgs boson with spin 0.

The particles composed of three quarks either have spin 1/2, like the proton and neutron, or spin 3/2. Examples of the latter are the Δ resonances (↓). Finally, spin also plays a crucial role in proposed extensions of the Standard Model, such as supersymmetry (↓).

In addition to these theoretical aspects, spin also finds application in medical imaging, as described in the article on magnetic resonance (↓).

What Makes a Particle Elementary? → p. 18
Fermions and Bosons → p. 94
Resonances → p. 56
Supersymmetry → p. 278
Magnetic Resonance Imaging → p. 146

Scattering Experiments

Atomic Billiards

Scattering experiments are the main investigative tool of particle physics. Similar to a collision of two billiard balls, elementary particles collide in a scattering process. The involved particles are deflected or even new particles are created. Therefore, the results of scattering experiments allow for conclusions about the structure of matter and their interactions.

In the simplest version, an object to be investigated, referred to as the target, is bombarded with projectiles which bounce off the target. The deflection pattern of the projectiles allows for conclusions about the shape of the target. The basic idea can be illustrated using a simple mechanical model. When projectiles hit a round object, they are deflected in all directions, as shown in the figure. You get a uniform distribution of projectiles, i.e., regardless of the outgoing angle, you observe approximately the same number of scattered projectiles in the detector. For the square target shown below, on the other hand, the projectiles are only found at angles $\pm 90^\circ$ relative to the direction of incidence of the projectiles.

At the atomic level, projectiles and the particles that make up the target do not really collide. Here the laws of quantum mechanics apply (↓), so that projectiles and target particles are described by a wave function, which is altered by the scattering process. An interaction that leads to scattering, for example, occurs due to the electric charges of the particles involved. From the deflection of the projectiles, one can learn something about the charge distribution of the scattering partners.

A historically very significant scattering experiment was the Rutherford experiment, which was carried out on the suggestion of Ernest Rutherford by Hans Geiger and Ernest Marsden. A very thin gold foil served as the target, which was bombarded with helium nuclei (alpha particles).

Most of the alpha particles simply penetrated the gold foil. However, the surprising part were those that were scattered at large angels. A few were even observed at backward angles, i.e., they received a very large deflection. At the time of the experiment, there was only a

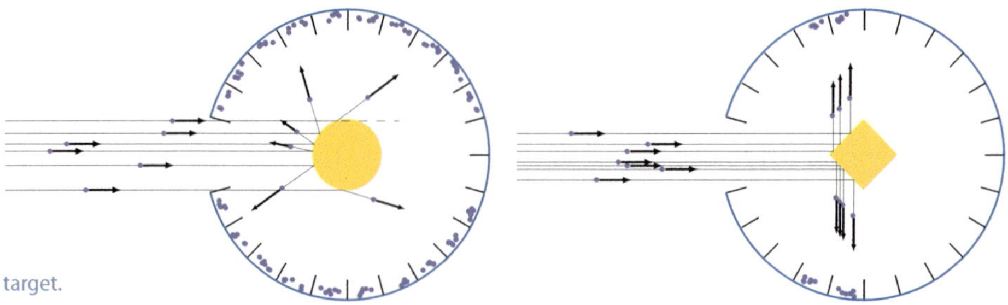

Projectiles are scattered at the target. The angular distribution of the projectiles allows conclusions about the shape of the target.

Concepts of Quantum Mechanics → p. 50

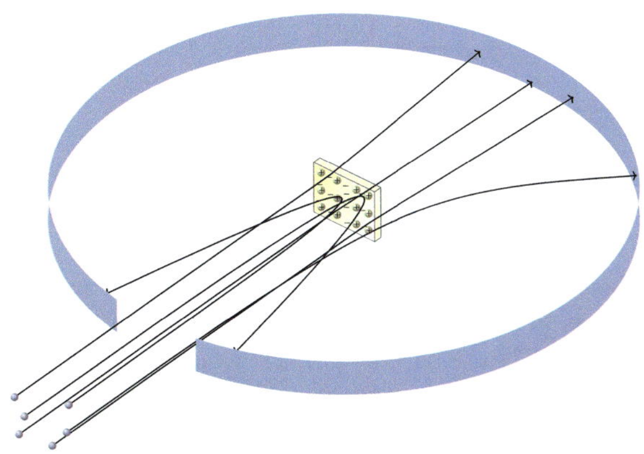

Principle of the Rutherford scattering experiment

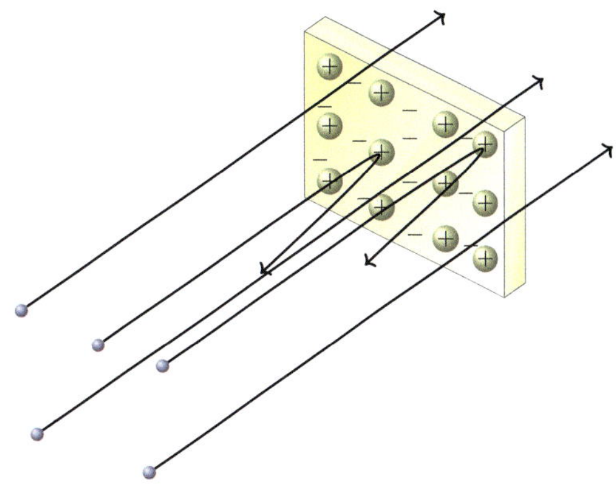

Scattering of α-particles on a gold foil

very vague idea of the structure of atoms. According to Thomson's atomic model, it consists of a uniformly distributed positive charge. Within it, negatively charged electrons are distributed like raisins in a cake, resulting in an overall electrically neutral object. Since the alpha particles are about 7000 times heavier than electrons, it is hard to imagine that these are responsible for backward scattering. Rutherford expressed his astonishment about this as follows: "It was as if you fired a 15-inch shell at a sheet of tissue paper and it came back to hit you."

From the experiment, Rutherford concluded that the gold foil is mostly "empty". This explains the unimpeded passage of the alpha particles. Merely a small part of the atoms consists of objects that are much heavier than the projectiles. Only in this way can the backward scattering be explained. Similar to how a coin is only scattered back in its original direction when it collides with a much heavier coin. With the Rutherford experiment, it was possible to exclude the Thomson model and establish the still valid concept of the structure of atoms: The positive charges are localized in a very small volume, the atomic nucleus (↓), which also carries the lion's share of the atom's mass.

In particle physics, scattering experiments play a crucial role both in investigating the structure of the target particles and in understanding the interaction between projectile and target. Thus, with the help of scattering experiments, the inner life of the proton can be investigated (↓) and, something can be learned about the structure of interactions from electron-positron scattering at high energies (↓).

A one-cent coin collides with a much heavier coin and is deflected.

On the Ranges of Interactions → p. 58
The Structure of the Proton → p. 212
Strong interaction: Quantum Chromodynamics → p. 204

Feynman Diagrams

Using Doodles for Computations

Feynman diagrams are graphical representations of reactions between elementary particles. The first figure below shows an example diagram in which an electron with momentum vector $\vec{p}_1$ and a muon with momentum vector $\vec{k}_1$ exchange a photon and thus scatter on each other

$$e^-(\vec{p}_1) + \mu^-(\vec{k}_1) \to e^-(\vec{p}_2) + \mu^-(\vec{k}_2)\,.$$

In all Feynman diagrams in this article, the time axis runs from left to right. On the left-hand side is the initial state, $e^-(\vec{p}_1) + \mu^-(\vec{k}_1)$, and on the right the final state, $e^-(\vec{p}_2) + \mu^-(\vec{k}_2)$. Typically, $\vec{p}_2$ will point in a different direction than $\vec{p}_1$, i.e., due to the interaction, the electron has changed its direction of flight, analogously for the muon. However, this is *not* shown in the diagram by the arrow directions, but only by the momentum labels.

Fermions such as the electron or the muon are shown as solid lines with an arrow. If the arrow points in the direction of time (to the right), it is a particle, in this case an electron or a muon. If the arrow points against the direction of time, it is an *antiparticle* (↓). In the second figure, we show electron-antimuon scattering: The arrow on the antimuon here points *against* the time direction, i.e., from right to left.

Photons are represented as wavy lines in Feynman diagrams. In both diagrams one photon is exchanged between the fermions. The point at which different lines come together, shown as red circles, is called the vertex. Which particles can form a *vertex* is determined by the interactions in the theory (Feynman Diagrams II ↓). In the simplest form of quantum electrodynamics, the theory of electromagnetic interactions, there are only vertices with a photon at a fermion line. Only here and in the article Feynman Diagrams II all vertices are highlighted in red for illustration.

There is an important rule: Fermion lines (↓) must always be continuous (as in our diagrams) or form a closed loop (↓). Boson lines, e.g., photons or gluons, can also end at vertices.

If a line starts at one vertex and ends at another, as with the wavy photon line, this is called a *propagator*, since the photon has to move from one place to another, i.e., propagate.

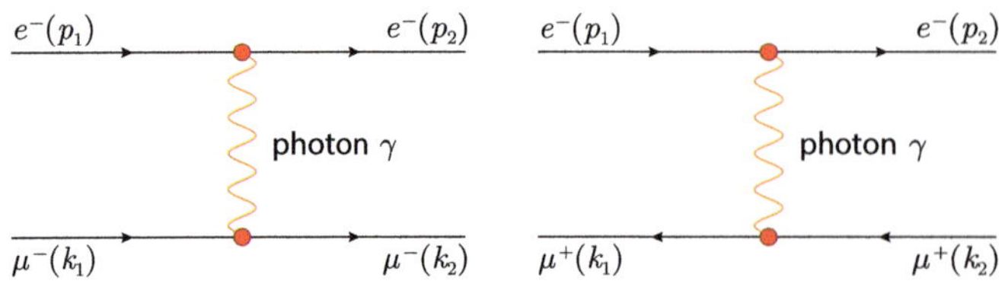

Scattering of an electron at a muon (left) and an antimuon (right)

What Is Antimatter? → p. 22
Feynman Diagrams II → p. 102

An up quark (u) and a down quark (d) interact via the strong interaction: they exhange gluons.

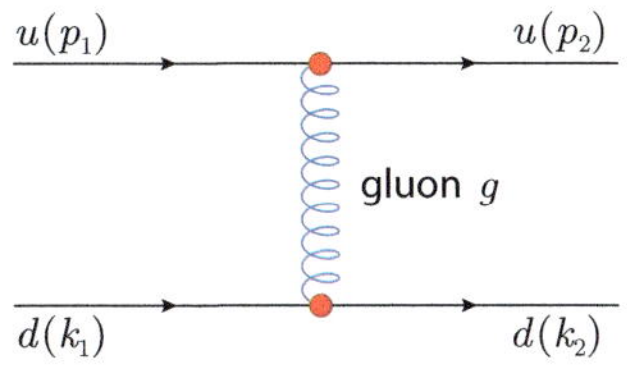

Like electrons, quarks can also exchange photons, but they can also exchange gluons. A diagram of quark-quark scattering through the exchange of a gluon is shown. The gluon is represented by a corkscrew line.

For physicists, Feynman diagrams are primarily a calculational tool. Every Feynman diagram can be translated into a mathematical expression using fixed rules. This expression is called an *amplitude*. The square of this amplitude can be used to directly derive the probability of a reaction taking place, the so-called *event rate*. The Feynman diagram can therefore be used to calculate the probability with which the outgoing electron has the momentum $\vec{p}_2$ if the initial state was $e^-(\vec{p}_1) + \mu^-(\vec{k}_1)$.

In the mathematical interpretation, it is important to note that the Feynman diagram is *not* a representation in normal (x, y, z)-space. Thus the two diagrams with the diagonal photon propagators are identical, i.e., they provide the same amplitude, although it looks as if in the left diagram the vertex V_1 takes place before V_2 and on the right after. Both variants are possible in quantum field theory and are summarized in *one* Feynman diagram.

Both Feynman diagrams are identical.

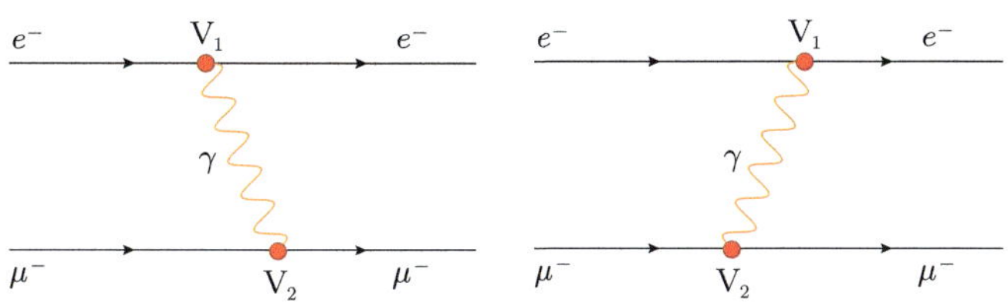

Finally, we show two independent Feynman diagrams for the scattering of a photon on an electron, so-called *Compton scattering*. (The electron line has a kink at the vertices, but only for aesthetic reasons. The momenta, including their direction, are determined solely by the labels). Physically we have an electron colliding with a photon. The photon is absorbed and then reemitted. Or, surprisingly, in the second diagram it is first emitted and *then* absorbed. The corresponding mathematical expressions of both diagrams must be added together to obtain the total amplitude, which is then squared to get the event rate. They are different because the outgoing photon with momentum k_2 couples once at the vertex V_2 with the outgoing electron, and once at the vertex V_1 with the incident electron. In both cases, we have an electron propagator between the two vertices.

Two contributions to Compton scattering ↳

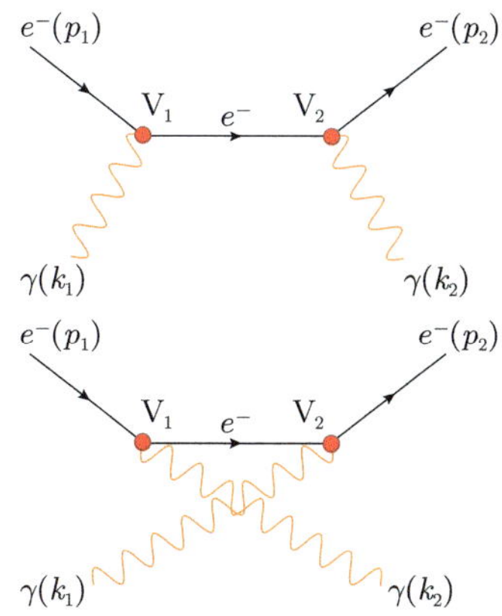

Fermions and Bosons → p. 94
Loops → p. 106

Concepts of Quantum Mechanics

The Trajectories of Arrows and Electrons

As already the legendary Robin Hood knew, if the launch speed and direction of an arrow are correct, then it will surely hit its target. In classical physics we can calculate this trajectory exactly, if we know the starting location and velocity exactly (as well as, for example, the wind conditions). However, on a microscopic level, things are quite different: Quantum mechanics does not allow us to know at the same time the location and velocity of, for example, an electron with arbitrary accuracy. All that is calculable is the *probability* with which an electron arrives at a certain location in a given experiment. Nevertheless, quantum mechanics and its more advanced version, quantum field theory, allow predictions of incredible accuracy, for example, for the energy levels of atoms.

The founding father of quantum mechanics, Max Planck, postulated the relationship $E = hf$ between the energy E and the frequency f for light or electromagnetic radiation, as this was the only way for him to explain the radiation frequency spectrum of bodies like molten metal or even the Sun.

The factor h is now known as *Planck's quantum of action*—it is, like the speed of light, one of the constants of Nature. Thus, the energy content of light of a certain frequency can only increase in steps of hf. Louis de Broglie picked up the idea and claimed that in reverse matter particles behave like waves, with a wavelength $\lambda = h/p$. Here, p is the momentum, which for slow (non-relativistic) particles is just their mass times their speed, $p = mv$. The wavelength of a particle thus becomes shorter as its momentum and thus its energy increases. As radiation and matter have wave-like properties, they are described by so-called *wave functions*. Computing the square of these wave functions gives the probability with which a particle can be found at a certain location. The temporal and spatial development of a wave function for slow (non-relativistic) particles is determined by the Schrödinger equation (see box).

But what does this mean for the properties of particles? Let us consider a plane water wave of a given wave length that propagates linearly. It has a well-defined velocity and thus also a well defined frequency, but is very widely spread in space along the wavefront. If this wave describes a particle, we know very well its velocity and thus momentum, or energy. It is the same as the wave propagation velocity. But the probability of finding the particle is the same along the entire wave front; the location is thus unknown.

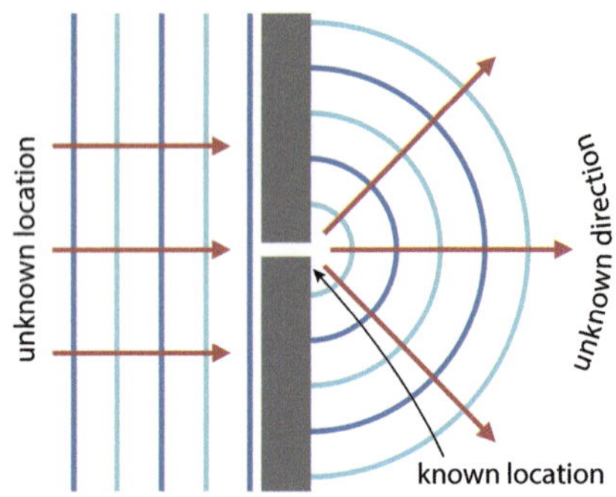

A plane wave squeezing through a gap. What is shown here for a water wave applies to all waves (including wave functions).

Now let's consider the instant when this wave passes through a narrow gap. If the wave as a wave function describes a particle, then we know its location very precisely at this point in time (within the gap)—but we no longer know much about its direction of movement, because when passing through the gap, the plane wave becomes a semicircular one, so that all forward-facing flight directions are equally likely.

Heisenberg has shown with his *uncertainty relation* that the product of the uncertainties in location and momentum must always be greater than $h/(4\pi)$. Planck's quantum of action thus controls whether quantum effects play a role—or not. More precisely, for every particle path, including that of the aforementioned arrow, its *action*[1] can be calculated. The arrow of Robin Hood therefore only flies on a well-defined path because the action assigned to it is much much larger than Planck's quantum of action, and thus quantum effects do not play a role.

But why is there a problem with simultaneous measurements of location and momentum? To determine the location of an electron, one can use photons—clearly it must be illuminated. However, these photons cannot provide a location measurement that is more accurate than their wavelength. In other words: To measure the location of an electron very accurately, one must use photons with large momentum, since, as mentioned above, a small wavelength corresponds to a large momentum. However, if a photon with large momentum hits an electron, a large momentum can also be transferred—thus information about the momentum of the electron is lost through the location measurement. A similar relationship as between location and momentum also exists between energy and time: $(\Delta E)(\Delta t) \geq h/(4\pi)$, where Δ (not to be confused with the derivative operator that appears in the box) describes the uncertainty (more precisely: variance) in the measured quantity (Statistics ↓).

The Schrödinger Equation

The time-independent Schrödinger equation in three spatial dimensions is

$$-\frac{h^2}{2\,M(2\pi)^2}\,\Delta\Psi(\vec{x}) = (E - V(\vec{x}))\,\Psi(\vec{x}) \ ,$$

where the operator $\Delta = \frac{d^2}{dx^2} + \frac{d^2}{dy^2} + \frac{d^2}{dz^2}$ and $\vec{x} = (x,\, y,\, z)$ denotes the position of the particle, $V(\vec{x})$ the potential energy of the particle at position $\vec{x}$, and E the total energy of the system; $\Psi(\vec{x})$ is the wave function.

Let us now consider an electron in a hydrogen atom as an example. Then $M = m_e$, and $V(\vec{x}) = -\alpha hc/(2\pi|\vec{x}|)$, with $\alpha = 1/137$ for the fine structure constant, the strength parameter of the electromagnetic interaction and $|\vec{x}|$ for the length of the vector $\vec{x}$, the distance between electron and proton. This is the quantum mechanical version of Coulomb's law.

For bound electrons, $E < 0$ applies. In this case, only wave functions are permissible that fall off quickly for large distances—after all, they are supposed to describe an electron that is bound to a proton. However, such solutions only exist for very specific energies. The binding energies are "quantized" and can only take the values $E_n = \alpha^2 m_e c^2/(2n^2)$, with $n = 1,2,3,\ldots$. For the ground state, this results in $E_0 = \alpha^2 m_e c^2/2 = 13.6$ eV.

[1] The action referred to here is a mathematical construct, namely the difference between kinetic energy and potential energy integrated over the entire flight time. This action is of order h for small atomic systems, but many orders of magnitude larger for macroscopic systems.

Statistics → p. 84

Chance, Interference and Predictions

Quantum Mechanics at Work

Consider a laser, which you shine onto two narrow, closely spaced slits. The image that appears on a screen behind them is not the simple linear projection of the slits. Instead you obtain a series of bright and dark stripes. The light waves from the two slits overlap and can either amplify (bright stripe) or cancel each other (dark stripe). This type of overlap is called *interference*, and the stripe pattern is an *interference pattern*. If one conducts the same experiment with electrons, an analogous stripe pattern emerges built from areas where many electrons hit and those where none arrive, because electrons also behave like waves. For that to work, the size of the slits and their separation should be roughly comparable to the electron wavelength.

Do the electrons perhaps influence each other, so that the pattern is created by their repulsion? To test this, you can reduce the electron flow such that only isolated electrons pass through the apparatus. Something surprising happens: The electrons do not fill the screen evenly, but hit mainly at the spots where many electrons had previously arrived. Furthermore, they avoid the spots that remained empty before. The observed pattern of electrons thus precisely represents the probability of individual electrons hitting certain spots, so it corresponds to the square of their wave function.

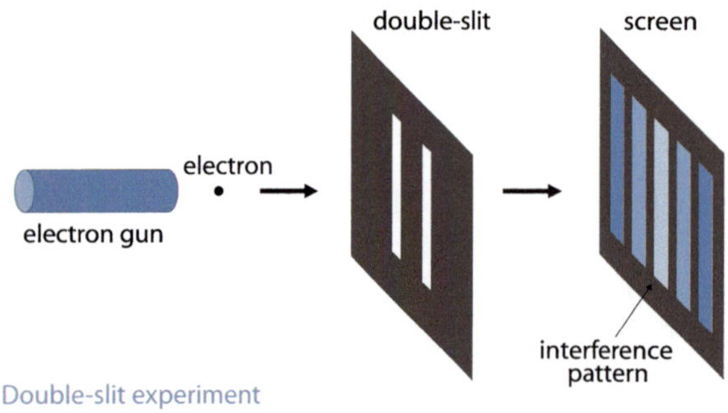

Double-slit experiment

But isn't it true that a single particle must fly through either one slit or the other? No, because, according to the laws of quantum mechanics, the individual electron is described by a wave function. When a plane wave approaches a double slit, it passes through both slits, and the now two waves from the separate slits interfere. If we, for example, illuminate the setup with light in order to observe the electrons and see through which slit they fly, we will influence the system in such a way that the stripe pattern disappears and the originally, naively expected images of the two slits appear on the screen. The reason for this is ultimately the same as the one explained in the previous article (↓) about the relationship between location and velocity measurements: In order to be able to decide, with the help of light, through which slit the electron flies, one must use photons with a wavelength of at most the size of the slit separation. However, the momentum that these photons give to the electron can be so large that the overlap of the waves passing through the different slits is completely destroyed.

It is true that the phenomena of quantum mechanics sometimes massively contradict so-called common sense. Many smart minds, such as Albert Einstein, have tried to find ways out of this contradiction by postulating that we perhaps simply do not know enough

Concepts of Quantum Mechanics → p. 50

about the propagation laws of electrons and that everything would be fine again, once we understood the system well enough. But all attempts of this kind have failed so far. So we probably have to accept: The world on a small scale works very differently from the world on a large scale. And yet, one can understand the transition from small to large: one cannot predict the outcome of a single roll with an ideal die, but can predict with high accuracy the average result of 600 000 rolls. 100 000 sixes will be present, with a deviation of plus or minus 300 being statistically permissible (Uncertainties ↓). In complete analogy, one cannot predict where on the screen a single electron will land in the double-slit experiment. However, one can calculate how the particles will distribute in the detectors after numerous repetitions of the experiment. Similarly, one cannot say in advance when a single, unstable atomic nucleus will decay. However, if one measures the decays of many nuclei, one finds a distribution of decays over time that is very precisely calculable—the so-called *decay law*. And the same applies to scattering experiments performed, e.g., at the LHC.

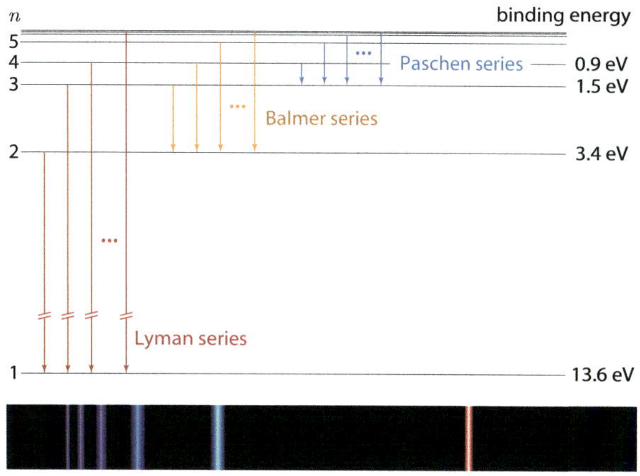

Top: Spectrum of hydrogen with transitions
Bottom: The visible lines of hydrogen (Balmer series)

As outlined in the previous article, quantum mechanics has two consequences for atoms: On the one hand, one can only specify the probability distribution of the electrons around the nucleus, but say nothing about their exact location—one cannot even speak of electron orbits. On the other hand, only very specific binding energies are permissible for the electrons. If an electron is in an atomic shell that corresponds to an excited state, characterized by a higher bound state energy than the ground state, it can transition to a lower state by emitting a photon (the photon must be emitted to conserve energy). Because of the separation of the binding energies, the energy spectrum of the emitted light shows characteristic discrete values, observed as stripes in the experimental setup (see figure). Each stripe corresponds to the energy difference between two bound states, that is released via the photon, when an electron transitions from a higher to a lower energy level.

In atoms, the electrons of the shell are bound to an atomic nucleus by electromagnetism. Particle physics also knows other bound states, e.g., atomic nuclei made of protons and neutrons (From Nucleus to Nuclei ↓). Protons and neutrons as well as many other particles are not elementary but composed, namely of quarks (↓). In both cases, in the nuclei, as well as in the nucleons (protons and neutrons), the constituents are bound by the strong interaction. Of course the laws of quantum mechanics also apply here. In particular, binding states are observed only at very specific energies, and transitions between these can also be measured.

Measurement Uncertainties → p. 86
From Nucleus to Nuclei → p. 226
The Quark Model: A First Look → p. 20

Particles' Births and Deaths

How Particles Can Create Other Particles

The Standard Model of particle physics (↓) allows the conversion of some kinds of particles into other kinds. The most well-known such process is the beta decay of the neutron, in which a free neutron is converted into a proton, an electron, and an electron antineutrino and which we will encounter frequently in this book. In this article, we look at the conditions under which such decays occur and what they can teach us.

In general, a particle is unstable and decays when

- the combined masses of all the final state particles is smaller than the mass of the initial particle and
- there is an interaction that allows the transition between the initial and final state.

The first condition is a direct consequence of the fact that the total energy must be conserved in every reaction: The total energy before a decay at rest is given by the mass of the decaying particle according to Einstein's famous formula $E = mc^2$. On the other hand, the total energy of the final state is at least determined by the sum of its masses, it's total mass, as any possible kinetic energy (↓) can only increase the energy. In the case of the neutron beta decay, this is fulfilled, as the mass of the neutron is greater than the sum of the proton, electron, and neutrino masses. And that's a good thing, because if the proton were heavier than the neutron and thus unstable, life as we know it would not be possible (Quark Mass Effect ↓).

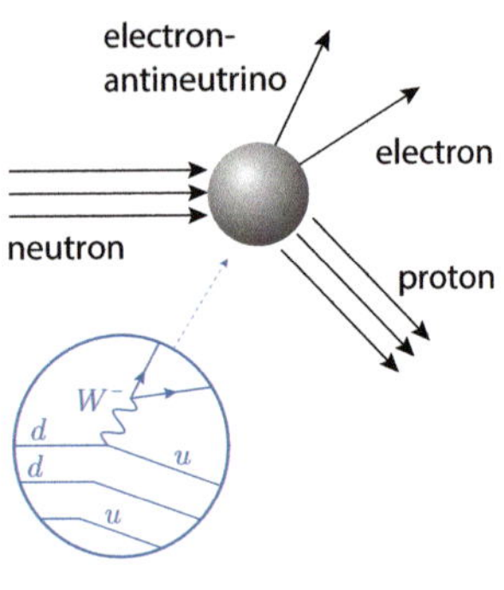

The β-decay of the neutron, both at the level of the composite protons and neutrons and at the level of the fundamental building blocks (up and down quarks, W exchange particles).

The second condition ensures that the process is allowed according to the rules of the Standard Model. A shorthand way of checking whether it is in principle allowed is using the fact that the structure of the Standard Model requires that certain properties of the overall system have the same value before and after all reactions. One such property is the electric charge. In our example: initially, the neutron has zero electric charge. After the decay the proton and electron charge add to zero, the neutrino is electrically neutral. In the Standard Model there are several other conserved quantities that will be discussed later in this book. Besides the above properties we need an interaction which enables the transition. In our example neutron beta decay is enabled by an interaction that converts neutrons (consisting of two down quarks and one up quark) into protons (consisting of two up quarks and one down quark). This is accomplished solely by the weak interaction (↓), which allows the transition from one down quark to an up quark and a W^- particle (which then decays to an electron and an electron antineutrino). Protons and neutrons would thus be stable, if there were only electromagnetism and the strong interaction.

Chapter 7 The Electroweak Standard Model of Particle Physics → p. 175
Kinematics → p. 40
The Quark Mass Effect → p. 224
The Weak Interaction → p. 178

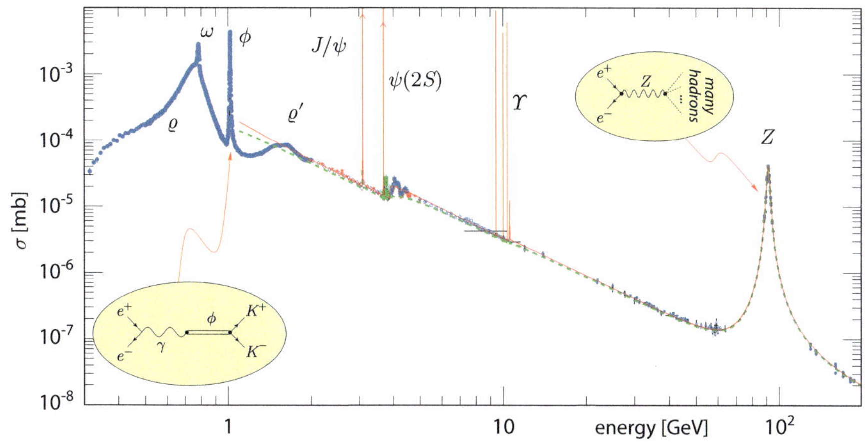

The total cross section (normalized count rate) for $e^+e^- \to$ hadrons

An important quantity that characterizes decays is the lifetime, denoted τ. It indicates how quickly a decay typically occurs. Since of course the laws of quantum mechanics (↓) also apply here, one can only make statements about the probability that a particle has decayed after a certain time. The more likely a transition is, the shorter the lifetime. Decays that are mediated only by the weak interaction are typically unlikely to happen, and thus have long lifetimes: The life time of the free neutron is 15 min; but even the lifetime of the μ^+ with $2 \cdot 10^{-6}$ s is long in the context of particle physics. In contrast, the lifetime of the so-called ϱ-resonance, whose decay is mediated by the strong interaction, is very short: 10^{-23} s.

In decays, particles can be transformed into others. The same applies to the scattering of particles: If two particles collide with sufficiently high energy, other particles can be produced in the course of the reaction. An example of this is shown in the image above. To measure the quantity shown here, electrons and positrons are shot at each other and it is counted how many strongly interacting particles can be found in the final state at a given energy. According to the Standard Model in this process the electron and the positron annihilate each other (this is possible with particles and antiparticles) and put all their energy into a photon or a Z^0 boson. Eventually, in some cases, strongly interacting particles are created from this energy, which are counted by a detector. At specific energies, the data show clear peaks, caused by so-called *resonances* (↓): These are nothing more than particles created by the photon or Z^0 and subsequently decay again. The figure highlights as an example diagram the creation of the ϕ meson, which decays into a kaon and an antikaon after its creation. Many of the particles named in the figure are discussed in the article on the quark model (↓). But away from the peaks there are also energy ranges in which the measured rate drops quite smoothly, without any structure. We will come back to these in the article on QCD (↓).

The peak marked with Z is something special: Here, instead of the photon, the Z boson (↓), the exchange particle of the weak interaction, is directly produced, which also shows as a resonance in the data. This process is also shown in the figure as a Feynman diagram.

Concepts of Quantum Mechanics → p. 50
Resonances → p. 56
The Quark Model Revisited → p. 202
QCD: Quantum Chromodynamics → p. 204
The Invisible Width of the Z Boson → p. 182

Resonances

When It's a Perfect Match

Resonance phenomena are widespread in all areas of physics. A resonance occurs whenever one system exactly matches another system which it tries to excite. In this way energy can be transferred very effectively from one system to the other. What appears very abstract at first glance, is actually familiar from your own experience. Take, for example, the pushing of a swing: If you push at the wrong moment, i.e., against the incoming swing, it will be slowed down and all the force you put in will be wasted. In contrast, pushing with the child on the swing is effective in the sense of "the energy is optimally transferred".

Resonance of a swing

The pushing must therefore occur exactly at the right time and with the right frequency. The latter is determined by the swing. The frequency that fits optimally is called the *resonance frequency*. It is the frequency at which the swing oscillates freely. If the swing is pushed exactly at this frequency, it will rise the highest. The resonance frequency of the swing is mainly determined (↓) by the length of the rope and gravity.

The maximal rise of the swing depends, of course, on how strongly it is pushed. But there is another limiting factor: air friction. The greater the friction, the less the swing rises—imagine trying to push a swing under water ... this connection is illustrated in the figure on the right.

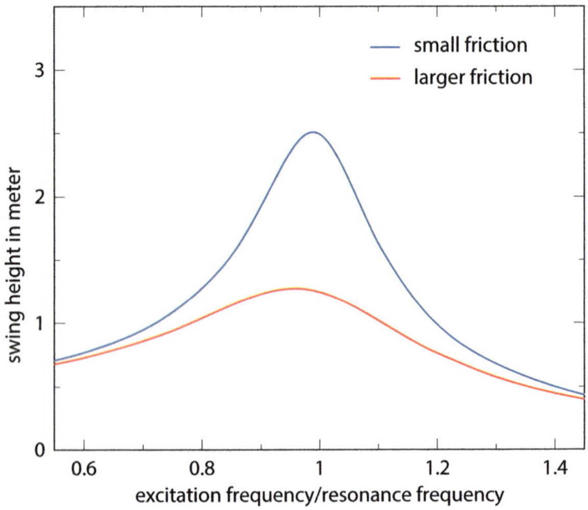

Illustration of the relationship between swinging height and push frequency for different levels of friction. The greater the friction and therefore damping, the lower and wider is the resonance curve.

Resonances also exist in particle physics. In scattering experiments (↓), for example, these manifest themselves in a strong increase in the count rates. Just like

Determining Measurement Uncertainties → p. 158
Scattering Experiments → p. 46

with the swing, a resonance in a particle physics scattering experiment only occurs when the frequency f of the incoming particle and, due to $E = hf$, its energy E (Quantum Mechanics ↓), closely match the properties of the target system. The total energy of the incoming particles must therefore correspond to the rest energy of the resonance. The analogue to air friction here is the decay probability: The more likely the transition from the resonance to other particles, the lower the mentioned increase in the count rate and the shorter the lifetime of the resonance (Particles' Births and Deaths ↓).

A beautiful example of this is the scattering of pions (Quark Model ↓) on protons. If you repeat the experiment often with slightly different pion energy and plot the number of scattered particles against the energy, you get the curve shown in the figure: The Δ-resonance (pronounced: Delta) is clearly recognizable.

The count rate is maximal, when the total energy corresponds to the energy of the resonance. The width of the curve reflects how likely the resonance, once produced, decays back into a pion and a proton.

We learn a lot from the measurement of resonances, from their position and their line shape. These parameters contain the information about the forces acting between the particles and between their components. In addition, one can perform the experiment with various probes—in the example above, one can also excite the Δ-resonance by bombarding protons with photons of suitable energy or frequency. This resonance can also be excited when cosmic ray protons hit CMB photons. The information we get from such experiments provide very important clues—they help us to find out how Nature works at the smallest distances. Another example of a resonance is the production of the Z boson (↓) at the LEP collider.

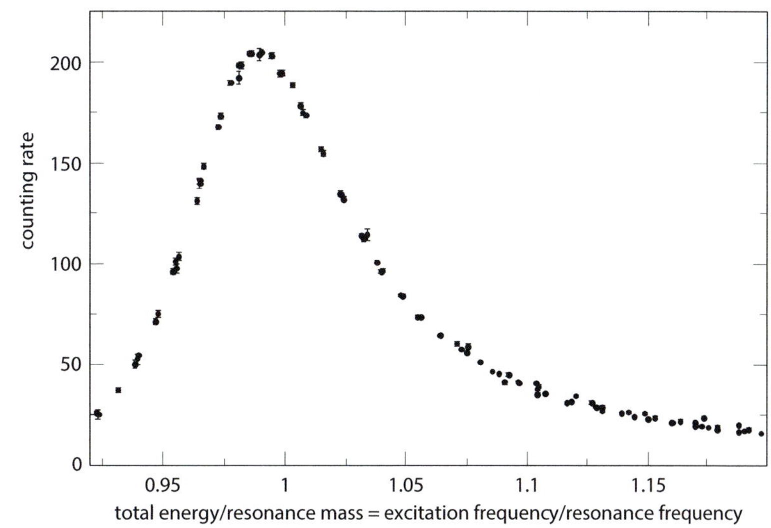

Measurement data for the reaction $\pi^+ p \rightarrow \pi^+ p$. The data clearly show a resonant increase, caused by the so-called $\Delta(1232)$ resonance.

Concepts of Quantum Mechanics → p. 50
Particles' Births and Deaths → p. 54
The Quark Model: A First Look → p. 20, The Quark Model Revisited → p. 202
The Invisible Width of the Z Boson → p. 182

On the Ranges of Interactions

From Intergalactic Distances to the Size of a Nucleus

The ranges of the various forces (↓) are very different: Gravitation has an infinite range and not only keeps us on the Earth and planetary systems and galaxies together, but also causes attraction between galaxies. On the other hand, the range of the electromagnetic interaction seems to be limited to centimeters—one feels it very directly when trying to bring two like magnetic ends closer to each other. The strong interaction, which holds atomic nuclei together, is effective only at unimaginably small distances of 10^{-15} m. The range of the weak interaction, which is responsible for decays like the beta decay of the neutron (↓), is yet much shorter. But what controls these ranges?

In particle physics, forces are described by the exchange of so-called *gauge bosons* (Quantum Field Theory ↓): The *photon* is the exchange particle of electromagnetism, the *gluons* are those of the strong interaction, and W and Z bosons mediate the weak interaction. Gravitation is attributed to the exchange of so-called *gravitons*, even though these have not yet been discovered. According to our current understanding, gravitons, photons and gluons are massless, while W and Z bosons have a mass of about 80 GeV and 90 GeV, respectively (about a factor 100 heavier than a proton).

In the box it is shown that the range R of an interaction or a force decreases as the mass m of the exchange particle increases. Thus we can understand that the short range of the weak interaction results from the large mass of the W and Z bosons and that the massless gravitons are still effective at cosmological distances.

But what about the other massless exchange particles, photon and gluon? The photon couples to the electric charge, i.e., the photon "senses" only charged particles. However, in an atom for example, there is a positive proton charge *and* the negative electron charge, of equal value but opposite in sign. Viewed from the outside, an atom or molecule is thus electrically neutral and is initially not "seen" by a photon at all. Two neutral atoms can only attract each other because the charge distributions of the nucleus and electron cloud can shift against each other due to quantum mechanical fluctuations. Although, viewed from a great distance, the atom is still neutral, at shorter distances the negative and positive charges are noticeably separated in space, leading to an effective charge. The now "charged" parts can exchange photons, and attraction can occur, however, at long distances the electrically neutral objects cannot. This is called *shielding*. The attraction described is referred to as *Van der Waals force*. In contrast, gravity always acts attractively. There is no shielding as there are no opposite charges and accordingly the interaction is truly of infinite range.

The reason why the strong interaction is short-ranged, even though the gluon is massless, is a bit more complicated—but also more exciting. A central property of the strong interaction is *confinement* (↓): particles carrying a strong color charge cannot be observed as

The Four Fundamental Forces of Nature → p. 16
Particles' Births and Deaths → p. 54
Quantum Field Theory → p. 104

free, color-charged particles. They are only found as components of composite objects whose color charges neutralize each other (analogous to the electrically neutral atom). However, the gluon carries—unlike the photon—a strong color charge itself. Therefore, confinement does not allow it to mediate forces beyond the confinement radius.

By the way, the strong interaction is also responsible for the binding of the nuclear components, protons and neutrons, to each other. As mentioned above, these are not color-charged when viewed from the outside, so they do not interact with each other by exchanging gluons. The cause for the forces between the nuclear components and their range is that QCD produces light, color-neutral bound states of quarks with antiquarks that can mediate the strong force over somewhat longer distances. The lightest are the pions (symbol: π; Quark Model ↓). How these can contribute to the nuclear force at the quark level is illustrated in the figure: The pion has a mass of just under 200 MeV/c^2. Therefore, the strong nuclear force has a range of 1 fm = 10^{-15} m (see box).

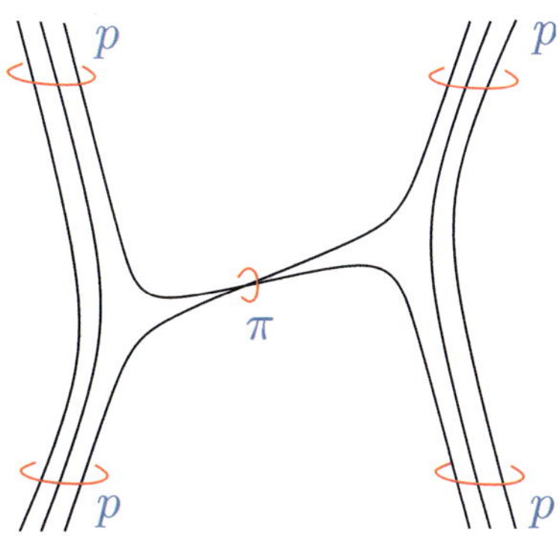

Derivation of the Range of a Force

For this we consider a matter particle, e.g., an electron with mass M, which is at rest, and emits a field particle, e.g., a Z boson with mass m.

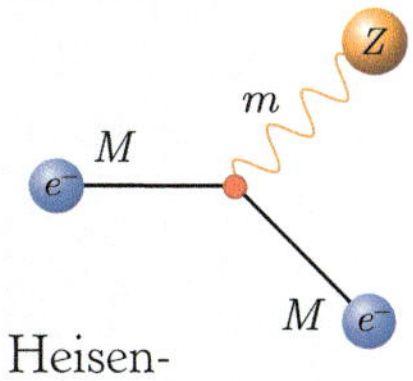

Before the emission, the energy of the electron is thus Mc^2, the energy of the final state is at least $(M + m)c^2$.

In quantum mechanics, the Heisenberg uncertainty principle allows for a violation of energy conservation (Quantum Mechanics ↓), however, only for the short time: $\Delta t \simeq \hbar/\Delta E \simeq \hbar/(mc^2)$. Thus, the larger the mass of the emitted particle, the more the energy conservation is violated, and the shorter the duration. Since a particle can fly no faster than the speed of light, c, the distance that such a virtual particle travels, i.e., its range R, is of the order of magnitude

$$R \simeq c\Delta t \simeq \hbar/(mc) = \hbar c/(mc^2)\ ,$$

where $\hbar c \simeq 200$ MeV fm (1 fm = 10^{-15} m).

Let's consider the weak interaction as an example. The exchange particles have a mass of about 90 GeV/c^2 = $90 \cdot 10^3$ MeV/c^2. This gives us a range of

$$\frac{200 \cdot 10^{-15}}{90 \cdot 10^3}\,\mathrm{m} \approx 10^{-18}\,\mathrm{m}\ .$$

← Visualization of the longest-range component of the nuclear interaction through quark lines: the range of the strong interaction is determined by the mass of the lightest strongly interacting particle—the pion (π).

Confinement → p. 206
The Quark Model: A First Look → p. 20, The Quark Model Revisited → p. 202
Concepts of Quantum Mechanics → p. 50

Symmetries

Beautiful and Practical

We speak of a symmetry when a system does not change through a transformation, for example a rotation or reflection. This applies, for example, to a sphere that is reflected about an axis passing through the center.

Symmetrical structures are often perceived as beautiful and pleasant. In addition to these aesthetic aspects, practical implications also play a role. For example, a fence of a fixed length encloses the largest area when it is built in the form of the highest symmetry, a circle.

In physics, symmetry considerations play an important role—for example, conservation laws such as energy conservation can be traced back to symmetries in spacetime (Noether's Theorem ↓). In the Standard Model of particle physics, both spacetime symmetries and so-called *internal symmetries* (↓) play a decisive role. Three symmetry transformations are of particular importance:

1. *point reflection*, also called *parity transformation*. Here, every point is reflected in the origin of the blue coordinate system. In our example, a right-handed screw becomes a left-handed screw,

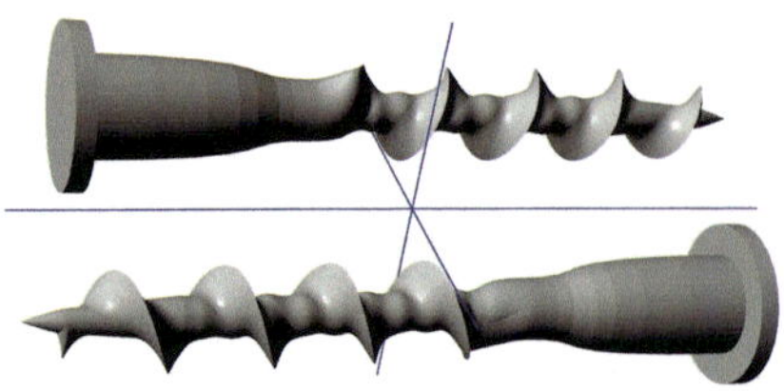

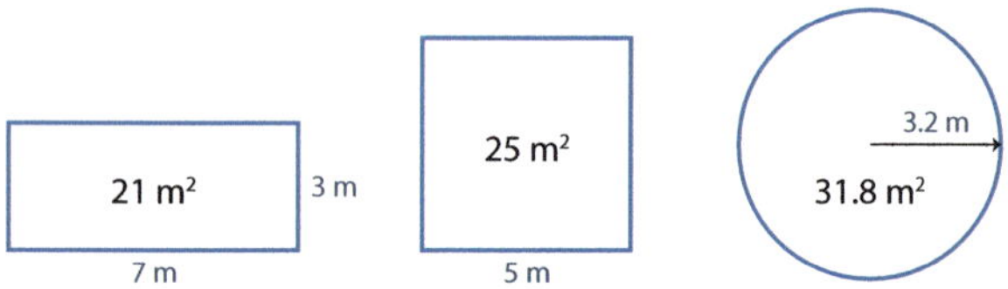

The higher the symmetry, the larger the enclosed area for the same circumference.

2. *charge transformation*, also called charge conjugation, in which each particle is replaced by its antiparticle (↓),

3. *time reversal*, in which time is run backwards.

Until the mid-20th century, it was assumed that the laws of Nature remain unchanged under these three transformations. For reflections, this means that a process and the corresponding mirrored process occur with equal probability. On a macroscopic scale, this seems to be fulfilled at least for the first two transformations. Whether we use left- or right-handed screws or on which side of the road we drive is pure convention. No law of Nature seems to prescribe this.

But are all laws of Nature really symmetric under the transformations mentioned above? On the suggestions of the two theorists Tsung-Dao Lee and Chen-Ning Yang, the Chinese-American physicist Chien-Shiung Wu was able to show in a groundbreaking experi-

Noether's Theorem → p. 100
External and Internal Symmetries → p. 96
What Is Antimatter? → p. 22

Clockwise from top left: Photo of a snowflake, ceiling of the octagon of Aachen Cathedral, ceiling of the Spanish Synagogue in Prague, Blue Mosque in Istanbul.

ment that the weak interaction (↓) violates the mirror symmetry. In the experiment, it was observed that electrons from the decay of a cobalt-60 nucleus were preferentially emitted in the opposite direction of the intrinsic angular momentum (Spin ↓) of the nucleus. The mirrored scenario, in which the electrons are preferentially emitted in the direction of the intrinsic angular momentum, is not observed in Nature. The world we live in, therefore, does distinguish between left and right on a microscopic scale and thus violates mirror symmetry.

Meanwhile, processes have also been found that proceed differently under the time transformation or under the combination of parity and charge transformation (↓). However, it is remarkable that the laws of Nature remain unchanged under very general conditions under the combination of the three transformations mentioned above. This law, known as the *CPT theorem*, is one of the pillars of the Standard Model. The three letters stand for the three transformations C (charge), P (parity) and T (time).

The American physicist Richard Feynman once described the violation of symmetry as follows: You can tell an extraterrestrial physicist your size as a multiple of a certain wavelength of a transition in the hydrogen atom. You can define your age using a transition frequency. Only because of the parity violation, as the violation of mirror symmetry in physics is called, is it possible to define what is right and left. If you then tell the alien that it is customary for us to shake the right hand in greeting and he extends his left hand to greet you, caution is advised. According to the *CPT* theorem, he is probably made of antimatter (Baryogenesis ↓)!

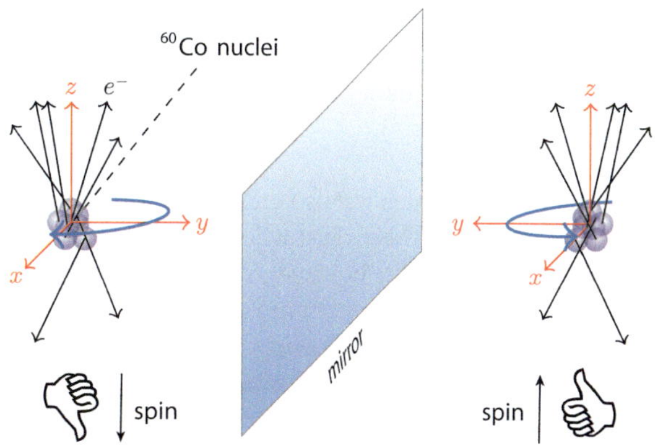

The principle of the Wu experiment: The process on the left, in which electrons, symbolized by the black arrows, are preferentially emitted against the direction of spin, is observed in Nature, the mirrored process (right) is not. The direction of spin is indicated by the right-hand rule through the thumb. The fingers indicate the direction of rotation.

The Weak Interaction → p. 178
Spin → p. 44
Discrete Symmetries: Violating *P* and *CP* → p. 190
CP Violation in Beauty → p. 258
Baryogenesis → p. 328

Spontaneous Symmetry Breaking

The Hidden Symmetry

Symmetries (↓) play a central role in particle physics. They often allow us to bring order to apparent chaos and lead to conserved quantities, which are important throughout particle physics (Noether Theorem ↓). Moreover, the construction principle of the entire Standard Model is based on the existence of certain so-called *gauge symmetries*.

Typically, a symmetry is directly reflected in the physics. For example, the proton, which consists of two up quarks and one down quark, is almost as heavy as the neutron, consisting of two down quarks and one up. The reason for this is that the strong interaction does not distinguish between quark types. In other words, the strong interaction is symmetric under the exchange of quarks. The very small but important mass difference comes from quark properties, not the interaction (Quark Mass Effect ↓).

But in particle physics we also encounter so-called *hidden symmetries*. These exist when the interaction has a symmetry, which is, however, dynamically or spontaneously broken. The meaning of this is explained in the following. An example of such a hidden symmetry is known to all of us: the permanent magnet. You can imagine this as a collection of many tiny little magnets. The material is magnetic, when all these little magnets are aligned in the same direction. This state is shown in the left half of the figure. If you heat the magnet and thus increase the kinetic energy of the little magnets, the alignment is lost. Then the material is no longer magnetic and in particular, there is no distinguished direction—it is symmetric. However, when it cools down again, below a certain temperature the material becomes magnetic again, thereby "choosing" a spatial direction—which, however, cannot be predicted. The symmetry is broken spontaneously.

This phenomenon is characterized by the fact that the interaction in the material is symmetric (it does not a priori prefer any direction for the alignment of the little magnets), but nevertheless, a direction is spontaneously distinguished when cooling down. Or more generally: Although the material does not specify a preferred direction and is therefore highly symmetric, the ground state distinguishes a direction and is therefore less symmetric. A clear illustration of this effect is the so-called *Mexican hat potential*, which is shown in the figure on the next page. Imagine a typical Mexican hat, which

Permanent magnet at room temperature (left) and heated (right)

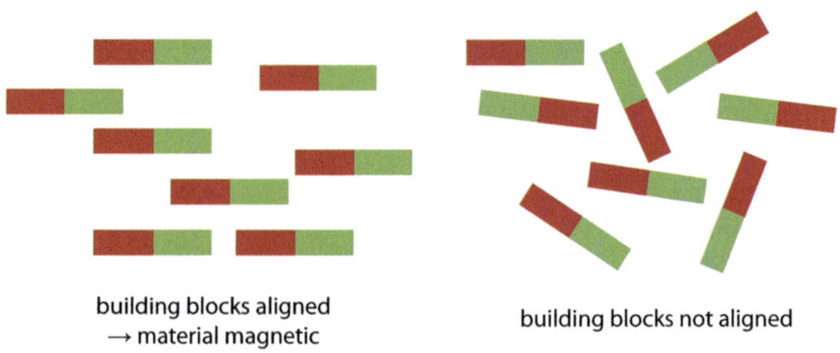

Symmetries → p. 60
Noether's Theorem → p. 100
The Quark Mass Effect → p. 224

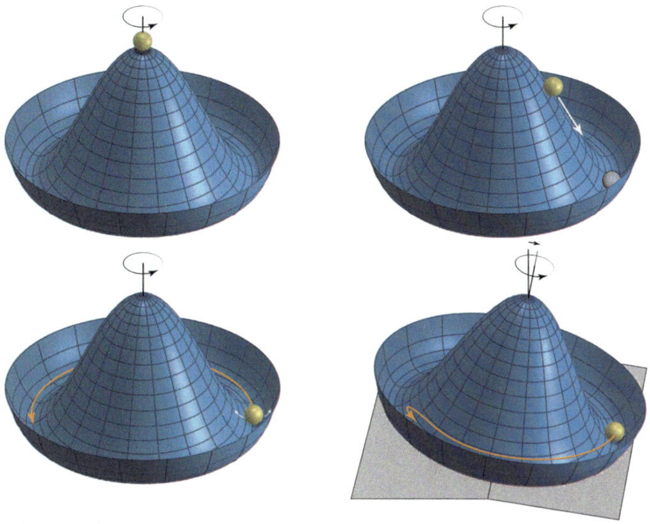

The Mexican hat potential

is symmetric under rotations around the central axis, as shown in the top left figure. The symmetry of the hat is still realised when a ball is placed on its tip. However, the ball will not stay on the tip for long, but will roll down in some direction to then lie in the brim of the hat. Once this has happened, the system as a whole, i.e., including the ball, is no longer symmetric under rotations. The symmetry is spontaneously broken, although the hat alone of course retains its symmetry.

As suggested in the lower left picture of the figure, it now costs no energy to move the ball around at the bottom of the brim. But it does cost energy, if you want to push the ball outwards or inwards. Thus the physics for these two actions is very different. And this remains the case even if the hat is slightly tilted, which corresponds to a small explicit breaking of the symmetry. The picture shows: As long as this breaking is not too large, it still costs very little energy for the ball to follow the bottom of the brim, but a lot more energy to move radially. The spontaneous symmetry breaking still determines the dynamics of the system.

In particle physics, spontaneous symmetry breaking underlies the Higgs and the Peccei-Quinn mechanisms (↓), to each of which an article is dedicated in this book. An example of a system controlled by spontaneous symmetry breaking, even though the symmetry is also explicitly broken (corresponding to a slightly tilted hat), is the phenomenology of the strong interaction at low energies. Here, the spontaneously broken symmetry is the one which would occur if the three lightest quarks (up, down and strange) had vanishing masses. The effect of the spontaneous symmetry breaking would then be that the pions, kaons, and the η meson, which consist of light quarks and which are discussed in the article on the quark model (↓), would also have no mass. (This is the quantum field theoretical analogy to the fact that it costs no energy to move the ball along the bottom of the brim of the sombrero.) However, up, down and strange quarks are not massless: the masses (↓) of up and down quarks are very small, that of the strange quark is 20 times larger. These non-vanishing masses ensure that the "hat" of the strong interaction is indeed tilted, but only very slightly. Accordingly, pions are not massless, but have a much smaller mass than all other particles participating in the strong interaction. Kaons and the η, which also contain strange quarks, are heavier, since for those the hat is tilted more. But they are still much lighter than all other particles with strange quarks.

The Discovery of the Higgs Boson → p. 246
Peccei-Quinn mechanism: The Strong *CP* Problem → p. 290
The Quark Model Revisited → p. 202
Where Does Our Mass Come From? → p. 220

3 Experimental Basics

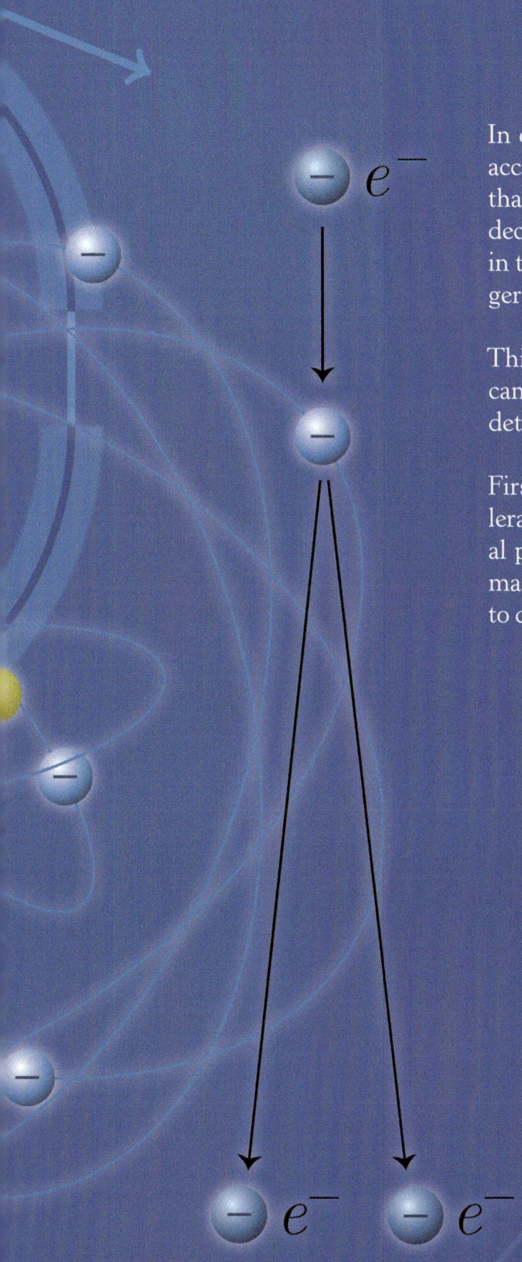

In order to learn about elementary particles, it is necessary to very accurately measure their properties and reactions. This is easier said than done. For example, in the human body about 10 000 radioactive decays occur per second. About 10^{14} neutrinos cross the human body in the same time period and a few muons from cosmic radiation trigger reactions. We notice nothing of all this.

This shows that, unlike objects of everyday life, elementary particles cannot be readily observed. A certain technical effort is required to detect them. How this is achieved is discussed in this chapter.

First, we describe how elementary particles are generated and accelerated. Then various methods are discussed for detecting individual particles and determining their properties. These include charge, mass, spin, and momentum—this is exactly the information needed to draw conclusions about the reactions that have taken place.

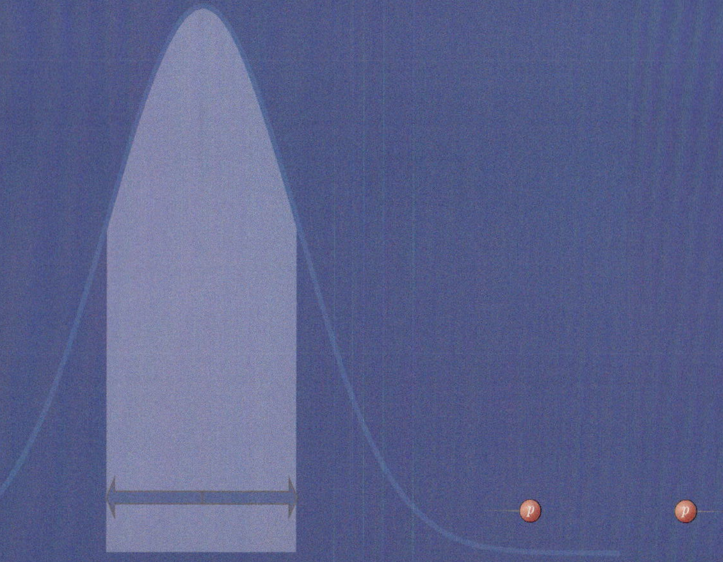

How Does a Particle Accelerator Work?

With Electric Fields to High Energies

One of the main tools of elementary particle physics are particle accelerators. As the name suggests, they are used to accelerate particles to a certain energy. When a collection of many such particles, a particle beam, hits matter or when the particle beams collide, one can learn many interesting things from the resulting processes (Scattering Experiments ↓).

But how can one generate high-energy particle beams? Similar to how a body is accelerated when falling in the Earth's gravitational field, a charged particle experiences acceleration in an electric field, which is created, for example, between two electrically oppositely charged plates. This is the basic principle of every accelerator. A cathode ray tube, formerly used in every TV-set, is essentially a particle accelerator! After passing through the acceleration path, the particles hit, for example, a target, as the object under investigation is called.

When a charged particle passes through a voltage difference, it gains energy, which is equal to the product of the magnitude of its charge and the voltage it passes through. In particle physics, the corresponding unit electron volt (eV) is commonly used as the unit of energy. If a particle with an elementary charge e passes through a voltage of one volt, it gains an energy of 1 eV $\approx 1.6 \cdot 10^{-19}$ Joule (Energy and Power ↓). One cannot accelerate a particle to arbitrarily high energies with this method, as electrical discharges (flashes) occur between the two plates if the voltage difference is too high.

One way to circumvent this limitation is to use alternating electric fields. In the upper part of the figure below, the negatively charged particles are attracted by the positively charged first and third cylinder. Once the particles are in the shielded inner area of the cylinders, the voltage is reversed and acceleration occurs be-

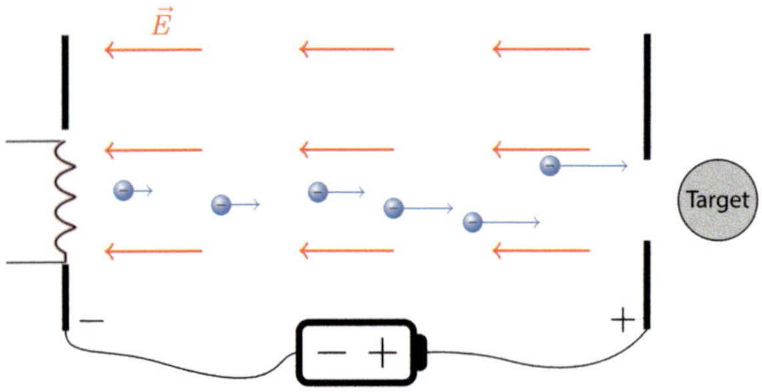

Principle of an accelerator. Electrons emerge from a cathode on the left, are accelerated in an electric field, and then hit a target on the right.

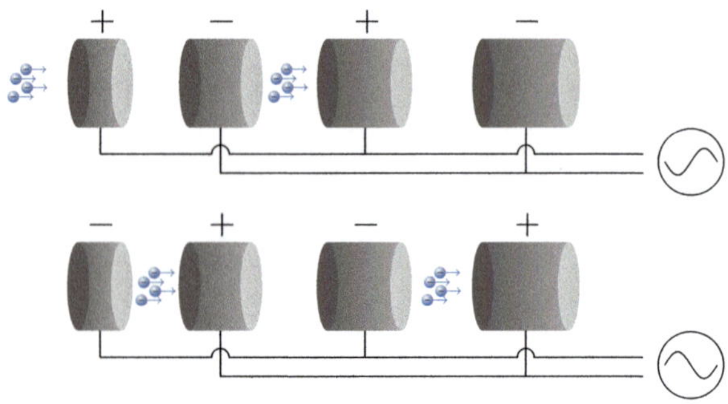

The principle of a linear accelerator with alternating electric fields according to Wideröe. The particles are accelerated from left to right. The upper image shows the situation at time $t = 0$, the lower one at a later time.

Scattering Experiments → p. 46
Energy and Power → p. 36

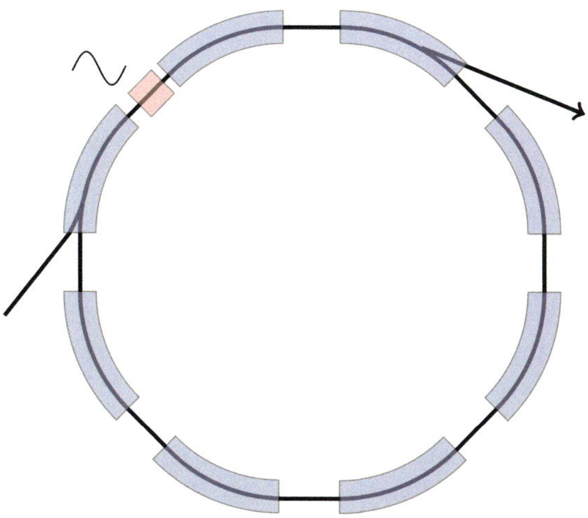

Construction of a synchrotron. The accelerator section is in red, the deflection magnets are in blue. The name synchrotron is derived from the fact that, as the particle energy increases, the field of the deflection magnets must be increased synchronously in order to keep the radius of the particle path constant.

tween the first and second or the third and fourth cylinder. In principle, this method allows for arbitrarily high particle energies, but the accelerator becomes longer and longer. An example of such a linear accelerator is the Stanford Linear Accelerator in the USA, which is more than three kilometers long.

To achieve even higher energies, the idea was to have the particles pass through such an accelerator section several times. This leads to a ring-shaped design. The beam is injected by a pre-accelerator and passes through the acceleration ring thousands of times. Deflection magnets are used to keep the particles on a circular path. The actual accelerator section only makes up a small fraction of the ring. As the particles become more and more energetic, the magnetic fields are increased correspondingly to keep the particles on the circular orbit. After reaching the desired energy, the particle beam can then, for example, be directed onto a target. Here too, one eventually reaches limits, as either the radius of the circular accelerator or the maximum magnetic field must be increased to reach higher energies. One way to further increase the particle energy available in a scattering process is to let two particle beams collide, as is done at the Large Hadron Collider at CERN (LHC ↓).

Accelerators are not only tools of particle physics, but are used in other areas, such as materials research or medicine (↓).

The COSY synchrotron at the Jülich Research Centre. The accelerator has a circumference of about 180 m. It allows protons and deuterons (bound state of proton and neutron) to be accelerated. The orange elements are deflection magnets to keep the particles on a circular path.

Video on the Synchrotron COSY at Jülich Research Center:
https://www.youtube.com/watch?v=MXNOVTF3utE

The LHC Accelerator Complex at CERN → p. 116
Material research: Synchrotron Radiation → p. 126
Accelerators in Medicine → p. 122

Interactions Between Particles and Matter

How to Detect Particles?

To detect a particle and determine its properties in a detector, two conditions must be met: They must interact with the detector material, and the reaction of the detector material to this interaction must be registered. The latter typically occurs through electronic signals (Semiconductor Detectors ↓). In this article, we deal with the interaction between particles and the material.

Of the four known interactions (↓), only electromagnetism (↓) and the strong force (↓) are suitable for creating a noticeable interaction between particles and detector material. Gravity and the weak force are much too weak to generate electronic signals in the detectors. However, the weak force can be studied with detectors when, for example, in neutrino physics experiments a neutrino produces a charged particle in the detector material through the weak interaction, which can then be further investigated using electromagnetism. For all electrically charged particles, electromagnetism can be used as an interaction between the particles and the detector.

How exactly these interactions occur depends on the material of the detector and the type of particle to be investigated, as well as on its energy. This means that a complete particle physics experiment is typically made up of different types of detectors. As a particle flies through the detector, depending on its nature it interacts in different ways with the different parts of the detector. The various resulting signatures can then be used to experimentally distinguish between different types of particles (↓).

The detector must be designed in such a way that it covers the full energy range of the particles under investigation. As an electrically charged particle passes through the detector, its energy and mass (Special Relativity in Particle Physics ↓) determine exactly how it electromagnetically interacts with the detector material. We now take you through the various possibilities.

The first possibility is that the particle can simply be deflected in the electric field of the atomic nuclei and thus "nudges" the relevant nuclei, transferring some of its energy to them. This can happen at any energy of the particles, but it is only the dominant process at a very low energy, at energies smaller than its rest mass. It usually cannot be used for measurements, as the change in nucleus momentum is difficult to detect. The situation is different when the particle lifts an electron from the lowest energy level in an atom to a higher one. The excited electron then returns to its original energy level by emitting a photon. If the material is transparent to the wavelength of the photon, the photon can be electronically registered as a light signal. This process is called *scintillation*. Although this process is very important for the electronic detection of particles, it contributes little to the energy transfer from the particle to the material. Their energy often is so far above the binding energy of the electrons in the atom that they are hardly slowed down.

When the particle has a kinetic energy significantly higher than the ionization energy of the atoms of the material, but below about a hundred times its own rest

Semiconductor Detectors → p. 70
The Four Fundamental Forces of Nature → p. 16
Electromagnetism → p. 176
Strong force: Quantum Chromodynamics → p. 204
Signatures of Particles → p. 80

Two examples of different detector types: A silicon vertex detector uses ionization in silicon. In a drift chamber, a particle ionizes a gas along its track.

mass (SRT ↓), the *ionization* of the detector material is the dominant process. Here, the particle does not raise the electrons in the atoms to a higher energy level, but knocks them completely out of the atom. This creates a charged ion and a free electron, which can be amplified as an electrical signal and registered electronically.

If the energy of the particle exceeds about a hundred times its rest mass, *radiation* becomes the dominant process of energy transfer. The incoming particle is decelerated in the electric field of the nuclei (↓) and emits high-energy photons, to which it gives about half of its energy. This process continues in a cascade until the energy (↓) drops below the $100 \cdot mc^2$ threshold. For very light particles such as electrons, this process plays a major role. For very heavy particles, this process rarely occurs in practice.

For electrically neutral particles, other physical processes must be used for detection, which will be further elaborated in the following sections, using individual detector types as examples. In contrast to the photons with energies in the visible spectrum created in the scintillation process, high energy photons, for example, cannot be detected directly. They first must be "converted" into electrons and positrons before they can be registered. Electrically neutral hadrons such as neutrons can be detected using the strong interaction between the particle and the protons and neutrons in the atomic nuclei of the detector material. Another special case applies to the antiparticles (↓) of the building blocks of matter: positrons, antiprotons, or antineutrons can also be measured at very low energies when they annihilate with their partners.

With the help of all these processes, the high-energy particles in the material generate either light in the form of photons or electrical signals in the form of free electrons, which are then converted into an electronic signal.

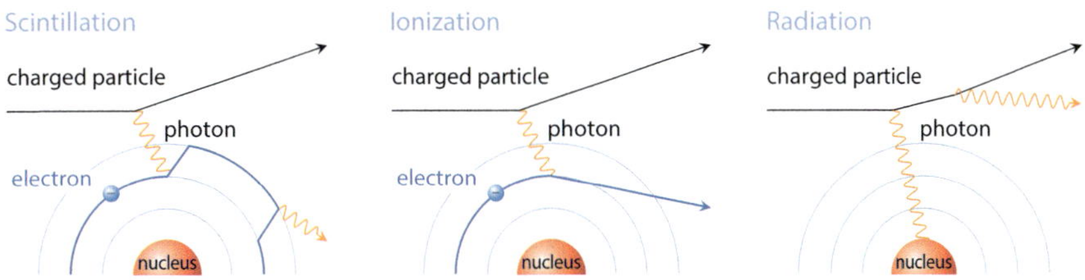

The various types of energy transfer important for the detection of particles in the detector in schematic representation: scintillation, ionization and radiation

Special Theory of Relativity in Particle Physics → p. 92
SRT: Special Relativity → p. 38
Acceleration and Radiation → p. 42
Energy Measurement → p. 78
What Is Antimatter? → p. 22

Semiconductor Detectors

The All-Rounders

Most people carry millions of tiny semiconductor detectors in their mobile phone cameras. This type of detectors also plays a significant role in the field of particle physics. The ability of a material to conduct electrical current is determined by the strength with which electrons are bound in atoms.

In solids, the wave functions of electrons on individual energy levels overlap to such an extent that the likelihood of finding electrons throughout the solid remains nearly constant. Therefore, rather than discussing individual wave functions, we refer to energy bands within the solid.

Nevertheless, unrestricted movement of electrons and the resulting flow of electrical current cannot occur if all available energy levels within a band are already occupied. In such a scenario, the Pauli exclusion principle (↓) prevents the transfer of electrons from one level to the next. Metals are electrically conductive because they contain partially occupied energy bands known as *conduction bands*. In contrast, completely filled energy bands are referred to as *valence bands*. The latter is the case in insulators.

Commercial germanium detectors mounted on vessels for liquid nitrogen.

Germanium detectors of the LEGEND neutrino experiment

Semiconductors basically are insulators. Yet, a relatively modest energy input (typically around three electron volts) can facilitate the transition of electrons from the valence band to the available and empty conduction band. In the most straightforward scenario, this transition occurs through thermal excitation, explaining why semiconductors become progressively more conductive as their temperature rises, leading to a corresponding decrease in their electrical resistance. By measuring this resistance, the temperature in turn can be accurately measured, making semiconductors invaluable as thermometers in various applications. The excitation of electrons into the conduction band can also be induced by photons or electrically charged particles. The quantity of liberated free electrons is directly proportional to the energy deposited by these particles. When an electrical voltage is applied, and the free electrons are gathered, it becomes possible to measure the energy of the initial particle by assessing the recorded electrical charge.

The energy gap between the valence band and the conduction band is the energy required to produce a detectable free electron. It is just a few electron volts in semiconductors. However, in gas-based detectors (↓) or scintillators (↓), an ionization energy of several tens of electron volts must be supplied to generate a signal. Consequently, for the same energy deposition,

Pauli principle: Fermions and Bosons → p. 94
Gaseous Detectors → p. 72
Scintillators and Photomultipliers → p. 74

semiconductor detectors produce significantly more charge carriers, resulting in a significantly better sensitivity (Statistics ↓).

Semiconductor technology is highly advanced, enabling the mass production of exceptionally compact semiconductor detectors. A mobile phone camera consists of a chip with several million pixels, each with a semiconductor detector only micrometers in size. They are used for measuring optical photons, i.e., photons visible to the eye.

Semiconductor detectors find versatile applications, with their dimensions ranging from the miniature detectors in camera chips to large germanium detectors weighing several kilograms. They are employed in various settings, including the compact sensor chips found in mobile or digital cameras, as well as larger versions used in astronomical telescopes. Similarly, silicon detector chips serve as imaging cameras on satellites dedicated to X-ray astronomy.

In the very center of large particle detectors such as ATLAS, CMS, ALICE, and LHCb at the LHC accelerator are *vertex detectors*. These comprise multiple layers of highly pixelated silicon detector chips, facilitating three-dimensional measurements of particle tracks. A very broad field of application for larger individual detectors is also the spectroscopy of X-ray and gamma radiation. Gamma spectrometers are frequently used in environmental analysis for detecting radioactive contaminants and analyzing trace elements. In order to minimize interference caused by thermally excited electrons, semiconductor detectors are cooled to enhance their sensitivity in spectroscopic applications. Typically, this cooling is achieved using liquid nitrogen, which lowers the temperature to approximately 77 K. Looking at a semiconductor spectrometer, one mainly sees the nitrogen container (see images).

The largest semiconductor detectors, weighing a few kilograms each, are made from high-purity germanium. One prominent example is the Advanced Gamma Tracking Array (AGATA), designed for the study of the structure of atomic nuclei. AGATA comprises a spherical shell housing 180 germanium detectors. These detectors not only accurately measure the energy of gamma quanta emitted during nuclear reactions, but also precisely determine their emission directions. To enhance the precision of pinpointing the locations where gamma photons are absorbed, each of these detectors, weighing approximately 2.4 kilograms, features 36 read-out contacts distributed across its surface for collecting the generated charges.

A part of the AGATA array consisting of hexagonal germanium detectors

Installation of the ATLAS track detector consisting of many silicon detector chips

Statistics → p. 84

Gaseous Detectors

Measuring With Seemingly Nothing

Gas-filled detectors were the first particle detectors to operate with electrical signals. The first of this kind was the *counter tube*, developed by Hans Geiger and Walther Müller in 1908. The construction of the Geiger-Müller tube is simple: A cylindrical tube with as thin a metallic wall as possible is filled with a noble gas. A thin wire runs along the central axis of the tube. A voltage of about a thousand volts is applied between the outer wall and the central wire via a resistor so that the wire is positively charged and the outer wall is negatively charged. When an ionizing particle (e.g. an electrically charged particle from a radioactive decay, in which particles with several thousand electron volts (eV) or more of energy are produced) penetrates the wall into the tube, it can ionize the noble gas atoms (↓), i.e., knock an electron out of its atomic shell. These electrons are called "primary electrons". They are produced along the entire path of the particle in the tube.

Due to the electric field, the primary electrons drift inward towards the wire. In doing so, they experience an increasing electric field because the field lines become denser as the wire comes closer. This can be understood from the drawing of the wire and the electric field lines (right page), which run radially from the wire outward to the wall. Close to the wire, the drifting electrons are thus increasingly accelerated and become highly energetic particles themselves, which can in turn cause ionization—a self-amplifying process. This results in an avalanche of electrons.

A charged particle traverses a Geiger-Müller counter.

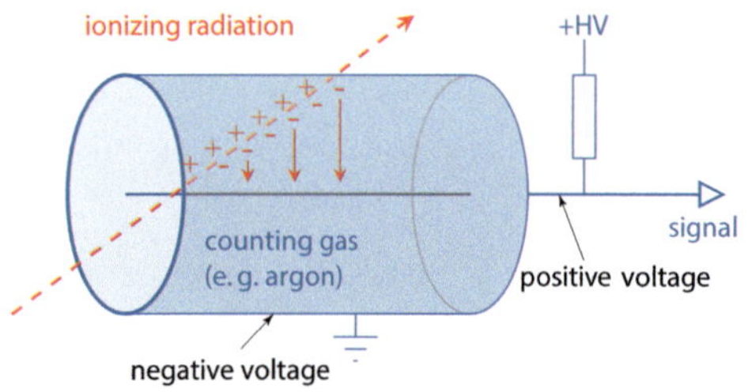

Ionization of an atom by a high-energy electron

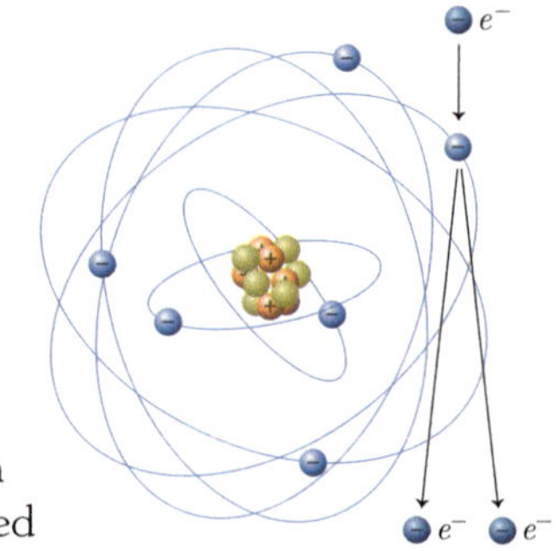

Depending on the field strength, about 100 billion electron-ion pairs can be created per initial particle passing through. The process comes to a halt after a short time because the electron avalanche and ions generate a measurable electric current between the wall and the wire. This current closes the circuit over the external resistor. The voltage drop over the resistor at a high current reduces the electrical voltage in the counter tube. Thus, no electrons can be accelerated anymore, and the avalanche stops. But why this elaborate construction? Why set free about 100 billion electrons from a few hundred primary electron-ion pairs that the original high-energy particle has generated? The reason is simple: The signal generated by the current of 100 billion electrons can generate an audible cracking noise

Interactions Between Particles and Matter → p. 68

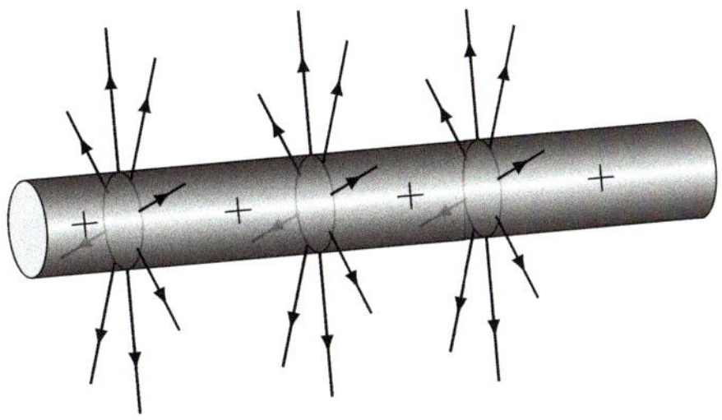

The electric field around a thin wire

in a loudspeaker. A few hundred primary electrons are not enough to generate an audible signal. This challenge existed not only in 1908, but essentially persists until today. Now computers read out the detector, not a loudspeaker and our ears. Still, modern gas-based detectors have to amplify the electrical signal in the gas to make it measurable, even if only by about a factor of 1000 and not 100 billion.

The great advantage of gas-filled detectors is also their biggest weakness: It is desirable that a detector is made up of as little material as possible because then the high-energy particles are hardly disturbed by the lightweight detector. Therefore, the detector—in this case both the detector structure as well as its gas—has little impact on the particle properties. However, the lack of material is exactly why in the initial ionization only a relatively small number of primary electrons are created and why good tricks for amplifying the signal must be found—such as the avalanche around the signal wire. Another major disadvantage of gaseous detectors is that they are quite slow: The time intervals between individual particle generations in modern experiments are sometimes in the range of a few nanoseconds. However, the primary electrons take several microseconds to drift through the detector, which is up to a thousand times longer than the time between interactions.

Despite these disadvantages, the advantages often outweigh them. Modern gas-based detectors, such as the *time projection chamber* no longer need to be tubular, but can also be thin plates. They can track the entire particle path in three dimensions with an accuracy of about 0.1 mm. They no longer measure the signal of the entire avalanche at just one wire, but are measuring the avalanches of each individual primary electron separately on small pads. The smallest of these pads are only about 50 µm in size. A microscope is needed to see them. The principle of gas-based detectors is over a hundred years old. It has been continuously modernized and still enables one of the most important types of detectors today.

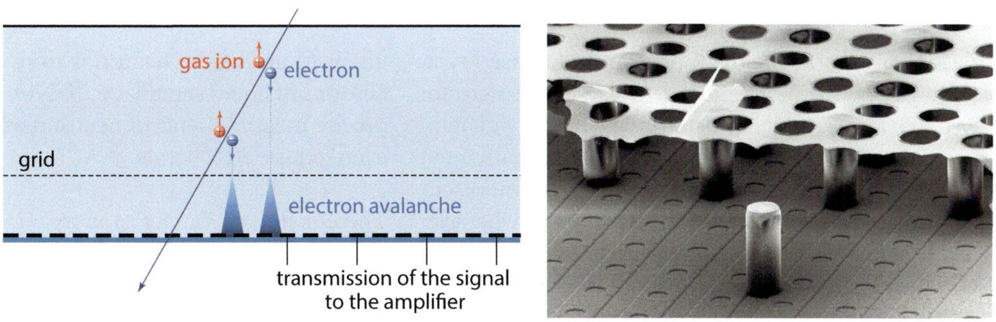

A modern microscopic readout structure of a gaseous detector

Scintillators and Photomultipliers

The Big Eyes of Particle Physics

Scintillation means twinkling or flickering. It describes a process crucial for particle detection. Without it, most of the experiments described in this book would literally be in the dark. In many materials, the excitation of molecules or atoms can be triggered by the passage of charged particles or photons, leading to the partial release of excitation energy in the form of light. Materials manifesting this phenomenon are called *scintillators*. They serve as a key detection material in particle physics.

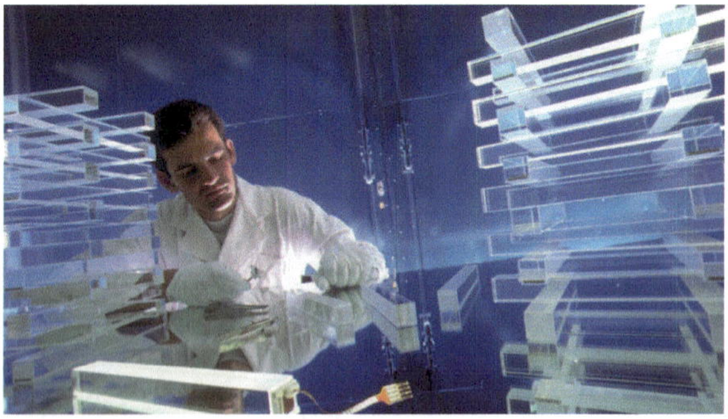

Test of the $PbWO_4$ crystals for the CMS detector at CERN.

The applications span from scintillation detectors in the scale of centimeters for positron emission tomography (PET ↓) to calorimeters in the scale of meters for measuring particle energies in the detectors at CERN (↓) and up to detectors with size of several 10 meters in neutrino physics (↓). Many of the materials used for scintillation are cost-effective to produce, making them particularly well-suited for large-scale detection setups. Depending on the material, the light emission can happen extremely quickly, which plays a major role when high time resolution is important, such as in PET or in the detectors at CERN.

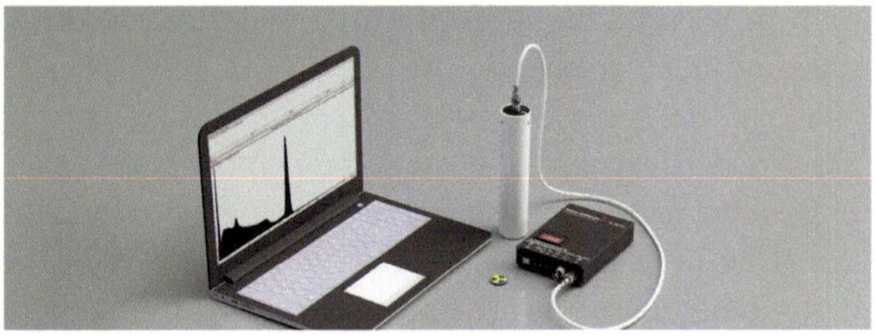

Commercial radiation detector with a cesium iodide scintillator, tightly packed together with a photomultiplier in the white cylinder. The black box contains the power supply and read out electronics.

Plastic scintillators are the most commonly used type. Large plates can be employed for large-area detector systems, for example, to determine whether radiation from outside penetrates into a sensitive measurement setup and possibly generates false signals. Liquid scintillators, mostly organic liquids, are used whenever large-volume detectors are required. This is particularly the case in neutrino physics. Borexino, an experiment to investigate neutrinos from the Sun (↓) used 800 tons of liquid scintillator. JUNO, an experiment for precise measurements of neutrino oscillations (↓), will even require 20 000 tons.

There is a wide variety of inorganic scintillators available. Crystals of NaI or CsI are commercially very widely used as gamma detectors. Other crystals offer

PET: Positron Emission Tomography → p. 144
CERN: Typical Detectors at Accelerators → p. 128
Large Detectors Deep in a Mountain → p. 132
Neutrinos From the Sun → p. 344
Neutrino Oscillations → p. 198

the flexibility to precisely tailor the detector material to meet specific experimental requirements. $PbWO_4$ crystals, which contain the heavy elements lead (Pb) and tungsten (W), have a very high efficiency in slowing down high-energy particles. This is why they are installed in large numbers in the CMS detector at CERN for energy measurements. With materials like $LuAlO_2$, an extremely short time to light emission can be achieved, which is necessary when extremely high time resolution faster than nanoseconds is required.

All these beautiful scintillators would be useless without the ability to measure the emitted photons. The most sensitive light sensors by far are photomultipliers. They convert photons into measurable electrical signals. On the front surface of an evacuated glass bulb, a photon generates an electron through the photoelectric effect. Behind this surface, a series of electrodes, known as *dynodes*, are arranged in a chain. These dynodes are subjected to an increasing electrical voltage, with each dynode connected to a higher voltage than the previous one. Consequently, the electron is accelerated toward the first dynode, where it causes the emission of additional electrons upon impact. This process of electron multiplication repeats itself several times, resulting in the generation of a measurable signal composed of several million electrons. Photomultipliers can be produced in various shapes and sizes fitting to specific applications. They exhibit exceptional speed and sensitivity. Even in the case of very large photomultipliers, such as those with a diameter of 50 centimeters used in the Super-Kamiokande (↓), experiment, one can detect individual photons.

At the photocathode, photons induce the release of electrons through the photoelectric effect. These electrons are subsequently accelerated as they traverse from one dynode to the next, with each collision causing the ejection of additional electrons from the metal. Finally, at the far end of this process, a measurable signal comprised of millions of electrons can be extracted.

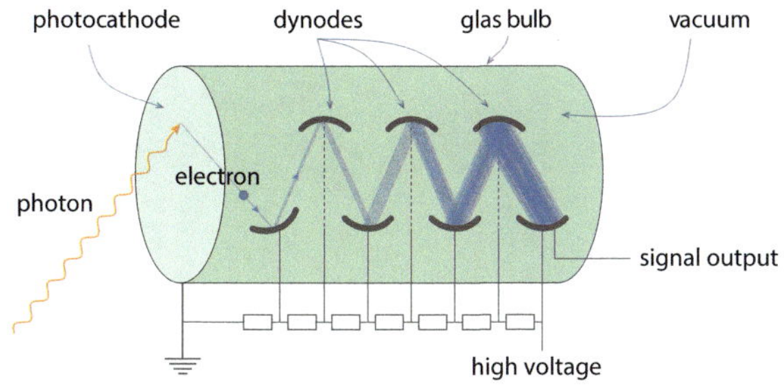

Examples of photomultipliers of different sizes

The world's largest photomultipliers are used in the Super-Kamiokande neutrino experiment.

Super-Kamiokande: Neutrinos → p. 196

Position and Momentum Measurement

Whither and Hither?

Where do particles come from, where are they going, and what is their momentum (Kinematics ↓)? These often are amongst the central questions of particle physics experiments. A detector for such measurements has to meet the following criteria: First, it should be easily penetrated by particles, as typically the path of the particle should be disturbed as little as possible by the detector material. Therefore, gas-based detectors (↓) are suitable for many applications. This is not always necessary; sometimes a detector measures the origin, path, and energy of the particles all at once, and absorbs the entire particle energy. Second, it should usually be able to measure very accurately the position of the particle along its path. Typical uncertainties of the location measurement along the particle path are between 0.1 mm and 0.05 mm! This is important because some particles decay (↓) into others and fly only a fraction of a millimeter before they decay.

Third, the detector often needs to be embedded in a magnetic field. This enables the determination of the particle momentum. For electrically charged particles the momentum is measured from the curvature of the particle path on a helix, whose axis is parallel to the magnetic field. The magnetic Lorentz force, which pushes a moving particle onto a circular orbit, is always exactly balancing the centrifugal force of a particle (see sketch). Particles with high momentum feel a high Lorentz force, but an even much higher centrifugal force would be needed to keep them on the same radius as a low momentum particle: Therefore, a large measured radius corresponds to a large momentum, and the other way around. The sign of the particle charge is also measured by the curvature, since the positively charged particles are curved in the opposite direction to the negatively charged ones. This can be seen in a picture of an event of the ATLAS experiment, where there are tracks curved to the left and right.

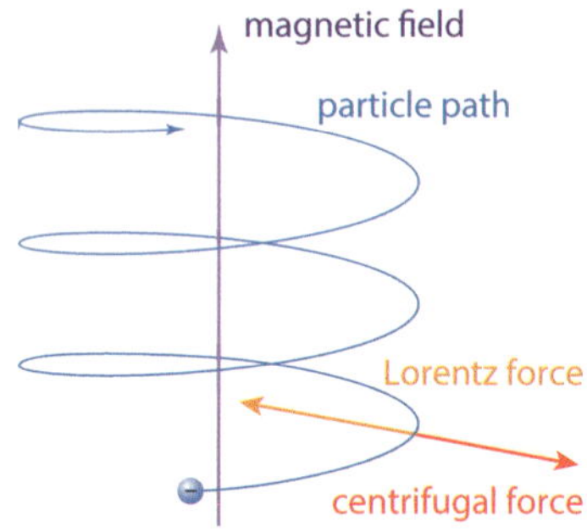

The spiraling path of a charged particle in a magnetic field

The curved tracks of many charged particles in the ATLAS detector

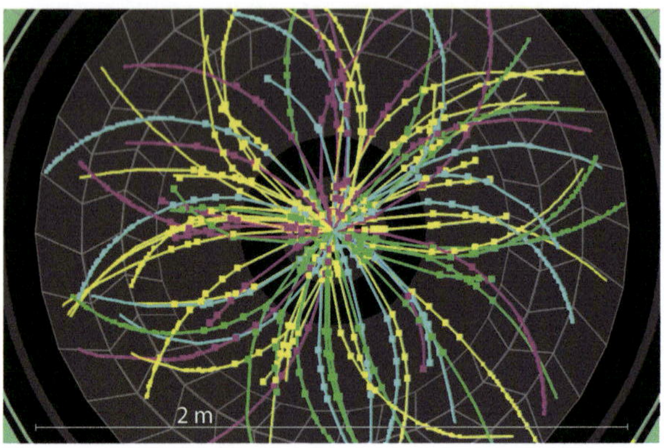

Kinematics → p. 40
Gaseous Detectors → p. 72
Particles' Births and Deaths → p. 54

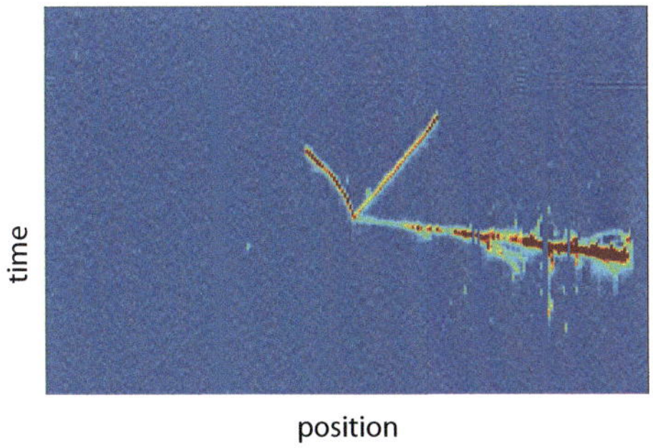

An image of a neutrino interaction in a liquid argon time projection chamber

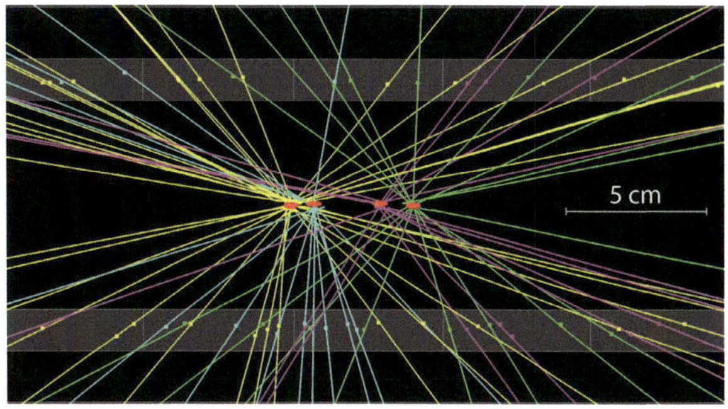

The reconstruction of particle collisions (red points) in the vacuum of the beam pipe from reconstructed particle tracks

All detectors for location and momentum measurement use ionization to generate a signal (Interactions Particles Matter ↓). This means that typically the particle path can only be continuously tracked for electrically charged particles, as neutral particles must first convert into other particles before they cause ionization.

To meet these properties, detectors can either measure the particle path continuously, or only at selected points along the path. An example of the first type is the time projection chamber, TPC. A TPC is capable of providing a continuous three-dimensional image of a particle path or even a particle interaction with multiple particles. This is shown in the left picture on top of this page, where two electrons and a positron can be found in a TPC filled with liquid argon from the ArgoNeuT Collaboration (Direct Search for Dark Matter ↓).

The vertex detector of the ATLAS experiment is an example which only measures discrete points along the path of a particle. This can be seen in the second image of the same collision event as before (top right). Here, the view is rotated 90°, so that the beam axis of the accelerator (LHC ↓) runs horizontally through the image from left to right. In the middle of the image, the particle packets collide. Apparently four of the protons coming from the left collided with four other protons from the right. Each pair of protons created bunches of new particles. Colorful lines are supposed to represent the tracks of the charged particles (Trigger and Data Processing ↓). However, only the small rectangular points correspond to real measurements. Only there, parallel to the beam axis, are layers of silicon detectors (↓), each layer being only about 0.1 mm thick. These are divided into small pixels, like in a normal camera, and measure the particles passing through the thin silicon material. The colorful lines are computer generated reconstructions of the particle tracks, generated literally by "connecting the dots". The innermost of these layers in the image is about 5 cm from the center of the beam pipe. The lines meet at only four points exactly on the beam axis. There, the four pairs of protons collided. These points are inside the vacuum of the beam pipe. Such an origin of particles is called a *vertex*. This is how a measurement can be made virtually, in a vacuum, where no detector is placed at all!

Interactions Between Particles and Matter → p. 68
Direct Search for Dark Matter → p. 310
The LHC Accelerator Complex at CERN → p. 116
Trigger and Data Processing → p. 150
Silicon detectors: Semiconductor Detectors → p. 70

Energy Measurement

How Many Calories Does a Detector Eat?

The detectors used for measuring the energy of particles are called *calorimeters*. In a calorimeter, the kinetic energy of the particles is fully or partially converted into another form of energy. In this process, the original particle is captured or destroyed. Therefore, energy is measured usually at the end of a chain of various measurements (Position and Momentum Measurement ↓).

The name "calorimeter" is, not without reason, taken from the Latin word *calor* for "heat": Some calorimeters indeed measure the temperature increase that a high-energy particle generates in the dense detector material when it literally pushes on many other particles that make up the detector material (Interactions between Particles and Matter ↓). This can be imagined exactly as when a massive body slides over a rough surface: its kinetic energy is converted into vibrations of atoms and molecules through the friction between the surfaces. These vibrations can be felt as heat.

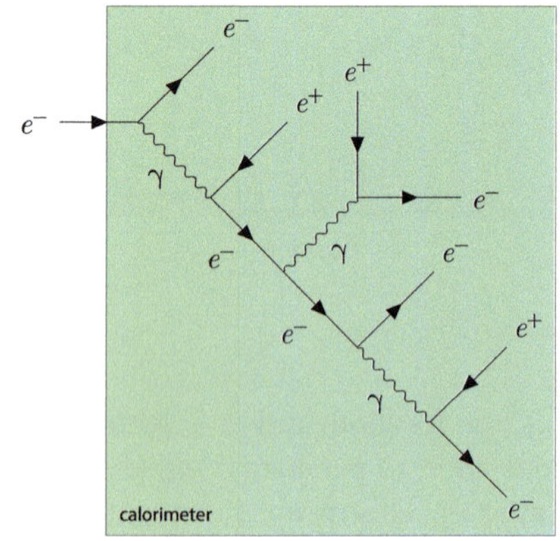

An electron and its interactions with the material of the calorimeter

The calorimeter typically consists of a high-density material, so that the particles transfer their energy to the detector in a confined space. As an example, let us consider an electromagnetically interacting particle like the electron. It undergoes a cascade of interactions with the electric fields of the nuclei and the electrons of the atoms of the material. In each of the steps of this cascade, new photons or electron-positron pairs can be formed, which on average each carry half the energy of the particle in the previous step, and then in turn trigger a new cascade. This chain stops when the photons carry less than twice the electron

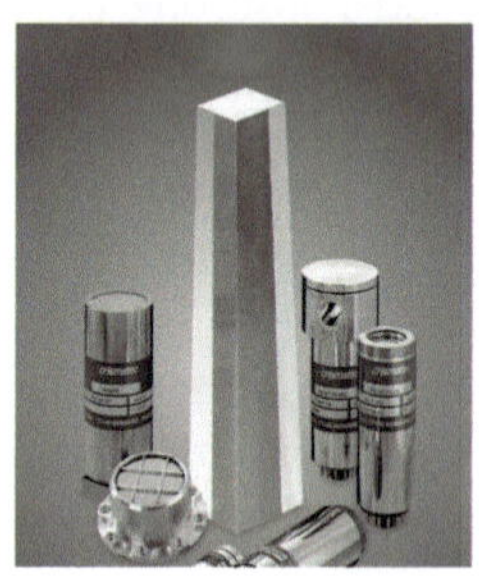

Typical calorimeters

Position and Momentum Measurement → p. 76
Interactions Between Particles and Matter → p. 68

mass as energy and can no longer form new electrons and positrons. The total amount of particles produced in this cascade is called a *shower*.

In most modern calorimeters, a direct electronic signal is measured instead of the increase in temperature. For this purpose, mainly two different methods are used: Either the particles are stopped in a dense scintillator (↓) and the energy is measured via the amount of light generated by the scintillation in a photomultiplier. The other possibility is to directly generate an electrical signal by incorporating semiconductor detectors (↓) or liquid noble gas detectors (↓) into the calorimeter.

A calorimeter can be used not only for the energy measurement of individual particles but also to support the identification of particle types. In an experiment, a wide variety of particles can be produced, such as only electromagnetically interacting particles with (e.g. electrons, muons) and without (e.g. photons) electric charge, as well as particles interacting with the strong force, which in turn can be electrically charged (protons, charged pions) or electrically neutral (neutrons, neutral mesons).

Some calorimeters allow for a high spatial resolution, i.e., they are divided into many small segments, each of which can be read out individually, or otherwise form a three-dimensional structure. Then the spatial structure of the energy deposits in the calorimeter can be used for particle identification (↓): Particles that interact with the strong force form a highly irregular shower (top). Heavy particles interacting only electromagnetically, such as muons, form a well-defined track, as they only slowly generate new particles (middle). Electrons or photons form a compact shower (bottom).

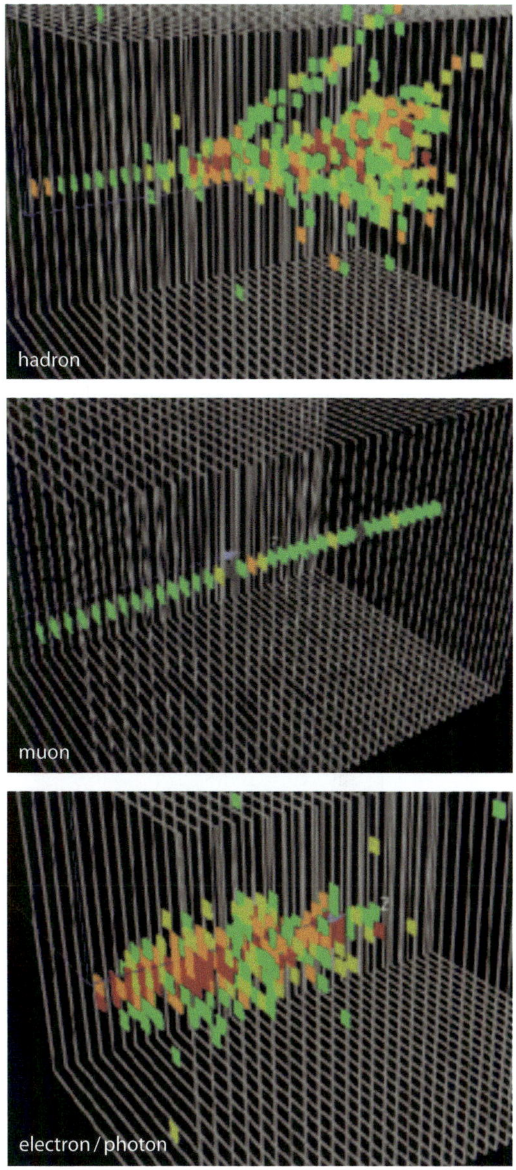

Particle identification by differently shaped showers in the calorimeter

Scintillators and Photomultipliers → p. 74
Semiconductor Detectors → p. 70
Liquified Noble Gases as Detectors → p. 138
Signatures of Particles → p. 80

Signatures of Particles

Footprints of Particles in the Detectors

In particle physics experiments we encounter a wide variety of different particles. Often, the presence of specific types of particles is key to identifying the underlying physical process. Therefore, an important aspect of measurements is the ability to precisely determine the type of particles that are traversing the detector. The experimentalists benefit from the fact that the different particles exhibit distinctly different combinations of interactions with the detector material (↓).

Since different densities of materials produce different signatures, many particle detectors, especially at high energy accelerators, are composed of various sub-detectors (Typical Detectors ↓). The aim is to use as little material as possible in the vertex and tracking detector during measurements of the origin, direction, and momentum of electrically charged particles (↓). This layer is followed by the calorimeters (Energy Measurement ↓), which are made of dense material. Behind the calorimeters, typically only one type of particle can be detected: muons, heavy partner particles of the electrons. For them, there is often another tracking detector system behind the calorimeters.

Different particles and their specific interactions in layers of detector components

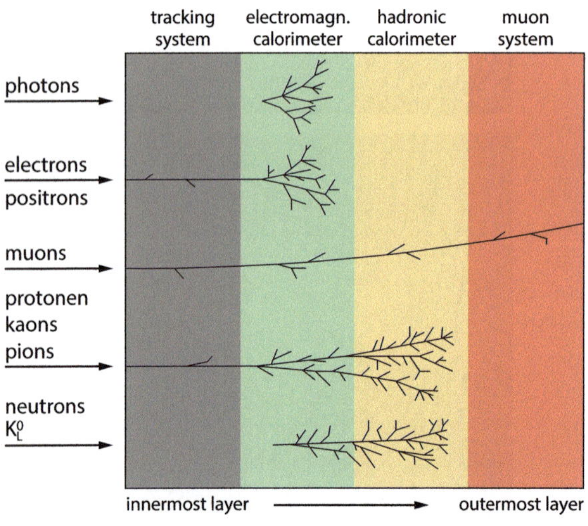

The detection of particles in the detector always uses either the electromagnetic interaction, the strong interaction, or a combination of the two. The relative proportion of the two is determined by the electromagnetic charge of the particle one wishes to identify, whether it is a hadron held together by the strong interaction, and by its mass. The most commonly identified particles in accelerator-based experiments are as follows:

A *photon* is neither electrically charged nor subject to the strong interaction and, therefore, typically does not leave a trace in the tracking detector. However, it is the exchange particle of electromagnetism; although it is electrically neutral, it can still interact with electrically charged particles. For example, electrons are present in large numbers in dense matter. The photon thus transfers its energy to electrons and newly formed electron-positron pairs. These electrically charged light particles release their energy in a compact, dense process into the detector material. A photon is therefore reconstructed as a compact "cluster" without a track directed to it from the original interaction point.

Interactions Between Particles and Matter → p. 68
Typical Detectors at Accelerators → p. 128
Semiconductor Detectors → p. 70, Gaseous Detectors → p. 72 and Position and Momentum Measurement → p. 76
Energy Measurement → p. 78

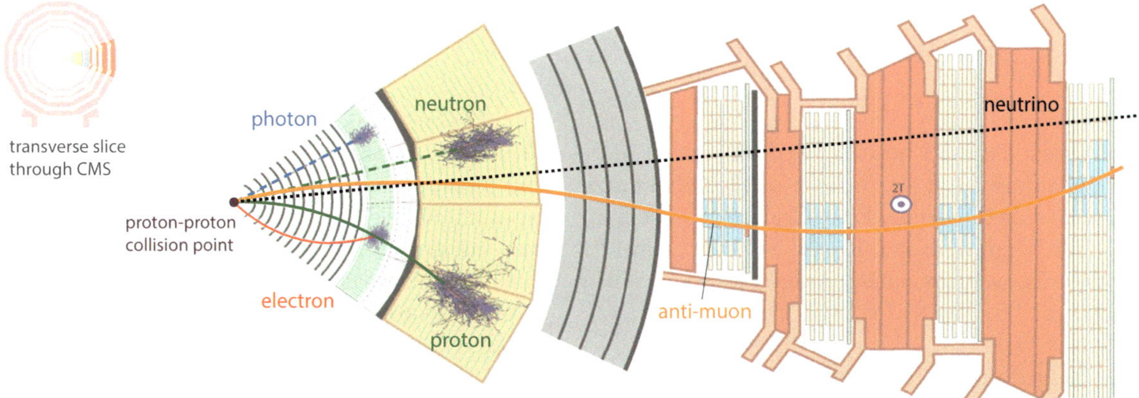

Different types of particles on their path through the CMS detector. The colors correspond to those in the figure on the left page.

An *electron* behaves in almost the same way but is not very massive, and electrically charged. Therefore, it ionizes the material that it penetrates along its path. A track is created in the tracking detector. In the calorimeter, it behaves like the photon.

A *muon* is the heavy partner particle of the electron. However, it does not interact identically to the electron: This is because its mass is 200 times higher. Therefore, it has more "penetrating power" than an electron. It penetrates calorimeters and muon detectors as a track. At the usual energies of muons in particle physics experiments it has so much energy that it decays only far outside the detector (SRT ↓). This also means that the energy of those muons is not measured directly, but that it is inferred indirectly from their momentum, and from knowing their mass.

Charged hadrons such as protons or charged pions differ from muons in their ability to interact via the strong interaction. Unsurprisingly, the strong force is very strong, thus causing intense interactions. However, hadrons are color-neutral combinations of quarks. The strong force acts only between hadrons when their distance is in the order of magnitude of their own size (↓). Because of this interplay, they create irregular clusters, which differ in their shape from those created by electrons or muons and show large statistical variations in deposed energy density. The same applies to neutral hadrons, such as a neutron or a neutral kaon, but without a pointing track. Also here, the difference from the photon lies in the shape of the cluster in the calorimeter.

The only particles that escape most detectors are *neutrinos*. They only interact via the weak interaction, mostly far outside the detector or even far outside Earth. In rare cases, they can be found in a dedicated detector (FASER ↓), though.

Studying all the resulting patterns, physicists begin a kind of puzzle game: From which interaction in the vacuum of the beam tube might the observed combination of particles in the detector have originated?

SRT: Special Theory of Relativity in Particle Physics → p. 92
On the Ranges of Interactions → p. 58
FASER → p. 312

Cherenkov Radiation and Particle Identification

Faster Than Light

This article discusses an effect that allows one to determine the speed of a particle. It is called *Cherenkov effect* after its discoverer. Pavel Cherenkov[1], Ilja Frank and Igor Tamm were awarded the Nobel Prize in 1958 for the "discovery and interpretation of the Cherenkov effect".

If a charged particle moves through a medium at a speed v that is greater than the speed of light in this medium, it emits a characteristic electromagnetic radiation. This is referred to as *Cherenkov radiation*. It occurs at an specific angle ϑ relative to the path of the particle.

The cosine of this angle is given by:

$$\cos(\vartheta) = \frac{c_{\text{medium}}}{v} = \frac{c}{nv} ,$$

where c_{medium} is the speed of light in the medium, which results from the ratio of the vacuum speed of light c and the refractive index n known from optics. A similar effect occurs for sound waves, when a supersonic aircraft moves faster than the speed of sound in the medium air.

There are various ways to use the Cherenkov effect. If one considers a particle beam consisting of different types of particles such as kaons and pions with the same momentum (Kinematics ↓), they inevitably have different speeds, as they have different masses. If you now let this particle beam pass through a gas volume, you can adjust the refractive index n using the pressure so that a pion generates a signal, but the heavier kaon does not, as its speed is still below the speed of light of the gas. In this way you can distinguish between pions and kaons in the particle beam.

If you also measure the angle ϑ of the photons relative to the path of the particle, you can determine the speed. A particularly elegant method to do this is a so-called *ring Imaging Cherenkov (RICH) detector*. Here a set of mirrors is arranged spherically so that, regardless of where photons are emitted along the track, they are projected onto a ring in the detector (see figure).

The radius of this ring provides information about the angle ϑ and thus about the speed. If the momentum of the particle is known as well, e.g., from the deflection

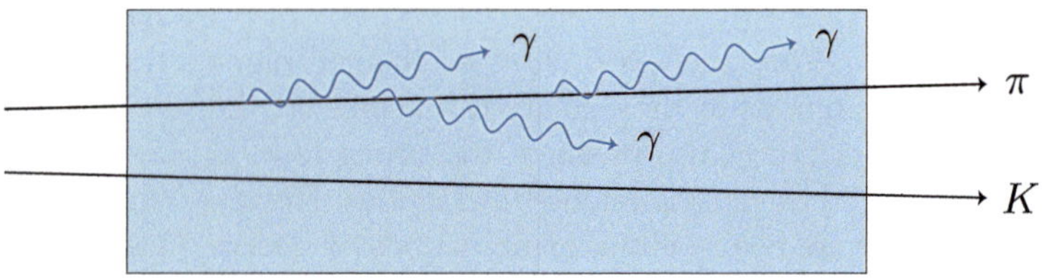

Principle of a threshold Cherenkov counter: The refractive index of the gas in the blue volume is chosen so that the faster pions emit Cherenkov light, while this is not the case for the slower kaons.

[1] You can also find the spelling "Tscherenkow".

Kinematics → p. 40

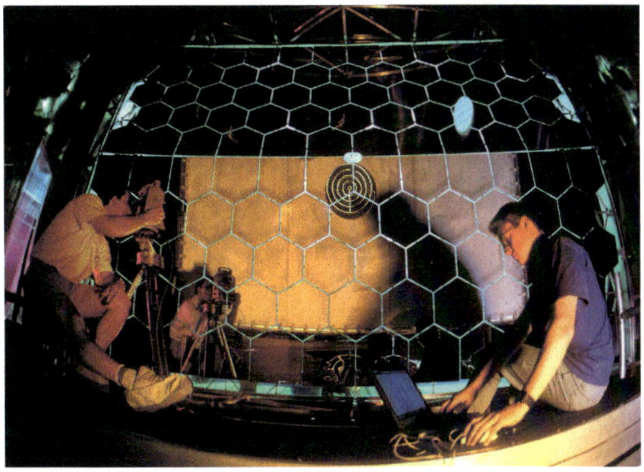

The RICH (Ring Imaging Cherenkov) detector of the COMPASS experiment at CERN during the construction phase. The spherical mirror arrangement made up of hexagons is visible.

in a magnetic field in another part of the detector, one can infer the identity of the particle, since the mass can be calculated from the knowledge of momentum and speed.

Principle of a Ring Imaging Cherenkov Detector. Regardless of where a photon was emitted along the particle track, it always hits the detector at the same distance from the particle track. In two dimensions, this results in a ring.

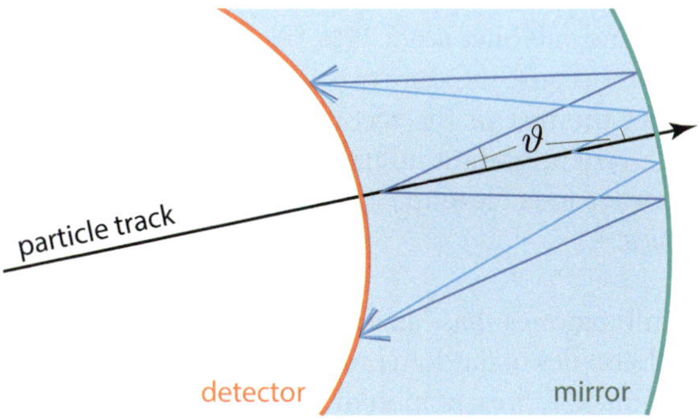

One can even use the ice in Antarctica as a Cherenkov medium. This is exploited at the IceCube detector (↓) near the South Pole. Photomultipliers (↓) embedded in the ice are used to detect the Cherenkov light generated from high energetic cosmic ray particles passing though the ice.

Bluish Cherenkov light in a nuclear reactor caused by fast electrons in water

The IceCube Detector → p. 134, Large Detectors Deep in a Mountain → p. 132
Scintillators and Photomultipliers → p. 74

Statistics

Mathematically Describing Randomness

For several reasons particle physics always has to deal with chance or randomness: On the one hand, quantum mechanical processes (↓) in Nature are obviously fundamentally guided by chance. On the other hand, many uncertainties (↓) that occur in quantitative predictions and experiments can be described by random processes. Uncertainties always exist when observing the world through measurements: no instrument always measures exactly the same when the measurement is repeated, and the calibration is never exact. The mathematical study of processes characterized by randomness is called *statistics*.

A characteristic of randomness is that the *exact* value of the next random number—e.g., the lottery numbers or the measured value for the next measurement—cannot be predicted. However, this does not mean that scientists know nothing about chance. The processes that lead to results that are subject to random uncertainties may be random, but they are not arbitrary. They generally follow the laws of probability, so that a definitive statement can be made about the probability of obtaining a certain result. Almost all fundamental measurements are based on this principle.

We can perform calculations with probabilities. A simple example is a coin toss: heads or tails each have a probability of 50 % with a perfect coin. The sum of all possibilities must always add up to 100 %, because there is always one definite outcome in each toss. For two coin tosses in succession, it follows that the probability of two heads is $50\,\% \cdot 50\,\% = 25\,\%$.

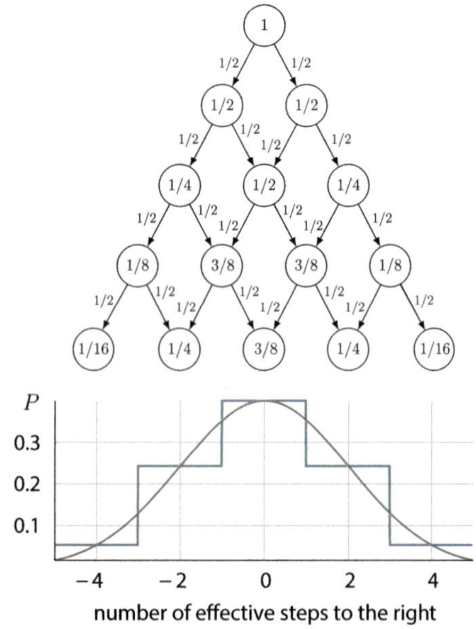

The Galton board and the Gaussian distribution

It is a little less intuitive that the probability of heads once and tails once is not 25 %, but $25\,\% + 25\,\% = 50\,\%$. This is because there are two ways to achieve this result: Either the first *or* the second throw can show heads! This very important influence of the number of different ways to achieve the same result is called *combinatorics*.

Combinatorics has an important influence on the probabilities of random processes, as can be seen in the schematic representation of a *Galton board*: Imagine

Concepts of Quantum Mechanics → p. 50
Measurement Uncertainties → p. 86, Determining Measurement Uncertainties → p. 158

a ball rolling downwards from above, which can turn left or right once on each level. The probability of rolling to the left or right at each turn is exactly 50 %. The superposition of many of these independent processes leads to a completely new probability distribution. While each individual branching (left vs. right) has an *equally distributed* probability (50:50), the combinatorics of the superposition of many equally distributed processes results in a new probability distribution in which the probability P of obtaining a certain result is similar to the Gaussian distribution. There are many different paths for the ball to reach the center, but only one for the very left or right. Incidentally, this would also apply if we tilted the board at an angle, i.e., the probability for left and right were not exactly the same anymore.

Regardless of the probability distribution of the individual random processes—if we superimpose enough of them, the result is a Gaussian distribution. This is due to combinatorics: In a symmetrical Galton board, every single decision on the way from top to bottom happens with 50 % probability to the right or left. However, there are many more possible paths that end in the center (i.e., vertically close to the starting point) than far away to the sides. Many measurements are not only influenced by a single random process, but by many different ones: For example, if we follow the path of a particle through a detector (Position and Momentum Measurement ↓), it is sometimes pushed slightly to the left, sometimes to the right when it collides with atoms on its path—just like the ball in the Galton board.

Poisson distribution

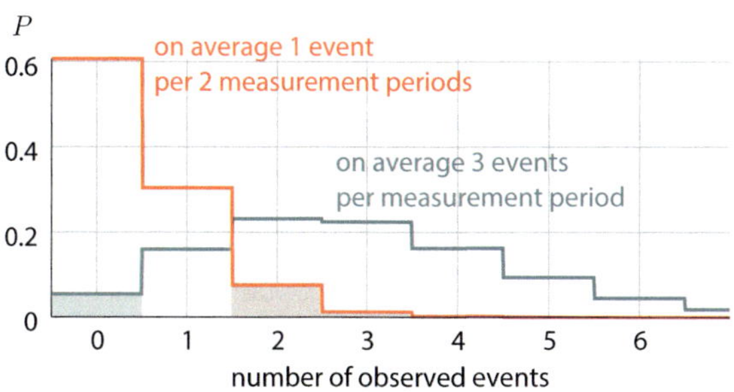

If the random process is not a superposition of many independent random processes, but is dominated by a small number, other distributions are needed. Among these, it is worth mentioning the so-called *Poisson distribution*. This distribution plays a big role in counting experiments with small numbers of observed counts. An example would be the observation of rare radioactive decays, where only a handful of counted decays occur within a certain timeframe. The Poissonian probability distribution shown in the last figure represents the probability P that the number of observed events is 0, 1, 2, 3 or more, if on average 0.5 events per measurement period (red) or 3 events per measurement period (blue) are expected. An important observation: This distribution is very broad compared to the mean or expected count rate. This is especially true for small expected count rates. For example, even if one expects 3 events to occur per experimental run, in about 5 % of the repetitions of the experiment, no event occurs at all. And if it is expected that on average only one event will occur in every second repetition, two or more events may also occur. This is shown in the colored areas in the graph of Poisson distributions. This property of probability is taken into account when calculating the *significance* (↓), and is very very important for determining whether a particle or process can be discovered or not.

Position and Momentum Measurement → p. 76
Significance: Drawing Conclusions From Statistics → p. 160

Measurement Uncertainties

Everyone Makes Mistakes

The result of a measurement is often given in the form measured value ± uncertainty This implies that a repetition of the experiment will not yield exactly the same result, but a value that fluctuates around the (unknown) true value μ. More quantitatively: About 68 % of the measured values lie between $(\mu - \sigma)$ and $(\mu + \sigma)$. The Greek letter σ is often used for the uncertainty.

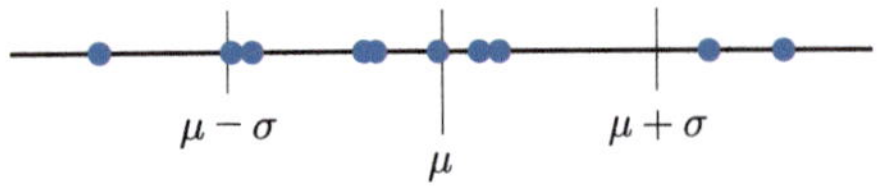

In repeated measurements, about 68 % of the measured values (blue dots) fall within the range $\mu - \sigma$ to $\mu + \sigma$.

The value of 68 % is pure convention and results from the properties of the Gaussian normal distribution (see box and Statistics ↓), which plays a major role in the consideration of measurement uncertainties. Measurement results usually follow a Gaussian normal distribution around the true value of the quantity to be measured.

Uncertainties are unavoidable in measurements. In colloquial language, a measurement uncertainty is often also referred to as an error or measurement error. However, this expression is an unfortunate choice, as no one has done anything wrong.

A distinction is made between *statistical* and *systematic uncertainties*. The difference can be well illustrated using the example of election polls. Typically, 1000 eligible voters are surveyed. In a simple example, let's assume that there are only two parties, A and B. If 400 people vote for party A and 600 people for party B in the survey, it is easy to predict vote shares of 40 % and 60 % for the total population. The statistical uncertainty of these shares is easy to calculate (see box) and becomes smaller the more people are surveyed.

Systematic uncertainties are far more difficult to handle. In our example, these can occur if the respondents were not selected representatively. If, for example, only people in rural areas are surveyed, this leads to a distortion of the result, when it is extrapolated to the entire population. In the diagram shown at the beginning, the values then no longer scatter around the true value μ, but around a value shifted by the unknown distortion. This systematic error or distortion cannot simply be reduced by surveying more people in the same area. Something has to be changed in the methodology, here specifically in the selection of the respondents.

Let us now move on to an example from physics. If you systematically press "Stop" on a watch too late when measuring the oscillation period of a pendulum, you cause a systematic uncertainty, which does not decrease by measuring *one* period duration multiple times. The only remedy here is a change in methodology. In this simple example, this can be done by measuring ten period durations and dividing the result by ten. This will be discussed in more detail in the article on the determination of measurement uncertainties (↓).

Statistics → p. 84
Determining Measurement Uncertainties → p. 158

Carl Friedrich Gauss and his distribution function were featured on the 10 Deutsche Mark bill in Germany until 2001.

As a concrete example of reducing the statistical error by repeating the experiment, we consider tossing a coin. To check whether there is a probability of $1/2$ for either side to be on top, you could, for example, toss it ten times. If heads comes up four times and tails six times, this is certainly not sufficient to make a statement about whether the coin is biased. However, if in 1000 tosses, heads comes up 400 times and tails 600 times, the statistical error is $\sigma = 0.015$ (see information box). The difference between the measured value of 0.4 and the hypothesis 0.5 is therefore $0.1 = 6.5\,\sigma$. This is referred to as a $6.5\,\sigma$ effect. The probability of such an effect occurring randomly is about 10^{-10}. Thus, we can practically rule out that both sides of the coin appear with a probability of $1/2$.

Measurement result and uncertainty for the probability that a coin toss results in heads for 10 and 1000 tosses

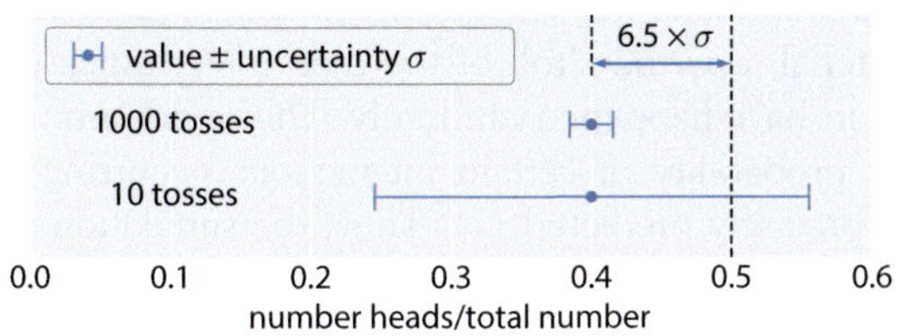

Binomial Distribution in Coin Tossing and Election Surveys

If in N coin tosses, k times the result „heads" occurs, the probability for the event „heads" can be estimated by k/N and the measurement uncertainty by

$$\sigma = \sqrt{\frac{\frac{k}{N}\left(1-\frac{k}{N}\right)}{N}}.$$

Accordingly, in the survey of N people, where k voted for party A, their share of votes can be estimated by k/N and the uncertainty with the formula above.

Gaussian Normal Distribution

Measurement results x are often distributed according to a Gaussian normal distribution

$$f(x) = \frac{1}{\sqrt{2\pi}\sigma} e^{-\frac{(x-\mu)^2}{2\sigma^2}}.$$

About 68 % of the measured values fall in the range of one standard deviation σ around the mean value μ, which corresponds to the true unknown value of the measured quantity.

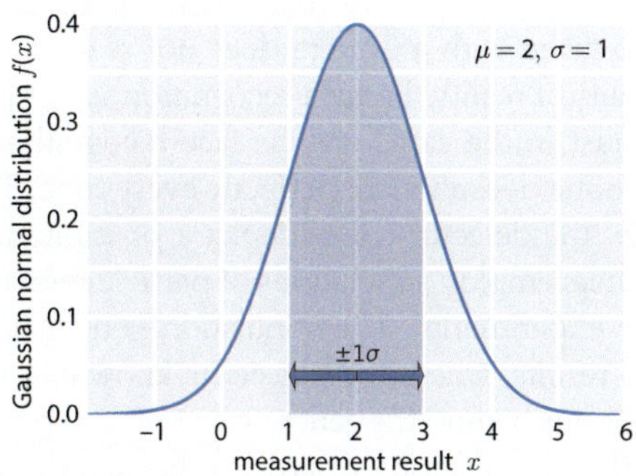

Simulations in Experimental Physics

Which Observation is *Really* Predicted?

A detector (↓) is a device that creates an "image" of the particles in an event. The image sensor of a camera, such as that of a smartphone, is such a detector for light particles. The sensor in the camera converts photons, the light particles, into electrical signals, which in turn are converted into an image on the touchscreen. But how do we know which electrical signal corresponded to which photon energy or photon number? Or how the electrical signal spreads laterally in the sensor? Here is where the simulation comes in. If you are interested in how electrical signals are interpreted as physical quantities, here you can dig deeper into this topic.

It would be highly desirable if the detector could simply recognize each particle exactly, similar to an imagined ideal camera with very accurate imaging and built-in face recognition that recognizes each person exactly. An example of the detector's output would then be: "Exactly at location x, a photon with energy E was created and it was moving in an exactly known direction". Such a detector does not exist. The detectors do not produce perfectly "sharp" images, just as a photo taken with a camera does not really show all the details of reality. Particle recognition is not perfect either, just in the same way as face-recognition on a smartphone doesn't work perfectly every time. To find out how the detector could depict a possible particle, researchers employ simulations of particle interactions in their experiments. The simulation generates hypothetical results, where the physicists know exactly the assumptions behind the generated results, in particular about the propagation of particles through the matter and electric or magnetic fields of the detector. Now, the same processes in the detector with the same energy and momentum can be simulated many times. Each time, slightly different non-ideal processes are simulated. The distribution of different simulated results for the same true particle then provides a quantitative estimate of the expected uncertainty of the measurement. In the smartphone analogy, this corresponds to taking, e.g., 100 photos of the same person under different conditions, lighting and facial expressions and then checking for how many of them the smartphone recognizes the person correctly.

For a complete detector, such a simulation can be very complex. The detector consists of several layers, each measuring different properties of the particles. At each place in the detector, either electromagnetism or the strong force can transfer energy from the particles to particles in the detector material, or create new particles. The fundamental interactions are described by quantum field theories (↓). Therefore, they make probabilistic statements: The prediction of the theory is not that something specific always happens, but with what probability something specific happens and with what probability something else occurs.

In our simulations, when a particle enters the detector material, a virtual "roll of the dice" determines what could have happened randomly. This is not arbitrary; the probability of certain interactions occurring can be precisely predicted (↓). Thus, the simulation rolls the dice for a specific realization of a possible event in

Typical Detectors at Accelerators → p. 128
Quantum Field Theory → p. 104
Interactions Between Particles and Matter → p. 68

A high-energy electron enters dense detector material. In each square, the simulation "rolls the dice", deciding what should happen in the simulated event with each of the generated particles.

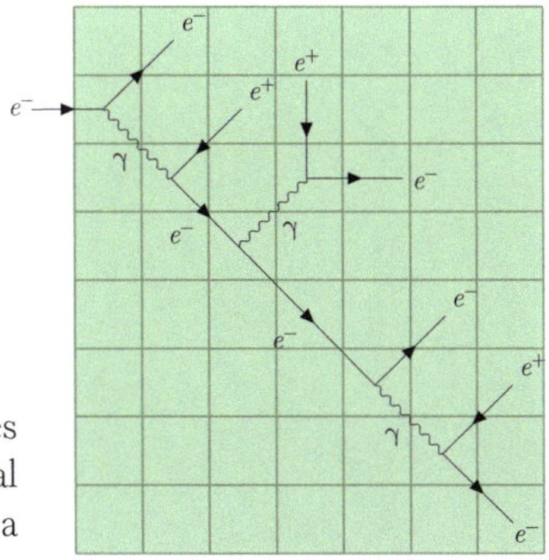

the detector. In the drawing, the detector material is divided into small squares that represent each individual roll of the dice. For example, a high-energy electron e^- enters the material on the upper left. Now, in each square, it is determined by chance what happens—for example, the electron can emit a high-energy photon under the influence of the electromagnetic fields near the atomic nuclei, which can in turn transform into an electron and a positron e^+, and so on. This process continues until the energy of the initial electron is distributed among so many particles that each individual one has too little energy to create new particles. This results in a complex picture, as can be seen from the representation of a real simulation. There, the high-energy electron flies from the upper left into the detector material, and thousands of new particles are created, which in turn stimulate the detector material (this generates an electrical signal, which is also simulated in the "digitization" step) and create new particles. The shape of this energy deposition can be measured in the detector, as well as the total amount of the generated signal. Both measurements can then be compared with the simulation, in order to cross-check the understanding of the underlying process encoded in the simulation.

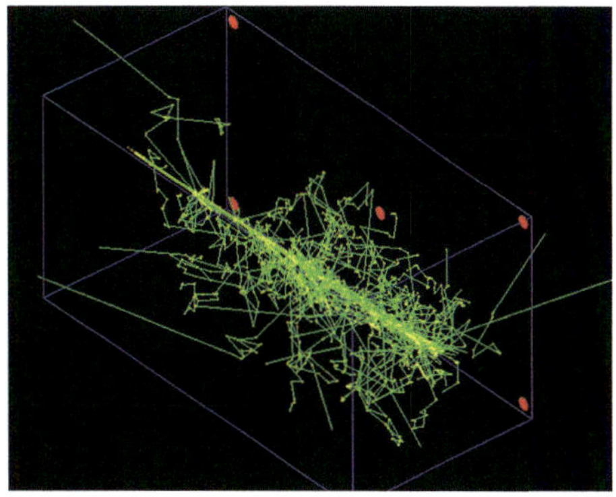

The simulation of the energy transfer of a high-energy electron to the calorimeter

Measured energy fraction for simulation and test beam experiment

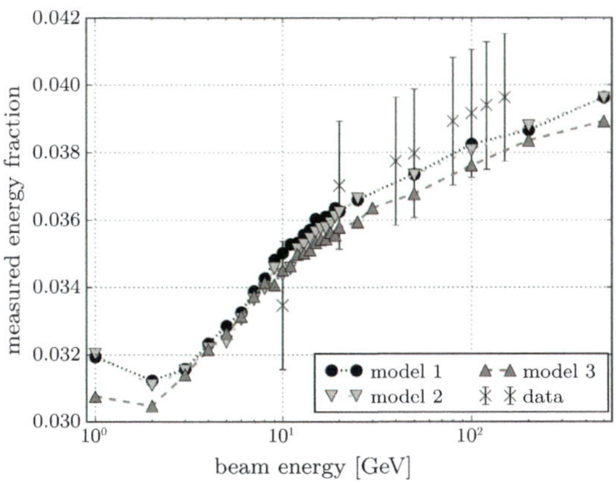

Therefore, the next step has to be a check whether this simulation is accurate enough. For this purpose, a particle accelerator shoots high-energy particles directly into the detector in test beam experiments. The known energy and type of particle from the accelerator can be directly compared with the measurement. In the example shown, this works quite well: The measurements (crosses) of the visible energy in the detector agree well with the simulations (dots and triangles) within their uncertainties.

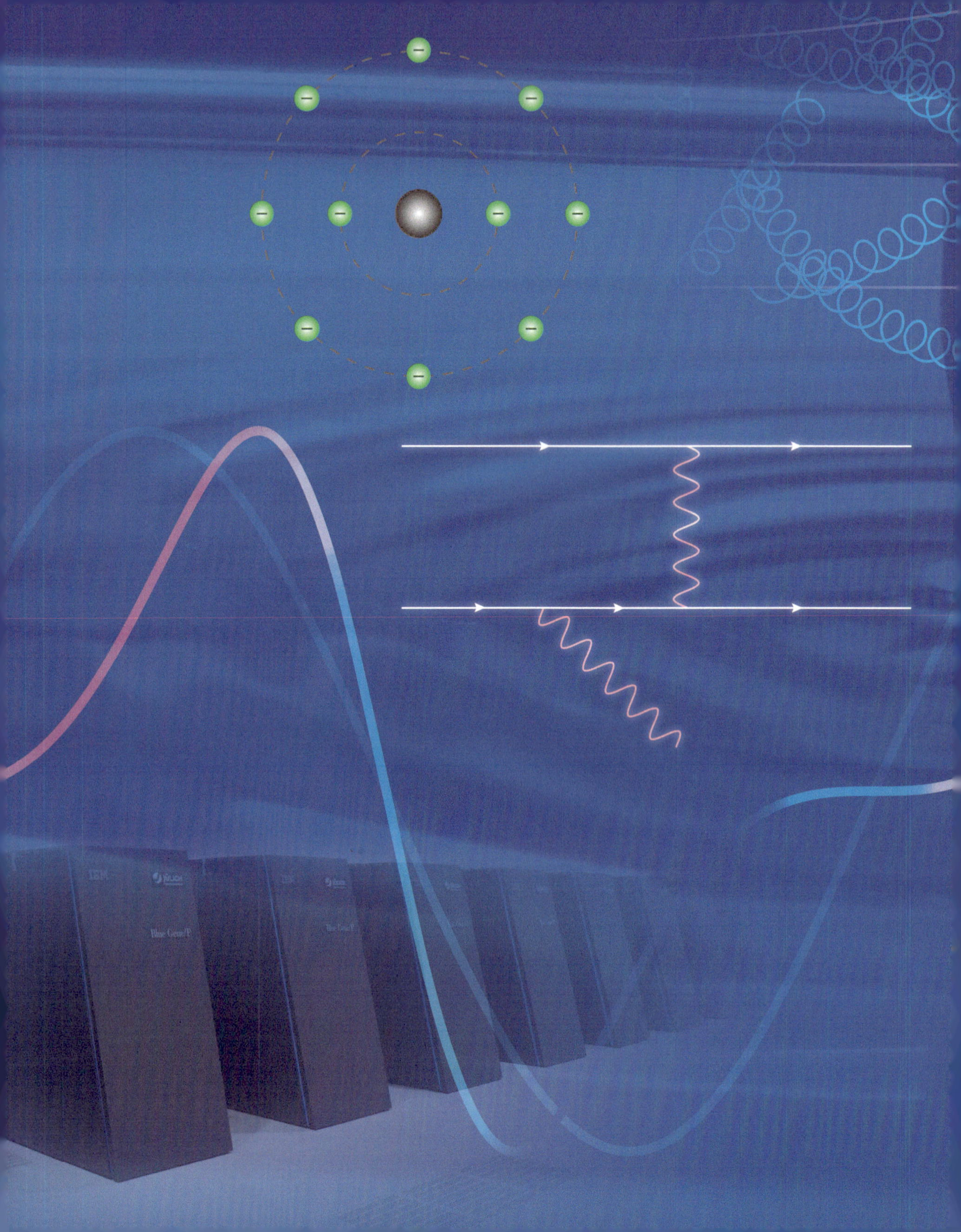
IBM
Blue Gene/P

4 Basic Theoretical Concepts

On the one hand, elementary particle physics relies on complex experiments conducted on both large and small scales. On the other hand, it is built upon a rich foundation of theoretical concepts, which span all areas of the field and are thus relevant to nearly every chapter of this book. Here we explore these concepts in more detail.

The basis for describing all moving particles is the theory of special relativity. Here, we present concrete examples from everyday particle physics. Symmetries play a central role, particularly local symmetries, known as "gauge symmetries", which are essential for describing the forces in the Standard Model, the fundamental framework of elementary particle physics. Noether's theorem reveals a fascinating connection between symmetries and conservation laws, offering us an intuitive understanding of the reactions observed in particle experiments. Quantum field theory serves as the fundamental mathematical framework for our understanding of particle physics. It allows us to make predictions based on the Standard Model, which can then be tested through experiments. This forms the crucial link between the theoretical model and the observable world.

Special Theory of Relativity in Particle Physics

The Speed of Light in Everyday Particle Physics

In this article we explain the relevance of the special theory of relativity (↓) for particle physics using a few examples.

Example I: The protons in the beam at the LHC have the energy $E = 6{,}8$ TeV. In the special theory of relativity, the following holds for the total energy of a proton in the beam

$$E = \gamma m c^2 = \frac{m c^2}{\sqrt{1 - \frac{v^2}{c^2}}}$$

With the known proton mass $m = 0.938\ \text{GeV}/c^2$, and inserting the energy $E = 6.8$ TeV we obtain for the speed $v = 99.999999\,\%$ of the speed of light. Furthermore, the boost factor γ is very large, about $\gamma = 7250$. The special theory of relativity is therefore directly relevant to know the relationship between the velocity and the energy of the protons. For the particle collisions under investigation the energy is decisive, as it is conserved. For the flight curves of the protons in the accelerator and thus for the control of the magnetic and electric fields, however, the *exact* velocity or the momentum $p = \gamma m v$ is essential.

Example II: A free neutron decays into a proton, electron and an electron antineutrino:

$$n \rightarrow p + e^- + \bar{\nu}_e\,.$$

Energetically, this is possible because the neutron is heavier than the sum of the masses of the proton, electron and antineutrino. Since energy is conserved and according to the special theory of relativity, the following remaining energy

$$E_{\text{kinetic}} = (m_n - m_p - m_e - m_{\bar{\nu}_e})\, c^2$$

must be distributed amongst the motion of the produced particles: the proton, electron, and antineutrino. The electron energy distribution for many decays is predicted by the theory of the weak interaction, see the figure on the right side. This is in complete agreement with experiments. Decays can only happen if the mass of the decaying particle is greater than the sum of the masses of the decay products (Particles' Births and Deaths ↓).

Example III: Our Earth is bombarded with high-energy cosmic rays, mainly protons and alpha particles (helium nuclei, see From Nucleus to Nuclei ↓). Coming from outer space, these hit the troposphere at an altitude of around 30 km. Here the atmosphere is so dense that violent collisions with the atomic nuclei of the air molecules take place immediately. Among other things, these collisions produce charged π mesons.

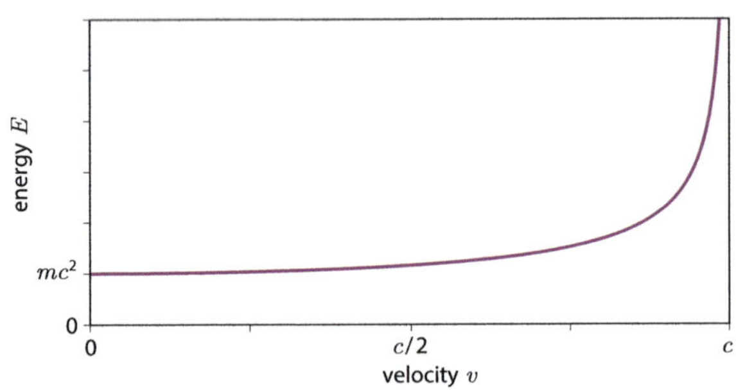

The relation between energy and velocity

Special Relativity → p. 38
Particles' Births and Deaths → p. 54
From Nucleus to Nuclei → p. 226

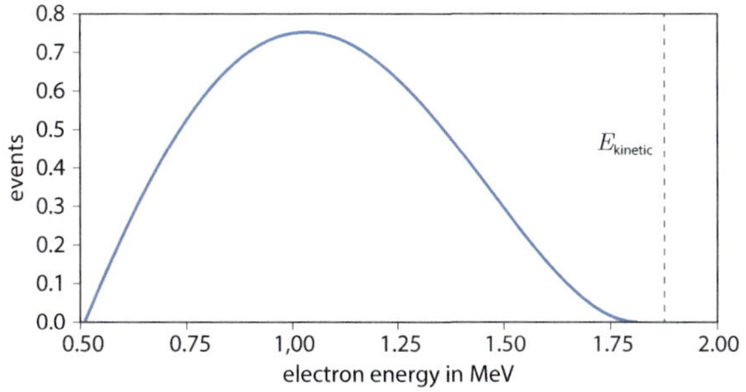

The electron energy distribution in neutron decays

These decay rapidly into muons and muon (anti)neutrinos. These collisions are similar to those we generate and observe at the LHC.

As measured in the laboratory, the muon decays with a very short half-life in such a way that after about 10^{-5} seconds 99 % of the muons have decayed. Even if they flew at the speed of light, 99 % of the muons should decay within about 3 km. They would therefore never reach the Earth's surface. Nevertheless, we very often see cosmic muons at the surface, even with simple detectors. This is how they were originally discovered! How can this be?

The answer is again the special theory of relativity. The muon is always in its own rest system, with an internal clock. According to this clock, 99 % decay within 10^{-5} seconds. In the muon's rest system, the Earth is hurtling towards the particle. The flight distance seen from the Earth is thus contracted by the boost factor gamma (γ) for the muon. If $\gamma \approx 50$, corresponding to a muon energy of about 5 GeV, which often occurs in these high-energy collisions, then the muon sees the 30 km as only 600 m. A distance that it often overcomes before it decays.

Viewed from the Earth's surface, the muon flies at a very high speed. We on Earth see the time of the muon stretched by the boost factor gamma (γ). For $\gamma \approx 50$, the *lifetime* is effectively 50 times longer, 99 % of the muons decay within $5 \cdot 10^{-4}$ seconds. This corresponds to a distance of 150 km. Thus, thanks to the theory of special relativity, we now understand why muons reach the Earth's surface so copiously.

But muons are also a nuisance for particle physics: there are so many of them that they cause background noise in our experiments, when we search for the new physics (see, e.g., The Mystery of Dark Matter ↓). This is why many very sensitive experiments are carried out deep underground in mines or tunnels to shield them from the muons.

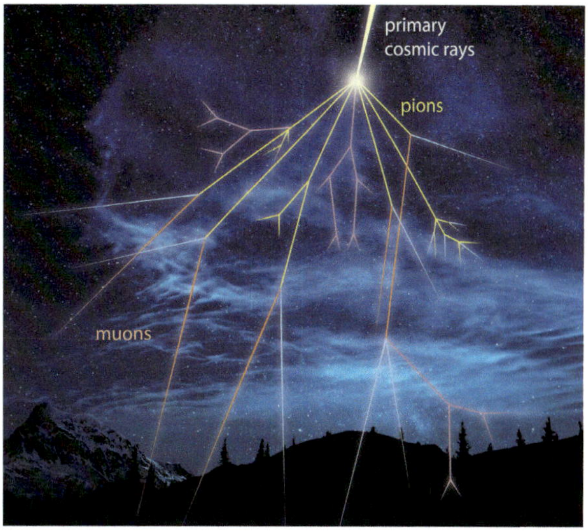

Muons are created in the upper atmosphere through cosmic rays. Due to the special theory of relativity they can reach the Earth's surface before they decay.

The Mystery of Dark Matter → p. 308

Fermions and Bosons

The Basis not Only for Chemistry

The chemistry of atoms is very diverse, as different substances behave very differently. This is mainly due to the fact that the electrons in their cloud around the nucleus do not all cluster in the lowest energy level (↓), but fill up the energy levels one-by-one. These levels are grouped into shells. In other words, no two electrons can occupy the same state. This is known as the *Pauli Principle*, which implies that the electron cloud of atoms becomes increasingly complex with an increasing charge number of the nucleus. Thereby certain similarities in the chemical properties of those elements emerge that have the same number of electrons in their outermost shell. In contrast, the many photons emitted together by a laser are all in exactly the same state—this is what gives the light of the laser its special properties, such as its very precisely determined frequency and large energy density.

But why is this so? Why do electrons and photons behave so differently? The underlying reason is the quantum mechanical principle of indistinguishability, which ultimately also leads to the Pauli Principle. To understand this principle, we would like to invite you to consider a thought experiment. Imagine two identical balls moving from A and B, respectively, to C and D. Our experience tells us that at any given time it is clearly known which ball started where and thus also which ball arrived where. But this is not the case anymore in quantum mechanical systems, e.g., when two electrons come close to each other. To be able to distinguish them then, we need to know the location and speed of both electrons very precisely. According to the uncertainty principle, however, this is not possible. So for the electrons, we only know that one started at A, the other at B, and that one electron arrived at C and another at D. But we do not know which is which. The two electrons cannot be distinguished and this is referred to as the *indistinguishability* of the electrons.

But if we cannot distinguish paths (1), where one electron moves from A to C and the other from B to D, from paths (2), where one moves from A to D and the other from B to C, then the two cases must occur with exactly the same probability. In the article on quantum mechanics (↓), we saw that probabilities are calculated from squares of wave functions. Now let us call the wave

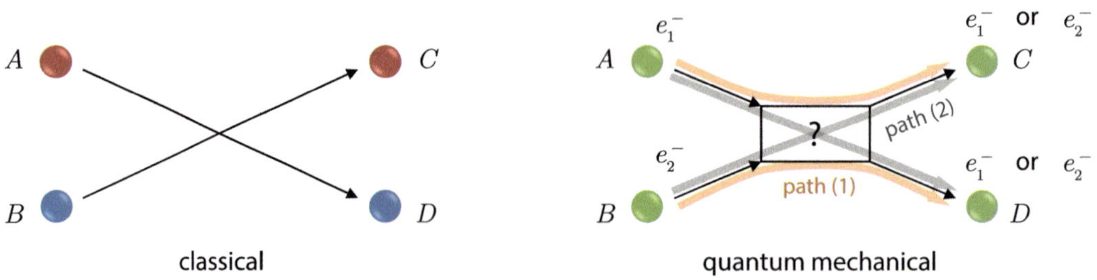

Comparison of classical with quantum mechanical scattering of indistinguishable particles

Energy levels: Chance, Interference and Predictions → p. 52
Concepts of Quantum Mechanics → p. 50

function for paths (1) $f(C, D)$ and for paths (2) $f(D, C)$ (since the difference between the two cases is exactly that the two electrons are swapped at the destination). That the probability must be equal for the two scenarios thus translates to $|f(C, D)|^2 = |f(D, C)|^2$. Since, in addition, two exchanges of the paths must return the original wave function, there are only two possibilities: Either it remains the same when the two electrons at the destination are swapped, $f(C, D) = f(D, C)$, or $f(C, D) = -f(D, C)$ applies, since the sign is lost when squaring. As we will see, systems where the minus sign occurs behave very differently from those where it does not.

In relativistic quantum field theory, it can be shown that wave functions for two indistinguishable particles with half-integer spin (1/2, 3/2, 5/2, ...) must change their sign when the two are interchanged. In contrast, the wave function of two particles with integer spin (0, 1, 2, ...) remains the same in this case. This rule can also be extended to systems with more than two particles. The two types of particles have been given their own names: particles with half-integer spins are called *fermions* and those with integer spin are called *bosons*.

The indistinguishability of two elementary particles and the resulting consequences naturally does not only apply to wave functions that describe paths, but also, for example, to the wave function that describes two electrons in an atomic shell. Here, each energy level carries its own identifier Q (which collects the so-called *quantum numbers*). Since electrons are fermions, (elementary particles ↓), the wave function of two electrons, which are in the energy levels with the quantum numbers Q_1 and Q_2, must change its sign when the two are interchanged. So, $f(Q_1, Q_2) = -f(Q_2, Q_1)$ must hold. This means in particular that the two cannot be in the same energy level, because $f(Q_1, Q_1) = -f(Q_1, Q_1)$ is only solved by $f(Q_1, Q_1) = 0$. Here it is: the Pauli Principle.

The behavior would of course change completely, if electrons were bosons. Simplified and neglecting the mutual repulsion of the electrons, in every atom the state would be preferred in which all electrons were in the ground state. Chemistry would be dramatically different—and poorer. Only thanks to the Pauli Principle do different atoms behave very differently. It is the reason why atoms can be classified according to how full their outer shell is.

In a laser, on the other hand, it is exploited that bosons, in this case photons, particularly like to be in the same state, i.e., in this case, have exactly the same energy and even oscillate in the same way.

Comparison of atomic structure in Nature (electrons as fermions, left) with a world where electrons would be bosons (right)

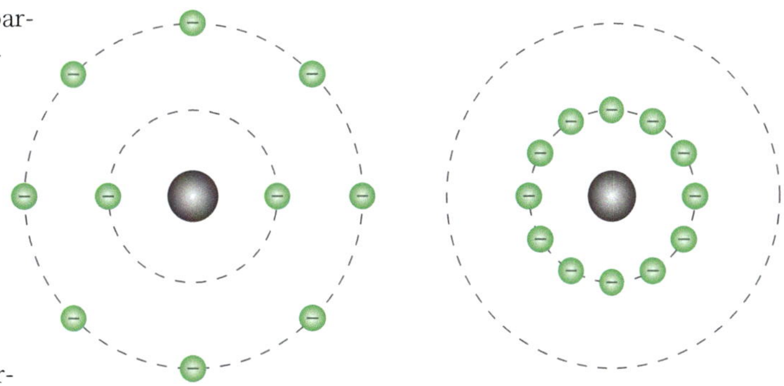

The Elementary Particles → p. 14

External and Internal Symmetries

On Display or Hidden Away?

In the article on elementary particles (↓) we saw that an elementary particle is initially defined by its mass and its spin (↓). Mass, spin, and momentum, together with the coordinates x, y, z (space) and t (time) are the so-called *outer* properties of a particle. They are directly related to spacetime. Recall, spin is the intrinsic angular momentum of a particle, i.e., part of its total angular momentum.

A particle is furthermore defined through its charges with respect to the three gauge forces in Nature. The electron, for example, has an electric charge and a charge with respect to the weak force, but not the strong force. These charges are the *internal* properties of the particle, as they exist independently of space and time. The particle carries these internal properties with it like a passport in its pocket. Only the gauge bosons of the electromagnetic, weak, and strong forces force can "see" or interact with the charges. The associated symmetries are the *gauge symmetries* (↓).

Let us take a closer look at what external and internal *symmetries* are. As an example, first consider a perfect sphere. It is symmetric under any rotation around the three independent axes in space in a coordinate system that has its origin in the center of the sphere. If you close your eyes and reopen them, you do not know whether the sphere has been rotated or not. We say: There are three *external* (rotational) symmetries. It becomes a little more abstract if you imagine the sphere floating in the empty Universe. Then there is no reference point and you cannot tell if the sphere is moved in the x-, y- or z-direction. We call this *translation symmetry*, which is also an *external* symmetry. If the system is invariant in time, then shifts in the time-coordinate t are a further external symmetry.

Thus we have seven independent *external* symmetries (three rotations, three translations, time). In the special theory of relativity, every system can also be viewed in a coordinate system which moves with uniform velocity $\vec{v} = (v_x, v_y, v_z)$, which has three independent directions. There are therefore a total of ten independent *external* symmetries in Nature.

In quantum field theory (QFT ↓) individual particles are described by functions $\Psi(x, y, z; t)$ over the whole space and time. These are the wave functions of quantum theory (↓). This is similar to writing down the wa-

A sphere is symmetric with respect to rotations in all three space directions.

What Makes a Particle Elementary? → p. 18
Spin → p. 44
Gauge Symmetries → p. 98
Quantum Field Theory → p. 104
Concepts of Quantum Mechanics → p. 50

Propagation of a water wave

ter height for each point in a swimming pool in the two dimensions of the water surface: $H(x, y; t)$. If the water is disturbed, waves spread out and the height at the respective points varies with time. The wave functions $\Psi(x, y, z; t)$ also change in space and time. In QFT, they also contain information about the spin and the internal charges.

Interactions between particles are described mathematically as products of wave functions. In quantum electrodynamics (QED ↓), the quantum field theory of electromagnetism, for example, an electron (e^-) can emit a photon (γ), as shown by the Feynman graph (↓). At the point of radiation three particles or functions come together: the incident electron, Ψ_{e1}, the outgoing electron, Ψ_{e2} and the emitted photon, A_γ. This is represented mathematically as the product of the three: $\Psi_{e1}\Psi_{e2}A_\gamma$.

In QFT interactions are symmetric if the products of the wave functions in the interaction, e.g., $\Psi_{e1}\Psi_{e2}A_\gamma$, do not change under transformations of all fields at the same time. If these are transformations in space and time, i.e., displacements, rotations or a transformation into a moving coordinate system, they are *outer* symmetries. All of these are realized in QED.

The wave functions of QFT have further *internal* mathematical properties, so-called *phases*, similar to the phases in the three-phase current in electricity. These phases in QFT are *internal*, i.e., not externally observable. In the case of the electromagnetic force it is one phase per field, for the other forces there are several phases. There are abstract transformations in the mathematical space of these phases. If the interaction does not change under these transformations, it is invariant or symmetric. This corresponds to an *internal* symmetry. Every gauge symmetry is invariant under these phase transformations.

In the Standard Model, the gauge symmetries are internal symmetries. Baryon and lepton number, not discussed here, are also internal symmetries. The above ten external transformations are the external symmetries of the Standard Model. Supersymmetry, a hypothetical symmetry transformation between the spins of particles, would be another *external* symmetry. As described in article on supersymmetry (↓), it is the last possible external symmetry.

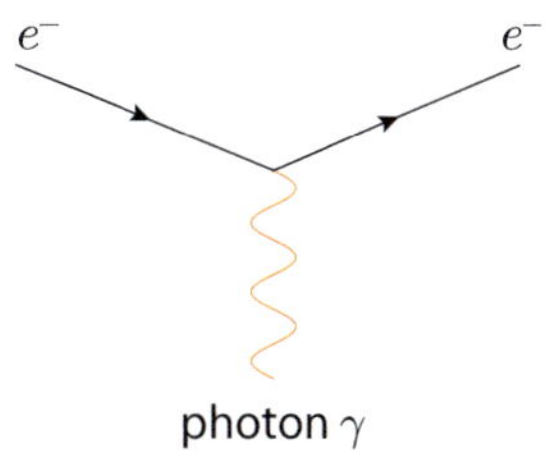

An electron radiates a photon.

QED: Electromagnetism → p. 176
Feynman Diagrams → p. 48
Supersymmetry → p. 278

Gauge Symmetries

Symmetries as the Origins of the Forces

A *symmetry transformation* is a transformation that leaves a system unchanged. So far we have only encountered *global* symmetries (↓). In that case the system is transformed uniformly as a whole. These are the standard symmetries, such as mirroring or rotation of an entire object. However, there is another very important class of symmetry transformations: the *local* symmetries, also called *gauge symmetries*. These are more general and lead directly to the forces of the Standard Model. How this works is described in this article.

With a global transformation, the entire object is transformed in the same manner, as with the circle which is displaced unaltered to the right in the figure below (left). Under a local transformation, each point of the circle can be shifted differently (figure right). This leads to a displacement *and* a deformation of the circle.[1]

In quantum mechanics (↓), electrons are characterized by wave functions $\Psi_e(x)$. The top left figure on the next page shows two waves described by the functions $y_{b/r} = A_{b/r} \sin(\pi(x - \Delta_{b/r}))$ (y_b: black, y_r: red). Each wave is characterized by an *amplitude* (maximum height of the wave, here $A_b = 1$, $A_r = 1$), a *wavelength* (distance between two wave crests, here for both waves 2), and a *phase* $\Delta_{b/r}$. The phase of the black curve is zero, $\Delta_b = 0$, as it has height zero at the origin ($x = 0$). The phase of the red curve $\Delta_r = \Delta \approx 0{,}32$ determines by how much the red curve is shifted in the x direction. A *global* phase transformation of the wave shifts the entire wave in one piece, as shown in the figure, but retains the shape. Amplitude and wavelength are unchanged.

A *local* phase transformation of the wave shifts the wave differently at each point, i.e., $\Delta(x)$ depends on x. This allows the wave to be deformed, as shown in the figure at the bottom right of the next page. Now comes the important step: In quantum electrodynamics (QED) we postulate that the system of electrons should remain unchanged under *local phase transformations* of the electron wave function $\Psi_e(x)$. This initially strange postulate has far-reaching consequences. It allows us to derive QED in its entirety and ultimately the complete Standard Model. We now try to explain how this works.

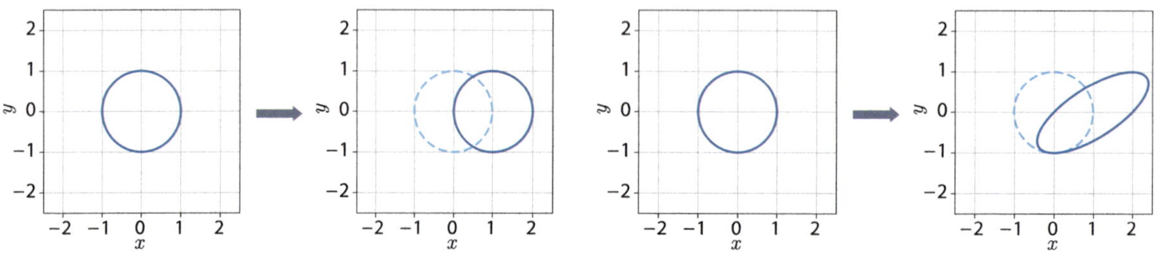

Global transformation ... ... and local transformation of a circle

1 For the global transformation, each point of the circle is shifted: $(x, y) \rightarrow (x+1, y)$. For the local transformation: $(x, y) \rightarrow (x+y+1, y)$, which is different for each location coordinate y of the circle.
Symmetries → p. 60
Concepts of Quantum Mechanics → p. 50
QED: Electromagnetism → p. 176

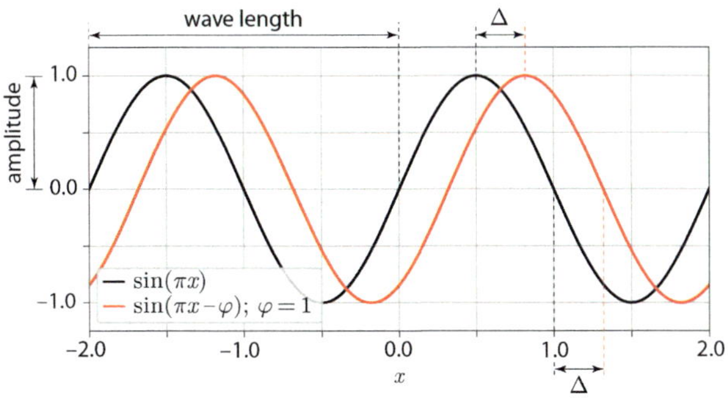

Global transformation of a wave

The deformation of a wave function costs energy. For the system to remain unchanged, this must be compensated. This can only be achieved by introducing another field $A_\gamma(x)$. The phase transformation behaves in such a way that this compensating field *must* have spin 1. In order for the the field $A_\gamma(x)$ to be able to compensate the phase deformations of $\Psi_e(x)$, it must also interact with the electrons. We identify the spin-1 field $A_\gamma(x)$ with the photon field of QED. The resulting interactions from our postulate (not shown here) are exactly those of QED!

Finally, to guarantee the invariance of the whole theory under local phase transformations the $A_\gamma(x)$ must also transform in a certain way when the electron field is transformed. These transformations of the $A_\gamma(x)$ are called *gauge transformations*. They leave the (observable!) electric and magnetic fields $\vec{E}$ and $\vec{B}$ unchanged.

All in all, we have constructed QED from a simple mathematical principle: we require invariance under local phase transformations, extended by necessary gauge transformations of the photon field $A_\gamma(x)$, and we automatically obtain the complete QED. Fantastic!

In QED there is only one such phase and therefore only one gauge boson, the photon. The associated symmetry is one-dimensional and is denoted: $U(1)_{EM}$. In the weak interaction there are three phases and thus three gauge bosons $W^\pm$, W^0 (Spontaneous Symmetry Breaking I ↓). The symmetry is denoted: $SU(2)_L$. For the strong interaction, quantum chromodynamics (QCD ↓), there are eight phases and therefore eight gluons, and the symmetry is denoted: $SU(3)_C$.

Interestingly, the coupling of the gauge fields in these higher symmetries, $SU(2)_L$, $SU(3)_C$, is necessarily universal. The W bosons, for example, couple with equal strength to electrons or muons, and the gluons couple to all quarks with the same strength. This is because the phase transformations of the W fields or the gluons can only be defined once. This has been confirmed experimentally with high accuracy.

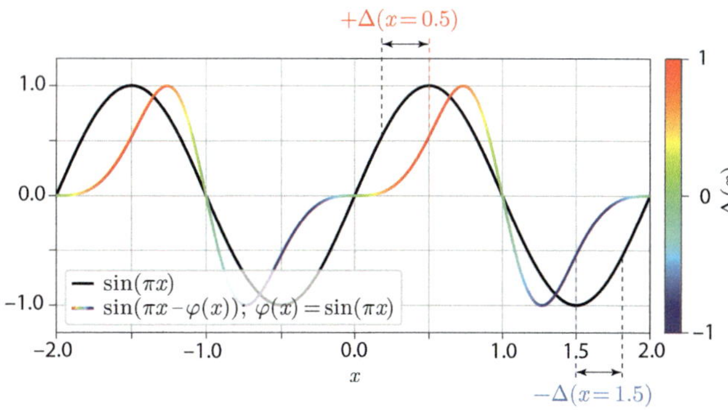

Local transformation of a wave

Spontaneous Symmetry Breaking I → p. 184
Quantum Chromodynamics → p. 204

Noether's Theorem

Symmetries Imply Conserved Quantities

Noether's theorem is a mathematical statement with profound implications for physics. It was proven in 1915 by the German mathematician Emmy Noether. It says: if a system has a *continuous* symmetry, then there are associated quantities that do not change over time, so-called *conserved quantities*. The conservation of energy, momentum, electric charge or the number of baryons and leptons are very important concepts in particle physics.

If a system is invariant under rotations, it has a *continuous* symmetry, since the angle of rotation can be changed *continuously*. In contrast, a mirroring is a *discrete* symmetry. Either everything is mirrored completely or nothing at all, there is no continuously changing parameter. Noether's theorem does *not* refer to discrete symmetries, such as parity (P), charge conjugation (C) or time reversal (T) (see discussion in the article Discrete Symmetries ↓).

As an example of Noether's theorem, let us consider a water tank at a height of 10 m, connected to a pump, like in the figure on the left.

A water tower stores energy in the form of the potential energy of the water.

If you pump 100 liters of water up into the tank, you have to do physical work, namely $W = m\,g\,h$.. Here $m = 100$ kg is the mass of the water, $g = 9.81\ \mathrm{m/s^2}$ the gravitational acceleration and h the height of the water tank. This performed work $W = m\,g\,h$ is then exactly the potential energy of the elevated water.

If the water is allowed to flow out of tank down to the ground, then the potential energy is converted into kinetic energy of the flowing water, $E_{\text{kin}} = 1/2\, mv^2$, where v is the velocity of the water at the ground. This can be used, for example, to operate a turbine and generate electricity. Without frictional losses, a 100 % efficient system returns exactly as much energy as was put in. The energy is *conserved*. But why?

Noether's theorem provides a simple answer. The system is invariant in time, since the gravitational acceleration g does not change over time. At the beginning, when we pump up the water, $g = 9.81\ \mathrm{m/s^2}$, and later, when it flows down, the same applies. The elapsed time t is a continuous parameter and therefore we have a continuous symmetry. The conserved quantity according to Noether's theorem is energy. That's a great connection!

If gravity were weaker at night than during the day, the water could be pumped up in the evening with an energy input E_N. It could flow out again during the day retrieving the energy $E_D > E_N$. The system would no longer be invariant in time and energy would no longer be conserved.

Discrete Symmetries: Violating P and CP → p. 190

The wagon with the pendulum is invariant under horizontal translations—linear momentum is conserved.

Emmy Noether's theorem tells us that this relationship between the invariance of the system and conserved quantities is general. If a system is invariant under spatial displacements, the associated momentum of the overall system is conserved. An example is a pendulum on a wagon on a flat table, like the on in the figure. The combined system of wagon *plus* pendulum is invariant with respect to displacements in the direction of travel of the wagon. It does not matter where on the long table the wagon stands. Thus the total momentum is conserved in this direction, of course, only if friction is neglected. If the pendulum swings back and forth, the wagon moves accordingly against it, so that the total momentum is conserved.

Analogously, if the whole system is invariant under rotation about an axis, the angular momentum (Kinematics ↓) about that axis is conserved. A figure skater uses this in a pirouette. He turns faster when he pulls his arms in and slower again when he extends his arms. The total angular momentum is maintained.

In each example we have used some geometric invariance to derive a property of the whole system: the corresponding conserved quantity. This is a statement of fantastic generality. In particle physics, energy, momentum and angular momentum are conserved in the collisions at the LHC because the system is invariant with respect to the associated symmetries. It does not matter, for example, whether the protons collide now or two nanoseconds later (time invariance leads to energy conservation).

As discussed in the article on external and internal symmetries (↓), in quantum field theory there are also *inner* symmetries, in addition to the above-mentioned geometric symmetries. Quantum electrodynamics is invariant under so-called *local* changes in the phase of the wave function. These changes are continuous. Thus, according to Noether's theorem, there must be a conserved quantity. This is precisely the electric charge. In the article on gauge symmetries (↓) we saw how the requirement of *local* phase invariance immediately led to the structure of quantum electrodynamics. Now we see that we also automatically obtain charge conservation.

A figure skater uses the rotational invariance and the resulting angular momentum conservation.

Kinematics → p. 40
External and Internal Symmetries → p. 96
Gauge Symmetries → p. 98

Feynman Diagrams II

Simple Handicraft

Here we provide simple instructions on how to assemble Feynman diagrams (↓) from the individual parts of the theory. We look at quantum electrodynamics with only electrons and photons. There are three different components, as shown in the first figure below: firstly, an electron line with an arrow; secondly, a wavy photon line. Both lines are so-called *propagators* .

The third ingredient is called a *vertex*, where a photon couples to an electron line at a red dot. A given Feynman diagram consists of these individual parts, each element can occur more than once. The chosen elements must be connected to form a complete diagram; they are connected by vertices.

In the diagrams, time runs from left to right. For particles (electrons) the arrow points in the direction of time, for antiparticles (positrons) against the direction of time.

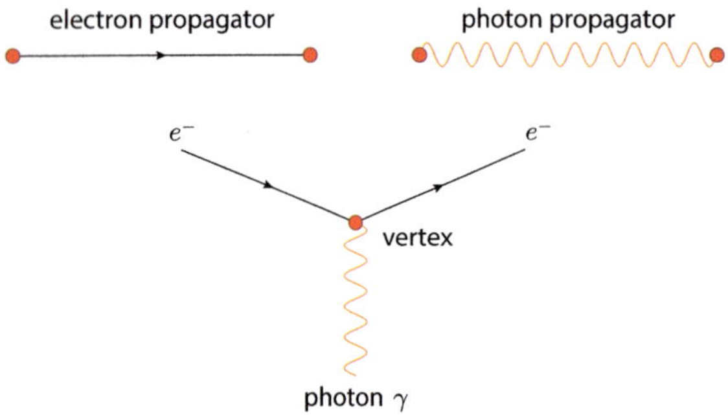

The ingredients for a Feynman diagram

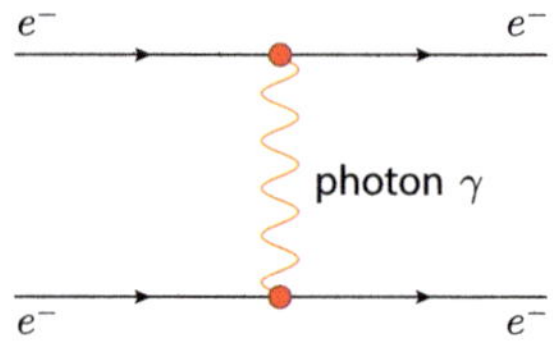

Scattering of electrons on each other

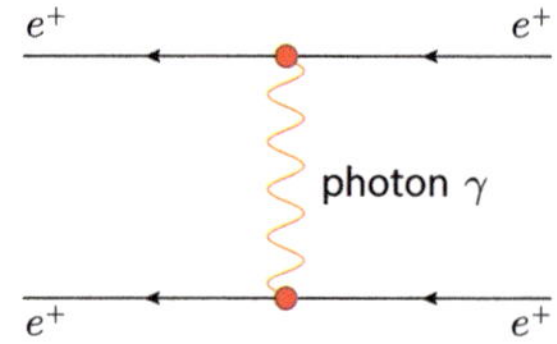

A rotation by 180° results in the positron-positron scattering.

In the figure on the top we join two vertices by connecting the wavy photon lines together to form a photon propagator. So we take one vertex from the first figure and a second one, which we turn upside down. The two wavy lines are combined into one, and we obtain the diagram for electron-electron scattering. We can rotate the entire figure by 180° and thus obtain a new diagram for positron-positron scattering (top right). The arrow on all lines points from right to left, against the direction of time and thus denotes a positron.

In the figure below we glue two vertices together via a fermion line. It is important that the direction of the arrow is continuous across all fermion lines. This diagram

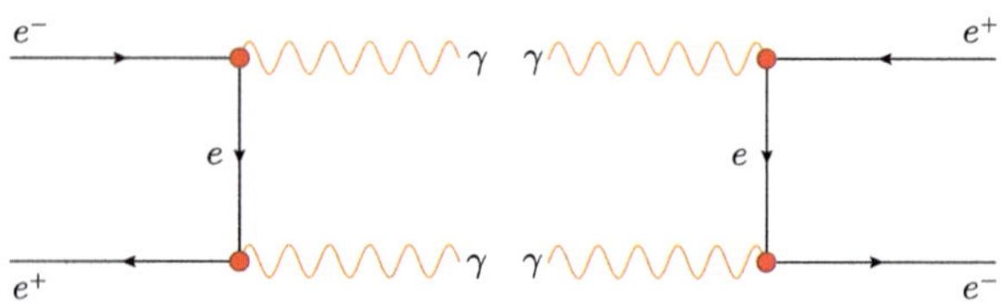

Left: Electron and positron annihilate.
Right: Two photons create an electron-positron pair.

Feynman Diagrams → p. 48

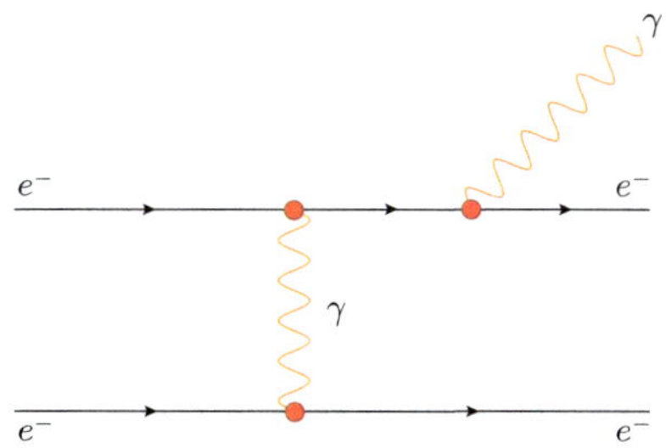

Two electrons scatter on each other. After scattering one electron emits a photon.

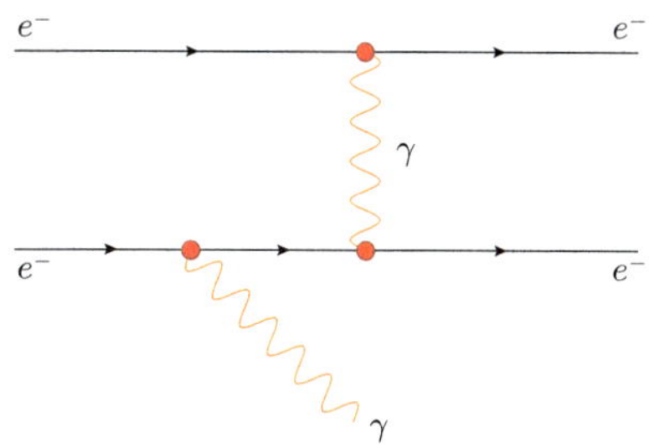

The same reaction as the left diagram, except here the lower electron emits the photon. There are two further such diagrams contributing, not shown here.

shows the annihilation of an electron and a positron to two photons (left), or, rotated by 180°, the creation of an electron-positron pair from two photons (right).

In the figure at the top right we glue three vertices together, once via the photon line and once via an electron line. Two electrons scatter and at the end one of the electrons emits a photon. There are three more independent diagrams leading to the same final state, because the photon can be emitted from each electron line. Another such diagram is shown in the top right figure. All these diagrams must be added to a total amplitude before the probability of the reaction can be calculated by squaring the sum.

As a last example, we combine four vertices in a special way in the right figure. We take twice the entire diagram for $\gamma\gamma \rightarrow e^+e^-$ from the left page, rotate one by 180°, glue them together via two of the fermion lines and obtain the last diagram. Now we have a closed box of fermion lines. These fermions never fly away freely, but are only created and destroyed in the process. They are called *virtual*. The physical process is the scattering of two photons on each other. A diagram with such a closed line is also called a one-loop diagram (Loops ↓).

This is a pure quantum phenomenon; in classical electrodynamics, electromagnetic waves do *not* scatter on each other. In quantum field theory however, this does occur for example via such one-loop diagrams, albeit suppressed.

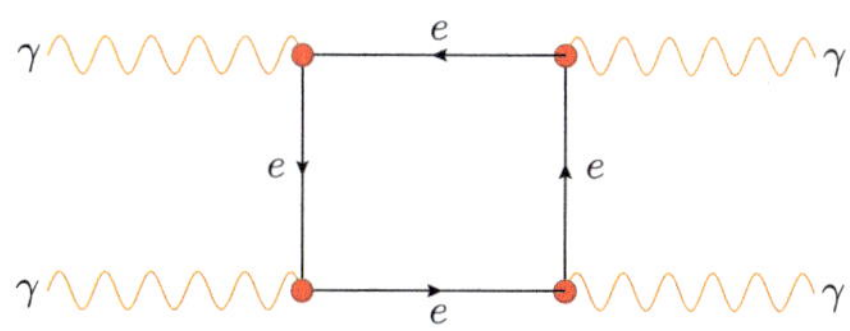

Scattering of photons on each other. Virtual particles can be created and destroyed within the closed electron line (loop).

Loops → p. 106

Quantum Field Theory

Creating and Annihilating Particles

Quantum field theory (QFT) is an extension of relativistic quantum mechanics (↓). As the name says, it combines the concepts of the *quantum* and *field theory*. Here we take a closer look at this.

In classical mechanics, the electron is a point particle, with spatial coordinates x, y, z, and time coordinate t. Furthermore, at any moment in time an electron has a specific velocity $\vec{v}$. This velocity can be arbitrarily small or large. In *quantum* mechanics, an electron is characterized by a wave function $\Psi(x, y, z; t)$. The square of the wave function $|\Psi(x, y, z; t)|^2$ indicates the probability of finding the electron at a specific time t at the location $\vec{r} = (x, y, z)$. In particular, the electron definitely exists, i.e., at any point in time it is *somewhere* in space. The electron never disappears, and no new electrons with new wave functions appear. QFT goes beyond quantum mechanics and allows for the creation and annihilation of particles.

Electromagnetism is an example of a field theory in classical physics. It has electric $\vec{E}(x, y, z; t)$ and magnetic *fields* $\vec{B}(x, y, z; t)$, both functions of the space and time coordinates. These fields can be generated by accelerated charges, propagate in space and transform into each other in the process (Acceleration and Radiation ↓). They can also be absorbed by electrical charges, e.g., in an antenna, and thus disappear again. In this process, any small amount of energy can be converted into radiation or be destroyed. Interestingly, even classical electromagnetism, as Einstein noted early on, is a theory in accordance with the special theory of relativity. It turns out that a (renormalizable ↓) quantized field theory is necessarily relativistic (↓).

In the quantum field theory of electromagnetism, also known as *quantum electrodynamics* (QED), the physics of the electron and that of the fields $\vec{E}(x, y, z; t)$ and $\vec{B}(x, y, z; t)$ are brought together in a quantum theory. The quantization of the electromagnetic field leads to the fact that at a certain frequency f of the electromagnetic wave, the field only exists in fixed energy packets, the quanta, with energy $E = h \cdot f$. Here h is Planck's constant. These light quanta, also known as *photons*, are generated by accelerated charges, but can also be absorbed by electrons and thus disappear, just as previously the classical $\vec{E}$ and $\vec{B}$ fields. The simplest

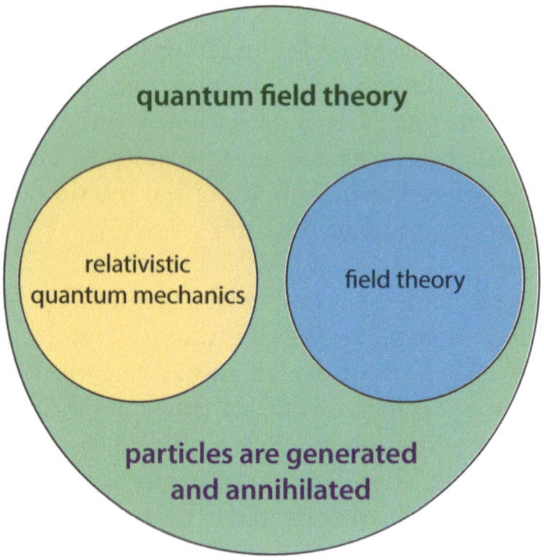

Concepts of Quantum Mechanics → p. 50
Acceleration and Radiation → p. 42
Renormalization → p. 108
Special Theory of Relativity in Particle Physics → p. 92

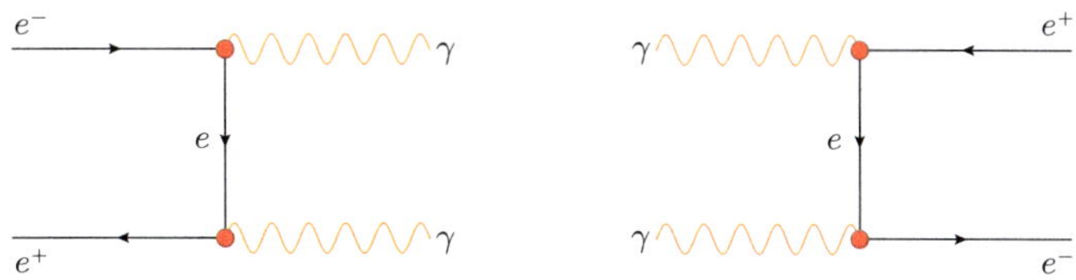

In quantum field theory it is possible that pairs of particles are annihilated (left) or created (right).

example is the light in a room: when it is switched on, the light source continuously generates photons, which fly through the room and are absorbed or scattered by the walls and objects. If it is switched off, no new photons are generated. All previously generated light particles are absorbed and disappear, it becomes dark.

In QED, the wave function of the electron also becomes a *quantized* field. The electron must therefore be part of a relativistic theory. As we saw in the article on antimatter (↓), the positron, the antiparticle of the electron, *must* also exist. The quanta of this one field are the electrons *and* the positrons. These quanta can be created and destroyed like the photon. This means that electrons and positrons can be created and annihilated, but only in pairs, so that electric charge is conserved. The lepton number must also be conserved: all leptons ($e^-, \mu^-, \tau^-, \nu_e, \nu_\mu, \nu_\tau$) have lepton number + 1, their antiparticles have lepton number − 1.

The top left Feynman diagram (↓) shows the annihilation of an electron and a positron. Energy and momentum can only be conserved if at least *two* photons are created. In the Feynman diagram, the time goes from left to right, i.e., the initial state is on the left, the final state on the right. In the top right Feynman diagram we show the reverse process, the production of an e^+ e^- pair.

Quantum chromodynamics (QCD ↓) is the quantum field theory of quarks, antiquarks and gluons. The gluons are the quanta of the color fields of the strong interaction, which can be created and annihilated. Two gluons can annihilate each other and create a quark-antiquark pair. Because of their electric charge, quarks also couple to photons, like electrons and positrons above, and can thus also be created and annihilated via photons.

The electroweak theory is more complicated, since the associated gauge symmetry is spontaneously broken (↓). The quantized fields before and after the breaking of the symmetry are different. At high energy, before the breaking, they are the massless gauge fields W^0, $W^\pm$ of the $SU(2)$ symmetry, as well as the B^0 of the $U(1)_Y$ symmetry. At low energies, after symmetry breaking, the W^0 mixes with the B^0 resulting in two new fields: the massive Z^0 and the massless photon γ. Furthermore the charged bosons $W^\pm$ are now massive. At high and low energy, the quarks and leptons are added as quantized fields. A massive W^- boson can, for example, decay into an electron and an electron antineutrino. In the process the W^- is annihilated and the electron and antineutrino are created, in accordance with the conservation of lepton number (Feynman Diagrams II ↓).

What Is Antimatter? → p. 22
Feynman Diagrams → p. 48
Quantum Chromodynamics → p. 204
Spontaneous Symmetry Breaking I → p. 184
Feynman Diagrams II → p. 102

Loops

Special Effects in Quantum Mechanics

For particularly curious readers, we want to delve a little deeper into quantum mechanics and quantum field theory (↓) in this and the next article. For illustration we will focus on the example of positron-muon scattering (similar to Feynman Diagrams II ↓). If we want to improve the accuracy of the calculation, we need to evaluate diagrams with more than one photon in addition to the one-photon exchange described there. All diagrams with one and two photons that contribute to the scattering process are shown in the figure on the right. The diagrams with two or more photons have an interesting new feature: The conservation of energy and momentum does not determine anymore the energy and momentum of all particles in the diagram (see box). In such a case, the rules of quantum mechanics require that all possible intermediate states must be taken into account—formally, this requires integration over *all* momenta and energies (collected in the four-momenta) of the particles. Diagrams in which this is necessary are referred to as *loop diagrams*.

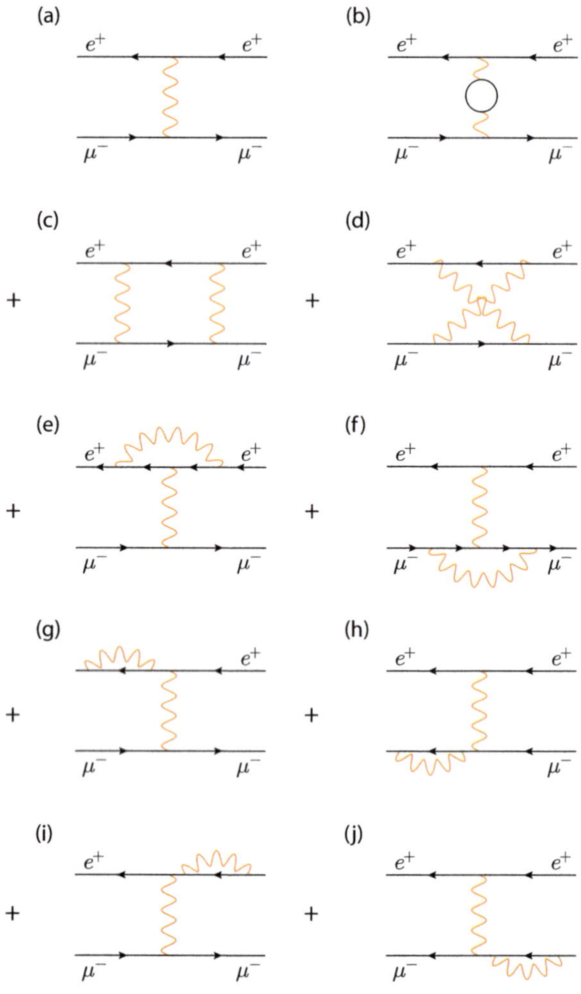

Contributions to $e^+\mu^-$ scattering, including one-loop corrections

How important are such loop contributions quantitatively? As can be seen from the figure, each photon line or better propagator comes with two vertices that capture the coupling of the photon to the leptons. Each vertex contains the electric charge e as a factor. In addition, the evaluation of the relevant integrals shows that loop diagrams typically generate prefactors of $1/(4\pi)$. All together one finds that each additional loop in electrodynamics comes with a factor of $\alpha_{\text{QED}} = e^2/(4\pi) = 1/137$. Therefore, it is expected that the diagrams $(b) - (j)$ provide a correction of only 1 % com-

Quantum Field Theory → p. 104
Feynman Diagrams → p. 48
Feynman Diagrams II → p. 102

pared to the one-photon exchange without a loop (diagram (*a*)). The explicit evaluation also shows exactly this for the diagrams (*b*) and (*d*) – (*j*) in the figure (for some of these there is another special feature, which is discussed in the next article). However, the calculation of the integral that belongs to the loop diagram (*c*) yields an additional factor of 1/(relative velocity of e^+ and μ^-), which cancels the above-mentioned suppression of the loops by α_{QED} for very small velocities. Because of this all diagrams of the mentioned type (i.e., ladder-like with exchange of three, four, five, etc. photons—as shown in the figure below) become equally important, so that all of them must be summed up. This is exactly what the Schrödinger equation (Quantum Mechanics ↓) does. This summation then leads, for example, to bound states—in this case, it would be an $e^+\mu^-$ bound state, but the calculation for an e^-p bound state, i.e., the hydrogen atom, would proceed analogously.

In the strong interaction, the corresponding parameter α_{QCD} is at low energies of the order of 1 and then generally there is no suppression of any loop diagram—with drastic effects on the properties of strongly interacting systems and the methods of their theoretical description. We will return to this topic in the chapter on the strong interaction.

Momentum Flow in Loop Diagrams

Let us now take a closer look at a specific loop amplitude for

$$e^+(p_1) + \mu^-(p_2) \to e^+(p_3) + \mu^-(p_4) \ ,$$

shown in diagram (*d*) of the first figure. According to the rules of Feynman diagrams, the momentum must be conserved at each vertex. That means the sum of all momenta entering a vertex must equal the sum of all momenta leaving it. This is achieved, for example, by the choice of momenta, as shown in the figure below. From the conservation of momentum at the lower right vertex it follows that

$$p_4 = p_1 + p_2 - p_3 \ ,$$

so that the total momentum conservation is ensured. However, the momentum k is not determined by this—momentum conservation is ensured for any value of k—and therefore, as described in the main text, an integration must be performed over all values of k. This type of integration is a characteristic of every loop diagram.

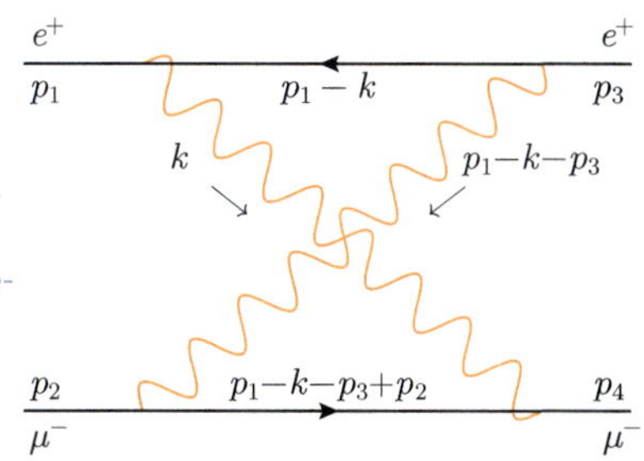

Momentum flow in the Feynman diagram: The conservation of four-momentum does not determine the internal momentum *k*.

Sum of ladder diagrams. The Schrödinger equation sums *all*, which leads to bound states like atoms.

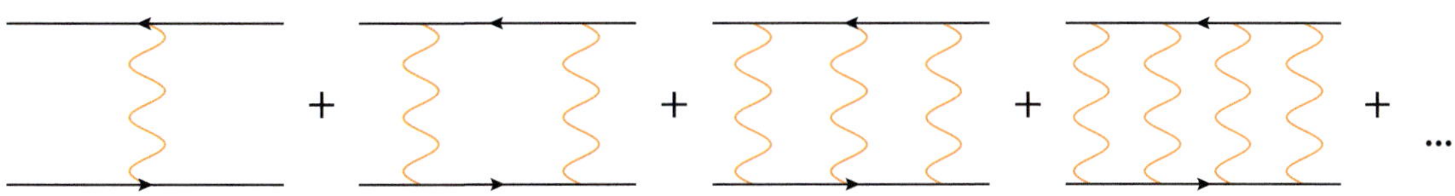

Concepts of Quantum Mechanics → p. 50
Chapter 8 The Peculiar Strong Interaction → p. 201

Renormalization

Taming the Infinite

We would like to point out that this article is quite technical: To get even a rough understanding of renormalization, we have to work through some abstract subtleties—which require a bit of perseverance. But: It's worth it, because it gives you an insight into a central concept of modern field theory.

There is an important aspect in the evaluation of the loop diagrams (↓): The integral over all energies and momenta in the intermediate state does not necessarily yield a finite result. There are cases where one gets an infinite result (in the diagrams of the first figure of the previous article, this is the case for the diagrams (*b*) and (*g*)−(*j*)), since one formally integrates over all momenta and energies—without any upper limit. One could now argue that one should not consider arbitrarily large momenta, but only those that are smaller than some given value, which we want to call P_{max}—after all, infinitely large momenta are not physically meaningful. But then the result depends on the value of P_{max} and is thus to a certain degree arbitrary. But does this mean that our theory is useless?

Of course not: Whenever loop diagrams with undefined results occur, the theory provides another diagram with an adjustable parameter that allows one to eliminate the arbitrariness described above. To see how this works let us take problematic diagrams of the previous article as an example. Those are summarized again in a new figure here: diagram (a) has no loops and is therefore not problematic. The other five have initially undefined loops. Taking all these diagrams together, the incoming positron in diagram (a) is modified by the emission and absorption of a photon in diagram (*c*).

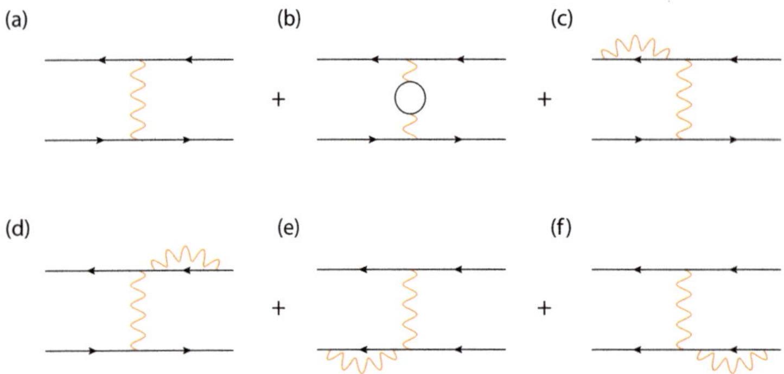

Single-photon-exchange diagrams that contribute to mass and charge renormalization

This is shown again in isolation in figure on the right page: The first diagram describes a positron propagator. This contains the electron mass as a parameter. The second diagram is the one-loop correction to the positron propagator, which on its own does not yield a clearly defined result. However, the positron property described by the sum of the two diagrams is the mass; one speaks of *mass renormalization*.

Loops → p. 106

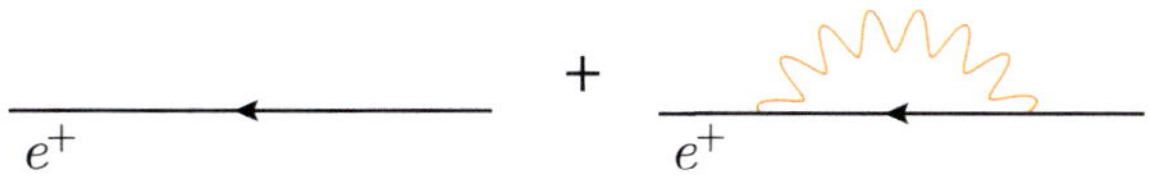

The positron propagator including its one-loop correction.

It is therefore sufficient to demand that the sum of the original, in fact unobservable, mass parameter of the positron and the loop correction corresponds to the physical positron mass. And indeed: It works! The dependence on P_{max} has disappeared. The concept lead to a well-defined, finite answer. In the box we provide more details about this method. The argument for the last three diagrams of the first figure runs analogously. The second diagram—the fermion loop—contributes to *charge renormalization* and leads to the so-called *running coupling constant*, to which we return later.

In this way, all undefined parts from the amplitudes can be absorbed into a few parameters. And even more: It can be shown that, for example, in the case of quantum electrodynamics with only electron, positron, and photon fields, the two mentioned parameters are sufficient, namely electron charge and mass (the positron has exactly the same mass as the electron and the opposite charge), no matter how complicated the diagram is. The general procedure is referred to as *renormalization* and a theory that has only a finite number of such parameters is called *renormalizable*.

In 1971 Gerard 't Hooft proved in his Ph. D. thesis that the complete Standard Model, with 21 free parameters, is renormalizable. He jointly received the 1999 physics Nobel Prize with his advisor Martinus Veltman for this work.

On the Renormalization of Mass

We denote the original mass of the positron (without loop corrections) as m_0 and the loop contribution as $\Sigma(P_{max}, E)$. Here, E is the energy of the positron that we assume to be at rest and we allow ourselves to neglect the effects of the electron spin. The parameter P_{max} is the previously mentioned, arbitrary cutoff parameter for the integral. Then the mass term in the positron propagator results in

$$M(E) = m_0(P_{max}) + \Sigma(P_{max}, E) \; .$$

Now the renormalization prescription is that we demand for the physical positron mass m_e

$$m_e = m_0(P_{max}) + \Sigma(P_{max}, m_e c^2) \; ,$$

such that one obtains the positron mass, when the original mass m_0 is added to the loop correction, evaluated at the positron mass (i.e., $E = m_e c^2$). Clearly, this can only hold for all values of P_{max}, if we demand that m_0 also depends on P_{max}. This results in the mass term,

$$M(E) = m_e + [\Sigma(P_{max}, E) - \Sigma(P_{max}, m_e c^2)] \; .$$

The expression in the square bracket is finite and well defined (i.e., independent of P_{max}) and disappears for $E = m_e c^2$. In fact, we have absorbed the dependence of the loop correction on P_{max} into a (P_{max})-dependence of m_0.

Formal Considerations on the removal of arbitrary components lead to the so-called renormalization group equations, which we will encounter again in the running coupling article (↓).

The Running Coupling → p. 210

Accurate Calculation of the Unsolvable

How to Cunningly Unlock Nature's Secrets

The title of this section sounds a bit like cheating—but it is not. However, the concept is quite abstract and this text is primarily aimed at the particularly curious readers. As a warm up, we want to approximate a complicated function with a simple one. To do this, we show in the box and the figure below that it is indeed possible to represent the sine function, $\sin(\beta)$, by a sum of a finite number of simple terms. For small angles β, a straight line is a good approximation. If you also want to describe the first curvature, you have to subtract from the straight line $(\pi\beta/180)^3/6$, and so on. Of course, there is a catch: The larger the range in which you want to describe the function, the more terms are needed. Thus, the first term describes the sine function well only up to about 30°, the next one works up to 70°. If you go up to β^{13}, you get a good description up to 290°—but this already requires seven coefficients. More concretely, for $\beta = 1°$ the first term in the expansion reproduces the correct result with an accuracy of half a percent.

Example: Sine as a Power Series

For illustration, we consider the function $\sin(\beta)$, where β denotes an angle in degrees. The function is shown in the bottom left figure as the red curve. The other curves show that the sine can be approximately described by expressions of the type

$$\sin(\beta) = a\,\beta + b\,\beta^3 + c\,\beta^5 + d\,\beta^7 ...$$

By the way, only odd powers of β appear, so that the property $\sin(-\beta) = -\sin(\beta)$ is also reflected in the expansion. For the dotted curve (blue), only the first term with $a = \pi/180$ is considered, for the dashed one (orange) the β^3 term with $b = -(1/6)\,a^3$ is added. Accordingly, the terms up to β^7 and β^{13} were added for the two further curves—of course with increasingly complicated coefficients (for example, the one from β^{11} is $-(1/39916800)\,a^{11}$).

The coefficients of the powers of β can be calculated with a well known mathematical rule. The expansion is called *Taylor series*.

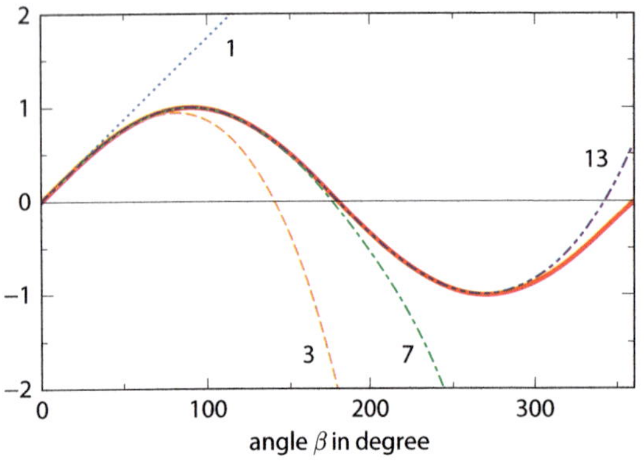

Adding in the next term improves this accuracy to 0.0008 %. At $\beta = 100°$, on the other hand, the first term is 77 % off, adding in the second improves the result to 13 % and with the next the full result is missed by less than one percent.

← Approximation to the sine function (red curve) with increasing powers in the angle β. The highest power considered in the series is noted next to the curve.

Similarly, modern theoretical particle physics proceeds to calculate experimental quantities such as experimental count rates in principle arbitrarily accurately. An example of such a calculation is the one for the so-called anomalous magnetic moment of the electron, described in the article on precision experiments (↓). What was the angle in the sine approximation is in this calculation the strength of the electromagnetic interaction, $\alpha_{\text{QED}} = 1/137$. Since this has such a nice small value, the contributions from the following terms, which have to be evaluated in increasingly complicated calculations, quickly become smaller (Loops ↓).

However, if you want to calculate effects of the strong interaction with the same methods, this only works in the range of very high energies, because here the parameter analogous to α_{QED}, α_{QCD}, also becomes small (The Running Coupling ↓). At low energies, however, where α_{QCD} becomes large, this kind of approach does not work, because one would have to sum up infinitely many terms to get a meaningful result. Then there are only two possibilities: Either one carries out complex numerical simulations, as described in the article on lattice QCD (↓), or one uses the so-called *effective field theory*. The picture below illustrates the difference: A direct QCD calculation has to compute extremely many diagrams of the type shown on the left page. If, on the other hand, one uses the effective field theory, for example, for the calculation of pion-nucleon scattering at low energies, the ratio of either the pion mass or the pion momentum to nucleon mass is used as the expansion parameter instead of α_{QCD}. The first term of this expansion consists of a single diagram, and this can also be calculated very simply. The corrections contain higher powers from the ratio of the pion energy and the proton mass. At low energies, this ratio is about 15 % and thus rather small. However, with increasing energy the effective field theory gets more complicated as well and eventually looses its efficiency.

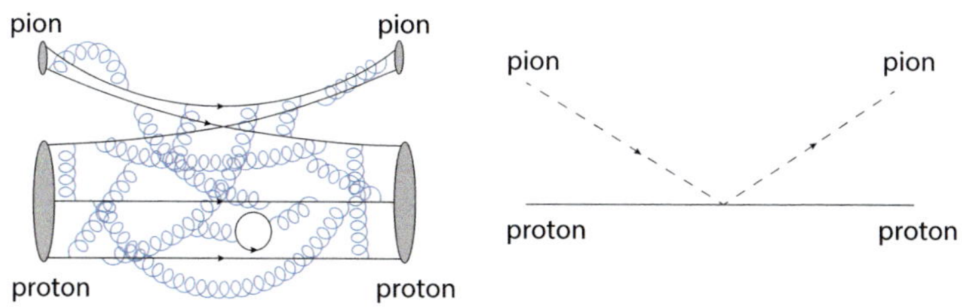

typical QCD diagram

leading contribution in the effective field theory

Typical QCD diagram for pion-nucleon scattering at low energies compared to the leading contribution in the effective theory.

The effective field theory just outlined captures the physics of the light quarks. It is also possible to construct an effective field theory for the physics of the heavy quarks. This so-called *Heavy Quark Effective Field Theory* even allows an analysis based on quarks and gluons (taking advantage of asymptotic freedom). The construction principle of this theory for heavy quarks is different, but the idea behind the concept is exactly the same. The principle is even more general: One can even regard the Standard Model as the leading expression of the effective field theory of an even more general theory, which might then allow one to answer the open questions of the Standard Model. This idea can also be quantitatively tested by constructing the next order of this effective theory and testing whether the predicted effects then occur in experiments. Such approaches are pursued, for example, in the search for electric dipole moments (↓).

Precision Experiments → p. 250
Loops → p. 106
The Running Coupling → p. 210
Lattice QCD → p. 222
The Search for Electric Dipole Moments → p. 318

Simulations in Theoretical Physics

How Computers Reach the Unreachable

Computers have become indispensable in our daily lives: They strongly influence how we communicate, travel, and work. Accordingly, it is not surprising that they also play a very important role in modern theoretical physics.

Starting from the laws of Nature, theoretical physics provides a mathematical apparatus that, in principle, allows for the prediction of the further development of a system, if the initial state is known. As an example, we can consider a spring pendulum, as described in the box: Although we can describe the motion for small deflections and without friction with known functions, the motion of a realistic pendulum is significantly more complicated—here a numerical simulation is required.

For the example from the box, it works like this: Let us assume we know the velocity and location of the mass at the beginning of the motion. Then we get the velocity a small time Δt later by adding its change Δv. According to Newton's law, Δv is given by the product of the force F and the time span Δt, divided by the mass m: $\Delta v = (F(x)/m)\Delta t$. To find the corresponding location, proceed analogously by adding to the location the change of location, given by the just calculated speed times Δt. Now we know the speed and location a time Δt later and can go through the steps given above again to create another Δt step. For a "sufficiently" small but not too small Δt, one can thus navigate through the complete evolution of the system (a somewhat more complicated example is discussed in detail in the scientific article given below ↓)

The supercomputer Jugene at the Jülich Research Centre

The outlined program can be very easily implemented on a computer. It will describe the real motion more accurately, the more accurately we know the force (including friction) at a location and the smaller we can choose Δt (the latter is, for example, determined not only by how long one is willing to wait for the result but also by the numerical accuracy of the computer)—this is called a *numerical simulation* of the physical system. In this way, one can also describe much more complicated systems, such as the development of the weather, where of course the complexity of the system presents an additional complication. Then very large computers or the networking of many computers into so-called supercomputers are needed to meet the numerical challenge.

Numerical simulations also play a central role in theoretical nuclear and particle physics. Here, in contrast to the classical equations of motion discussed above, the

C. Körber et al. 2018, *A primer to numerical simulations: The perihelion motion of Mercury,* Physics Education, 53, 055007, doi:10.1088/1361-6552/aac487

Lattice QCD → p. 222

equations of quantum mechanics need to be employed. One typical application is to solve the sometimes very complicated equations of the Standard Model; this is further elaborated in the article on lattice QCD (↓). The basic concept here is similar to what has already been said, even if the practical implementation is much more complicated.

Furthermore, simulations are necessary to predict the expected experimental signals in complicated reactions. Let us consider a typical reaction at the LHC: (↓) Here, two protons collide at extremely high energies. The energy is so high that many new particles can be produced in the reaction (Special Relativity ↓), as energy can be converted into mass. This can also happen in many stages, with unstable intermediate states initially being created, which ultimately decay into the particles that are measured in the detectors (Particles' Birth and Death ↓). Each of these decay chains occurs with a certain probability, and the sum over all chains then gives the distribution of the predicted signal, which is to be compared with the actual measurements. This type of simulation, in which the computer actually "rolls the dice" on the various possible event sequences, is only possible with the use of supercomputers.

How do discoveries and tests of the theory work? The first step is to calculate the fundamental reaction, which one wants to test using the methods of quantum field theory (↓) to get the amplitude predicted by the Standard Model. Then one computes how an experimental signal arises from this, using the methods described above. A discovery is then made, when the experiment shows significant deviations from the predictions based on the simulations using known physics (New Particles at LHC ↓)

Example: Spring Pendulum

Newton's laws of motion state that the acceleration of a body a is determined by its mass m and the force F acting on it

$$F = ma\,.$$

The force exerted by a spring with spring constant D on a mass is proportional to the deflection x, so $F = -Dx$. In addition, the acceleration corresponds to the change in speed v with time, and the speed itself is the change in location with time:

$$a = \frac{\mathrm{d}v}{\mathrm{d}t} \quad \text{and} \quad v = \frac{\mathrm{d}x}{\mathrm{d}t}\,,$$

where the "d" symbolizes small changes. Thus, we now have a relationship between the change in location and the force, which depends on the location:

$$m\frac{\mathrm{d}}{\mathrm{d}t}\left(\frac{\mathrm{d}x}{\mathrm{d}t}\right) = -Dx\,.$$

This is a so-called *oscillation equation*. By direct evaluation, one can convince oneself that

$$x(t) = A\sin(\omega t) + B\cos(\omega t)$$

solves the oscillation equation for any A and B, as long as $\omega = \sqrt{D/m}$ holds. The parameters A and B are determined by the initial conditions (how far the spring is initially stretched and whether one gives it an additional push or not). However, these equations describe a real spring well only for short times and small deflections—if one wants to be more precise, for example, friction must be taken into account. Then one is in general dependent on numerical simulations.

The LHC Accelerator Complex at CERN → p. 116
Special Theory of Relativity in Particle Physics → p. 92
Particles' Births and Deaths → p. 54
Quantum Field Theory → p. 104
Searching for New Particles at the LHC I → p. 302

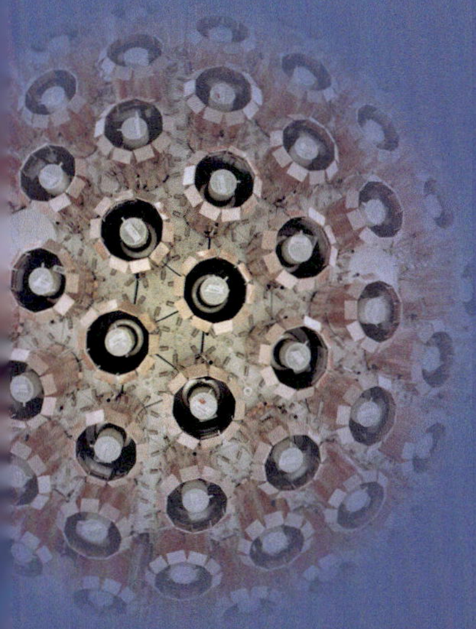

5 Accelerators and Detectors

Particle accelerators are one of the main tools of particle physics. They accelerate particles to a certain energy. With increasing energy particle collisions on the one hand can resolve ever finer structures and, on the other hand, can also produce yet undiscovered heavier particles. However, merely accelerating particles to high energies does not in itself yield any new knowledge. Therefore, detectors that measure the properties of particles after a reaction took place play an equally important role.

This chapter covers the way accelerators work and how detectors are developed by combining various experimental techniques that can jointly determine all important properties of the particles. Accelerators and detectors have also found their way into our daily lives, especially in medicine.

The LHC Accelerator Complex at CERN

From a Hydrogen Bottle to Collisions

In several articles, we demonstrate that accelerators play an important role in discoveries in particle physics. Here we introduce the LHC accelerator complex at CERN. The Large Hadron Collider (LHC), with its 27 km circumference, is currently the largest accelerator in the world. Protons or lead ions are accelerated according to the demand of the experimenters. Before the protons reach their final energy of 6.8 TeV (teraelectron volts, see Energy and Power ↓), they pass through several pre-accelerators. Such a series of accelerators is used, because a given accelerator can only be operated in a certain narrow energy range. In this article, we follow the stages the protons go through, from their source to the experiment.

The first question is: where do the protons later accelerated in the LHC come from? Surprisingly they come from a simple hydrogen gas bottle. With the contents of a standard gas bottle, one could theoretically fill the LHC with protons for several million years. However, the bottle only contains electrically neutral hydrogen molecules (H_2) consisting of two protons and two electrons.

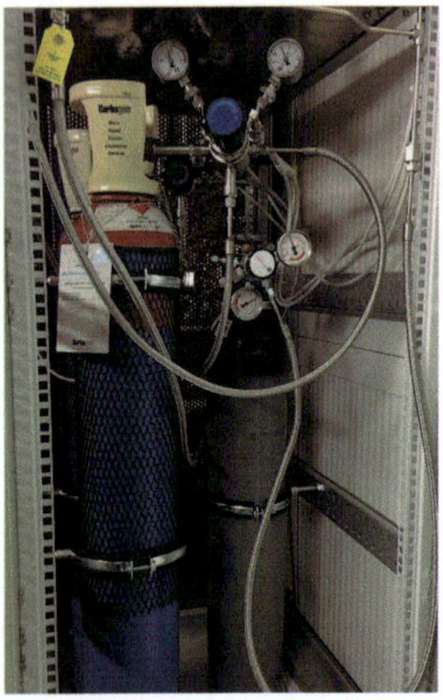

The LHC's hydrogen bottle, where the journey begins ↵

The neutral molecules are first stripped of a proton. This creates negatively charged hydrogen ions, H^-, consisting of one proton and two electrons. These H^- ions, and not protons, are then accelerated in a 86-m long linear accelerator to a kinetic energy of 160 MeV before they hit a thin carbon foil, which is referred to as a ***stripping foil***. Here the two electrons are stripped off the H^- ions. What remains are "naked" protons, which are then fed into the first ring accelerator where they are further accelerated.

You may wonder why initially electrically negatively charged H^- ions are accelerated instead of the desired positively charged protons. As can be seen in the figure, below the protons already injected into the magnetic holding field of the ring accelerator take a right bend at the point where the new particles are injected. If you want to inject the next packet into this ring, you realize that it is more advantageous if these particles describe a

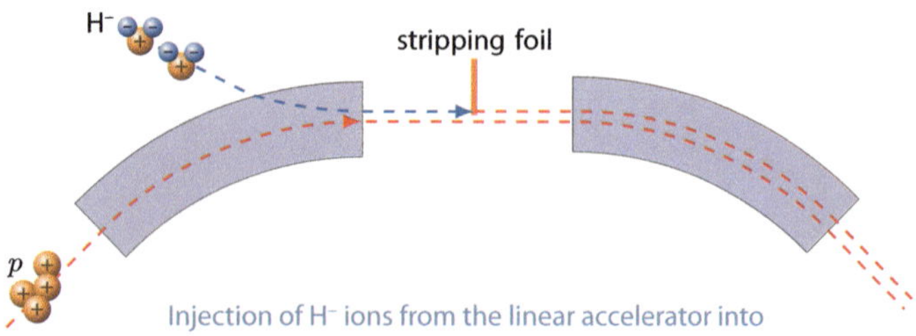

Injection of H^- ions from the linear accelerator into the first ring accelerator

Energy and Power → p. 36
How Does a Particle Accelerator Work? → p. 66

left bend in this magnet. This makes filling the ring accelerator with proton packets easier and results in more protons being stored overall.

In the first ring accelerator, the Proton Synchrotron Booster (PSB), the protons are accelerated to a kinetic energy of 2 GeV. In the next step, they are accelerated up to 25 GeV in the 628 m circumference Proton Synchrotron (PS). The subsequent Super Proton Synchrotron (SPS) already has a circumference of almost 7 km. Here, the protons are accelerated to 450 GeV. From there, proton bunches are fed into the LHC in both clockwise and counterclockwise directions. Each of these over 2000 bunches per direction of rotation contains 10^{11} protons. In the LHC, the protons are accelerated to the final energy of 6.8 TeV. This entire process takes about 20 minutes.

The two opposing beams are then brought to collision at four collision points, where the four experiments ALICE, ATLAS, CMS, and LHCb are located. The image at the top shows the result of a proton-lead collision, which was registered at the ALICE experiment.

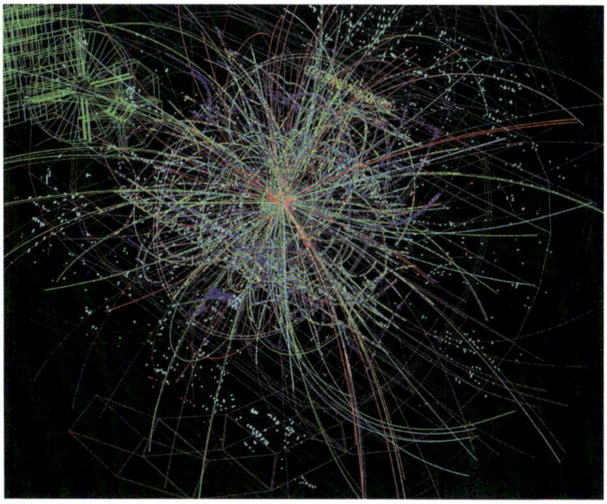

Particle tracks of a proton-lead collision at the ALICE experiment

There are about one billion collisions per second. Since this is only a small proportion of the stored particles, the LHC only needs to be refilled after about 10 hours. The remaining particles make 400 million rounds in the LHC during this time. This corresponds to a distance of 10 billion kilometers.

Linear accelerator for H^- ions at CERN

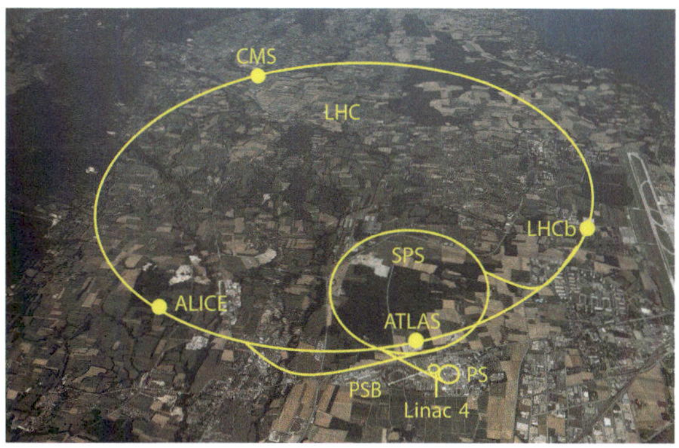

The LHC accelerator complex at CERN near Geneva

Beam Optics

Don't Go Off the Rails

In a circular accelerator, a particle beam consisting of many billions of particles performs several million revolutions over a period of several hours (LHC ↓). It is quite astonishing that the beam remains focused for such a long time. This is no coincidence. Many effects can cause particles to be lost. For example, they could collide with remnant air molecules in the beam pipe. To minimize this, it is evacuated. To have a high-quality beam, the pressure in the beam pipe has to be reduced compared to the atmospheric pressure by up to 13 orders of magnitude. Furthermore, there are particles that do not have exactly the energy for which the accelerator's magnetic field is designed, or particles that are not moving exactly in the right direction. Eventually, these will hit the wall of the beam pipe and are lost. In addition, we know since Galileo: "All bodies fall at the same speed." This also applies to elementary particles in accelerators. But why don't the particles fall downwards in an accelerator?

In accelerators, so-called holding fields force the particles of a specific energy onto a circular path with a fixed radius. In order to counteract the above mentioned problems, further electric and magnetic fields are applied which we now discuss in more detail.

The magnetic field of an accelerator is designed for a specific energy, which is called the nominal energy. How does one ensure that particles which deviate from this nominal energy are not lost?

For the acceleration, sinusoidal electric alternating fields (Accelerators ↓) are used. The trick is to place the particles slightly after the maximal available field strength at a slightly lower value. For simplicity, let's assume that all particles have already been accelerated to 99.99... % of the speed of light. Further acceleration then no longer leads to a significant increase in speed, but only to an increase in energy (Special Relativity ↓). A particle whose energy is too low compared to the nominal energy is deflected more strongly in the next magnetic field section and therefore moves on a circular path with a smaller radius. Similar to a runner on the inner track of a stadium, it travels a shorter distance and therefore arrives earlier at the accelerator section. There it sees a higher field towards the peak of the sinusoidal field, as shown in the figure below, and is accelerated more strongly. Correspondingly, particles whose energy is too high travel a longer distance, arrive later and experience a smaller field. This ensures that all particles have the same energy on average. This process is referred to as *phase focusing*.

Depending on the energy, the particles travel on different paths and are accelerated to varying degrees.

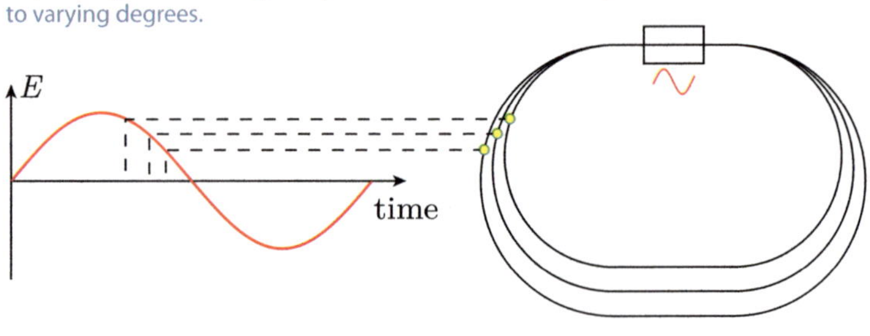

The LHC Accelerator Complex at CERN → p. 116
How Does a Particle Accelerator Work? → p. 66
Special Relativity → p. 38

Next, we look at what happens to particles that deviate to the left and right or up and down from their intended path, or that threaten to fall downwards due to the influence of gravity. In optics, lenses are used to focus light rays. Similarly, magnetic fields can be used to focus particle beams. This is done with so-called *quadrupole magnets* (see pictures below). The magnetic field is designed so that a particle moving through the center, i.e., on the intended path, does not feel the magnetic field. Particles that move too far to the left or right feel a field that is stronger the further they move away from the center. This field causes a deflection of the particles towards the center.

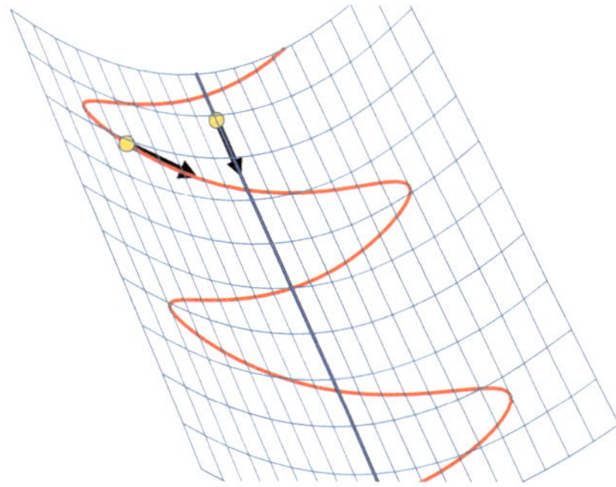

Just as a sled is repeatedly deflected back to the intended path by the shape of the toboggan run, particles are deflected by quadrupole magnets.

Unfortunately, focusing a particle beam does not work in both horizontal and vertical directions simultaneously, like with a light beam. If a particle is too far up or down, it is deflected even further up or down. The solution is a sequence of two quadrupoles rotated 90° relative to each other. The particle that was deflected further upwards in the first quadrupole is directed even more strongly towards the center in the second. The particle that was directed from the left towards the center in the first quadrupole receives only a small deflection outwards in the second. The combination of both quadrupoles then leads to a focusing effect in both directions.

Now it is also clear why the particles do not fall further down due to gravity. Initially, of course, they do. But then they feel the field of the quadrupole magnets, which ensures that the particles do not deviate too far downwards from the intended path. With these measures, conditions are created to store focused particle beams over long periods of time.

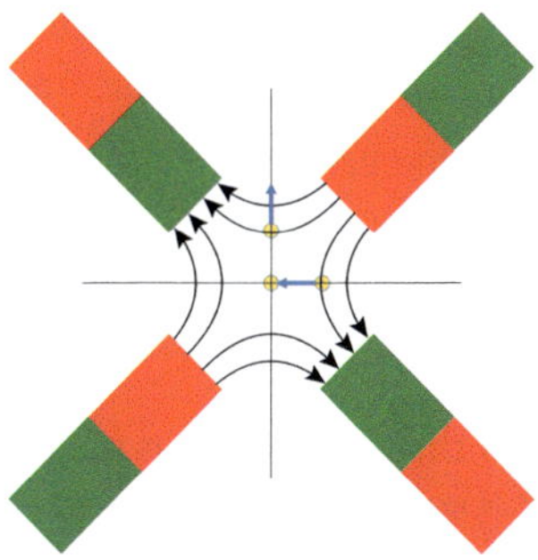

Principle of a quadrupole magnet. In accelerators, electromagnets are used instead of the bar magnets shown in the drawing.

Example of a quadrupole magnet

Plasma Acceleration

An Accelerator That Fits on a Kitchen Table

The essential ingredients for accelerating particles to energies as high as possible are strong electric fields. An electric field is measured in the unit volts per meter. If you apply a 9 V battery between two metal plates spaced one centimeter apart, this corresponds to an electric field of 900 V/m. Technically possible are electric fields of several megavolts per meter. If one tries to go beyond this, there are electrical discharges between the plates. So, if you want to achieve energies of a few gigaelectron volts (GeV), you have to arrange several acceleration stages in succession. This leads to very long linear accelerators or one accelerator section has to be passed through very often as in a circular accelerator (↓). But even with a circular accelerator, one eventually reaches limits, as the radius and/or the magnetic field must always become larger to keep particles of ever higher energy on a circular path.

However, the desire of physicists is to reach ever higher energies. For this, we need new methods. Is there a way to generate higher fields on short distances? One can ask oneself what electric fields occur within an atom. An electron in a hydrogen atom is exposed to the electric field of the proton. If you calculate its strength (see information box) you get an incredibly large value of 500 gigavolts per meter, far beyond the fields that can be generated in the laboratory in macroscopic structures. Unfortunately, this does not help to accelerate particles, as in normal matter positive and negative charges of protons and electrons neutralize each other and we do not perceive these high fields. But maybe there is still a way how to use such fields.

Electric Field Strength in an Atom

How large is the electric field that an electron in a hydrogen atom experiences from the proton? On average, an electron is at a distance from the proton equal to the so-called *Bohr radius*

$$a_0 = \frac{\varepsilon_0 h^2}{\pi m_e e^2} \approx 5.3 \cdot 10^{-11}\,\mathrm{m}\,.$$

Here, ε_0 is the electric field constant, e is the elementary charge, m_e is the rest mass of the electron, and h is Planck's constant.

The force between the electron and proton is calculated from Coulomb's law

$$F = \frac{1}{4\pi\varepsilon_0}\frac{q_1 q_2}{r^2}\,.$$

The electric field then results from dividing the force by the magnitude of the electron's charge $|q_1| = e$. Note that the proton also carries an elementary charge. Using for the distance $r = a_0$, one finds

$$E \approx 500\,\mathrm{GV/m}\,.$$

This is where plasma acceleration comes in. In addition to the known states of matter: solid, liquid and gaseous, matter can also exist as a *plasma*. It is characterized by the fact that the individual electrons are no longer assigned to a single atom. The atoms are thus fully or partially ionized. Such plasmas exist, for example, in a fluorescent lamp, also called neon tube, or in the Sun.

How Does a Particle Accelerator Work? → p. 66

The boat corresponds to the laser pulse, the water to the plasma, the surfer to the particles to be accelerated.

A separation of electrons and ions can, for example, be achieved by a short laser pulse, providing strong electric fields. Behind the laser pulse, a wakefield is created, similar to the wake of a boat.

In the laboratory, a great deal of effort is required to convert matter into a plasma state. Even in a plasma, the electrons, like the ions, are initially uniformly distributed over the volume, so that no large electric fields occur. Electric fields only arise when charges with different signs are separated from each other.

Since the free negatively charged electrons have a much lower mass than the positively charged ions, the electrons can move much more easily. Thus, the uniform distribution can be disrupted by external influence and strong electric fields arise within the plasma, which can be used for particle acceleration.

In this field, electrons can be accelerated over a very short distance to several gigaelectron volts. With conventional accelerators much larger facilities are needed to reach such energies. In 2018, it was possible to accelerate electrons in a laser-plasma accelerator to 7.8 GeV over a distance of 20 cm. The laser pulse had a power (↓) of 850 TW.[1]

Even though plasma acceleration has so far only been successful on a small scale, this technique opens up possibilities to accelerate particles to very high energies with accelerators of manageable size.

A laser pulse (red) causes a wakefield (blue), which can be used to accelerate an electron beam (bright blue dot behind the laser pulse).

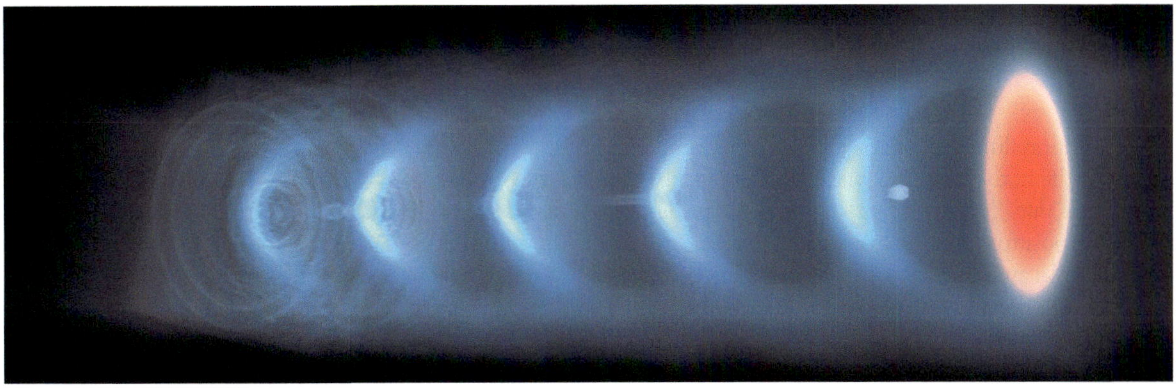

[1] A. J. Gonsalves et al. 2019, *Petawatt Laser Guiding and Electron Beam Acceleration to 8 GeV in a Laser-Heated Capillary Discharge Waveguide,* Physical Review Letters, 122, 084801, doi:10.1103/PhysRevLett.122.084801

Energy and Power → p. 36

Accelerators in Medicine

Diagnosis and Therapy

Particle accelerators are now indispensable in medical diagnostics and therapy. There is a wide range of applications, which will be discussed here.

The well-known X-ray unit (↓) requires a small electron accelerator. In positron emission tomography (PET, ↓), an accelerator several meters in diameter, typically a cyclotron, is used to produce short-lived radioactive isotopes which provide the positrons.

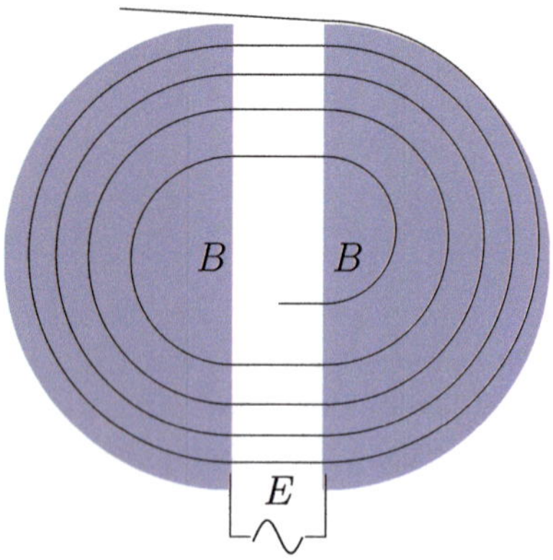

Principle of a cyclotron

In a cyclotron (see figure), the magnetic fields in the left and right blue semi-circles keep protons on a circular path. The path's radius increases, since the electric field applied between the two magnetic fields accelerates the protons. With a cyclotron, moderate energies of a few 100 MeV can be achieved, which are sufficient to produce the radioactive isotopes used in medical applications.

The protons are accelerated and directed onto a target. If there is oxygen in the target material, the following reaction occurs (From Nucleus to Nuclei ↓):

$$^{18}_{8}\mathrm{O} + p \rightarrow {}^{18}_{9}\mathrm{F} + n$$

The resulting fluorine isotope $^{18}_{9}\mathrm{F}$ is injected into a patient. In the body, it decays with a half-life of 110 minutes back into $^{18}_{8}\mathrm{O}$. In this process a positron is emitted. The positron eventually encounters an electron in the body. In an electron-positron annihilation process, photons are created and detected outside the human body. To keep the radiation exposure for the patient as low as possible, only radioactive substances with a lifetime on the order of the treatment duration, i.e., a maximum of a few hours, are considered. Therefore, these cyclotrons are usually located directly in the hospital, as most of the nuclei would have decayed after a longer transport.

Isotopes, produced in an accelerator, are also used for scintigrams, e.g., of the thyroid. All these methods are diagnostic methods, for example to determine the exact position of a tumor in the body.

However, accelerators are not only used for diagnostic purposes. In addition to surgery and chemotherapy, they are also used to fight tumors. The goal is to deposit as much energy as possible at the site of the tumor. The patient is irradiated in order to damage the tumor cells as much as possible, while the surrounding healthy tissue should remain intact.

X-Rays → p. 124
Positron Emission Tomography → p. 144
From Nucleus to Nuclei → p. 226

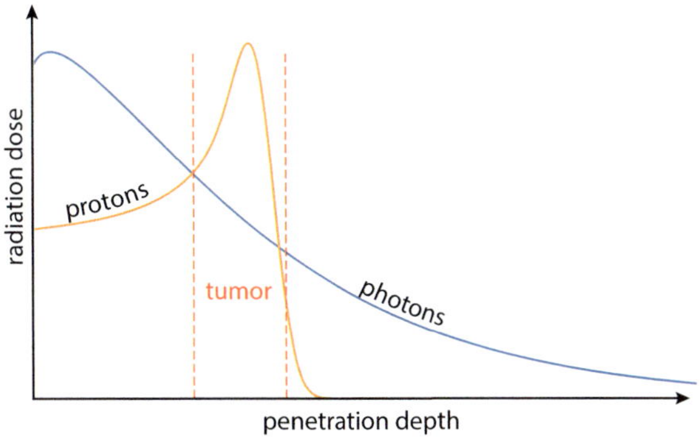

The deposited radiation dose as a function of penetration depth for photons and protons

Here, photons or charged hadrons, mostly protons, are used. Irradiation with photons has the disadvantage that, as can be seen from the figure, most of the energy is released shortly after entering the human body. As the penetration depth increases, the energy deposition decreases.

Siemens cyclotron for the production of radioactive isotopes

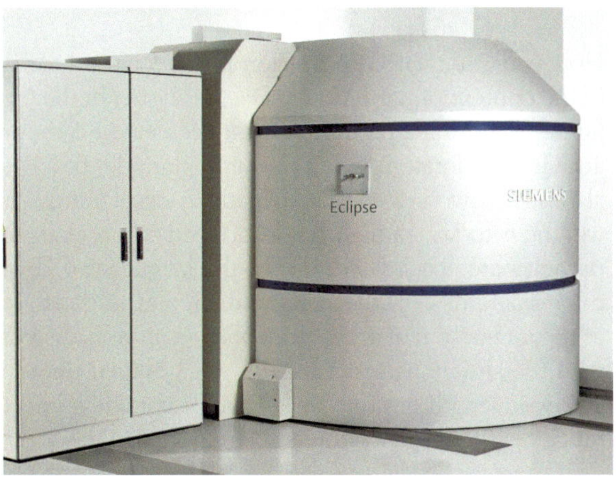

The situation is different with hadron beams: they deposit most of their energy at the end of their trajectory. There they slow down and linger longer. Since photons are massless and fly at the speed of light in the medium, they do not show this effect.

The interaction between protons and matter strongly depends on the energy of the protons. This allows the penetration depth of the protons, where they deposit most of their energy, to be set accurately by varying the beam energy. Thus the tumor, whose position has been precisely determined by imaging methods, can be irradiated to within a millimeter.

It is now even possible to use imaging methods simultaneously with the irradiation. This way, a tumor moving with the body, e.g., due to breathing, can always be optimally irradiated.

A movable beam guide (gantry) at the Paul Scherrer Institute PSI in Switzerland. The apparatus allows the proton beam to be directed at a patient from various directions.

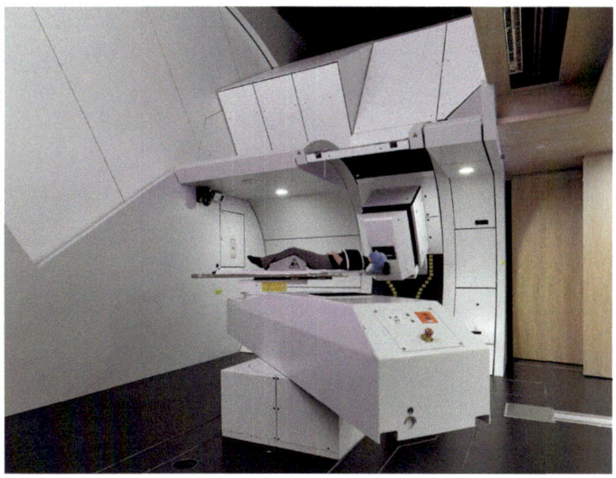

X-Rays

They Go Under Your Skin

The first Nobel Prize in Physics was awarded in 1901 to Wilhelm Conrad Röntgen for the discovery of a new type of radiation. In the German-speaking world, it is named in his honor as ***Röntgen radiation***. In English, these rays are referred to as ***X-rays***. The term X-rays was also originally chosen by Röntgen, as he did not know what he had found. He experimented with cathode ray tubes (Origins of Particle Physics ↓), as were later used in television sets. The aim of his investigations was actually the measurement of electrical conductivity in gases. In his darkened laboratory, however, Röntgen observed that a luminescent screen at a great distance brightened up. An unknown radiation must have been generated in the glass bulb, passed through the glass and a cover, and finally excited molecules on the screen to glow.

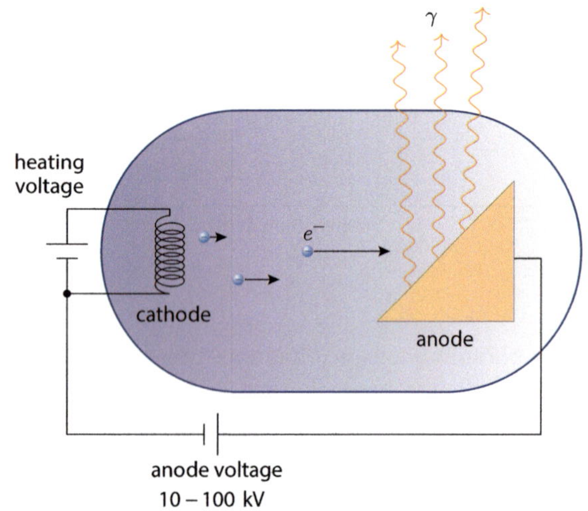

Principle of the generation of X-rays in a cathode tube. Accelerated electrons hit an anode, are slowed down there and emit X-ray radiation.

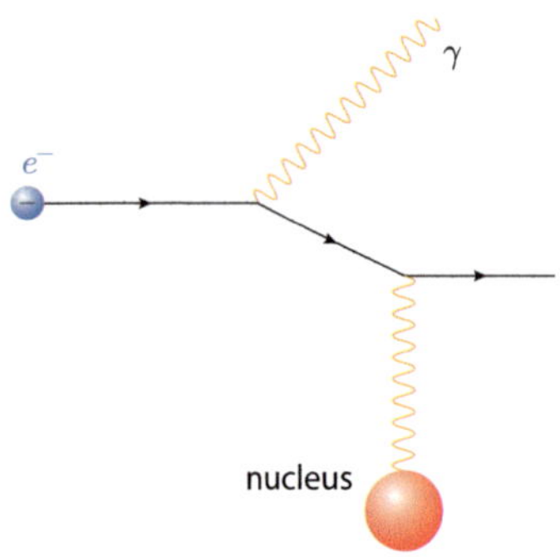

Feynman diagram for Röntgen radiation. In the interaction with an atomic nucleus in the anode, the electron emits a Röntgen photon γ.

How can Röntgen's observation be understood today in the context of particle physics? As shown in the top figure, electrons are emitted from the heated cathode and are accelerated. Upon hitting the anode, they are abruptly slowed down, i.e., negatively accelerated. As we know today, radiation is produced when charged particles are accelerated (↓). In the language of Feynman diagrams (↓), the X-ray radiation is understood as the interaction of an electron with an atomic nucleus with the simultaneous emission of a Röntgen photon. The wavelength of the photon is approximately in the range of 10 nm to 10 pm.

Origins of Particle Physics → p. 10
Acceleration and Radiation → p. 42
Feynman Diagrams → p. 48

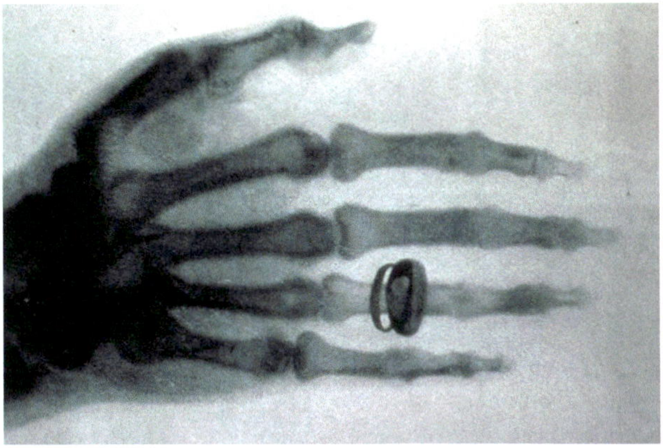

Historical X-ray image

separation of neighboring atoms in the crystals. One then observes diffraction patterns on a screen, as is also known from optics. From these patterns, one can draw conclusions about the structure of the irradiated crystals or molecules. Thus, X-ray radiation contributed to the resolution of crystal structures and also to the deciphering of the structure of DNA molecules. Further applications are discussed in the article on synchrotron radiation (↓).

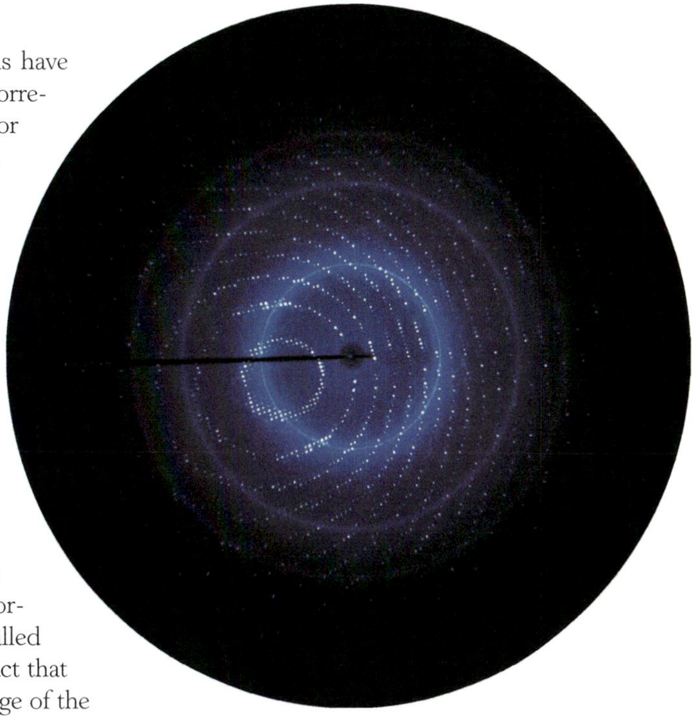

Diffraction pattern of a lysozyme crystal obtained using X-ray radiation.

In contrast to visible light, the X-ray photons have a wavelength 1000 times shorter and, correspondingly an energy higher by the same factor (Quantum Mechanics ↓). Therefore, Röntgen photons can partially penetrate matter. Depending on the material, the X-rays either go through or get absorbed. The probability of absorption increases with the atomic number Z of the material. Therefore, bones (calcium, atomic number $Z = 20$) appear as shadows in X-ray images, compared with the softer tissue in the body (carbon $Z = 6$, nitrogen $Z = 7$, oxygen $Z = 8$ and hydrogen $Z = 1$).

The use of Röntgen radiation is not limited to creating a shadow image of an object, as in medical imaging. When X-rays hit an ordered structure, such as a crystal (↓), so-called diffraction effects occur. This is due to the fact that the wavelength of X-ray photons is in the range of the

Concepts of Quantum Mechanics → p. 50
The Crystal → p. 4
Synchrotron Radiation → p. 126

Synchrotron Radiation

Sources of Special Light

When charged particles are accelerated (↓), they emit electromagnetic radiation, which consists of photons. Precisely this effect is exploited in an X-ray machine (↓), where previously accelerated electrons are decelerated to generate radiation.

If a particle is on a circular path, it is constantly *accelerated* towards the center of the circle and therefore constantly emits radiation. On a circular path, the particle does not necessarily get faster, but it constantly changes its direction, which also corresponds to an acceleration, as the velocity vector of the particle changes.

Such radiation is produced in a circular accelerator (↓), more precisely a *synchrotron*, and is therefore referred to as *synchrotron radiation*. For particle physics, synchrotron radiation is an unwanted side effect, as the particle loses energy through the radiation, which must be constantly replenished.

To get an idea of the order of magnitude, let us consider the largest electron accelerator to date, LEP, which was operated at CERN until 2000. In it, electrons were accelerated to an energy of 100 GeV. The energy loss per revolution due to synchrotron radiation was 2.9 GeV, which had to be constantly replenished. At the LHC (↓), which was later installed in the same tunnel as LEP, the energy loss is only 6.7 keV per revolution. This is due to the type of particles being accelerated. For light particles like electrons, the effect is much larger than for the heavier protons that are accelerated in the LHC.

While synchrotron radiation is a nuisance for particle physicists—after all, you want to give as much of the energy as possible to the accelerated particles—it is a blessing for other areas of physics. For example, to examine materials, the synchrotron radiation is directed onto a material sample. The scattered photons then provide information about the structure of the material (Scattering Experiments ↓).

Synchrotron radiation in a deflection magnet

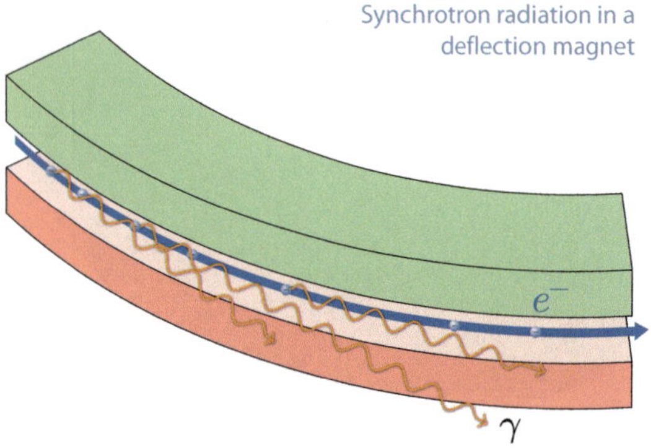

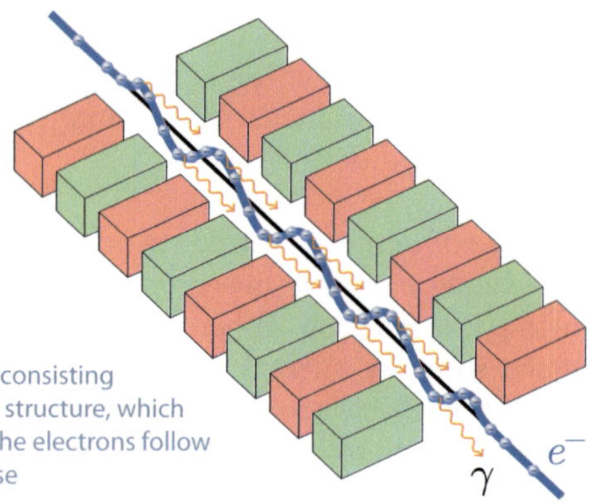

An undulator consisting of a magnetic structure, which ensures that the electrons follow a slalom course

Acceleration and Radiation → p. 42
X-Rays → p. 124
How Does a Particle Accelerator Work? → p. 66
The LHC Accelerator Complex at CERN → p. 116
Scattering Experiments → p. 46

Initially, the emitted synchrotron radiation was a kind of by-product. Later accelerators were built specifically as synchrotron radiation sources. In this case, so-called *wigglers* or *undulators* are used. These consist of a magnetic field arrangement that forces the electrons to wiggle on their flight path (see figure). The electrons then emit photons in the same direction multiple times. This results in a more finely bundled and intense radiation, which also has a better defined energy. The higher intensity allows for a more precise study of spectra or diffraction patterns, which are characteristic of the material being examined.

A face is hiding behind the excerpt of the van Gogh painting "Patch of Grass" in the top picture. It can only be vaguely recognized with a regular X-ray image (bottom left). With fluorescence excited by synchrotron radiation, the face can be clearly reconstructed (bottom right).

Applications range from measurements on individual molecules to the examination of paintings. The right figure shows the result of measurements on a painting by Vincent van Gogh. In an X-ray image, one can only vaguely recognize that there is another picture under the actual painting. With the much more intense synchrotron radiation source, more possibilities open up to unravel the mystery of the picture behind the picture.

Synchrotron in Grenoble, France

The further development of synchrotron radiation sources leads to a Free Electron Laser (FEL). Through increased intensity and short pulse duration, chemically fast reactions such as photosynthesis can be observed. Such devices are operated at DESY in Hamburg, Germany and at SLAC in California.

Typical Detectors at Accelerators

The Big and Small of Detectors

Particle physics detectors come in a great variety. Some are located beneath kilometers of rock or ice. Or they might be huge water tanks and constantly penetrated by particles from the cosmos (↓). However, despite this variety, the experiments at particle accelerators (↓) can be roughly divided into two categories: *collider* or *fixed-target* experiments. An example of each of these types is presented here.

The CMS detector is one of the two large general purpose collider detectors at the LHC (↓) at CERN. This means that the experiment was designed to cover a very wide range of physics topics. About 4000 researchers from around the world have built the detector and operate its components. Each year, they collectively publish about one hundred contributions in scientific journals that deal with a wide range of topics in particle physics. CMS stands for *Compact Muon Solenoid*, which means that it is a compact detector with a solenoid magnet and has something to do with muons. A solenoid magnet can be imagined as a spool of yarn, with the current flowing through the yarn. This creates a uniform magnetic field inside the winding of the spool, also called *coil*. Here, "compact" is relative: The detector has a diameter of about 20 meters and weighs 14 000 tons. It is however in volume much smaller (a factor of almost 7) than for example the ATLAS detector, which is shown in the next article.

CMS is a *collider detector*, which means that it surrounds the point where the accelerator collides bunches of protons with each other with a center-of-mass energy of 13.6 TeV. The detector surrounds the collision point from all sides like a barrel lying on its side. At its center, new particles are created in the collision of protons, which then penetrate the detector in all directions

The CMS detector: a collider detector

Particle collision in the CMS detector

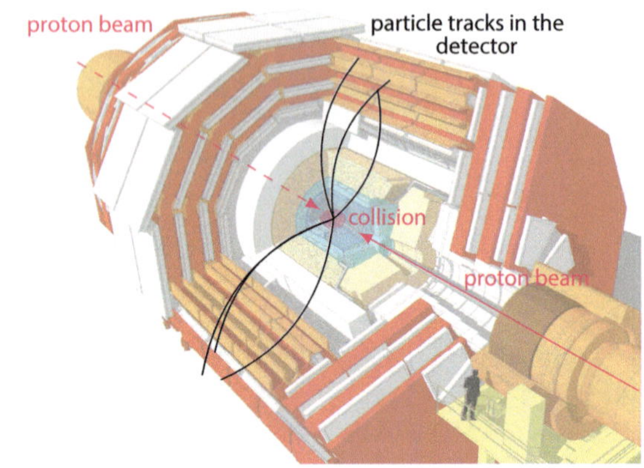

Direct Search for Dark Matter → p. 310, Ultra-High-Energy Cosmic Rays → p. 338, Large Detectors Deep in a Mountain → p. 132 and Cosmic Rays → p. 336
How Does a Particle Accelerator Work? → p. 66
The LHC Accelerator Complex at CERN → p. 116

and are measured. Because the colliding protons meet from both sides with the same momentum, the result of the collision must also be symmetrically distributed according to the conservation of momentum—hence the symmetrical and cylindrical shape of the detector.

In the drawing, the CMS detector has been cut open, so that a typical particle reaction becomes visible (see Signatures of Particles ↓). Due to the onion-like layered structure of the detector, the outgoing particles drawn in black are measured from the inside out in all their properties: First, the origins and tracks of the electrically charged particles are measured (small magenta and dark cyan detector parts). The particle energy is measured next (cyan and yellow), then comes the magnet coil (gray). Finally the least interacting detectable particles, the muons, are measured in the red and white/yellow detector parts, which are sandwiched between steel plates (red).

Due to the strong magnetic field and the possibility to examine the muons once inside the magnet coil and then again outside with a track detector, the detector also got its name: *Compact Muon Solenoid*, the "compact magnet coil for muons".

By far not all experiments are as large as CMS, or are operated by so many physicists. And by far not all are located at colliders. An example for a completely different type of detector is the BGO-OD experiment in Bonn, Germany, at the ELSA accelerator, where electrons with energies of "only" 3.2 GeV are shot at a *fixed target* to produce high-energy photons. These are then brought into interaction with a second fixed target of polarized matter, where a magnetic field has been used to align the nuclear spins. The detector is placed behind the collision point and subsequently examines the resulting products of the reaction between polarized photons and the polarized nucleons.

Not all accelerators are optimized for the same properties. The ELSA accelerator, for example, is one of the few facilities in the world from which individual electrons—on average—can be extracted in each orbit of the accelerator. This allows the experiment to be provided with a sequence of individually measurable particles one after the other instead of a whole bunch at a time. This, and the polarization of the electrons and the target, offers completely different possibilities for measurements than other accelerators.

Since the target, the material at which the electrons are shot, does not move, the momentum of the electrons ensures a bundling of the reaction products: All particles continue to fly forward like in a lying bouquet of flowers, hence the experiment is rather built like a cone opening downstream from the collision point, and not like a barrel. Even though the experiment is only about four meters wide and ten meters long, it still consists of specialized layers of individual sub-detectors, all of which together fulfill the common mission: to measure all important properties of all particles in the event.

The BGO-OD detector: a fixed-target experiment

Signatures of Particles → p. 80

The ATLAS Experiment at the LHC

A Graphical Overview

Particle physics experiments come in a wide variety of shapes and sizes, according to the measurements they are intended to perform and the environment they are exposed to.

Here, the currently largest experiment at a collider is shown in detail: The ATLAS experiment (**A** **T**oroidal **L**HC **A**pparatu**S**[1]) at the Large Hadron Collider LHC.

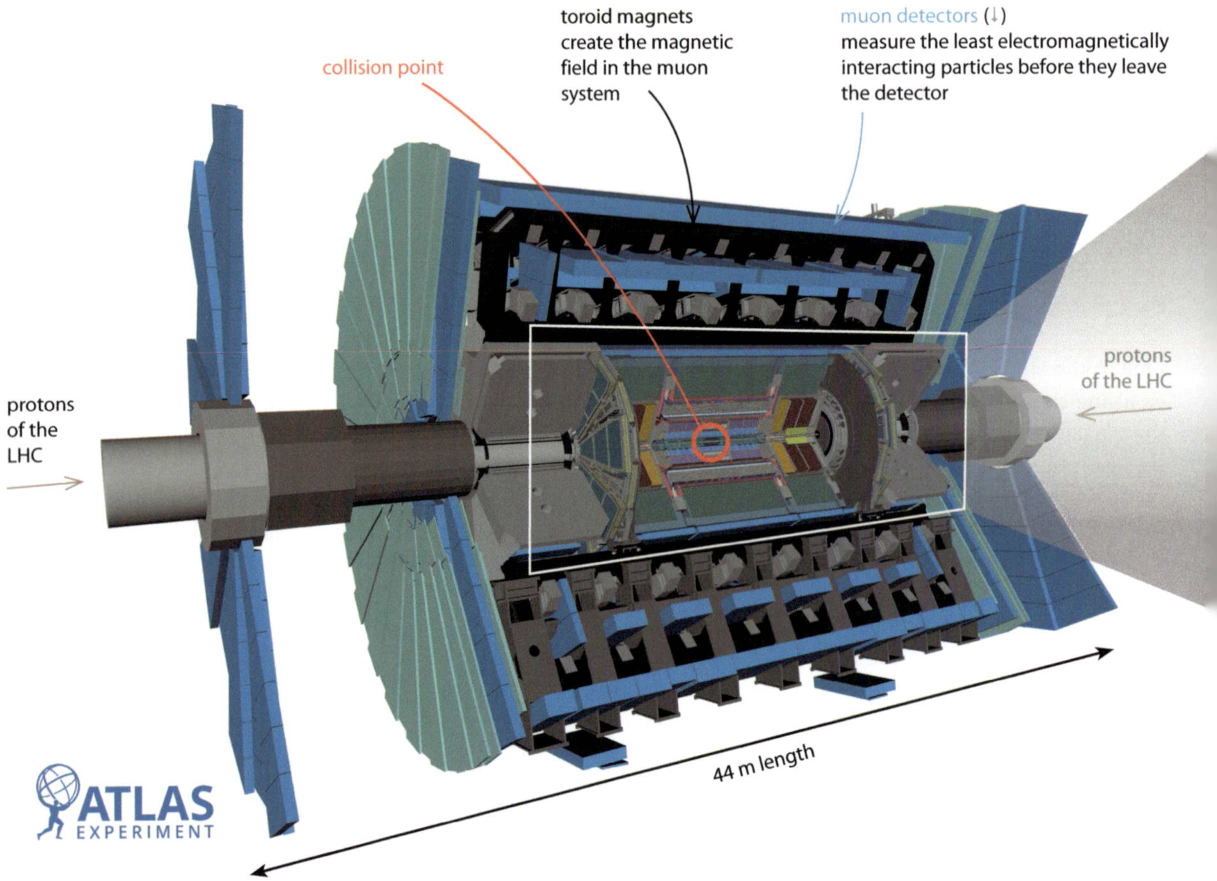

[1] Sometimes, the acronyms chosen for particle physics experiments are not based on the most insightful backstory. Often, a nice sounding acronym is more important than its "meaning"

Muon detectors: Identifying the Particle Type via the Mass → p. 156

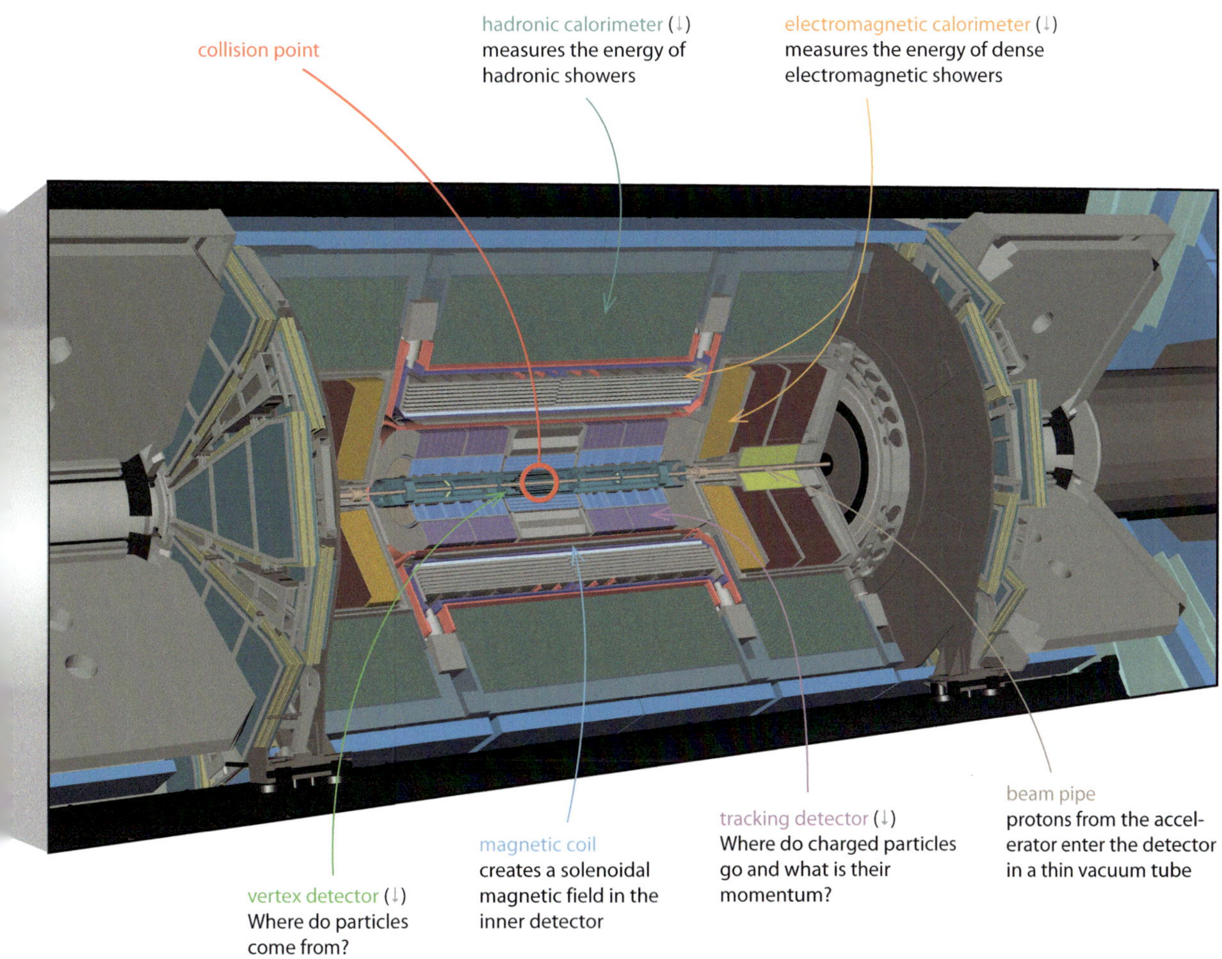

The ATLAS detector at the Large Hadron Collider LHC at CERN

Calorimeters: Energy Measurement → p. 78
Vertex detector: Semiconductor Detectors → p. 70
Tracking detector: Position and Momentum Measurement → p. 76

Large Detectors Deep in a Mountain

Hunting for Neutrinos in Caves

Neutrinos (↓) are highly elusive. Thus their investigation requires lots of neutrinos and/or very large detector volumes to have a chance of seeing anything. The detector used for the first discovery of neutrinos in 1956 was still quite small with approximately 400 liters of water. This was compensated for by the enormous flux of neutrinos produced by the reactions in a nuclear reactor, so that very close to the reactor (about 11 m) one neutrino reaction was observed about every half hour. The largest neutrino detectors built today contain many thousand cubic meters of detection volume. Even this is surpassed when sea water or Arctic ice is used as the detection material. All these detectors work according to the same principle: an incoming neutrino reacts via the weak interactions (↓) and is either converted into a charged particle or transfers energy to a charged particle which then can be observed. In a weak charged current reaction for example a neutrino can be converted into an electron or a muon, or, if it was an antineutrino, into a positron or antimuon. As charged particles, these are observed in the detectors, e.g., via Cherenkov light (↓).

Inside the still empty Borexino detector for investigating solar neutrinos. The picture shows the photomultipliers for measuring the scintillation light.

Driven by the first surprising results of the measurement of solar neutrinos (↓) there was increased interest in analyzing the oscillations of solar neutrinos, of neutrinos from nuclear reactors or neutrinos from cosmic radiation—so-called atmospheric neutrinos—in more detail. Initially, pre-existing large detectors were employed. These had been built to search for another rare, hypothetical event, namely proton decay (↓).

The Fréjus detector installed in a side chamber of a road tunnel in the French Alps, was such a detector. It consisted of alternating layers of iron and scintillator plates and weighed around 900 tons. Because of the high density of the detector material, the neutrinos mainly react in the iron and produce charged particles whose tracks can be traced in the alternating scintillator layers. Another example was the IMB detector in a salt mine in the USA consisting of a tank filled with 10 000 tons of water, which were surrounded by 2048 photomultipliers (↓). Water was chosen because it is the most cost-effective material with a very high proton proportion.

The Kamioka Nucleon Decay Experiment, or Kamiokande experiment for short, was installed in the Kamioka mine in Japan. It consisted of of 3000 tons of water and 1000 photomultipliers for the detection of Cherenkov light. It was expanded in 1996 to Super-Ka-

Neutrinos → p. 196
The Weak Interaction → p. 178
Cherenkov Radiation and Particle Identification → p. 82
Neutrinos From the Sun → p. 344, Neutrino Oscillations → p. 198
Proton Decay → p. 320

miokande with 50 000 tons of water and about 13 000 photomultipliers. With Kamiokande, neutrino oscillations were clearly detected for the first time, and with Super-Kamiokande the solar neutrino deficit of an earlier experiment was confirmed to be caused by neutrino oscillations (↓).

The MINOS detector, consisting of iron plates and scintillation detectors, examines neutrinos from an accelerator 720 km away.

The KM3Net detector uses the water of the Mediterranean Sea. Photomultipliers are installed at a depth of around 3000 meters.

The Kamiokande detector normally observed a maximum of about one neutrino event per day. Then something spectacular happened on February 24, 1987. On this day, the supernova 1987A from the Large Magellanic Cloud, a satellite galaxy of our Milky Way, produced an unusal burst of eleven neutrino signals within just thirteen seconds registered in Kamiokande. At the same time, the IMB detector in Ohio, USA, recorded eight events while two detectors in France and in the Soviet Union each measured five more neutrino events respectively. This was the first and so far only detection of neutrinos from a supernova (↓). It confirmed the idea that the energy of a supernova is mainly carried away by neutrinos.

Today, detectors are getting bigger and bigger in order to further increase sensitivity. At 20 000 tons, JUNO (Jiangmen Underground Neutrino Observatory) will be the largest liquid scintillator detector that has ever been built. It has been constructed in China, and from 2025 on will measure neutrino oscillations from nuclear reactors 50 kilometers away. The scintillator liquid contained in an acrylic sphere will be observed with 53 000 photomultipliers. Also under construction is DUNE (Deep Underground Neutrino Experiment) in the USA, a detector for measuring neutrinos generated 1300 kilometers away in the Fermi National Laboratory. DUNE will consist of 70 000 tons of liquid argon, which corresponds to the volume of a cube with an edge length of 37 meters.

The largest detector volume built by humans will be Hyper-Kamiokande in Japan. 260 000 tons of water will provide about ten times more usable volume than Super-Kamiokande. This will not only cover a broad neutrino physics program, but also continue the search for proton decay. It is scheduled to start in 2027.

The ingenuity of physicists has made even larger detectors possible. The idea is to use naturally occurring detector material such as seawater or the ice of the Antarctic. The largest detectors by far are KM3NeT in the water of the Mediterranean, and IceCube (↓) in the ice sheet at the South Pole. Volumes of one cubic kilometer, corresponding to a billion tons of water or ice are equipped with photomultipliers to detect traces of very high-energy neutrinos from astrophysical sources (↓).

Scintillators and Photomultipliers → p. 74
Verifiying Neutrino Oscillations → p. 254
Supernovae as Particle Physics Laboratories → p. 324
The IceCube Detector → p. 134
Multimessenger Astronomy → p. 340

The IceCube Detector

Searching for Neutrinos at the South Pole

IceCube Laboratory
data is collected at the South Pole and sent by satellite to the University of Wisconsin–Madison.

Digital Optical Module (DOM)
DOMs are light sensors consisting of photo multipliers (↓) and related electronics

The last Digital Object Module (DOM) was signed by the people working on the construction of IceCube.

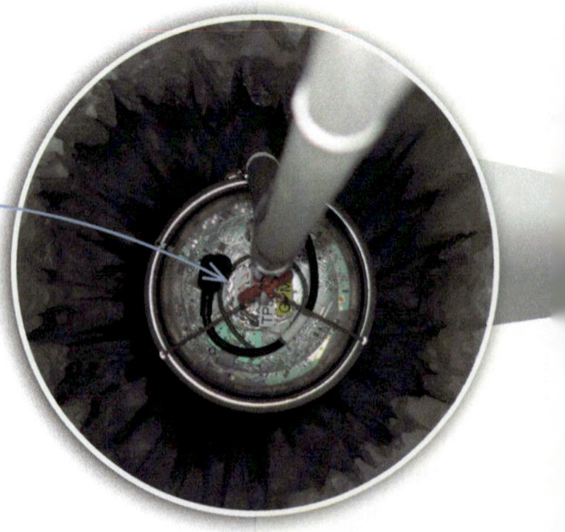

A DOM is sent on its way deep into the ice, where it will record data. The hole is "drilled" by melting the ice.

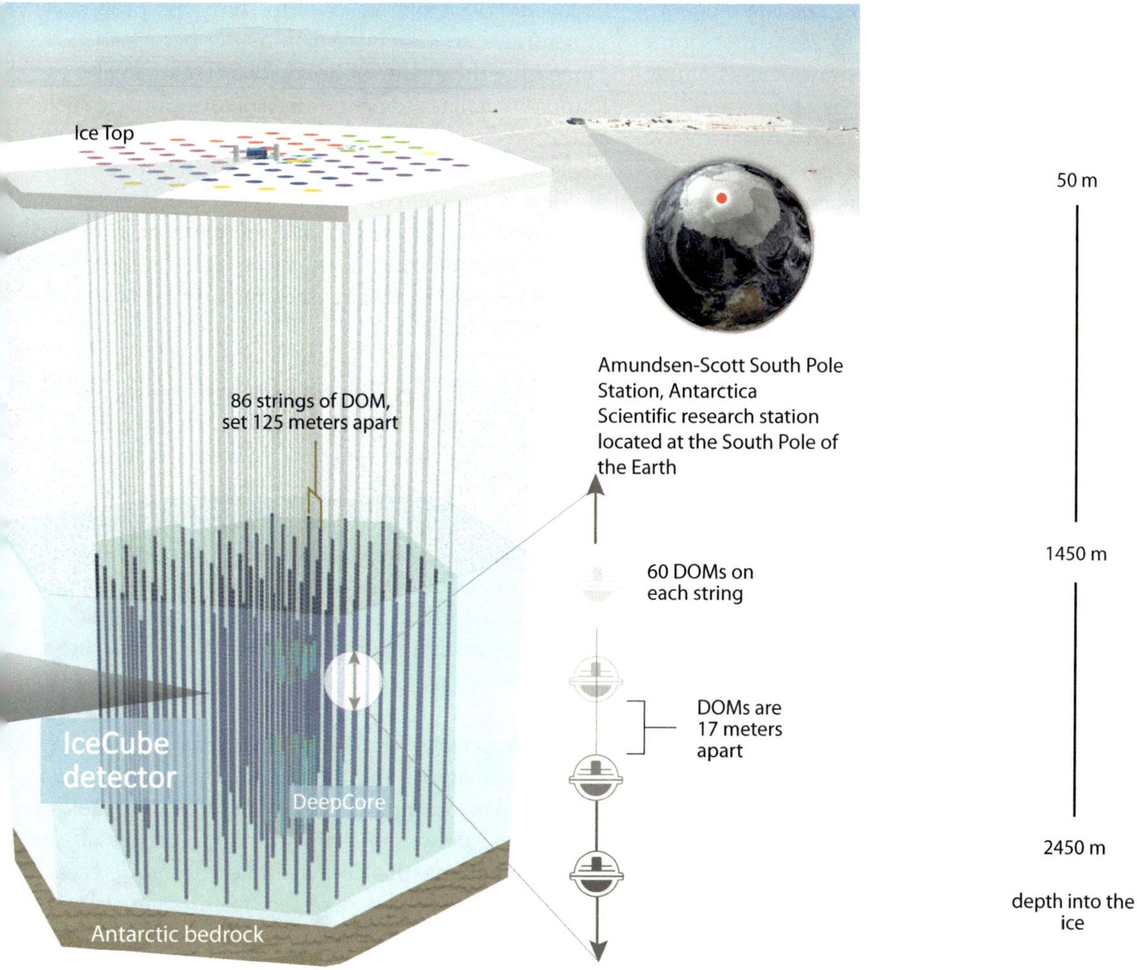

Scintillators and Photomultipliers → p. 74

Cryogenic Detectors

High Demands at Low Temperatures

Particles are measured by detecting their interactions. In doing so, they transfer energy to the matter of the detector. In most cases, it is the electromagnetic interaction (↓) that generates free charges or light and thus a measurable signal for detection (Semiconductor Detectors ↓, Scintillators ↓). But how can one detect particles that do not interact electromagnetically, or only very inefficiently?

Regardless of the type of interaction, every energy transfer ultimately leads to an increase in temperature. Temperature is a measure of the average kinetic energy of a material's atomic vibrations. If energy is transferred from an incoming particle to just one atom, the energy is gradually distributed evenly to all atoms and we can speak of a temperature increase. The measurement of this temperature increase as a signal for particle detection is the principle of so-called *cryogenic detectors*.

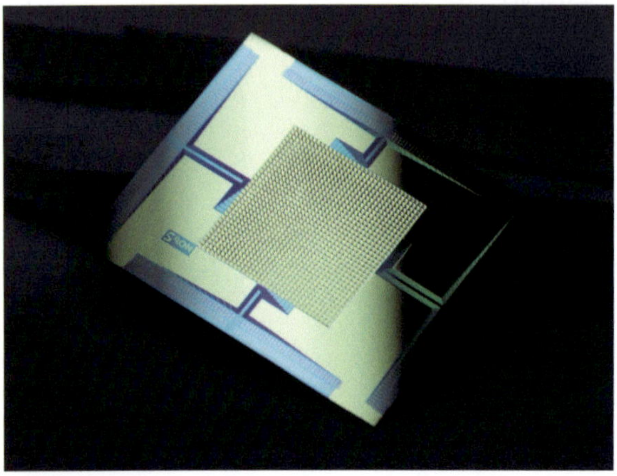

Cryogenic detectors are used to study the properties of neutrinos, to search for dark matter particles (↓) or to measure the energy of heavy ions. They can also be found on satellites for some areas of astrophysics, such as the precise investigation of the cosmic microwave background or high-resolution spectroscopy of astrophysical X-ray sources. The detectors are operated at low temperatures below about one tenth of a degree Kelvin above absolute zero, in order to achieve the greatest possible relative temperature increase for a given energy transfer.

Dark matter in the Universe is most commonly assumed to consist of so far unknown elementary particles. The only thing we know for sure about them is that they do not interact electromagnetically. If they exist, they have been created in the early Universe by some other interactions. These non electromagnetic interactions would not create measurable charges but a temperature rise in a detector material. Cryogenic detectors are therefore particularly suitable for the direct search for dark matter.

Cryogenic detectors can be produced from almost any solid-state material and then adapted to the physical problem. For example, in the search for neutrinoless double beta decay (↓), only a limited number of isotopes can undergo this decay. With cryogenic detectors it is possible to realize detectors from materials with

Camera with 1200 pixels from very small cryogenic detectors. Such cameras are used to generate images in X-ray light on satellite-based X-ray telescopes.

Interactions Between Particles and Matter → p. 68
Semiconductor Detectors → p. 70
Scintillators and Photomultipliers → p. 74
Dark Matter → p. 282, Direct Search for Dark Matter → p. 310
Neutrinoless double beta decay: Searching for a New Experiment → p. 296

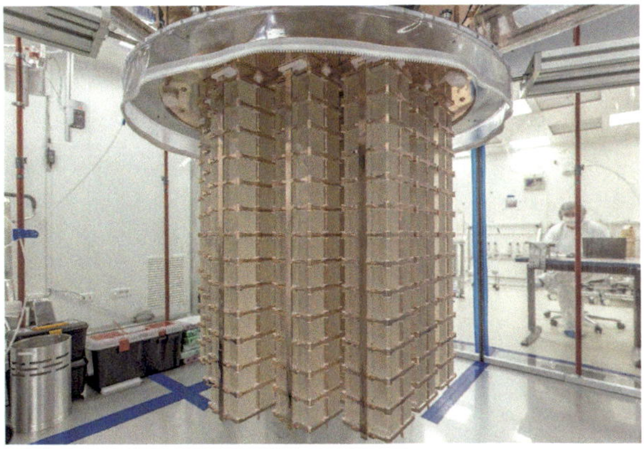

In the CUORE experiment—here during installation—around 1000 cryogenic detectors, each made from a TeO_2 crystal weighing one kilogram, are operated at a temperature of 20 mK.

some of these isotopes in which case source and detector material coincide. A prominent example of this is the CUORE project in the Gran Sasso underground laboratory in Italy, which operates the world's largest system of cryogenic detectors to date with around one ton of tellurium oxide (TeO_2).

Significantly lower energy thresholds and better energy resolutions are possible with cryogenic detectors than with conventional detectors. This is the decisive advantage that lead to the growing importance of cryogenic detectors in astrophysics. They are therefore also being investigated as a way of determining the neutrino mass using high-resolution beta spectroscopy (↓). Almost any measurable parameter that changes with temperature can be used to accurately measure the temperature increase. Most widely used is the temperature dependent electrical resistance of a semiconductor or even of a superconductor. The latter are metals that change their phase below a critical temperature to so-called *superconductivity*, whereby the electrical resistance disappears completely. The change in resistance is extremely large with small temperature fluctuations, which can be used for a very precise temperature measurement.

When attempting to detect dark matter particles directly, combinations of the detection principle of cryogenic detectors with conventional detection methods have become established. For example, the amount of light or charge generated is measured simultaneously with the temperature increase. The ratio of temperature increase to the amount of light or charge varies when measuring different particles, which can thus be distinguished. This greatly expands the possibilities for investigating particles and increases the sensitivity to previously unknown particles. It is not about playing the individual detection principles off against each other—as is so often the case in life, the combination makes the team stronger.

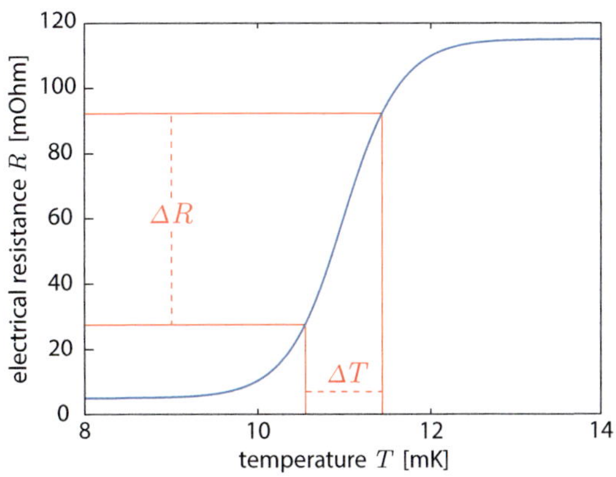

At the critical temperature, in this case around 11 mK, the resistance of a superconductor changes significantly. A small change in temperature leads to a very large measurable change in resistance.

Measurement of the Neutrino Masses → p. 256

Liquified Noble Gases as Detectors

Rare Liquids to Search for Rare Particles

Researchers in particle physics often need very large detectors. This can be necessary in order to effectively stop and analyze highly energetic particles that are generated in reactions at accelerators, or simply because the reactions they are looking for are extremely rare. For example, to detect neutrinos or to search for reactions of hypothetical dark matter particles, the chance of seeing a signal is greater the larger the detector (↓).

Measurable signals of particle reactions are generated in most detectors by ionization, i.e., the generation of free electrons and ions, or the generation of light (↓). Free charges drift in an electric field to electrodes, where they are measured as a current signal, while the photons of light are detected by photomultipliers. However, it is not always possible for free charges to drift over long distances. Light, on the other hand, propagates uninhibitedly. Thus large-volume detectors are usually scintillation detectors (↓).

In many cases, they are made of organic liquids, as this makes it possible to realize large detectors at relatively low cost. For many applications, however, the elements of organic scintillators, mainly hydrogen, carbon and oxygen, are not a good choice to effectively stop the particles. Liquid noble gases are used instead, in particular argon and xenon. When cooled to about −200° C, argon and xenon become liquid and are very good scintillators. Due to their high density and, in comparison to organic liquids, high atomic number, high-energy particles can be stopped over a shorter distance in order to measure their energy via the scintillation light.

Check of the photomultipliers to measure the scintillation light from the XENON detector to search for dark matter particles

In the direct search for dark matter (↓), one method is to detect the scattering of the particles from atomic nuclei. During scattering, the dark matter particle transfers part of its kinetic energy to the atomic nucleus. As in billiard, the energy transfer during scattering is larger if the masses of the particles and the atomic nucleus are similar. With xenon detectors in particular—the atomic mass of xenon is 130 GeV/c^2—it is therefore possible to search very effectively for relatively heavy dark matter particles in the range around 100 GeV/c^2, as is done with the XENON experiment. Liquid argon with an atomic mass of 40 GeV/c^2 is another excellent detector material for the search for somewhat lighter dark matter, as for example in the DEAP experiment.

Large Detectors Deep in a Mountain → p. 132
Interactions Between Particles and Matter → p. 68
Scintillators and Photomultipliers → p. 74
Dark Matter → p. 282
Direct Search for Dark Matter → p. 310

The noble gases xenon and argon are extracted from the atmosphere. A special feature of argon is that in addition to the main component ^{40}Ar, traces of ^{39}Ar are also obtained. ^{39}Ar is formed by cosmic radiation in the atmosphere. It is unstable and decays with a half-life of 269 years through beta decay.

For the detection of very rare events, such as in the search for dark matter, this represents an interfering background. For this reason, the argon for the DarkSide experiment is not obtained from the air, but from underground sources protected from cosmic radiation. Xenon also consists of a number of isotopes. Of particular interest is ^{136}Xe, as it decays via double beta decay. Therefore, liquid xenon detectors enriched in ^{136}Xe offer an excellent opportunity to search for neutrinoless double beta decay (↓), as is done with the nEXO experiment.

In most cases, the signal detection is not limited to the scintillation light, but also uses the charge generated simultaneously by ionization. The difficulty is to transport the electrons generated in the large-volume detectors over long distances to readout electrodes. To do this, the liquid noble gas must be very pure and the electric field very high. If the electrons that have been generated along the track of a particle in the detector drift to electrodes, a projection image of the original particle track is created on the wall of the electrodes, similar to a shadow.

Together with the prompt scintillation light as a timing signal, the particle track can be reconstructed in three dimensions. Such detectors are called TPC—*time projection chamber* (see Gaseous Detectors ↓). Such a TPC is being built for the neutrino oscillation experiment DUNE (↓). The neutrinos are generated in an accelerator at the Fermi Laboratory and detected 1300 kilometers away in a TPC detector. With the possibility of track reconstruction, particles that do not come from the direction of the accelerator beam can be sorted out. TPC detectors are a further development of the century-old principle of making particle tracks visible in cloud chambers (and later in bubble chambers), taking advantage of the fact that visible droplets or vapor bubbles form along the track of the ionizing particles under suitable conditions.

Liquid argon detector of the DEAP experiment. 255 photomultipliers enclose the volume with 3.6 tons of liquid argon.

Neutrinoless double beta decay: Searching for a New Experiment → p. 296
Gaseous Detectors → p. 72
Verifiying Neutrino Oscillations → p. 254

Particle Traps

Particles Locked Up in a Cage

Often one imagines that particle physics can only be conducted with huge accelerator facilities and collaborations involving several thousand members. However, exciting experiments can also be carried out with table-top experiments in which only a few researchers are involved. New particles and their interactions are not only measurable when they are produced at high energy in reactions at accelerators. They also leave traces in properties of particles well known to us, such as the proton and electron. These properties can be investigated in smaller apparatuses. This is due to the quantum field theory (↓) underlying the Standard Model.

An example of such experiments are particle traps, in which individual particles can be confined and studied in a small volume of only a few cubic centimeters in size. By a clever combination of electric and magnetic fields, it is ensured that the particle cannot leave a predetermined volume and can be studied for a long time.

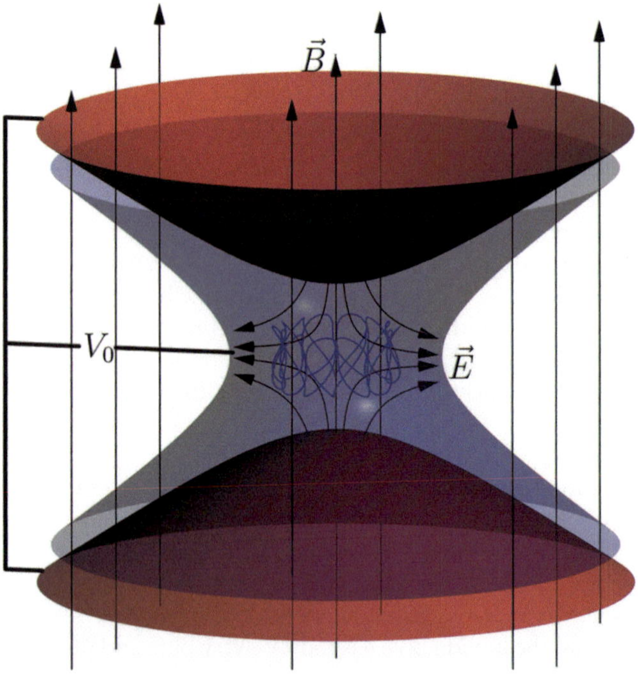

Particle motion (blue curve) in a Penning trap with electric and magnetic field. The entire arrangement is only a few centimeters in size.

Through the arrangement of an electric field as shown in the picture, which is realized by means of electrodes, and a magnetic field along the axis of symmetry (the vertical direction in the image), a charged particle follows a complicated trajectory in a small volume. Such an arrangement is called a ***Penning trap***, named after its inventor. Another way to store particles is the ***Paul trap***, named after the Nobel laureate Wolfgang Paul. It operates with alternating electric fields.

In particle traps, stored particles, similar to the electrons in atoms and molecules, can occupy various energy levels (see figure, Quantum Mechanics ↓) which can be measured very accurately. However, there is an important difference: While in an atom the atomic nucleus, which is responsible for the binding of the electrons, is predetermined by Nature, in a trap one can, so

Quantum Field Theory → p. 104
Concepts of Quantum Mechanics → p. 50

to speak, build one's own atom. One can arrange a field configuration adapted to the question at hand.

Particle traps are used in many areas of physics, for example in the construction of quantum computers. They allow one to build very accurate clocks. In particle physics, the precise determination of masses or the ratio of electric charge to mass is of interest. A precise knowledge of particle masses is important for our understanding of element formation (Nucleosynthesis ↓).

Individual antiparticles, such as antiprotons, can also be stored and measured in traps. The comparison with measurements on the proton then allows conclusions to be drawn about possible symmetry violations (↓).

In a particle trap, the anomalous magnetic moment of the electron could also be determined with great accuracy by measuring the transition of an electron between two energy levels in the trap. Since it is one of the best-determined quantities in physics, it is worth mentioning here the experimental numerical value with the measurement uncertainty (↓) given in parenthesis:

$$a = 0.001\,159\,652\,180\,62(13)\,.$$

Elsewhere (Precision Experiments ↓), it is explained that this quantity can be calculated with similar precision within the framework of the Standard Model. The agreement between theory and experiment is one of the great triumphs of the Standard Model of particle physics.

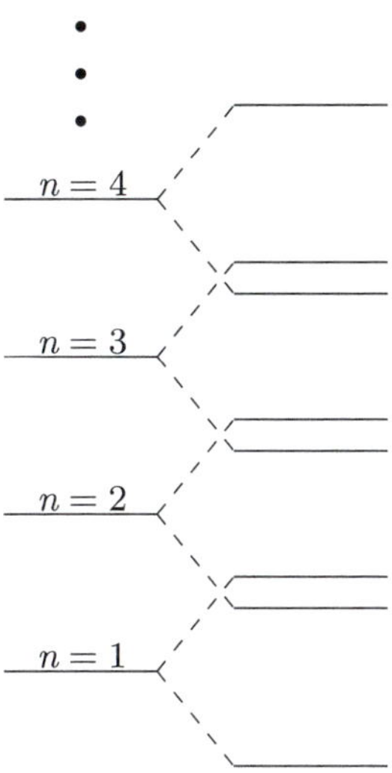

Simplified energy level diagram of an electron in a Penning trap

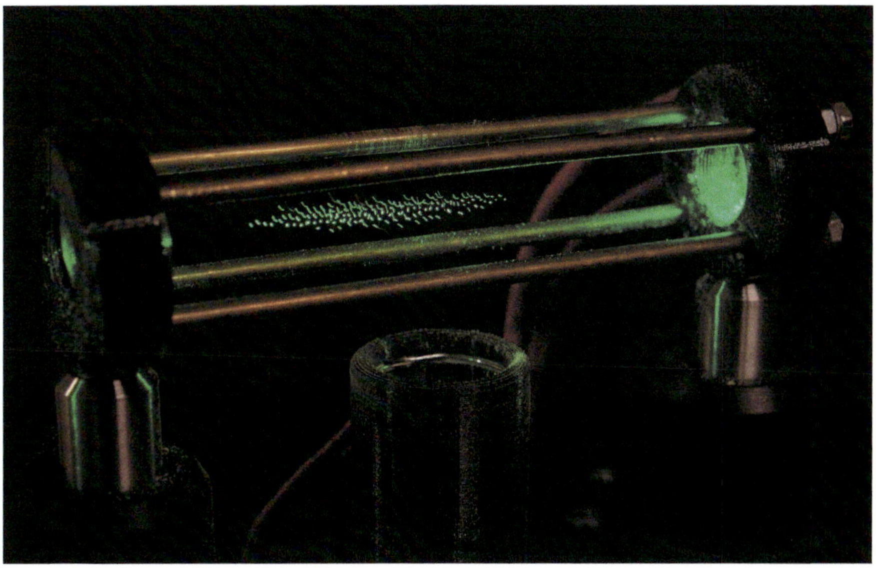

Flour grains stored in a Paul trap

Nucleosynthesis After the Big Bang → p. 330
Symmetries → p. 60
Measurement Uncertainties → p. 86
Precision Experiments → p. 250

A Penning Trap at CERN

Confining Single Particles in a Small Volume

The picture shows a schematic of two Penning traps designed to measure the so-called charge-to-mass ratio q/m of protons and antiprotons. The CPT theorem (Antimatter ↓, Symmetries ↓) predicts this ratio to be the same for a pair of particle and antiparticle like proton and antiproton. The impressive result of the experiment is

$$-(q/m)_p / (q/m)_{\bar{p}} = 1.000\,000\,000\,003(16)\,,$$

i.e., the ratios are indeed the same to 16 parts of 1 000 000 000 000. This device allows to store and analyze single antiprotons or negatively charged hydrogen ions (bound state of two electrons and a proton). From these two measurements the charge to mass ratios $(q/m)_{\bar{p}}$ and $(q/m)_p$ can be determined. Note that the size of the figure corresponds to more than twice the size of the actual trap.

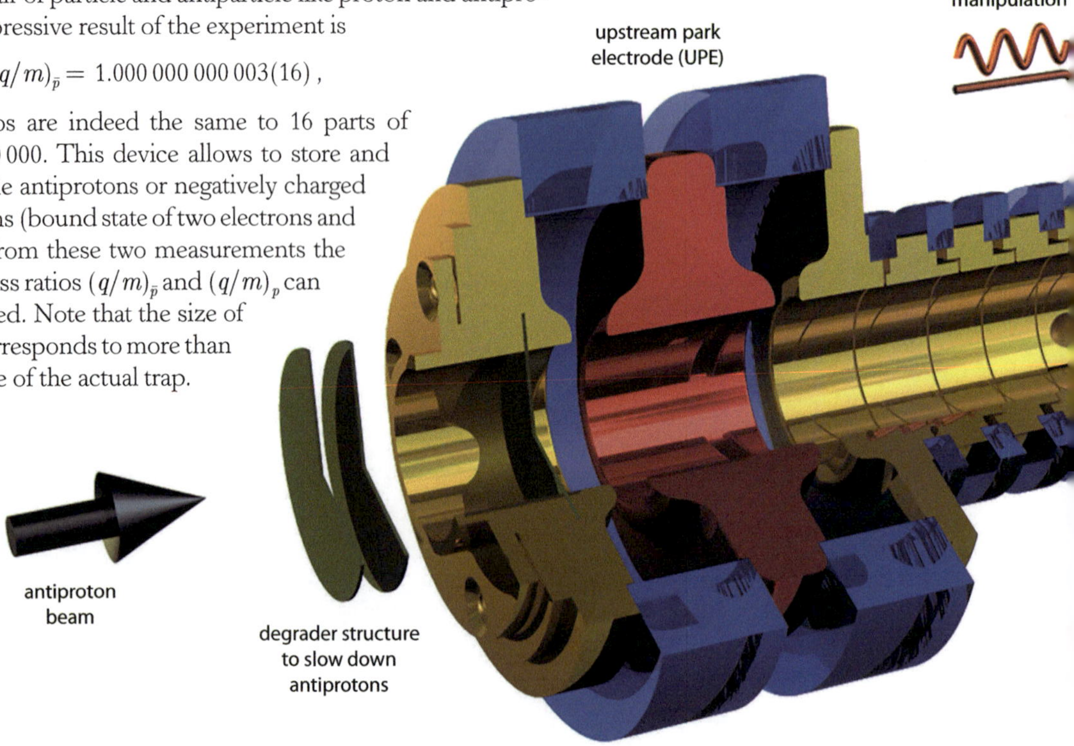

What Is Antimatter? → p. 22
Symmetries → p. 60

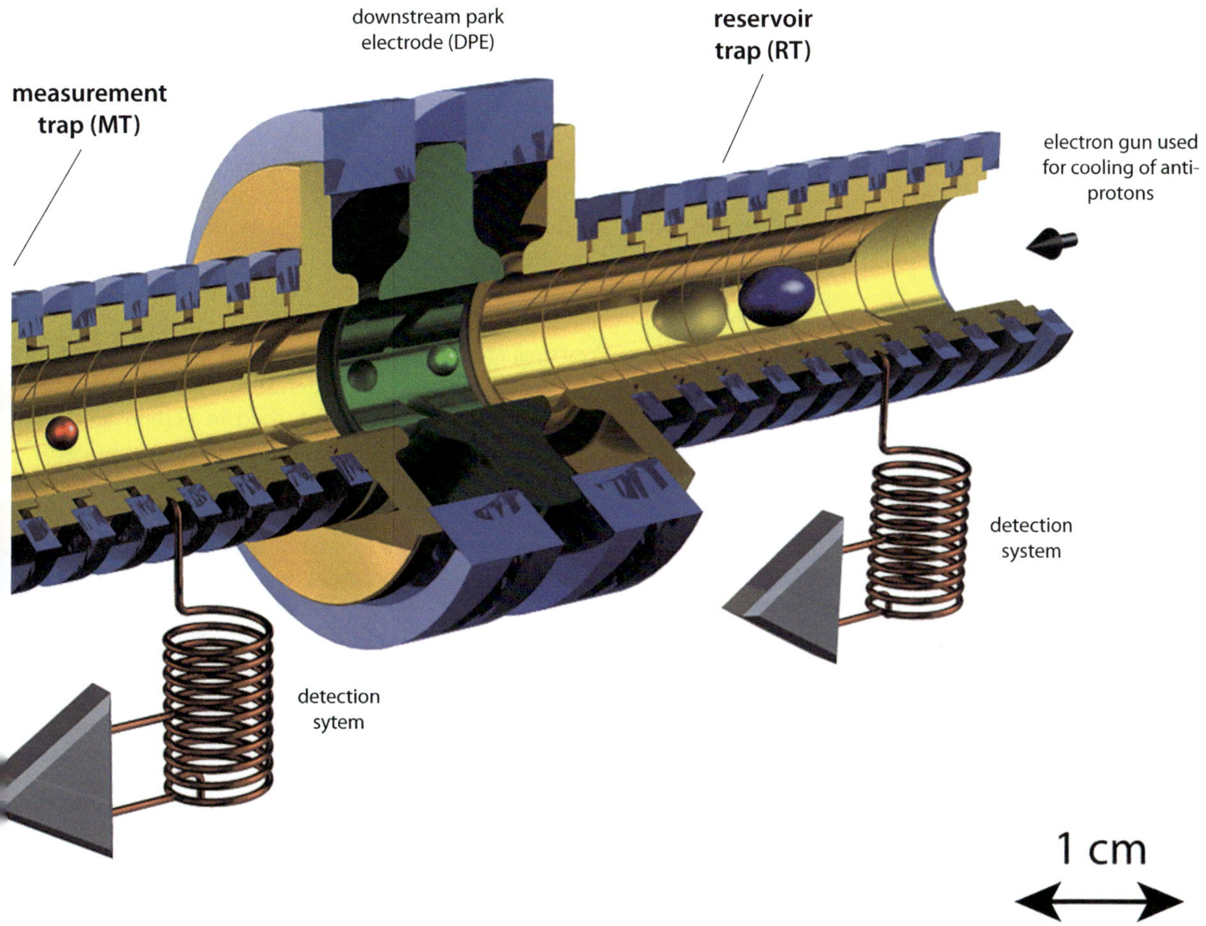

A Penning trap (↓) for antiprotons of the BASE collaboration operated at CERN.

Particle Traps → p. 140

Positron Emission Tomography

Matter-Antimatter Annihilation Inside the Body

Like everything in the world around us, we are made of matter, the components of which are essentially made up of protons, neutrons and electrons. However, in Nature there is also antimatter (↓) in the form of antiparticles, which exist for each of the known elementary particles. Fortunately, antimatter does not occur in large quantities, but only as individual particles in special reactions.

When antimatter meets matter, they annihilate to radiation (↓). As ominous as annihilation sounds, it can be used for three-dimensional imaging of physiological processes inside the body. Thus tumours can be localized at an early stage, or disorders of brain function can be detected.

The idea behind *positron emission tomography*, or PET for short, is to introduce antimatter in the form of positrons, the antiparticles of electrons, into specific structures in the body. As a charged particle, the positron interacts with its environment, is decelerated within a very short time, and then annihilates with one of the electrons present everywhere to form photons. Due to the conservation of energy and momentum, two photons are generated[1], which fly away from each other back-to-back. If one measures the two photons at locations A and B, one knows that the location of the positron's annihilation must lie on the line connecting A and B, as shown in the image below. Before it is annihilated with an electron, however, the positron must be slowed down. It typically travels only one to two millimeters. Therefore, the place where the positron is annihilated is not far from the place where it had been created. The accuracy of PET's location reconstruction is therefore limited to one to two millimeters.

Positrons are used here because they are the most easily available form of antimatter. If the usual balance of neutrons and protons is disturbed in certain atomic nuclei (↓), neutrons and protons transform into each other in β^- or β^+ decays until the number of neutrons and protons is balanced. As with all particle reactions, the conservation laws for charge and the symmetry of particles and antiparticles are observed via β^- or β^+ de-

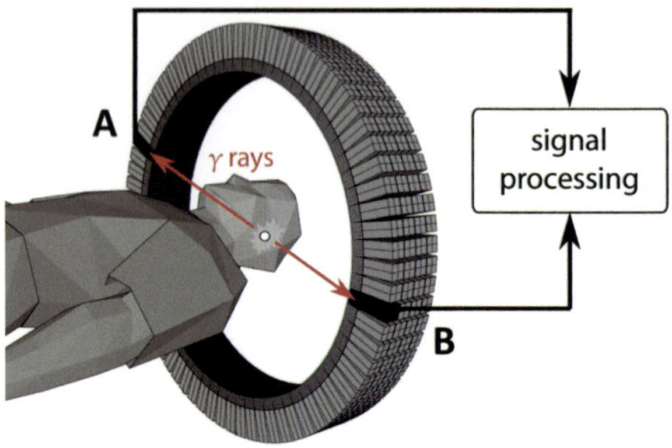

After the patient has been injected with the β^+ emitter, they are placed on a couch in the PET system , where they are surrounded by photon detectors. The two photons from the positron-electron annihilation lead to coincident, i.e., simultaneous, signals in two detectors separated by 180°.

[1] In less than 1 % of the cases, three photons are generated.

What Is Antimatter? → p. 22
Annihilation matter-antimatter: Feynman Diagrams II → p. 102, Quantum Field Theory → p. 104

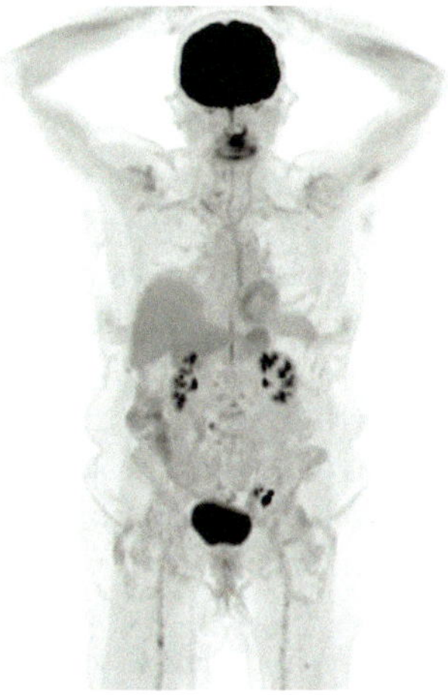

The PET image shows a patient after intravenous injection of glucose molecules containing a β^+ emitter. In addition to the normal accumulation of glucose in the heart, bladder, kidneys and brain, metastases of a tumor at the anus are also visible.

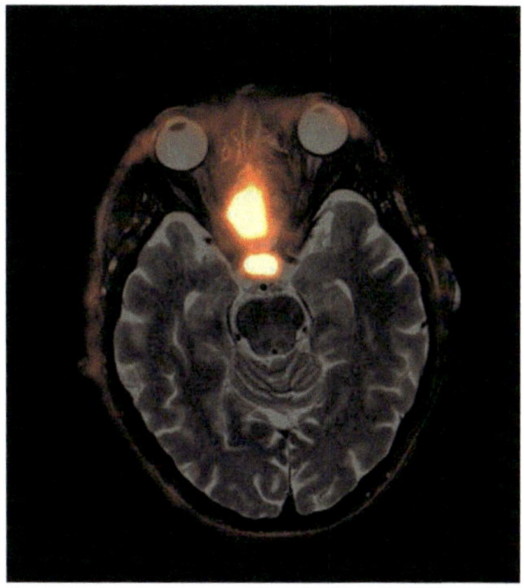

Overlay of a PET and magnetic resonance image of a brain with a tumor. While the MRI image provides a high spatial resolution and good tissue contrast, the PET image marks the tumor in bright orange.

cay. In the case of β^- decay, i.e., the transformation of a neutron into a proton, an electron and an additional antineutrino are emitted. As neutrinos are very difficult to detect (↓), not much can be done with antimatter in the form of the antineutrino. The situation is completely different with β^+ decay, i.e., the reverse transformation of a proton into a neutron. This produces a neutrino and a positron.

During a PET scan, the patient is administered so-called *radiopharmaceuticals* beforehand. These are molecules including a β^+ emitter that the body cannot distinguish from the non-radioactive versions, therefore they participate in normal metabolism. The β^+ emitter often used is ^{18}F, which decays to ^{18}O with a half-life of 110 minutes (↓). The molecules into which the ^{18}F is incorporated are selected depending on the body location to be examined via PET. If one takes FDG molecules (^{18}F-2-fluoro-2-desoxy-D-glucose), for example, which participate in glucose turnover, they will accumulate together with the ^{18}F contained therein at sites with high energy turnover. This allows tumors to be represented on the basis of their active cell growth or brain activity due to the energy consumption. Other molecules can be used to select uptake in certain organs or specifically in bones.

The patient is moved into the PET system on a stretcher, where he or she is surrounded by an array of detector elements for photon measurement. Photons registered simultaneously in opposite detector elements are interpreted as a positron annihilation signal on the connecting line between the two detector elements. By superimposing the connecting lines of many signals, the three-dimensional structure of the distribution of the radiopharmaceutical in the body can be reconstructed.

In more recent developments, PET and magnetic resonance imaging (MRI, ↓) are combined in order to make use of the advantages of both, functional labeling by PET and better spatial resolution and good tissue contrast with MRI.

From Nucleus to Nuclei → p. 226
Neutrinos → p. 196
Accelerators in Medicine → p. 122
Magnetic Resonance Imaging → p. 146

Magnetic Resonance Imaging

Looking Inside the Body With Spin

As abstract as the spin (↓) of particles may seem, it can nevertheless be used very concretely to look inside the human body. Since Wilhelm Conrad Röntgen first discovered the possibility of painlessly examining the inside of the body with X-rays (↓), a whole series of other methods have been developed, each with its specific advantages. One of these is *magnetic resonance imaging* (MRI). It has become indispensable in medical diagnosis. Whether for examining joints, the brain or abdominal organs, MRI has become a workhorse of medical imaging.

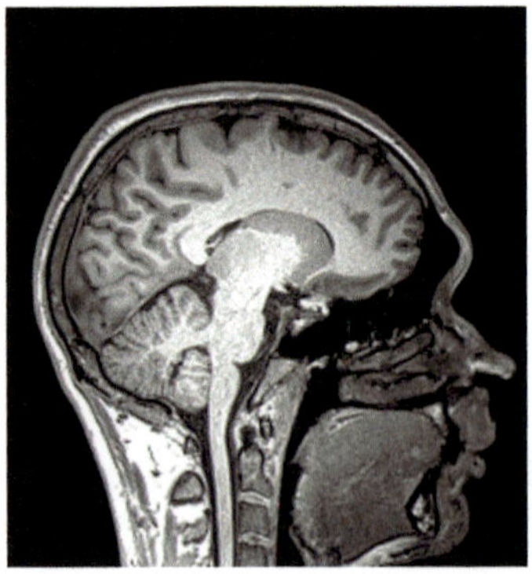

Sectional image of a head with MRI. The different shades of gray reflect the respective density of protons.

Our body contains atomic nuclei whose elementary building blocks have a spin. The vast majority of atomic nuclei in the body, namely 63 %, are hydrogen, consisting of only one proton. MRI is used to measure the spatial distribution of proton spins in our body. Depending on how the protons are bound in the body, in water molecules, proteins or fats, their density, which can be measured with MRI, varies. Therefore, the distribution of protons also provides a spatial image of the structures in the body. In addition to their spatial distribution, the environment of the protons can be explored, which allows conclusions to be drawn about the type of tissue. This allows various organs, vessels and bones to be differentiated and displayed in three dimensions.

The spin of protons behaves like a spinning top. A spinning top usually not only spins around its own axis, but often dances, with the axis of rotation performing a further circular motion. This movement, called *precession*, is caused by the weight acting on the top.

Weight does not play a role in elementary particles, but magnetic fields can be used to exert forces on them instead. The spin of the particles is linked to a magnetic moment, the orientation of which can be influenced by a magnetic field in a way similar to a compass needle. If a strong magnetic field is applied, a sufficient fraction of magnetic moments

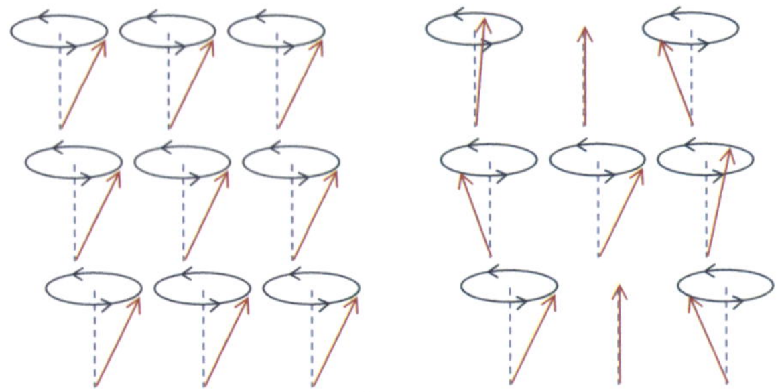

The applied magnetic fields cause the spins to move in precession in unison, emitting high-frequency signals. Over time, the spins get out of sync or stop moving, causing the signal to decay.

Spin → p. 44
X-Rays → p. 124

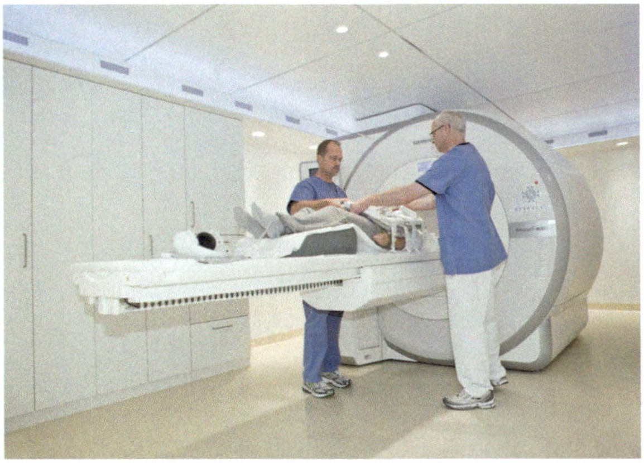

A patient to be examined is moved into the magnetic field. The coils for generating the magnetic fields and measuring the signals are located around the patient tube in the MRI machine.

and thus of spins is initially aligned in the direction of the magnetic field. Subsequently a short magnetic pulse of a high frequency is applied, which is perpendicular to the original field. This tilts the spins and thus initiates a precession movement. The proton spins rapidly orbit in unison around the direction of the original external magnetic field, which leads to measurable high-frequency signals.

The laws of quantum mechanics tell us that the frequency of the precession movement is proportional to the strength of the original magnetic field. The frequency of the captured signal therefore reveals how strong the magnetic field is at the location of the protons.

In MRI examinations, the magnets are arranged in such a way that the magnetic field has a different strength at each location, which means that the precession frequency varies from location to location. The intensity of the signal of a certain frequency thus reflects the number of protons at the corresponding location. In this way, the spatial density distribution of the protons or hydrogen atoms is measured and thus the distribution of the various molecules, such as water, proteins or fats, which contain hydrogen.

But there is more information in the signals. How quickly the signals decay again after a start pulse depends on the surrounding tissue structure. Two effects lead to the decay of the signals, which can be measured separately and thus contribute to the recognition of the tissue types. Firstly, the circulating spins become out of sync over time. If all spins no longer point in the same direction at all times, the generated signals become weaker. Secondly, the spins flip back to the original direction of the magnetic field over time and no longer contribute to the signal. The two effects occur at different speeds, which is why their respective contributions to the decay of the signals can be measured separately. The strength of these two contributions allows conclusions to be drawn about the type of surrounding tissue.

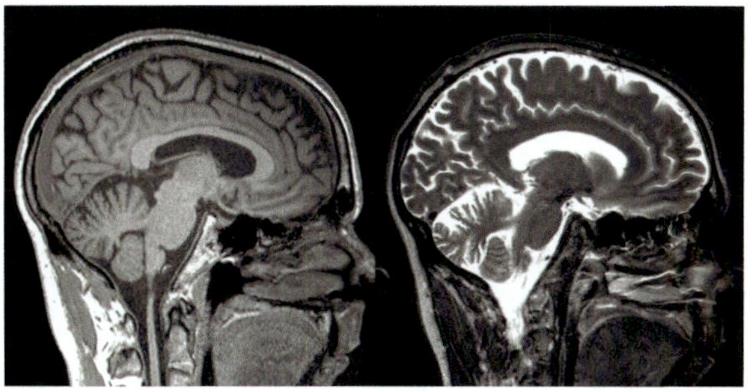

Sectional images of the brain. Same patient and same slice with different tissue contrasts, i.e., different weighting of the two decay processes, so that different structures and tissues stand out in the image.

6 Recording and Analyzing Data

The previous chapters explain the basics of particle interactions and the types of detectors that make up particle physics experiments. However, physical results do not simply emerge from the experiment on their own: detectors generate electronic signals that must be recorded, filtered, collected, and interpreted. These steps are collectively referred to as *data acquisition, reconstruction* and *analysis*. This chapter illuminates the path from recording individual interactions, over the treatment of statistical uncertainties, to the interpretation of the results in light of known or new hypothetical physical theories. Along the way, we discuss how events are selected to be stored, how the interesting events and particles for a specific measurement are chosen, and how elementary measured quantities of the individual recorded particles are determined.

Trigger and Data Processing

In Search of the Needle in the Haystack From the Detector to the Hard Drive

Particle physics experiments are used to record the interesting properties of particles and their interactions. The detector captures an image of each individual reaction between particles. However, usually there is either one of two opposite challenges for which a similar solution can be found: Either something interesting happens only very rarely (such as in the search for an interaction of neutrinos (↓) or dark matter), or something happens almost constantly, and it must be decided which event is interesting and which is uninteresting. In both cases, the question arises: When and of what should the detector record a picture? Is there a particle reaction at all, or is a given particle collision interesting enough to keep the reconstructed information about the measured processes on a storage medium? The device tasked with this decision is called a *trigger*.

A typical trigger consists of several stages and, in most experiments, is closely linked to the readout of the data from the detector and the subsequent processing of the data in computers. This step after the trigger is called *data acquisition*, that is, the extraction of the data from the detector of a given interaction or in a given timeframe. This is then followed by the calibration of the reconstruction algorithms, and the reconstruction of the recorded events. In an experiment like ATLAS or CMS (↓) at the LHC (↓) at CERN, about 1.7 billion collisions of protons occur per second. Data are recorded ideally almost 24/7, with only short breaks when the accelerator is filled with new particle bunches. This results in about one petabyte of data being created per second inside the detector. That would be the content of 1000 one-terabyte hard drives each and every second. This amount of data cannot be stored in its entirety. It cannot even be extracted sufficiently quickly from the chips built into the detector. However, a large part of the events are not particularly interesting for the evaluation: For example, it is about a billion times more likely that some already well-known interaction happens at the LHC than that a Higgs boson is produced, which is actually interesting for today's researchers.

The trigger is supposed to find and record as accurately as possible all events with interesting particles. Making sure that the trigger is well-understood is a very important task: It would be a tragedy if unexpected signs of yet unknown physics would be produced in the experiment, but discarded by the trigger, so that we would never know that we missed a major discovery (↓)! Only a few of the uninteresting ones are also recorded, those that were either misidentified or necessary to check the function of the trigger. In the end, a tremendous effort is made to produce, measure, and then almost completely discard 1.7 billion collisions per second, so that there are about 2000 "interesting" events per second which are selected, recorded, and can be used for a wide array of physics. Amongst them, there is about one Higgs boson per second.

Neutrinos: Large Detectors Deep in a Mountain → p. 132
Typical Detectors at Accelerators → p. 128
The LHC Accelerator Complex at CERN → p. 116
Searching for New Particles at the LHC II → p. 304

Illustration of the function of the trigger: the detection of the needle in the haystack (in the lower middle image) made easier by reducing the image to the essentials.

The photos of the search for the red needle in the haystack illustrate how the trigger does this: The upper row shows three different "events". The full information of the events (color, shape, resolution) is shown. This cannot be extracted from the detector because of the enormous amount of data. Therefore, in the first trigger stage, only the most relevant information is extracted. This is represented in the second row by only the grayscale information and the red color. Figuratively speaking, physicists expect their desired events to contain a red component. In reality, this could be, for example, information from calorimeters (Energy Measurement ↓) and special components of the muon detector.

Even in the middle row, the needle is still hard to find, and the events are still very detailed, so they still contain too much information to be extracted from the detector. Therefore, the data are further reduced, with the simplification optimized so that the properties of the real sought-after events are preserved as well as possible. This is represented by the bottom row: The contrast is reduced, and the resolution is greatly simplified. Lo and behold: In the lower-middle image, only the needle can still be identified as red and stands out. The trigger can now recognize that something interesting seems to have happened here. Therefore, this event is completely read out with all information from the detector and further processed, while the other two events are discarded. However, in the image seen by the trigger, the contrast is still not very strong. Therefore, once the event has been recorded, the full detector information is read out, the subsequent calibration and reconstruction further improve the image with a lot of effort, and ideally the needle in the haystack can be found.

Illustration of the fully reconstructed image: The stored images reveal the full information in the analysis.

Energy Measurement → p. 78

Data Analysis

Fishing for Discoveries in a Sea of Information

Particle physics experiments often record very large amounts of data. At the ATLAS and the CMS experiments at the LHC (↓) at CERN, about 2000 collisions of protons per second are recorded, which lead to about 60 Terabytes of data per second that are read out from the detectors. The evaluation of this huge amount of particle physics processes from these collisions is called *analysis*. The goal of an analysis is to gain insight into the physics of fundamental particles and their interactions from the recorded and reconstructed information. An example for such an analysis would be how often specific particles are produced in a given kinematic configuration. From this we can learn how particles interact with each other. The measured variables include the mass of a decaying particle (↓), the momenta of the final state particles or more complex information. In order to structure this information, particle physicists also use machine learning methods such as neural networks or deep learning (↓).

Data analysis involves many individual steps, of which only two important ones shall be highlighted here: the *selection*, i.e., the choice of the events that make the most significant contribution to the result, and the *statistical analysis*. Typically, the goal is to measure a mass or coupling, to calculate an exclusion limit, or a significance (↓). However, a very large part of the work involves the precise estimate of uncertainties (↓) using simulation and data. Often it is the task of one or more doctoral students to carry out one of these analyses with increasing responsibility during the preparation of their doctoral thesis.

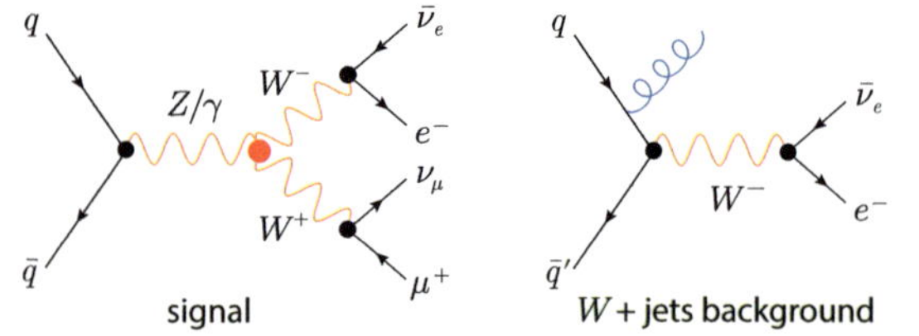

The generation of the signal (left) and the most common background (right). The examined coupling of three gauge bosons is indicated with the red dot.

As an example, a measurement by the ATLAS experiment is shown in the figures. It investigates whether events in which two W bosons are produced and decay into an electron plus electron antineutrino and a muon plus muon antineutrino, respectively, behave exactly as predicted by the Standard Model. The antineutrinos produced in this way can only be indirectly detected in the ATLAS detector through their "missing energy" (E_T^{miss})—that is, the energy not measured by the detector, which, however, must be carried by some product of the reaction due to the expected energy and momentum conservation. To select such WW-events, the researchers exploit that on average "signal" events (i.e., the events coming from the process under study) differ from "background" events (i.e., the events from any other process). Over several steps, they require that more and more properties of the retained events correspond to *all* of the expectations of the signal simulation. Each event that fails at least one measured property expected from the signal is removed. An example of a background process is the production of a single W boson (see the Feynman diagram on the right): This

The LHC Accelerator Complex at CERN → p. 116
Identifying the Particle Type via the Mass → p. 156
Neural Networks and Machine Learning → p. 170
Publishing a Non-Observation → p. 162 and Drawing Conclusions From Statistics → p. 160
Determining Measurement Uncertainties → p. 158

final state differs from the signal as it has only one charged lepton and that on average it has smaller E_T^{miss} values. The missing lepton might however be compensated by a misidentification of another particle as a lepton.

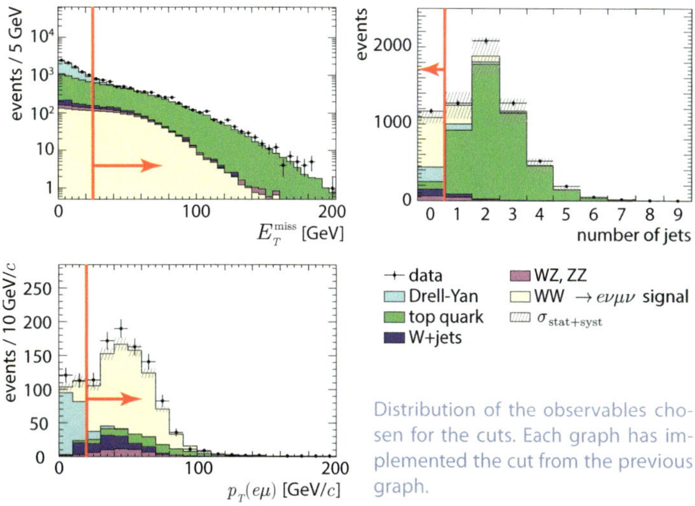

Distribution of the observables chosen for the cuts. Each graph has implemented the cut from the previous graph.

The figures to the right show such a selection step by step: The observed distribution (black data points; in color: expectations from simulations) of the missing energy, the number of jets (↓) and the momentum component of the electron and muon perpendicular to the accelerator's beam axis ($p_T(e\mu)$). The red arrows illustrate which parts of the events are selected. In each graph, only the events in the direction of the arrows are retained. There the signal shown in yellow is larger relative to the backgrounds, shown in other colors. This process is called *cutting* or *filtering*. Step by step, the huge amount of events is reduced to almost exclusively contain the signal events.

The second step shown below is the *fit*.[1] The goal is to calculate the significance at which a certain hypothesis can be distinguished from another hypothesis, based on the data. In the example shown, this is a variation of the number which parametrizes the coupling of three gauge bosons. Since the observed data (black points) match well with the sum of yellow signal and blue background expected in the Standard Model, large variations of the gauge boson coupling beyond the value expected in the Standard Model (↓) (purple lines) can be excluded. This is shown in the second graph: In the plane of parameters that indicate the possible strength of processes deviating from the Standard Model, the parameter range indicated by the green area is still allowed and the remaining area is excluded with $2\,\sigma$ significance. The point (0,0) in the parameter plane is included in the green area, hence, the Standard Model has not been excluded. At this point, the analysis of the data from the perspective of the properties of the production of two W bosons is completed and can be published.

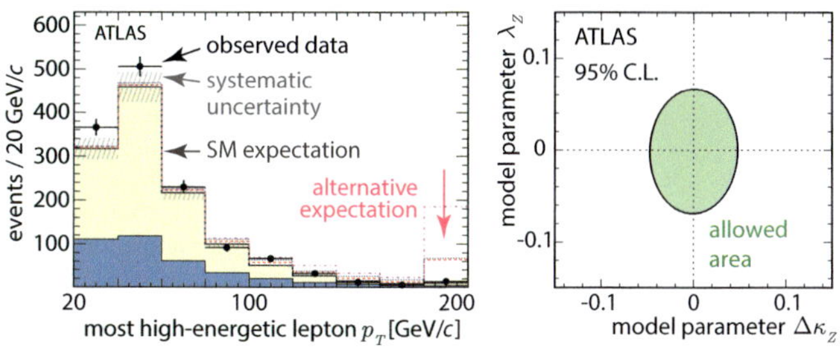

The fit of various signal hypotheses to the data (left) and the allowed range of theory parameters from the fit (green, right)

[1] G. Aad et al. (ATLAS Collaboration) 2013, *Measurement of W^+W^- production in pp collisions at $\sqrt{s}=7$ TeV with the ATLAS detector and limits on anomalous WWZ and WWγ couplings*, Physical Review D, 87,112001, doi:10.1103/PhysRevD.87.112001

Seeing Quarks and Gluons? → p. 216
Why the Standard Model? → p. 244

From Event to Result

A Graphical Overview

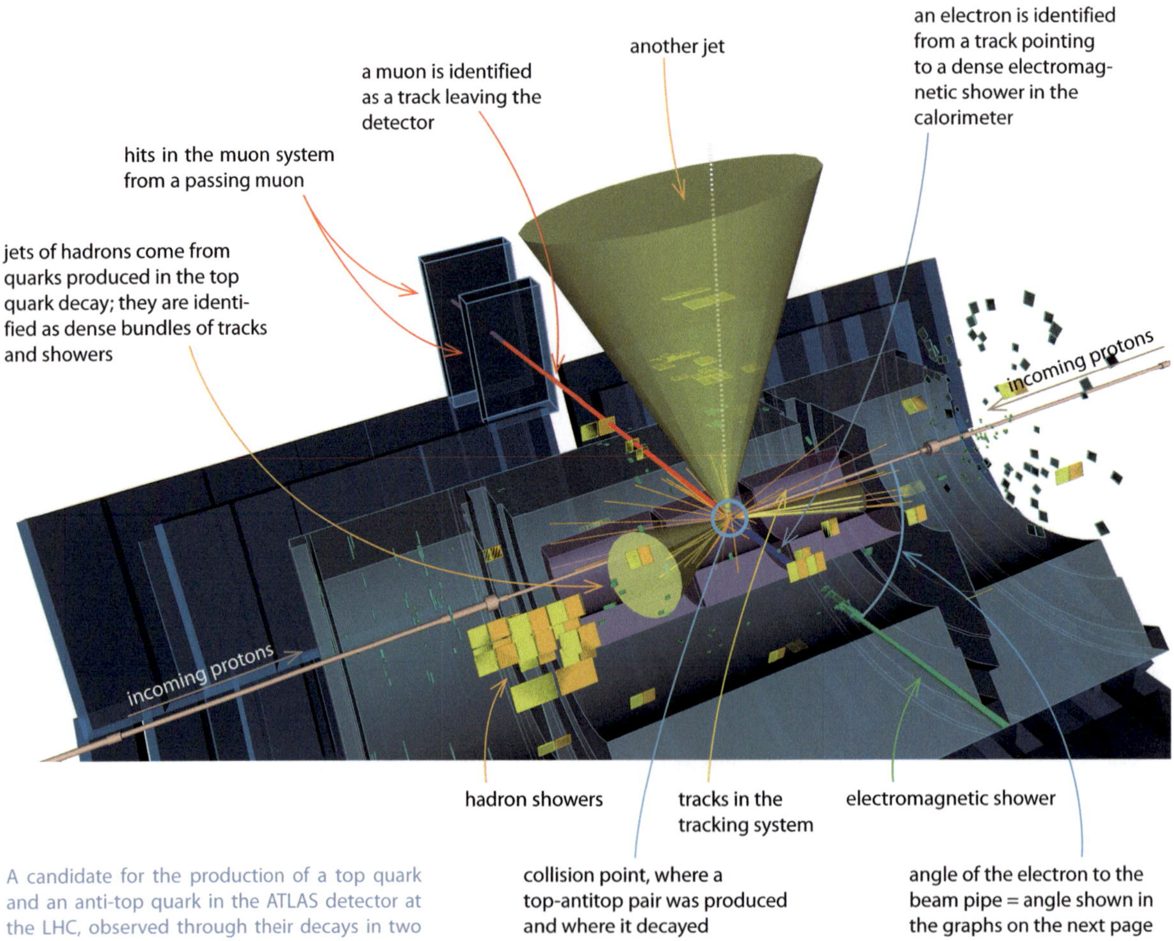

A candidate for the production of a top quark and an anti-top quark in the ATLAS detector at the LHC, observed through their decays in two jets (cones), a muon (red line) and an electron (blue line), respectively.

The ATLAS Experiment at the LHC → p. 130

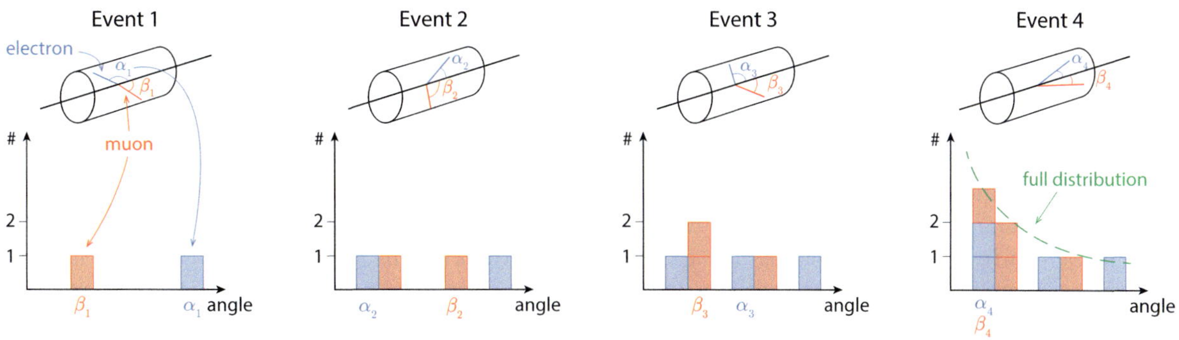

The reconstructed information from many events is combined in distributions of measured quantities. Here, the angle between a lepton (electron or muon) and the beam direction is measured.

Each event represents an entry in the histogram. Together, the distribution resulting from the combination of all events approximates the probability of finding an electron at a given angle.

The analysis of particle physics data can be complex, but much of it can be shown in graphical examples. The detector takes digital pictures of each collision. At the LHC, this happens 40 million times per second. The trigger (↓) selects the interesting pictures. Then, the pictures are analyzed (↓) by computer algorithms, and stored. The properties of the particles created in the events are then extracted, again by computer programs. Finally, all events that meet certain requirements—which may vary depending on the physics goal of the analysis—are used to create plots showing distributions of event properties in one or several dimensions. On this page, we show a simple example of the angle of a lepton (muon or electron) with respect to the beam direction in events that are compatible with the production of a top and an anti-top quark. The resulting distributions can then be compared to the probabilistic predictions of quantum field theory. Here, prediction and measurement agree: Thus, the Standard Model is preliminarily confirmed and the search for new phenomena continues.

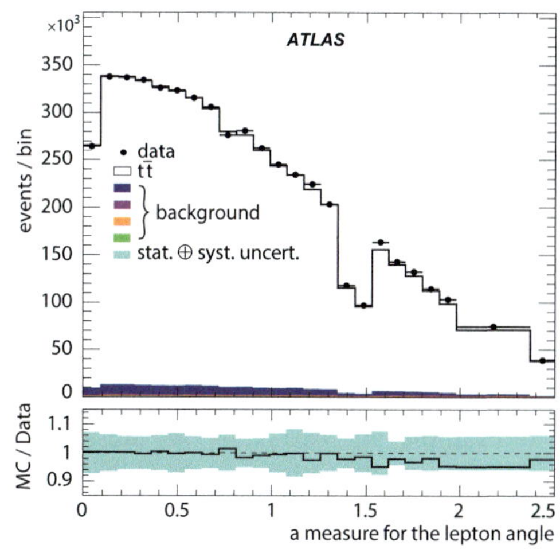

Combination of the information on properties of all electrons and muons in all candidate events: The cumulative distribution of a measure for the angle of all muons and electrons relative to the beam pipe is shown. Each lepton in each individual candidate event, as shown in the big event display, makes up just one out of hundred thousands of entries in this plot. All events together can be used to cross-check the prediction: The distribution of many events should approximately agree with the predicted probability to find a muon or electron at a given angle. Here this is the case: The data (dots) agree with the simulation based on the Standard Model theory (line).

Trigger and Data Processing → p. 150
Data Analysis → p. 152

Identifying the Particle Type via the Mass

Particles Playing "To Tell the Truth"

We have seen that Cherenkov detectors (↓) alone, under certain conditions, can identify the type of particle. However, there are many cases where they would be insensitive or where it is impractical to incorporate a detector optimized for Cherenkov radiation into the experiment. Many particles are also so short-lived that they decay into other particles before they reach a sensitive part of the detector.

One way to draw conclusions about the identity of a particle is through its mass. In combination with other measurable properties, such as charge, this is unique for each type of a particle. However, there is no "scale" in a particle physics experiment on which we could place the particle to determine its mass. So, how do we know which particle we are dealing with? This is where Einstein's famous formula $E = mc^2$ (Special Relativity ↓) comes into play. If a particle is not at rest, but has a momentum p (↓), this generalizes to

$$E^2 = (mc^2)^2 + p^2c^2 \text{ or}$$
$$m = \sqrt{E^2/c^4 - p^2/c^2} \, .$$

This means that if we can determine the energy E and momentum p of a particle, we immediately know its mass m. This method is also applicable if a particle decays before it reaches a sensitive part of the detector.

As an example, consider the K_S meson (Quark Model ↓): It decays after a short lifetime of $\tau_{K_S} = 9 \cdot 10^{-11}$ s, for example, into two pions (see illustration). The K_S is traveling at 99 % of the speed of light. The resulting time dilation extends the lifetime in the detector, leading to a decay length of two to three centimeters: too short to reach the detectors at LHC energies. In the image below, the illustration shows a collision event recorded at the LHC, in which a K_S has decayed into two charged pions. The red lines are the reconstructed particle tracks (Position and Momentum Measurement ↓) of the two pions, which originate from a K_S decay. They originate from a common point almost in the middle of the detector, in the vacuum of the beam pipe. The particle that decayed to the pair of pions has therefore never reached a sensitive part of the detector, yet as we now show, its mass can be determined and therefore it can be identified as a K_S.

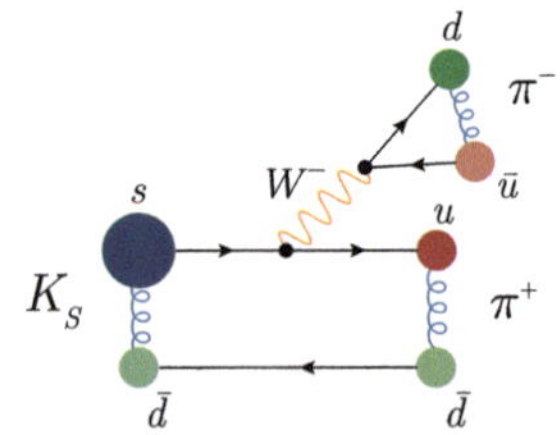

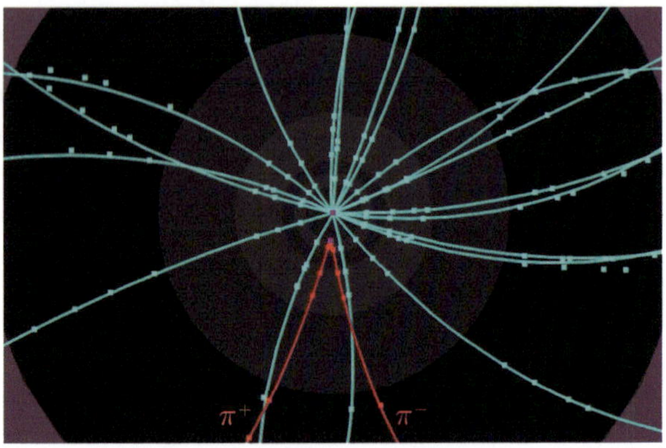

The decay of a K_S meson and its signature in the detector: Two tracks that do not meet where the other tracks meet.

Cherenkov Radiation and Particle Identification → p. 82
Special Relativity → p. 38
Kinematics → p. 40 and Special Theory of Relativity in Particle Physics → p. 92
The Quark Model Revisited → p. 202
Position and Momentum Measurement → p. 76

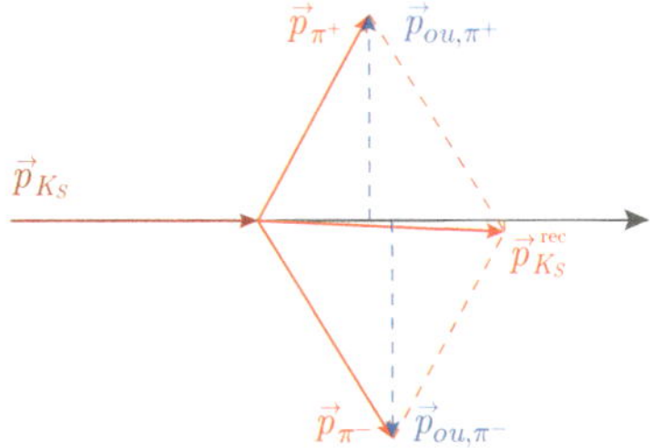

The calculation of the four-momentum of the K_S from the momenta of the decay products π^+ and π^-

The schematic drawing explains how this works: The K_S particle with momentum $\vec{p}_{K_S}$ decays into two pions with momentum $\vec{p}_{\pi^+}$ and $\vec{p}_{\pi^-}$. Since the momentum is conserved in the decay, the original momentum of the K_S, $\vec{p}_{K_S} = \vec{p}_{\pi^+} + \vec{p}_{\pi^-}$, can be determined from the sum of the pion momenta. The influence of measurement uncertainties (↓) is also illustrated, which results in the reconstructed momentum $\vec{p}^{\,rec}_{K_S}$ pointing in a slightly different direction and having a slightly different magnitude. The heavier the decaying particle compared to the pions, the greater the momentum of the pions can become. Since the K_S in our sketch only moves to the right and not up or down, the upward or downward momentum components p_{ou}, of the decay products cancel out: $p_{ou} = p_{ou,\pi^+} + p_{ou,\pi^-} = p_{ou,K_S} = 0$. This is also evident from the drawing with the measurement uncertainties: The sum of the observed pion momenta p^{rec} has no significant component upwards or downwards due to momentum conservation!

To calculate the mass of the K_S using our formula, the energies of π^+ and π^- and their momenta must be added separately. By combining the information from several different detectors, we can infer that the two particles are in fact pions. We can use the measured pion momenta and the known mass of charged pions (which is identical for positively and negatively charged pions) to get $E_{\pi^\pm} = \sqrt{(m_\pi c^2)^2 + p^2_{\pi^\pm} c^2}$. With this, we have all the ingredients for the mass determination of the decaying particle. The unknown particle has energy $E = E_{\pi^+} + E_{\pi^-}$ and momentum $p = |\vec{p}^{\,rec}_{K_S}|$. We can insert these two values into the equation for the mass. This mass is also called *invariant mass*, because it does not matter in which frame of reference we measure the momenta and energy: the invariant mass m is the same in all frames!

This method actually works: The measurement of the K_S mass from the ATLAS experiment at the LHC[1] shows that very many pairs of π^+ and π^-, which have a signature as in the event display shown here, are produced with a common mass of about 500 MeV/c^2. This peak is called the *resonance* (↓) of the K_S. The $\pi^+\pi^-$ pairs in the peak are exactly those from a K_S decay. The precise mass m_{K_S} is 497.611 ± 0.013 MeV/c^2. This method is indispensable for modern particle physics experiments: The Higgs boson, the top quark, the Z boson and many other particles and processes were first detected using exactly this mass reconstruction technique.

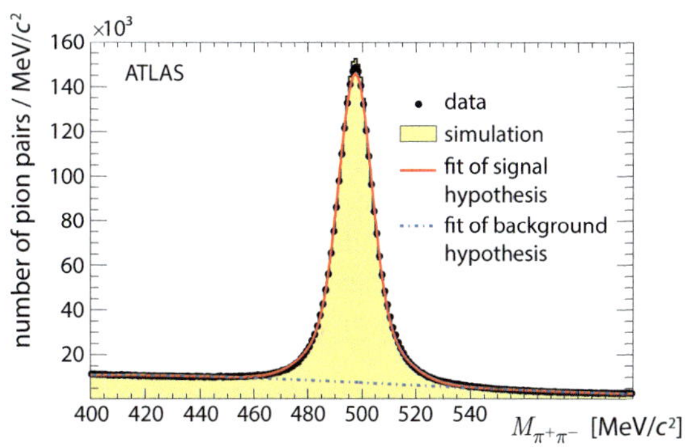

The mass distribution of the $\pi^+\pi^-$ pairs

[1] G. Aad et al. (ATLAS Collaboration) 2012, *K^0_s and Λ production in pp interactions at $\sqrt{s}$=0.9 and 7 TeV measured with the ATLAS detector at the LHC*, Physical Review D, 85, 012001, doi:10.1103/PhysRevD.85.012001

Determining Measurement Uncertainties → p. 158
Resonances → p. 56

Determining Measurement Uncertainties

Measure Better Rather Than Longer

Neither the prediction nor the measurement of a physical quantity is exact. The prediction is based on a hypothesis that probably does not exactly correspond to reality. The mathematical methods for calculating a measurement prediction may include approximations or imprecise input quantities. The measurement contains uncertainties (↓) that randomly fluctuate from measurement to measurement (statistical uncertainties) or the measurement setup is not exactly understood (systematic uncertainties). These uncertainties play an important role in the comparison of prediction and result (↓). Therefore, one of the most important criteria of a quantitative science is that not only something is measured, but that in particular the measurement uncertainties are determined as accurately as possible. The way this works will be explained here using a simple example: the oscillation of a pendulum (↓).

An important property of a pendulum is its period, that is, the duration of a complete oscillation from the left side to the right and back to the left. This duration can be predicted on the basis of "classical mechanics", as first developed by Isaac Newton, which is the theory about the relationship between forces and motion. The prediction for the period depends mainly on the strength of the gravitational force and the length of the thread. Weight, air resistance, and the initial maximum deflection also play a minor role. A measurement of the period is therefore a simple check of whether Newton's equations correctly describe Nature. The more accurately it can be measured, the more small effects (such as air resistance) must be taken into account in the calculation of the period to enable an accurate comparison between theory and experiment.

A stopwatch is a simple instrument to measure the period. A simple estimate of the statistical uncertainty of the measurement can be obtained from a repetition of the measurement: The distribution of the measurement results allows for the calculation of the *standard deviation*, a measure of uncertainty. In the example shown on the stopwatch, we measure 1.3; 2.0 and 2.3 s, resulting in an average of 1.8 s and a standard deviation of 0.4 s for the statistical uncertainty of this measurement. The standard deviation represents a quantitative measure of the uncertainty.

The main work of conducting a good experiment actually begins now: The goal is to reduce this uncertainty through more clever experimentation and to measure it more precisely! A simple solution would be to measure much more often, but this can take time and is not very exciting. A more promising approach is to improve the measurement method. So far, the measurement was always started and stopped at the right turning point when the pendulum briefly comes to a standstill. Alter-

The repeated measurement of the pendulum's period with the stopwatch. The measured times vary.

Measurement Uncertainties → p. 86
How Does Physics Make Progress? → p. 26 and A Deep Dive Into the Scientific Method → p. 30
Resonances → p. 56

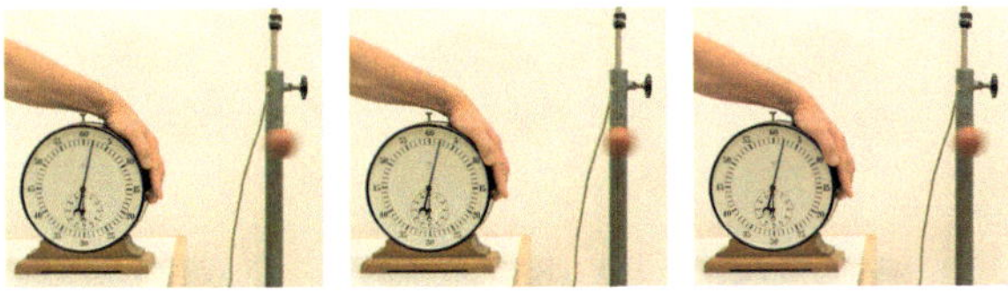

The improved measurement at the lowest point

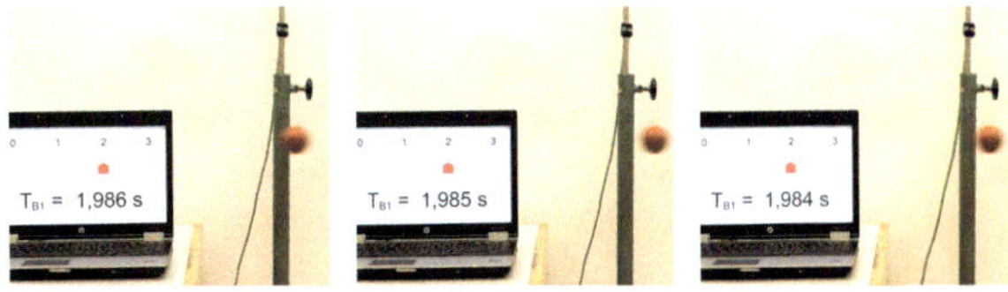

The measurement with a photoelectric sensor

natively, it could be better to measure at the lowest point of its movement, because there the pendulum is faster and the eye can more accurately determine the exact moment of the pendulum passing its support frame in the background. Whether this is really the case can be measured: The next series of measurements shows 2.0; 2.0 and 2.0 s as a result! The scatter is not visible. However, this does not imply that the standard deviation is zero, because the stopwatch only displays fixed steps of 0.33 s duration, within which the fluctuation of the individual measurement cannot be measured. The measurement must therefore be further improved!

This can be achieved by not measuring the duration of one period, but of 10 periods. Here, the result, divided by the 10 periods, is now 2.05 ± 0.03 s, where the uncertainty still only contains the statistical uncertainty. Thus, the precision of the individual measurement has improved by more than a factor of 10 overall.

However, measuring with a mechanical stopwatch by hand has many disadvantages. Is the stopwatch accurate at all? If it consistently measures time incorrectly, this would not be noticed in the dispersion of the individual measurement results used to estimate the statistical uncertainty. To test this, the stopwatch must be compared to a more accurate clock. Also, eye triggering is not very accurate. Instead, it would be better to have a photoelectric sensor connected to measurement electronics and a computer that synchronizes its clock with an atomic clock. In the next improvement of the setup, the photoelectric sensor is mounted at the top of the rope, where it moves slowly. Now, much more consistent individual measurements are achieved: The result now is a period of 1.9850 ± 0.0008 s.

While the improvements of the measurement have excluded many sources of systematic uncertainties and reduced the statistical uncertainty tremendously, the possibilities for improvement are not exhausted: Is the light beam, which is interrupted by the pendulum, as narrow as possible? Does the light sensor mount wobble? Wouldn't it be better to mount the light sensor at the bottom, where the rope moves much faster than at the top? With such improvements, the dispersion of the individual measurements continues to decrease and the result is an uncertainty of less than 0.0003 s. But even with this, the work is not finished: Can the light sensor be read out in even smaller time steps? Can the deflection of the pendulum be controlled more accurately? Can the complete measurement setup be compared with a complementary setup to determine the uncertainties even more objectively? The possibilities for improvement are almost unlimited.

In typical particle physics experiments, a large part of the work—in the authors' view one of the most creative and exciting parts—is exactly this task: finding creative ways of reducing uncertainties and determining them as accurately as possible.

Drawing Conclusions From Statistics

When Is Something Too Unlikely to Be True?

The previous article covered how measurement uncertainties can be determined and made as small as possible. For a very large number of experiments that measure a numerical value, such as the mass of a particle, making uncertainties as well-understood and as small as possible is the relevant task of the statistical method.

In the search for new particles, additional statistical questions arise. Given the data, when and how can we rule out the possibility that there is a new particle in a certain mass range? Or even better: When and how can we say we have discovered a new particle (↓)? For example, how was the Higgs boson discovered and what was different about the data in 2012 compared to previous searches for the same particle? Those interested in these questions will find exciting insights here: Does a certain predicted phenomenon exist at all in Nature? After all, the *discovery* of new particles is one of the most magnificent events in particle physics. And what happens if a predicted particle does not appear in the data? If a theory predicts that a particle should have been detected in the experiment but nothing significant was found, particle physicists speak of an *exclusion*.

To search for particles, a hypothesis is needed: a specific idea of what the fabric of Nature might look like, which particles contribute, and how they interact. Typically, the hypothesis is associated with a mathematical model of Nature (How does Physics make Progress? ↓) that makes concrete measurable predictions (↓). For example, such a prediction could be the mass of a particle and how often it is produced. Since all events in particle physics are probabilistic in Nature, the exclusion or discovery of such a hypothesis is based on complex statistical methods, the so-called *hypothesis tests*. However, these can be described in a simplified way in the categories of measurement uncertainties, as described in the previous pages.

The measurement required for discovery or exclusion is a *counting experiment*. One counts how often a process with certain properties was observed. This observation can then be compared with two different hypotheses: For example, the *signal hypothesis* may state that the Higgs boson exists with a certain mass, resulting in a certain number of events which pass the cuts. The *background hypothesis* states that only the particles known before the discovery of the Higgs boson exist, and correspondingly a different number of events should be observed. There can be any number of different signal hypotheses, each of which is individually compared with the background hypothesis. This evaluation, taking into account all possible hypotheses, as well as all systematic and statistical uncertainties, is often numerically very complex and requires large computing capacities.

Let us take the search for a generic new particle as a hypothetical example. The simulated frequency of observed events with a measured value of a certain mass is represented in the figures on the right side (black dots). The solid blue line represents the prediction of the background hypothesis, which states that on average about a hundred events should be counted per

The Discovery of a Particle → p. 164
How Does Physics Make Progress? → p. 26
A Deep Dive Into the Scientific Method → p. 30

measured mass value, and is thus flat. The signal hypothesis on the other hand is the expected frequency of events if the signal is present in addition to the background, and is represented by the red, dashed line. The data do not necessarily exactly match the hypotheses. This is due to the statistical and systematic uncertainties (↓) of the counting rate, which are marked by the vertical bars around the black dots. Whether they support a hypothesis with sufficient statistical certainty or contradict it is determined by the hypothesis test!

The schematic representations show various possible combinations of results and expectations for the signal hypothesis. The result of such a measurement might look like what is shown in the top figure: For most measured values, the data agree with the background hypothesis. For certain measured values, the data and its statistical description (dashed blue line) agree better with one of the signal hypotheses which in this case might mean that there is a new particle at a measured value of 65. The hypothesis test translates the degree of statistical agreement of the hypotheses with the data into a *significance* or *confidence*—the number of standard deviations σ that the result is separated from one of the hypotheses. This is exactly like determining a measured value on the previous pages: For example, if a measured value deviates three times the value of the standard deviation σ from a prediction, then the deviation has "3 σ significance". The center and bottom figure show other possible outcomes: the data cannot agree at all with a signal hypothesis (↓), which is called an *exclusion* of a hypothesis, or disagree significantly from the background hypothesis, which then is a discovery! In particle physics, high demands are placed on the significance of a discovery—typically, a 5 σ deviation from the background hypothesis is required! This corresponds to a probability of one in about three million that the observed data could have been generated by a statistical fluctuation of the background hypothesis. Therefore, the data shown in the top histogram with a significance of only 2.8 σ are not a discovery. With more data, the deviation either disappears (the signal hypothesis can be ruled out) or it becomes more significant—then the background hypothesis can be ruled out and a new phenomenon has been discovered!

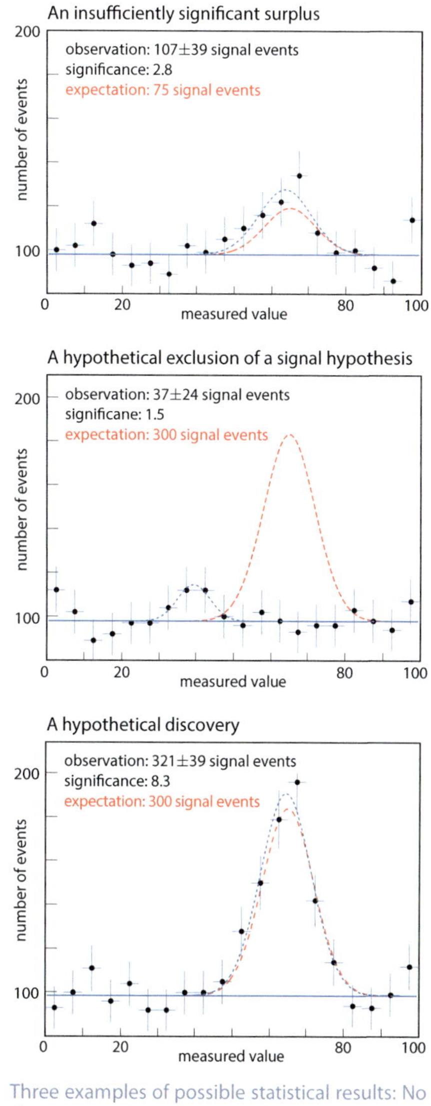

Three examples of possible statistical results: No discovery and no refutation of a hypothesis (top), refutation (center) and discovery (bottom)

Measurement Uncertainties → p. 86
Publishing a Non-Observation → p. 162

Publishing a Non-Observation

Finding Nothing Is an Important Result

After reading the previous article, one might think that the history of particle physics is simply the history of the discovery of particles. But this is not at all correct! On the one hand, because the precise measurement of the *properties* of particles and their interactions is one of the most important tasks of particle physics. On the other hand, because the *exclusion* of the existence of particles is also a very important result of research. Anyone deeply interested in this will find more details here about the probability that something does not exist.

Particle physicists constantly propose new ideas. They can explain phenomena in Nature that are not yet understood, or propose what unknown phenomena there might be (How does Physics make Progress ↓). Suppose experimental particle physicists had measured, but never published that a Higgs boson-like particle with the couplings close to what is predicted by the Standard Model, but half the mass of the discovered Higgs boson is experimetally excluded. Then new hypotheses could constantly be proposed in which such a second Higgs boson is predicted. This would be a huge distraction for the progress of science. Therefore, it is also important to publish results in which *no* signal was found. In the search for the Higgs boson, this approach allowed to gradually exclude more and more mass ranges. For example, a Higgs boson with a mass less than 114.4 GeV could already be excluded at the LEP Collider in the year 2000. This in itself represents a gain in knowledge as it excluded a multitude of hypotheses that predicted a lighter Higgs mass. The publication even includes what the significance (↓) is at which the proposed hypotheses have *not* been found. Unlike discoveries, where particle physicists have themselves set a threshold of 5 σ significance, weaker statements of 2 σ are also useful in case of exclusions. This significance corresponds to a maximal probability of 5 % to exclude a truly existing signal. This sounds like a lot but it is not that bad, because even in already excluded areas, the search for new phenomena continues.

In the figure labeled "Exclusions of different signal hypotheses", the hypothetical search for a particle with a measurement value around 65 is shown schematically. Let us assume that a theory predicts a count rate of about 75 events beyond the background as shown with the red dashed line in the graph. On the one hand, the data points do not suggest the existence of such a particle; on the other hand, the prediction of the theory is

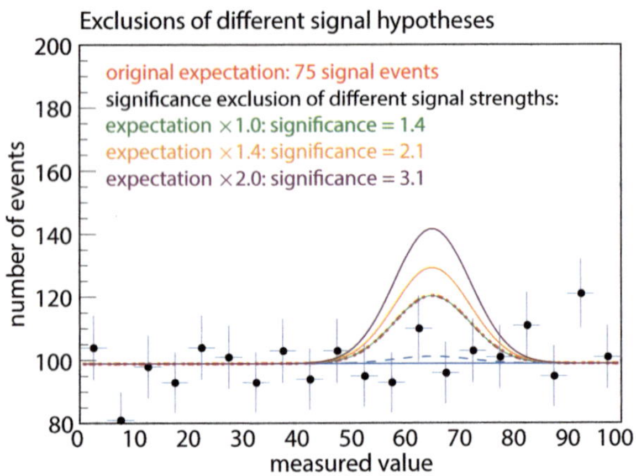

How Does Physics Make Progress? → p. 26
Significance: Drawing Conclusions From Statistics → p. 160 and Measurement Uncertainties → p. 86

not significantly different from the data—the original hypothesis can neither be confirmed nor ruled out, and the observed significance is 1.4 σ. Now, let us assume that in another theory the coupling of the particles to each other would be significantly larger, which significantly increases the predicted count rate. This is illustrated in the figure with the different colored curves which lead to exclusions with different significance. The hypotheses of the higher expected count rate match the data even less than the original hypothesis. Specifically, in this example, increasing the predicted frequency by a factor of 1.4 means that the new signal hypothesis does not match the data with a significance of more than 2 σ, namely 2.1 σ. This exclusion with more than 2 σ significance is then published. Now every physicist interested in proposing theories with particles consistent with those searched for knows that they are excluded if their production rate exceeds a certain published value, which is always the value which leads at least to a 2 σ exclusion.

A statistical significance of 2 σ sounds not too challenging. It means that there is an up to 5 % chance to accidentally exclude a real signal. Therefore, the same statistical error bar cannot be used for discoveries. An example of that is shown in the other plots. At one value, the measurement data is much higher than expected from the background, which can happen purely statistically. Moreover, these discoveries could appear not only where one has expected them, which in our example was a measurement value (e.g. an invariant mass) of 65, but also anywhere else in the measurement range! Because new discoveries always attract a lot of attention, the constant publication of statistical fluctuations of the data as alleged new discoveries would be a great continuous distraction for everyone, and the public trust in science and new discoveries would be gone. This is the reason for the very high threshold of 5 σ significance that particle physics has set for a discovery (Blind Analysis ↓).

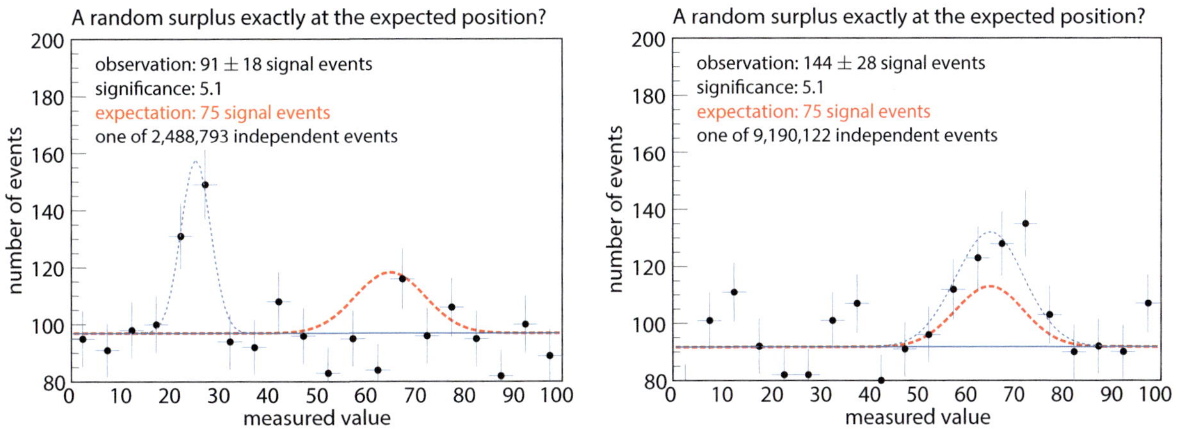

Three examples of statistical results: The exclusion of different hypotheses (left page), the "discovery" of a random fluctuation (left) and the discovery of an expected signal (right)

What Is a "Blind Analysis"? → p. 166

The Discovery of a Particle

Example: the Higgs Boson

The previous pages explain the discovery or exclusion of a new particle based on abstract criteria and hypothetical measurement data. The same process will be illustrated here using the real example of the discovery of the Higgs boson (↓) at the Large Hadron Collider LHC (↓).

We consider the search for the production of the Higgs boson and its subsequent decay (↓) into two Z bosons.[1] The results of this search at the time of the discovery of the Higgs boson in 2012 are summarized in the first graph on this page. This representation is very typical and recurs throughout this book: On the horizontal axis, a value is plotted that can be measured for each individual selected event. In this case, it is the reconstructed mass (↓) of the state that decayed into the two Z bosons, which then decayed into two charged leptons each. This mass varies for the counted events between a value of 80 and 250 GeV/c^2.

Two reasons cause the counted events to occur in a broad distribution of different invariant mass values: First, the invariant mass of the four leptons varies depending on the physical processes that led to the production of the Z pair. Secondly, due to measurement uncertainties (↓), the momenta of the four leptons are not measured exactly (↓), which smears out the reconstructed invariant mass.

On the vertical axis, the measured number of observed four lepton events with a certain invariant mass is plotted. For example, exactly two events were counted in

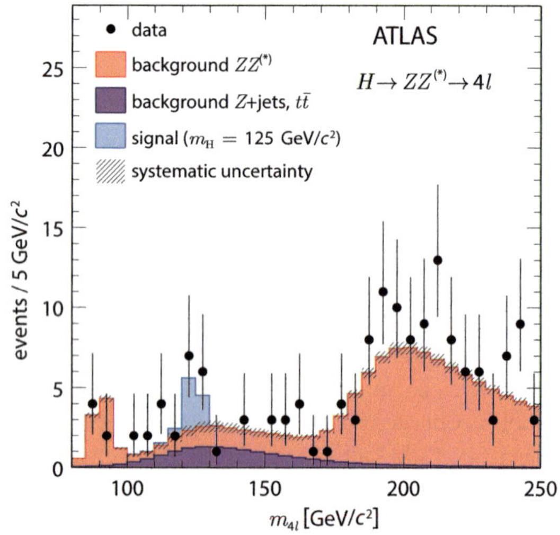

The distribution of the combined mass of four leptons

the mass window of 100 to 105 GeV/c^2. The measured data are again represented with dots. The statistical 1 σ uncertainty of the count rate is marked by a vertical bar, and the systematic uncertainty by a hatched region around the background expectation. The different expected background processes are separated graphically: Red or purple depending on whether they are real Z pairs that do not originate from a Higgs boson (red) or events that, due to uncertainties in the measurement process, appear to the detector to be a Z boson pair, but are not (purple).

The expected events that really originate from a decaying Higgs boson with a mass of 125 GeV/c^2 are repre-

[1] ATLAS Collaboration et al. 2012, *Observation of a new particle in the search for the Standard Model Higgs boson with the ATLAS detector at the LHC,* Physics Letter B, 716, 1, 1, doi:10.1016/j.physletb.2012.08.020

The Discovery of the Higgs Boson → p. 246
The LHC Accelerator Complex at CERN → p. 116
Particles' Births and Deaths → p. 54

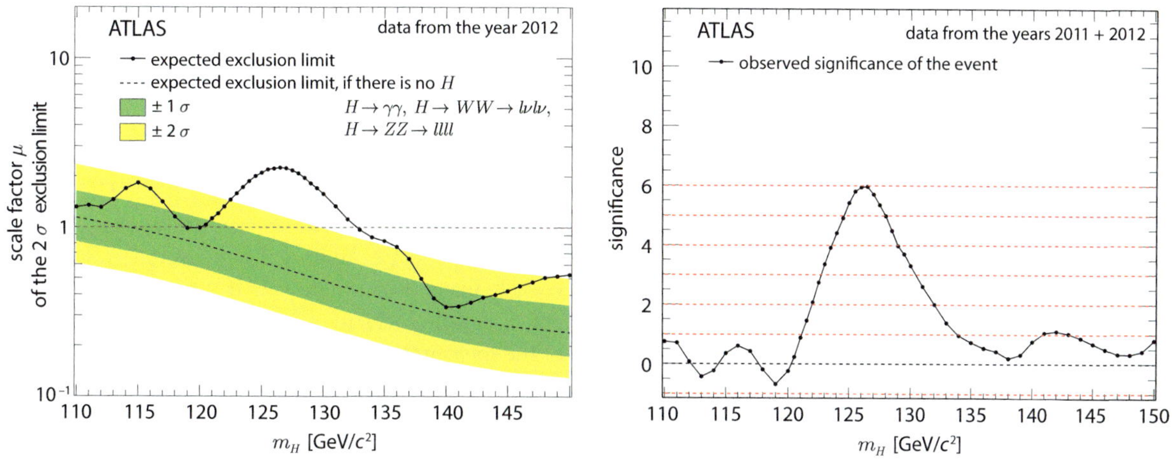

The exclusion of heavier Higgs bosons (left) and the discovery of a Higgs boson at a mass of 125 GeV/c^2 (right)

sented by the light blue area. The data agree very well with the sum of the expected background and signal, that is, with the signal hypothesis (↓). By itself, the deviation of the measurement data from the background hypothesis in this measurement is still clearly too small. It is less than the significance of 5σ required to find a deviation from the background hypothesis. Therefore, a joint statistical analysis of several searches for the Higgs boson was carried out. For being able to claim a discovery, decays of the Higgs boson into W boson pairs and into photon pairs were combined with the measurement shown here.

First, this joint statistical analysis was employed to analyze which hypotheses about the Higgs boson can be excluded with 2σ significance. This is plotted in the left graph on this page: All Higgs boson masses (shown on the horizontal axis) are excluded, where the black solid line is below the value 1 on the vertical axis. We can read from this graph that the Higgs boson of the Standard Model is excluded for masses heavier than 133 GeV/c^2. The green (yellow) region of the left graph indicates the area in which the 2σ exclusion would have been expected with 68 % (95 %) probability if there were no Higgs boson. Around the observed Higgs boson mass of about 125 GeV/c^2, there is no observed exclusion, as the black line is above 1. This is despite the fact that it would have been expected to be excluded if there were no Higgs boson at that mass: At the same point at around 125 GeV/c^2 where the observed exclusion is above 1, the expected exclusion for the case where there was no Higgs boson is below 1.

The picture on the right shows the significance of a discovery. The deviation of the data from the background hypothesis is more than 5σ significant around a mass of 125 GeV/c^2. Thus, the Higgs boson with a mass of about 125 GeV/c^2 was discovered!

Identifying the Particle Type via the Mass → p. 156
Determining Measurement Uncertainties → p. 158
Typical Detectors at Accelerators → p. 128
Signal hypothesis: Drawing Conclusions From Statistics → p. 160

What Is a "Blind Analysis"?

The Placebo Effect on Researchers

If an experiment has been carried out correctly and all uncertainties have been thoroughly scrutinized, the data itself should be the best guide to learn what discoveries are hidden in the data. However, caution is advised!

Measurement data and their precisely determined statistical and systematic uncertainties are essentially "judging" scientific hypotheses. Whenever measured data are used not as a judge, but rather as a "detective" to uncover new hypotheses, the probability of problems is high. This phenomenon is also known from other scientific fields: Studies measuring the effectiveness and safety of a vaccine are thus always conducted *blindly*. Neither the test subjects nor the administering medical staff know whether a particular subject is receiving the vaccine or a placebo; and those who evaluate the data have no contact with those who care for the subjects. The design of the study and all evaluation and termination criteria must be established before data collection begins. If these criteria are not strictly followed, the results of the study cannot be trusted because, for example, the perception of side effects can be consciously or unconsciously distorted by the knowledge of the underlying treatment.

Now, particles are not test subjects, and detectors are not medical personnel whose perception is inevitably influenced by subconscious processes, even in the case of the best motivation. So why should a *blind analysis* be necessary in particle physics? To demonstrate this, simulated measurements shall be used: Suppose that particle physicists conduct an experiment that can measure values between 0 and 1000. The already known background process has an average frequency of 10 events per unit of the measured value and is equally likely everywhere (↓). If no new signal is present, a sample measurement will result in a fluctuation of the measured rate (blue) around the value 10 due to the expected statistical fluctuations.

If the data are only to be used as a judge of a new hypothesis, but not as a detective, the decision about the method of statistical analysis of the data must not depend on the measured data itself. In this case: The hypothesis *where* in the data the signal is expected must not depend on the data, but must be made *blindly*. Here it is assumed that due to a new theory the predicted new signal should appear at a measured value of 500 in the form of a Gaussian distribution. In the example shown, no significant portion of the signal (↓) can be found at the value 500.

So far, the data have been used as a judge, but not as a detective. If the analysis is not performed *blindly*, but is guided by the data, then even without a "true" signal in the data, a statistically significant deviation is often found *somewhere*. This is simply caused by random statistical fluctuations of the data around their expectation value. Sometimes, these fluctuations can be large. The probability of large fluctuations *somewhere* is much higher than at exactly one spot.

Statistics → p. 84 and Measurement Uncertainties → p. 86
Drawing Conclusions From Statistics → p. 160

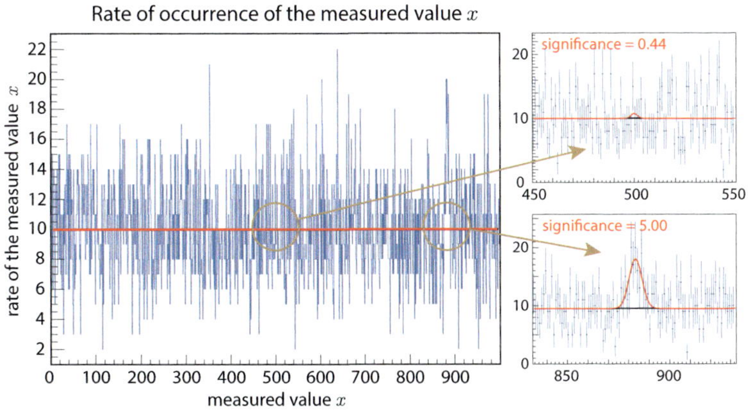

A simulated experiment. At a predetermined point in the middle of the distribution, a random excess is rarely found. But *somewhere* in the distribution, there is often a very significant excess created by purely random fluctuations.

In this simplified example, the correction is obvious: No new particles were discovered because a probability of random fluctuations creating a false signal in 2 out of 1000 attempts is way too high to claim a discovery. In practice, however, this correction is often much more complex because not only one measured value, but typically many possibilities of evaluating different regions of different measured values are involved. Then the correction shown becomes often impossible in practice. So, it is even more important to perform the analysis *blindly* and not to use the data as a detective.

In the example shown, this is approximately the case at the measured value 885: A signal with 5 σ significance was found, and thus apparently a new particle was discovered, although we know the simulated data do not contain any signal!

In everyday life, it is often not much different: Randomly noticeable events stick in our memory, while we tend to forget all the occurrences where nothing notable happened. Caution is also advised here!

In particle physics, this particular effect is also known as the *look-elsewhere effect*: You can always look elsewhere in the data and find something by chance. If one is aware of the exact extent of this process, it can be statistically corrected. If the expected data of the background are simulated many times, then the probability of achieving a high significance by chance somewhere can be calculated. Here, it turns out that a significance of 4 σ occurs randomly in 13 % of all repetitions and a significance of 5 σ in 2 out of 1000 of all repetitions of the experiment, instead of slightly less than 1 time in 10 million repetitions, as is actually required for a discovery.

The relative probability of high significances with predetermined evaluation (blue) and evaluation of the most significant point (red)

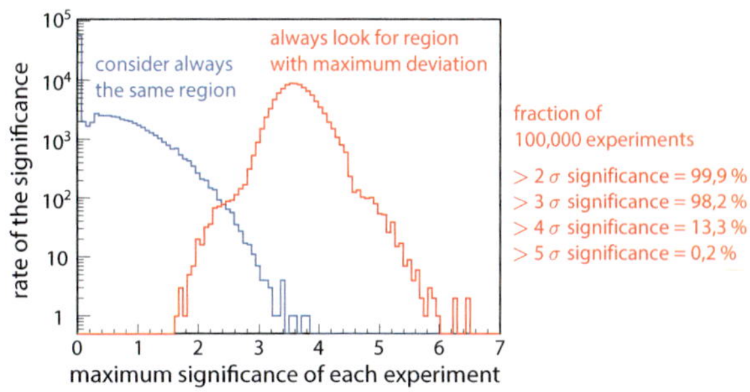

How to Fix False Discoveries

Correction Procedures in Particle Physics

It is the confusing characteristic of measurement uncertainties that they can be precisely known for the average of all measurements, while it can never be precisely known how far a single measurement result deviates from the unknown mean value of many other hypothetical measurements of the same quantity. If particle physicists knew this, then there would be no uncertainty!

Researchers can only know that they expect 68 % of the repetitions of a measurement to produce a 1 σ uncertainty interval that covers the unknown true value. 95 % of all results should cover the true value with their 2 σ uncertainty interval, and 99.7 % with the 3 σ interval. So there is always a small portion of all results left where something happens that is measured "wrong" by three standard deviations, even if nobody did anything wrong! This article will discuss how particle physics deals with this seemingly alarming situation.

The first example is a now refuted "discovery" of *pentaquarks*[1] from 2003 (Exotic States ↓). Pentaquarks were actually discovered in 2019, but with properties different from those of the "discovery" in 2003. At that time, the LEPS experiment observed a new particle with a significance of 4,6 σ, which is apparently less than the 5 σ threshold generally accepted in particle physics for an undisputed discovery. However, nine other experiments soon followed, each finding a similar particle in their data with sensitivities around or just below 5 σ. Each of these observations would almost constitute a discovery on its own. Together, they must surely be a secure discovery?

This could have been quite convincing. However, other experiments simultaneously demonstrated credibly that they should also be sensitive to this possible particle, if it did exist, but simply *did not* observe it. Here we see a prime example of why it is so important that negative findings are also published (↓), and not just forgotten or ignored.

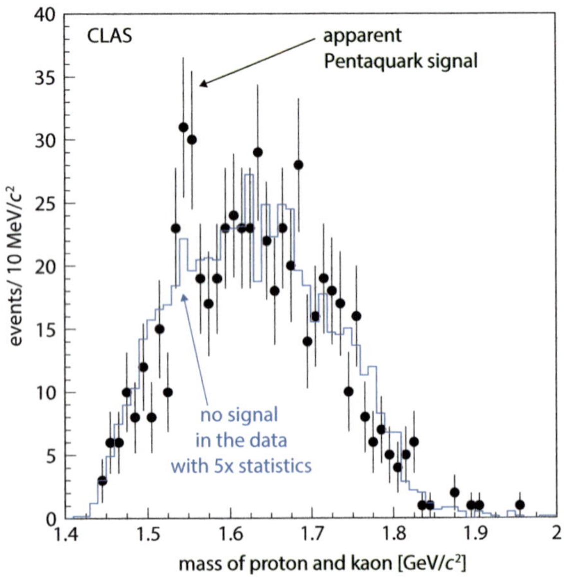

The apparent discovery of a pentaquark (dots with error bars) and the distribution of data with more statistics (solid line)

[1] K. Hicks 2007, *Pentaquark Searches*, Proceedings of the Hypernuclear 2006 Conference, doi:10.1007/978-3-540-76367-3_74

Exotic States → p. 230
Publishing a Non-Observation → p. 162

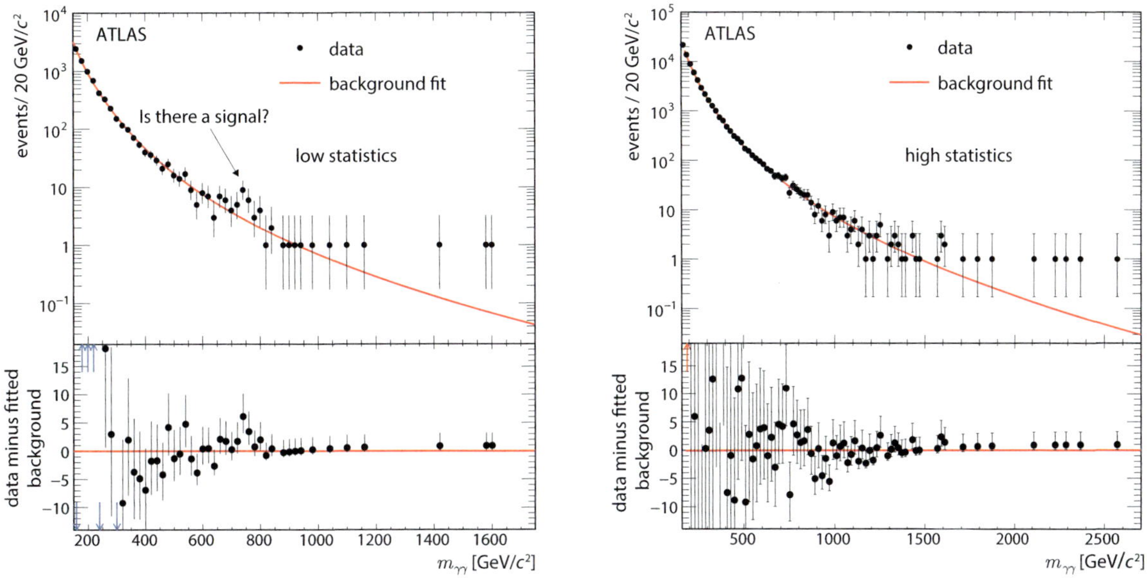

The apparent discovery of a new particle with a mass of 750 GeV/c^2 in the CLAS experiment (left) and its refutation with 10-fold statistics with the ATLAS experiment (right)

In case of such a controversy, it is important to compare the experimental methods accurately and to study the consistency of all observations. But even that is not always enough to clarify the question. A good idea is to continue measuring with increased self-criticism in the examination of all uncertainties. We see this in the graph of the CLAS experiment: The original "discovery" (data points) shows an apparent signal with a mass of approximately 1.5 GeV/c^2 on the horizontal axis. The data with a five times higher number of events, which were taken afterwards (shown as a blue histogram), do not observe this narrow peak. It is therefore reasonable to suspect that a deviation only appeared randomly in the selected region and that the calculation of the significance did not take into account that such a deviation could have occurred not only there but at any other measurement value in the data.

A similar phenomenon in a completely different measurement was shown in 2016 at CERN: CMS and then ATLAS saw a significant excess of up to 3.8 σ in events with two photons at a combined mass of 750 GeV/c^2. The graph on the left clearly shows the excess. This would have been a sensation. With ten times the amount of data, unfortunately no significant excess is visible in the right graph anymore.[2]

Misjudgments or intriguing statistical fluctuations can always occur when there are uncertainties. What is important is how they are dealt with: with an open invitation to constructive criticism.

[2] ATLAS Collaboration 2017, *Search for new phenomena in high-mass diphoton final states using 37 fb^{-1} of proton-proton collisions collected at $\sqrt{s}$=13 TeV with the ATLAS detector*, Physics Letters B, 775, 105, doi:10.1016/j.physletb.2017.10.039

Neural Networks and Machine Learning

Using a Simulation of the Brain to Get New Insights

How do we actually recognize objects? The answer to this question lies in our brain: large networks of neurons process external stimuli so that we can associate them with an object or a person. A fascinating fact: This complex process consists of a multitude of simple sub-processes. Signals are picked up by dendrites of a neuron (e.g. from the optic nerve or from another neuron) and processed by the nerve cell. If a stimulus threshold is exceeded, the cell transmits an electrical signal through the axon to the axon terminals, where another neuron picks up the signal with its dendrites and interprets it.

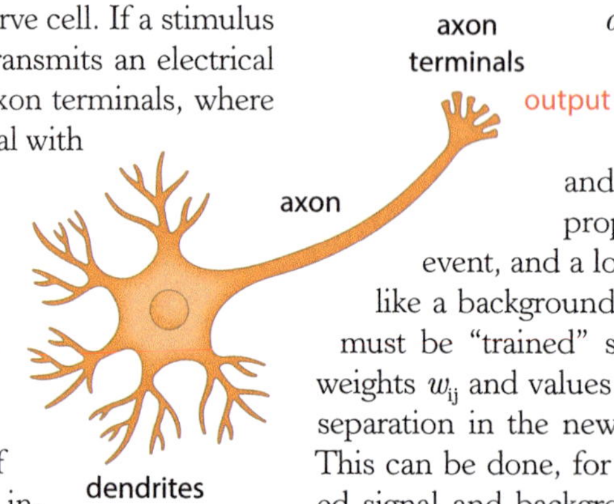

This biological process has inspired researchers to simulate neural networks on computers. The capabilities of neural networks and other methods of machine learning have steadily increased due to ever faster computers and are omnipresent in everyday life today. So it should not be surprising that such methods of machine learning have long been used in particle physics. Without these, many important discoveries would not have been possible at all (e.g. the discovery of the Higgs boson (↓) or decays of singly produced top quarks).

The illustration on the next page shows how an artificial neural network is used in the task of distinguishing an interesting signal process (red) from background processes (blue). For this purpose, various measured quantities (x_1, x_2, x_3) are used as input for a neural network with a single so-called "hidden layer". The network consists of individually simple mathematical functions that combine the inputs into a single new quantity, here called "new measurand". To do so, the measured quantities (x_i) are weighted (w_{ij}) in each neuron, and their sum, including a constant parameter a_j is evaluated in an activation function f. This gives a high value if an event with measured quantities x_1, x_2 and x_3 in the totality of its measured properties looks more like a signal event, and a low value if the event looks more like a background event. To do this, the network must be "trained" so that it can learn and adjust weights w_{ij} and values for a_j that allow for an optimal separation in the new combined measured quantity. This can be done, for example, with a set of simulated signal and background events, where one knows exactly which process is involved. At first glance, the illustration shows that the combined "new measurand" can distinguish the signal from the background much better than individual measurements x_1, x_2 and x_3 alone.

If we wanted to make a distinction without neural networks or other multivariate analysis techniques, we would have to distinguish events in each measured quantity individually. This is illustrated in the figure at the top of the next page using x_1 and x_2 as examples. Signal (red) and background (blue) are separated in the middle with two cuts. After the separation, we have

The Discovery of the Higgs Boson → p. 246

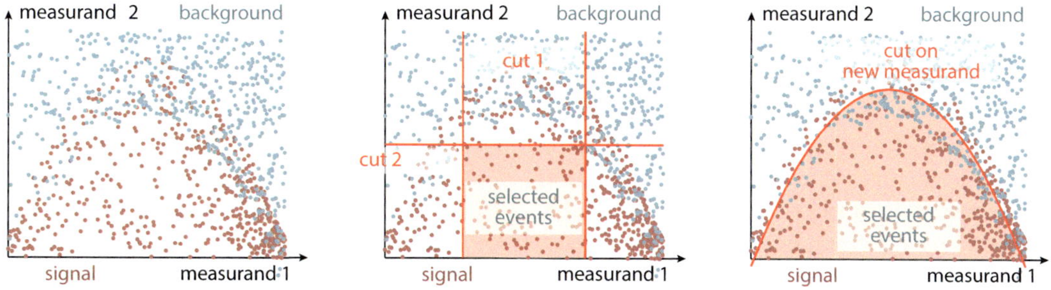

Each point represents the combination of the measurements x_1 and x_2 of individual events of a signal and a background. With cuts along the original measurements, the signal and background can be distinguished less well than with a cut on the new combined measurement of the neural network.

In the right figure, it can be interpreted such that the new combined measurement corresponds to the distance of a data point to the red cut line. If it is below the red line, it is more signal-like, and if it is above, of is more background-like.

excluded many signal events. If we use the neural network instead and make a cut in "new measurand", we can separate almost all signal events from background events. The neural network has learned to take advantage of the properties of the *conjugation* of x_1 and x_2 and can therefore separate at a different value of x_2 for each value of x_1.

Neural networks are also used in other tasks today. Networks can compete with each other: The first network learns to generate ever more faithful images of an object, and the second network tries to guess whether a true image or an artificially generated one was shown. This method is used to simulate events: A well-trained network can then simulate deceptively real-looking events (↓).

Today, the computational power has become so large that physical input quantities do not need to be prepared before given to the network. Instead "deep" information can be given directly to the network. The network then learns complex relationships from the unprocessed raw data (the "deep" data) of the detector, without the electronic signals of the detector being previously translated into physically meaningful measurement quantities such as momentum or energy (basically "lifted to a higher level"). This approach is called *deep learning* and is used, for example, in language models like Gemini, DeepL and ChatGPT.

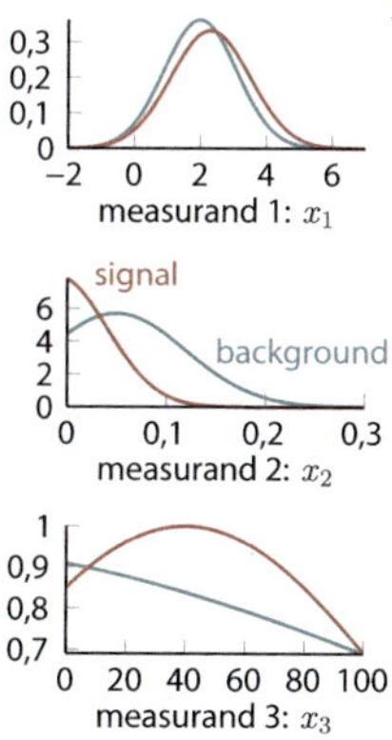

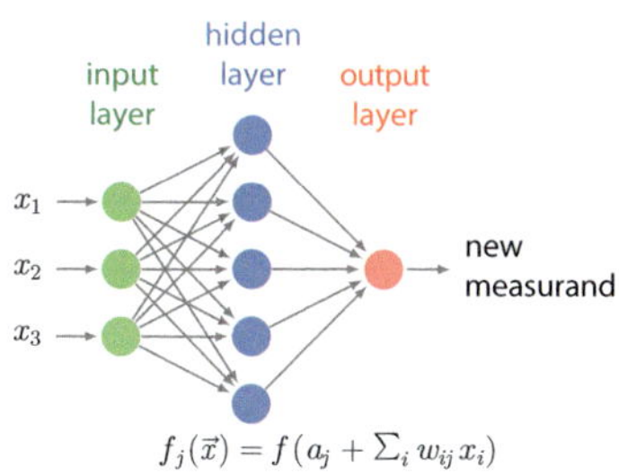

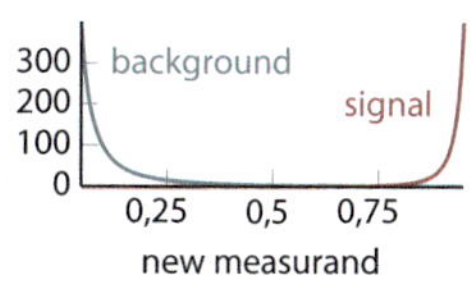

The principle of operation of a Neural Network using the example of a so-called *Feed Forward Neural Network*

The Invention of the World Wide Web

Global and Open Science for an Open Society

Particle physicists from all over the world work on projects that sometimes bring together thousands of experimental scientists in a joint effort on experiments and publications (↓). Furthermore, particle physics is concerned with the critical exchange of high-flying ideas between theorists from all over the world.All this work is based on the open exchange of ideas in an open society of science, across all borders. In the current "information age", this has fundamental consequences for all of us.

Tim Berners-Lee at CERN

Before the invention of the World Wide Web, the *internet* and its collection of protocols and information services, such as e-mails, was not a mass phenomenon. The origin of the wide-spead use of the internet lies in the year 1989. Tim Berners-Lee at CERN[1] provided the four experimental collaborations at the Large Electron-Positron Collider LEP (Why the Standard Model ↓) with a better opportunity for worldwide communication. He developed an information exchange service based on the visualization of text (and very soon images and videos), in which any content could be linked with any other: The World Wide Web (WWW) was born. CERN placed the new service under a free license, thus making the technological development as open and freely accessible to humanity as the scientific findings of the experiments.

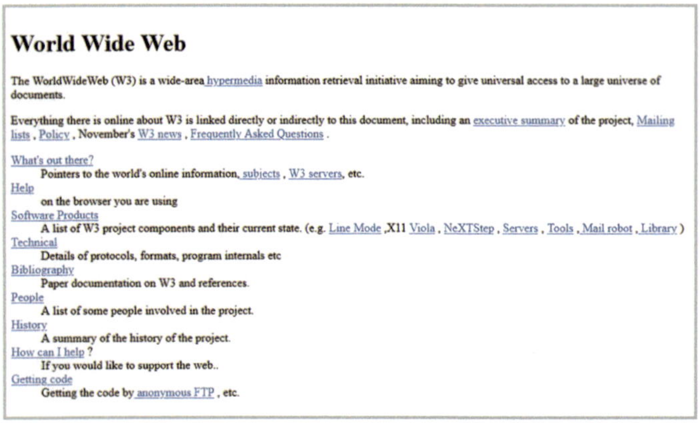

World Wide Web

The WorldWideWeb (W3) is a wide-area hypermedia information retrieval initiative aiming to give universal access to a large universe of documents.

Everything there is online about W3 is linked directly or indirectly to this document, including an executive summary of the project, Mailing lists , Policy , November's W3 news , Frequently Asked Questions .

What's out there?
Pointers to the world's online information, subjects , W3 servers, etc.
Help
on the browser you are using
Software Products
A list of W3 project components and their current state. (e.g. Line Mode ,X11 Viola , NeXTStep , Servers , Tools , Mail robot , Library)
Technical
Details of protocols, formats, program internals etc
Bibliography
Paper documentation on W3 and references.
People
A list of some people involved in the project.
History
A summary of the history of the project.
How can I help ?
If you would like to support the web..
Getting code
Getting the code by anonymous FTP , etc.

An image of the first website in the world can be seen in the figure on the left. However, the feature that makes the WWW a *network*—the ingenious idea of simple linking of content—and that shapes our information society today does not work in this book in printed form: tapping the links with your finger will not lead you to the linked content.

The world's first website at CERN from the year 1989

[1] CERN https://home.cern/science/computing/birth-web The birth of the Web

Experimental Research in Particle Physics → p. 28
How Does Physics Make Progress? → p. 26
Why the Standard Model? → p. 244

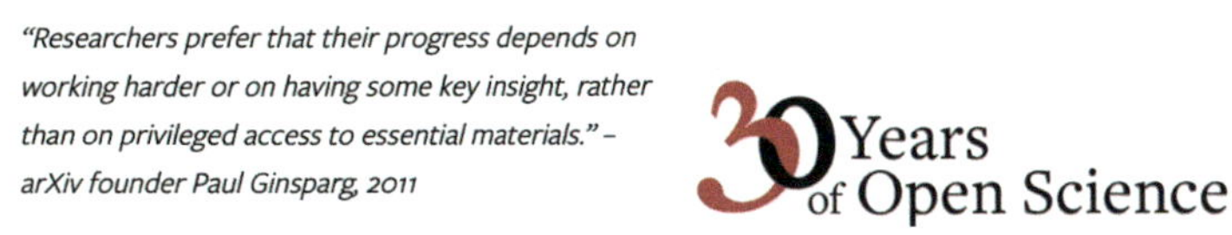

Statement by arXiv founder Paul Ginsparg, 2011

Since 1990, the year the WWW was published, information exchange on the internet has exploded. Already in 2014, every single person on Earth on average produced as much digital data transfer as was exchanged worldwide in 1984 in total—before the invention of the WWW. The network structure and the low-threshold accessibility of the WWW were crucial for this development. It is undisputed among critical and supportive voices that this led not only to more information exchange but also to societal changes of enormous magnitude.

The story of open data exchange does not end there: Today, the LHC collaborations and CERN operate a global analysis network for scientific data, in which more than ten gigabits of data per second are not only exchanged, but also processed for reconstruction and analysis (↓).

Open exchange includes free access to data and ideas, which also includes the free exchange of scientific publications without dependence on scientific publishers. This is also made possible by the World Wide Web: Today, a *preprint* server is a universally recognized platform for the free and accessible exchange of results independent of *peer reviews*, and the publication by a scientific publisher. The first and currently most active preprint archive, with about 16 000 new articles per month, is arXiv.org. It was founded by particle physicist Paul Ginsparg in 1991. Not only does this accelerate the exchange of results and ideas, it also allows all people without university affiliation to read the articles. Without the subscriptions, university libraries had to buy from publishers to access the findings of their own scientists, hardly any scientific article could be read. Thus, also the recent history of mathematics would have taken a different course without the arXiv: In 2003, Grigori Perelman published his revolutionary proof of the Poincaré Conjecture, one of the seven greatest unsolved puzzles of mathematics until then, only on arXiv.org and not in a journal. For this work he received the most important mathematics prize worldwide: the Fields Medal.

The international data flows of the particle physics data centers

Cornell University https://arxiv.org arXiv

Position and Momentum Measurement → p. 76, Energy Measurement → p. 78 and Signatures of Particles → p. 80
Analysis: Trigger and Data Processing → p. 150 and Neural Networks and Machine Learning → p. 170

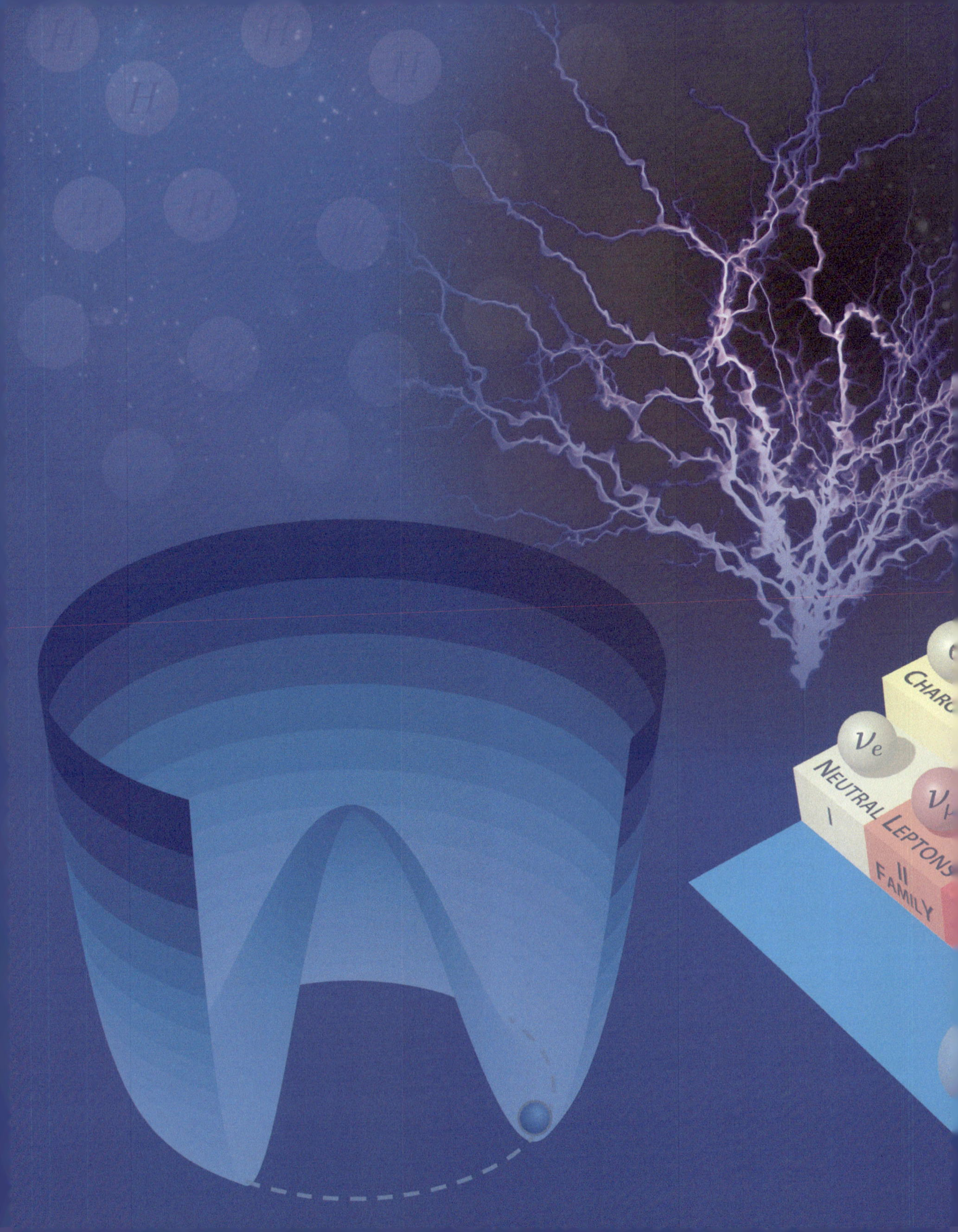
ν_e
Neutral
I
Leptons
II
Family

7 The Electroweak Standard Model of Particle Physics

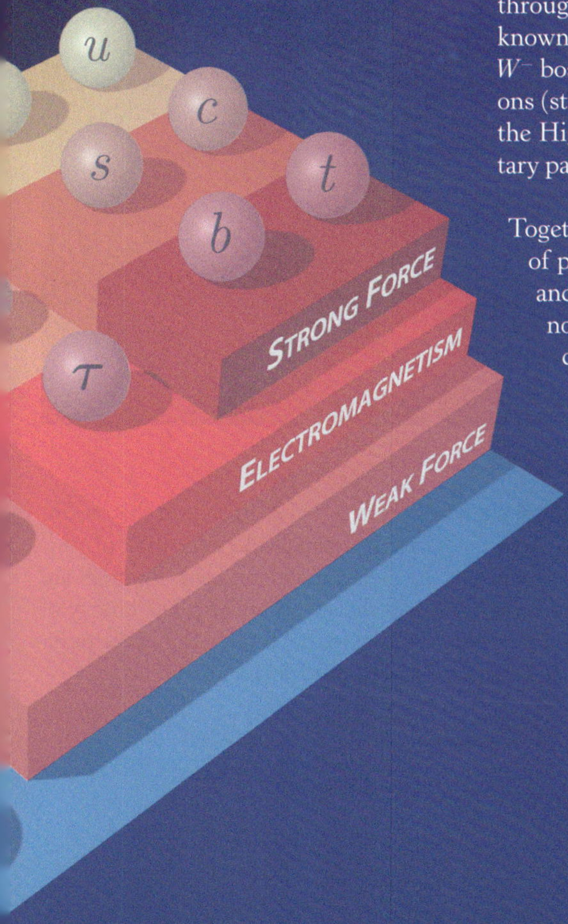

Already in antiquity, Greek philosophers asked themselves: What is matter made of? How many times can an object be divided? Are the elementary building blocks, originally called A-toms, indivisible? Today we know that the following building blocks of matter exist, which we believe are elementary: the three charged leptons (e^-, μ^-,τ^-), the three neutrinos (ν_e, ν_μ, ν_τ), the six quarks (u, d, s, c, b, t), and their respective antiparticles. These particles all interact with each other via the electromagnetic, the weak, and the strong force through the exchange of so-called gauge bosons. There are twelve known gauge bosons: the photon (electromagnetism), the W^+, and W^- boson, and the Z^0 boson (weak interaction), as well as eight gluons (strong interaction). This group of particles is supplemented by the Higgs boson, which is responsible for the masses of all elementary particles.

Together these ingredients constitute the so-called Standard Model of particle physics. The name may seem banal, but it is the best and most precise theory of Nature at present. It describes all phenomena from the hydrogen atom at an energy of 10 eV to the collisions at the LHC at 10^{13} eV.

This chapter provides an introduction to the basic concepts of the Standard Model, in particular the electromagnetic and weak interactions, and covers exciting topics such as symmetry breaking, parity violation and neutrino oscillations.

Electromagnetism

The Everyday Force

Electric charges attract or repel each other. *Moving* electric charges, e.g., an electric current, generate magnetic fields. Changing magnetic fields in turn generate electric fields. This can lead to spectacular discharges, as with the Tesla transformer in the figure. There, an electric resonant (↓) circuit is triggered at just the right frequency, which leads to a very high electric voltage. Electric and magnetic forces are inseparable and are unified in the theory of electromagnetism. This theory was the first great unification of natural forces. If quantum effects are included, i.e., taking into account that electromagnetic fields consist of photons, you get quantum electrodynamics (QED), a so-called *quantized gauge theory* (↓).

A Tesla transformer creates discharges.

Together with gravity, the electromagnetic force is the determining force in our everyday lives. All the light we see consists of electromagnetic waves, generated somewhere by accelerated electrical charges (↓) or atomic transitions (↓). All our electronic devices work with electromagnetism. Even the solidity of ordinary, neutral matter is based on electromagnetic forces. If you press your hand against a wall, or the wall presses back, you can feel the electromagnetic force. The charged electrons and nuclei within the atoms in the hand and in the wall repel each other if they come too close together.

At the level of elementary particles, electromagnetism is one of the three forces in the Standard Model of particle physics, alongside the weak and strong forces. Electromagnetism acts between electrically charged particles, such as between the electron and the proton in the hydrogen atom, through the exchange of photons, quanta of the electromagnetic field. Photons are massless and travel at the speed of light. The energy of a photon is proportional to the frequency f of the radiation (Concepts of Quantum Mechanics ↓).

Resonances → p. 56
Gauge theory: External and Internal Symmetries → p. 96, Gauge Symmetries → p. 98
Acceleration and Radiation → p. 42
Atomic transitions: Chance, Interference and Predictions → p. 52
Concepts of Quantum Mechanics → p. 50

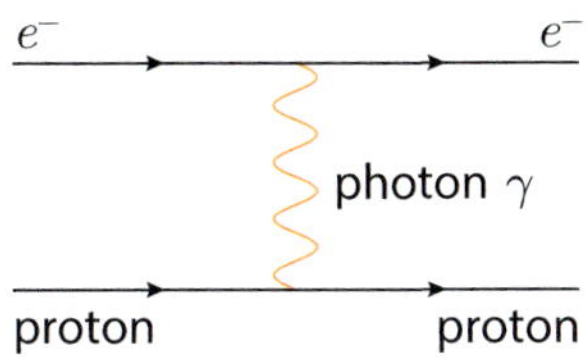

Electromagnetic interaction between an electron and a proton through the exchange of a photon.

The figure at the top shows a Feynman diagram for the interaction between an electron and a proton (↓).

The majority of devices used in high-energy experimental physics are based on electromagnetism. At the LHC accelerator, the protons are kept in a circular orbit by magnets and accelerated by electric fields. When the newly produced particles fly through the detector, they generate electromagnetic pulses, which are recorded to reconstruct their tracks.

The quarks are electrically charged and thus they also interact electromagnetically. When a very high-energy electron scatters off protons, the exchanged photon sees the quarks in the proton, as described in the article on the structure of the proton (↓).

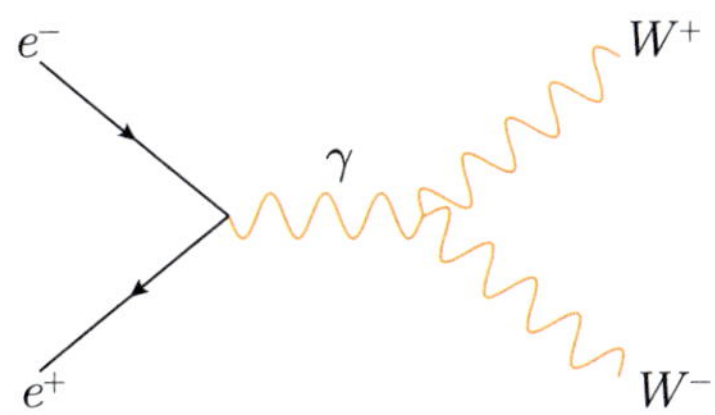

Feynman diagram for the production of two W bosons through the annihilation of an electron and a positron.

The W boson of the weak force is electrically charged and therefore also interacts with photons. For example, in the collision of an electron and a positron two W bosons can be produced. A photon is exchanged in the process, as shown in the Feynman diagram.

Neutrons are electrically neutral, but have a spin (↓), i.e., a magnetic moment. This allows them to be captured with magnetic fields and used for precision experiments. Neutrinos and gluons are also electrically neutral, so that photons do not interact directly with them.

Even though the Higgs boson is also electrically neutral, quantum effects can lead to interactions with photons. This is shown in the figure. Here, a Higgs boson decays into a top-quark-top-antiquark pair (the top quark has the strongest quark coupling to the Higgs boson, which annihilate and produce two photons). The top quark and the top antiquark are *virtual* here (see Range of Interactions ↓). Thus a Higgs boson can decay into two photons. This is how it was discovered in 2012 (↓).

The electromagnetic force is essential for understanding our everyday lives. It is a fundamental building block of the Standard Model of elementary particle physics and is used in all experiments.

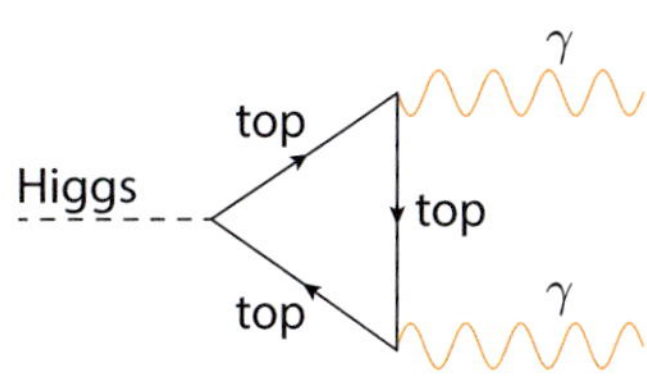

Feynman diagramm for the decay of a Higgs boson to two photons. This is only possible via a one-loop process.

Interactions Between Particles and Matter → p. 68
The Structure of the Proton → p. 212
Spin → p. 44
On the Ranges of Interactions → p. 58
The Discovery of the Higgs Boson → p. 246

The Weak Interaction

Metamorphosis Is Possible

The weak interaction is based on the exchange of $W^{\pm}$ and Z^0 bosons. In contrast to the massless photon of electromagnetism, these bosons have a large mass: $M_W = 80.4\ \text{GeV}/c^2$ and $M_Z = 91.2\ \text{GeV}/c^2$. This means that this force only has a very short range (↓), about 10^{-17} m to 10^{-16} m, smaller than the atomic nucleus radius of 10^{-15} m. The weak interaction therefore plays no role in our immediate everyday life. However, as we will see, it is essential for the Sun to work (and thus indirectly for our everyday life), for example, and plays a major role in particle physics experiments.

The weak interaction is fundamentally different from the electromagnetic and strong interactions in that it can convert particles into each other. For example in the beta decay of the neutron, its d quark is converted into a u quark (turning the neutron into a proton) and in the process a W^- boson is emitted. The W^- boson then decays into an electron and an electron antineutrino (Particle's Birth and Death ↓).

Further examples are the weak decays of the muon and the pion, which play an important role in cosmic rays, (see Special Relativity in Particle Physics ↓ and also the Feynman diagrams):

$$\mu^- \to \nu_\mu + e^- + \bar{\nu}_e, \quad \pi^- \to \mu^- + \bar{\nu}_\mu .$$

In the first case, we see in the Feynman diagram that a muon transforms into a muon neutrino and emits a W^- boson which again decays to an electron and an electron antineutrino. In the case of the pion, a d quark annihilates with a u antiquark to form a W^- boson, which now decays to a muon and a muon antineutrino. Since the W^- boson is charged, these reactions are also called *charged currents*. Note, however, that the muon mass (105.6 MeV/c^2) and the pion mass (139.6 MeV/c^2) are much smaller than the mass of the W boson. Therefore, in both cases the W^- bosons are virtual. The pairs of particles the W^- boson can decay into are summarized as:

$$\begin{pmatrix}\nu_e\\ e^-\end{pmatrix} \quad \begin{pmatrix}\nu_\mu\\ \mu^-\end{pmatrix} \quad \begin{pmatrix}\nu_\tau\\ \tau^-\end{pmatrix}$$

$$\begin{pmatrix}u\\ d\end{pmatrix} \quad \begin{pmatrix}c\\ s\end{pmatrix} \quad \begin{pmatrix}t\\ b\end{pmatrix}$$

We see six pairs of particles, grouped in three families. The top row contains three pairs of leptons, the bottom three pairs of quarks. The set of pairs arranged vertically above each other form a family, as indicated by the colored boxes. The three families (Up, Down, Left and Right ↓) are arranged from left to right by increasing mass. There are three types of neutrinos, or as we

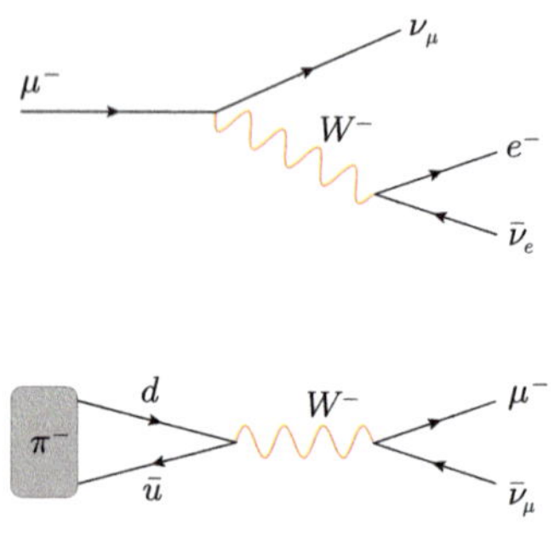

Feynman graphs for the decay of a muon (top) and a pion (bottom)

On the Ranges of Interactions → p. 58
Particles' Births and Deaths → p. 54
Special Theory of Relativity in Particle Physics → p. 92
Feynman Diagrams → p. 48
Up, Down, Left, Right → p. 180

say: *flavors*, one in each family: (ν_e, ν_μ, ν_τ). There are three charged lepton flavors (e^-, μ^-, τ^-) and six quark flavors (u, c, t, d, s, b).

As an application, consider the fusion of nuclei in the Sun. The energy radiated by the Sun is produced by a series of reactions, the most important is summarized as[1]

$$4\,\mathrm{H} \longrightarrow {}^4_2\mathrm{He} + 2\,e^+ + 2\,\nu_e\,.$$

Here, a u quark in two of the four protons transforms into a d quark, turning the respective protons into neutrons. Two W bosons are emitted, which each decay into $e^+ + \nu_e$ via the weak interaction. The strong force is responsible for generating the energy: the binding energy of the helium nucleus (two protons, two neutrons) is released as the kinetic energy of the positrons and neutrinos. Gravity holds the Sun together and through its pressure the nuclei are sufficiently close to feel the strong force and fuse. Since electromagnetism is responsible for the emission of photons, and thus the light of the Sun, we can only understand the Sun when taking into account all four known forces.

The Z boson is electrically neutral, like the photon. Particles do not change their identity when interacting with them. These interactions are called *neutral currents* (↓). The figure at the bottom right shows an example for a Z boson reaction. In any reaction in which a photon is exchanged, it could be replaced by a Z boson (see the figures in articles on Feynman Diagrams ↓), but not vice versa. The Z boson also couples to neutrinos.

Let us consider the electron neutrinos that are produced in the Sun. On Earth, some of them are converted in the detector by charged currents (W boson interactions) into electrons, which can be observed. Through neutrino oscillations (↓) some ν_e transform into ν_μ or ν_τ on their way to Earth. A ν_μ or a ν_τ cannot be detected via a charged current because the solar neutrinos do not have enough energy to produce a μ^- or τ^-, as their masses are too large, $m_\mu, m_\tau \gg m_e$. However, the SNO experiment has succeeded in detecting these neutrinos via their *neutral* current interaction. The SNO detector contained heavy water, D_2O, where the hydrogen is replaced by deuterium. Deuterium nuclei consist of a bound proton and a neutron. A passing solar neutrino can now exchange a Z boson with such a deuterium nucleus, as shown in the figure below. The energy of the neutrino is sufficient to split the deuterium nucleus. The resulting neutron can be detected.

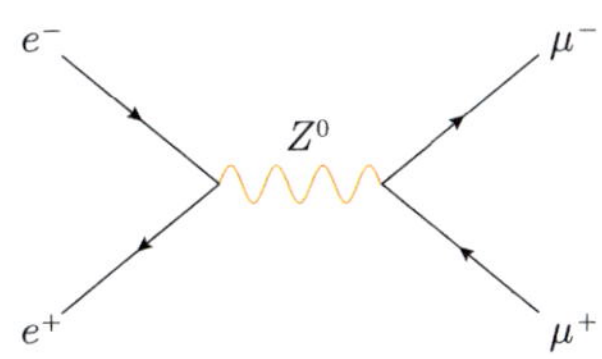

Feynman graph for the reaction $e^+e^- \to \mu^+\mu^-$ through the exchange of a Z boson

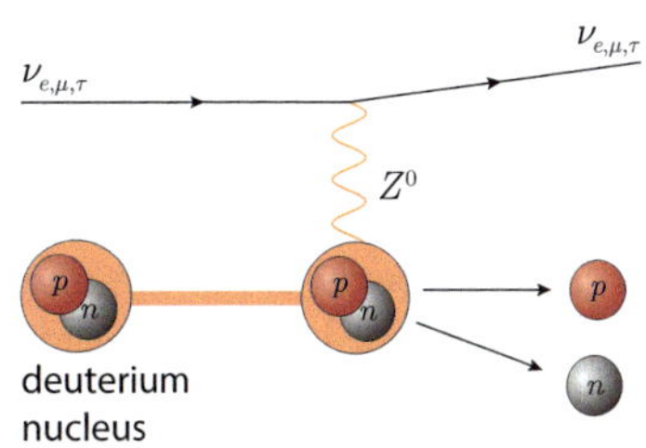

A neutrino splits deuterium through the exchange of a Z boson.

[1] The intermediate steps are described in the article The Birthplace of Nuclei → p. 342.

The Discovery of Neutral Current Interactions → p. 236
Feynman Diagrams II → p. 102
Neutrinos From the Sun → p. 344, Neutrino Oscillations → p. 198

Up, Down, Left, Right

Who Talks to Whom?

The Standard Model contains six leptons: the electron, e^-, muon, μ^-, and tau, τ^-, and the three associated neutrinos: ν_e, ν_μ, ν_τ. It also contains six types of quarks, so-called *flavors*: up, down, charm, strange, top and bottom, as well as the associated antiparticles for all quarks and leptons. Schematically, the particles are grouped into the following pattern:

$$\begin{pmatrix}\nu_e\\ e\end{pmatrix}_L, \; e_R \qquad \begin{pmatrix}\nu_\mu\\ \mu\end{pmatrix}_L, \; \mu_R \qquad \begin{pmatrix}\nu_\tau\\ \tau\end{pmatrix}_L, \; \tau_R$$

$$\begin{pmatrix}u\\ d\end{pmatrix}_L, \; d_R, \; u_R \qquad \begin{pmatrix}c\\ s\end{pmatrix}_L, \; s_R, \; c_R \qquad \begin{pmatrix}t\\ b\end{pmatrix}_L, \; b_R, \; t_R$$

The pairs with index L, e.g., $\begin{pmatrix}\nu_e\\ e\end{pmatrix}_L$, are the particles which can interact with the W and the Z boson (↓). The particles with index R can interact with the Z boson but *not* the W boson. In this article, we want to clarify the meaning of the indices L and R, as well as the connection with the structure of the Standard Model.

The spin (↓) is a form of angular momentum. For fermions, e.g., electrons, it is 1/2 and has an orientation in space. With regard to the particles momentum vector $\vec{p}$ it can be aligned either parallel or antiparallel. The thin orange arrow in the figure to the right shows the momentum direction, the broad blue arrow shows the spin direction. A particle with spin in the direction of momentum is called *right-handed*, with spin in the opposite direction *left-handed*.

One says: A right-handed electron has *helicity* +1/2, or positive helicity, a left-handed one −1/2, i.e., negative helicity.

In the special theory of relativity the direction of momentum depends on the frame of reference. If you move to a frame which is faster than the particle, the *direction* of momentum $\vec{p}$ reverses, as does the helicity. (The spin does *not* reverse direction.) Note this is only possible for massive particles which travel at less than the speed of light. Massless particles always travel at the speed of light, and there is no faster frame.

As an example, consider an observer on the ground, with a small propeller airplane flying by. The propeller turns, let's say, clockwise, as seen from the pilot. Then the corresponding angular momentum points in the direction of flight and the helicity is positive.

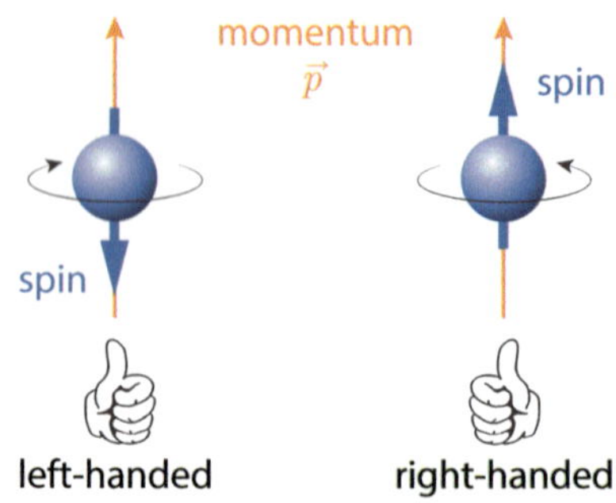

The helicity is determined by the relative direction of the spin and the momentum of a particle.

The Weak Interaction → p. 178
Spin → p. 44

Propeller airplane and jet

Seen from a much faster large jet the smaller airplane moves backwards, so the momentum vector is reversed. However, the propeller turns in the same direction for both, so the jet pilot sees a negative helicity. Overall, the helicity therefore depends on the reference system and is *not* an inherent property of the particles.

There is another particle property in relativity, it is called *chirality*. Unfortunately, it is not as intuitive as helicity. In the case of massless particles, which always fly at the speed of light, chirality is exactly the same as helicity. However, chirality *does not* depend on the reference frame. It is a property inherent to the particles. Somewhat misleadingly the particles are also called "right"- or "left"-handed, just as with helicity. For example, there are two chirality versions of the electron, a left-handed electron (e_L^-) and a right-handed one (e_R^-). In the LEP accelerator, the energies were so high that the electrons could be treated as if they were massless and flew at almost the speed of light.

At these speeds, the chirality and helicity happen to be practically identical. As we now discuss, however at any speed the chirality tells you who does or does not couple to the W boson.

The interesting thing is that right- and left-handed (chirality) particles behave differently under the special theory of relativity. They really are different particles! And surprisingly: Nature *knows* that these particles are different. It is *only* the left-handed electron that couples, i.e., talks to the W boson, not the right-handed one! This is shown in the pattern with the pairs above. The e_R^- is *not* in a pair. It does not couple to the W^- boson.

e_L^- and e_R^- both couple to the Z boson, but with different strengths, i.e., the Z boson senses the different chirality. Note, there is no right-handed neutrino ν_R in the Standard Model. The charges of a right-handed neutrino would be such that it would neither interact with the photon, nor the Z boson. This means that it does not interact at all in the Standard Model and is therefore omitted. We do not know whether right-handed neutrinos exist. If they exist, they could help explain neutrino masses (↓).

The photon couples with equal strength to e_L^- and e_R^-; the gluon couples with equal strength to u_L and u_R. It is only the weak interaction ($W^\pm$, Z^0) that recognizes this difference.

The particles in the orange box on the left of our pattern are together known as the first family. This structure is repeated with the other leptons and quarks shown as the red and dark red boxes. Together this results in the three families of the Standard Model.

QCD: Quantum Chromodynamics → p. 204
Discrete Symmetries: Violating P and CP → p. 190
Neutrino Masses → p. 272, Neutrino Oscillations → p. 198

The Invisible Width of the Z Boson

There Are Only Three Neutrino Flavors!

In this article, we take a closer look at the properties of the Z boson. From 1989 to 2000, CERN operated the *Large Electron-Positron Collider*, LEP, in whose tunnel the LHC was later built. At LEP, electrons and positrons were accelerated in opposite directions around the ring, and brought to collision at four specific points, where the detectors ALEPH, DELPHI, L3 and OPAL were located. The resulting events were examined in detail independently by each experiment and then compared.

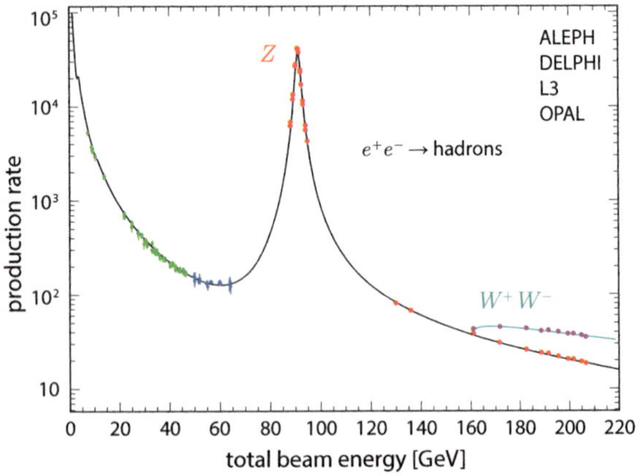

Observations (data points) and theoretical calculations (curve) of the production rate. There is a maximum at the energy of the Z boson.

We now take a closer look at one process: $e^+e^- \to q\bar{q}$, the creation of a quark-antiquark pair from an electron and positron collision. There are two corresponding Feynman diagrams, with either the exchange of a photon (γ) or a Z boson. If you add and then square the sum of the mathematical expressions of these two diagrams, the result is proportional to the production rate, i.e., the number of $q\bar{q}$ pairs produced per e^+e^- collision (Feynman II ↓). This production rate is shown in the figure as a function of the combined beam energy of the e^- and e^+ beam (black curve; note the scale on the left is logarithmic). The quarks are observed as hadrons (↓). At low total beam energy, 20 to 60 GeV, the photon exchange dominates and the rate decreases with increasing beam energy. From about 70 GeV the Z boson exchange dominates and the rate rises rapidly before falling again. Over the entire curve, the calculations agree very well with the observations.

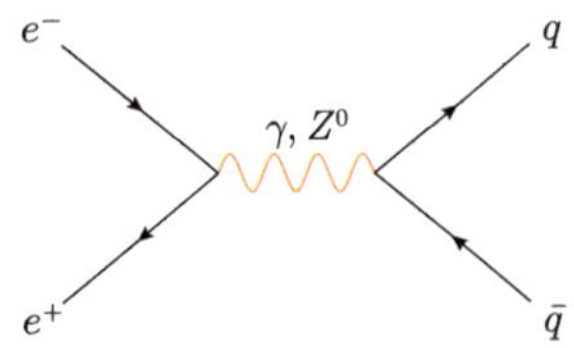

Feynman diagram for the generation of muons from the collision of an electron and a positron

The mass of the Z boson can be determined from this curve. To do this, the beam energy is varied in small steps in the accelerator, and at each value the production rate is measured. The peak of the resulting curve is very close to the Z boson mass: 91.2 GeV$/c^2$. The fact that there is such a peak is a *resonance phenomenon*. With the swing in the article on resonances (↓), pushing at the natural frequency of the swing, led to

LEP: Why the Standard Model? → p. 244
Feynman Diagrams II → p. 102
Seeing Quarks and Gluons? → p. 216

maximum deflection. At LEP, close to maximal production was achieved when the total beam energy was very close to the rest mass energy of the Z boson.

Further information can be obtained from the height of the maximum and from the width of the curve at half height. This width at half height is denoted Γ_Z (pronounced "Gamma Z") and is shown in the second figure for three values of the number of neutrinos. Γ_Z can also be calculated. It depends directly on the decay possibilities of the Z boson. There are many different ones: the Z boson can decay into pairs of charged leptons, pairs of neutrinos, or into quark- antiquark pairs.

$$Z \rightarrow \begin{cases} e^+e^-,\ \mu^+\mu^-,\ \tau^+\tau^-,\ \nu_e\bar{\nu}_e,\ \nu_\mu\bar{\nu}_\mu,\ \nu_\tau\bar{\nu}_\tau\ ; \\ u\bar{u},\ d\bar{d},\ s\bar{s},\ c\bar{c},\ b\bar{b}\ . \end{cases}$$

A decay into $t\bar{t}$ is not possible because the top quark is heavier than half the Z boson mass. The neutrino pairs cannot be observed.

The more such decays there are, the greater the width of the resonance curve (Γ_Z) and the lower the maximum. This experiment can therefore be used to determine the number of neutrinos that couple to the Z boson and with less than half the Z boson mass. And this, even though the neutrinos themselves are not observed. In the right figure, only the Z-resonance part of the previous curve is shown. There are now three curves: the middle curve in green is that of the Standard Model with three neutrinos, identical to the previous figure. The upper, red curve would be observed if there were only two neutrinos instead of three. The lower curve in orange would be measured if there were four light neutrinos.

We see that the upper, red curve at the maximum is significantly higher than the green curve. The lower, orange-colored curve is significantly lower at the maximum than the green curve. We have also drawn the respective widths at half the height of the maximum as dashed lines in the corresponding color. The data in red, especially the data point at the maximum, clearly favors the curve with three light neutrinos. If we consider the change in the number of neutrinos as a continuous parameter, the measured number of neutrinos from the LEP is given by $N_\nu = 2.994 \pm 0.012$, where 0.012 is the uncertainty of the measurement. The result is in excellent agreement with the three neutrinos of the Standard Model! This is astonishing: without ever having searched directly for these neutrinos, we can rule out the possibility that there are other light neutrinos, i.e., those with a mass below half the Z boson mass, and which couple to the Z boson.

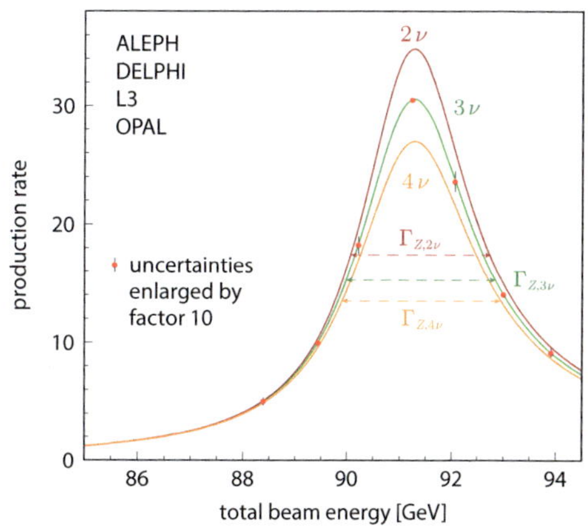

Extract from the previous figure (resonance part), with curves for 2, 3 or 4 neutrinos. Note the y-axis is now linear.

Resonances → p. 56
The Accidental Discovery of the J/ψ → p. 238
Measurement Uncertainties → p. 86

Spontaneous Symmetry Breaking I

The Higgs Boson Spoils the Party

In the article on spontaneous symmetry breaking (↓) we saw how a symmetry can disappear spontaneously. If a permanent magnet is cooled from a high temperature, when it is non-magnetic, to a low temperature, it turns magnetic with a certain orientation. The original rotational symmetry around three independent axes is reduced to a symmetry about only one axis. This way of breaking a symmetry also exists with the gauge symmetries (↓) in the Standard Model. Here, the initial state with high temperatures is the very hot plasma of elementary particles in the early Universe. At this early time, the gauge symmetry was the so-called $SU(2)_L \times U(1)_Y$, which corresponds to the weak interaction ($SU(2)_L$) together with another interaction ($U(1)_Y$) called *hypercharge*.

Since the Big Bang (↓), the Universe has been expanding, and it has cooled down in this process. As the temperature decreased, the potential of the Higgs boson deformed, so that below a temperature of about 246 GeV ($3 \cdot 10^{15}$ degrees Celsius) the symmetry was broken to the smaller symmetry $U(1)_{\text{EM}}$ of electromagnetism (QED, ↓). Thus in total we have

$$SU(2)_L \times U(1)_Y \xrightarrow{\text{temperature}\downarrow} U(1)_{\text{EM}}$$

We will now take a closer look at what happens to the Higgs bosons during symmetry breaking. Similar to the leptons, for example, the four Higgs fields at high energies form two $SU(2)_L$ pairs (Up, Down, Left, Right ↓)[1]

$$\begin{pmatrix} H^+ \\ H^0 \end{pmatrix}, \begin{pmatrix} H^- \\ \bar{H}^0 \end{pmatrix},$$

These couple to the W boson. The second pair contains the antiparticles of the first pair. The $\bar{H}^0$ has the opposite $U(1)_Y$ hypercharge to the H^0. Associated with the two fields H^0 and $\bar{H}^0$ there is a potential energy in the system. We show it schematically in three dimensions in the two images below. The axes in the flat area are the values of the Higgs fields H^0 and $\bar{H}^0$, the vertical axis represents the potential energy in the system, denoted V_{Higgs}. The larger the value of the Higgs field, the higher is the potential energy.

The left image shows the potential at very high temperatures, the right image shows the potential far below 246 GeV, e.g., at room temperature. At very high

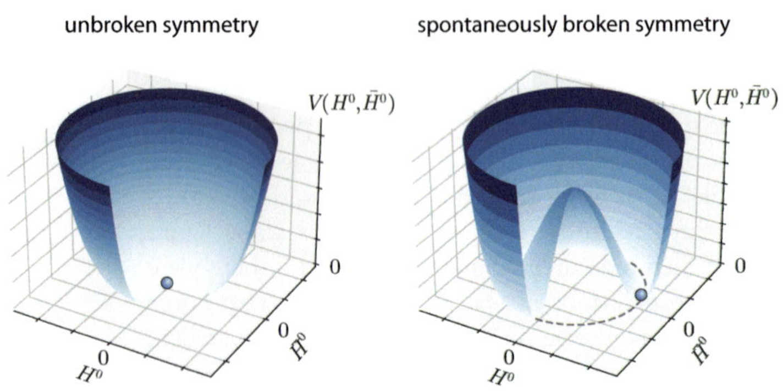

Higgs potential at high (left) and low (right) symmetries

[1] The charged Higgs bosons and the $\bar{H}^0$ get absorbed by the $W^\pm$ and the Z boson, respectively, when they become massive. A massless spin-1 boson has two, and a massive boson has three polarizations.

Spontaneous Symmetry Breaking → p. 62
Gauge Symmetries → p. 98

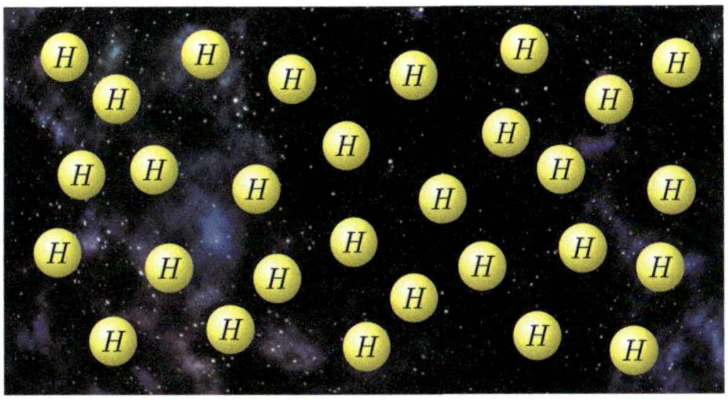

The Universe is evenly filled with Higgs bosons.

temperatures the potential looks like a salad bowl. It is rotationally symmetrical around the vertical axis and the minimum of the energy and thus the preferred ground state of the system is at the center, the origin, at the bottom of the salad bowl. This is where the entire Higgs system is located energetically, shown here as a small blue sphere. At this point, the ground states of the two fields H^0 and $\bar{H}^0$ are zero, so they disappear. At this temperature in the hot early Universe, the Higgs fields are created and annihilated in reactions, just like top quarks, for example, but there is no constant background of Higgs particles and therefore all elementary particles have a vanishing mass.

As the temperature drops, the potential deforms, as shown on the right. A bump forms at the center, which is surrounded by a "gutter". The bowl is still rotationally symmetrical, but the minimum is no longer at the origin. It is located in the gutter, some distance from the center. Therefore, the system shifts to a point in the gutter. All points in the gutter are energetically equal (rotational symmetry). However, the system must select a point as the ground state—the blue sphere must go somewhere—shown here along the H^0 direction.

In this state, the H^0 field has a constant value, the vacuum expectation value: $\langle H^0 \rangle = 246$ GeV. This means that the whole Universe is uniformly filled with a background of Higgs particles, as shown in the left picture The other particles, e.g., the electrons, necessarily scatter off these Higgs particles as they move through the Universe, and thus obtain their mass. The Feynman vertex for the electron-Higgs coupling is shown below: A left-handed electron e_L^- couples together with a right-handed electron e_R^- to the Higgs boson.

As always, all quantum numbers must be preserved at this vertex. Thus, on the left, an e_L^-, which is part of an $SU(2)_L$ pair, transforms into an e_R^-, which is *not* part of such a pair. This is fine because the two couple to the H^0, which is also part of a pair. The $SU(2)$ pair number is conserved, so to speak.

When an electron interacts with a background Higgs boson, we do not see the Higgs part of the reaction. This is shown in the right-hand image with the ⊗ for the Higgs boson. For us, it looks as if an e_L^- simply transforms into an e_R^-. The $SU(2)$ pair number is violated, the corresponding quantum number is *no longer* conserved. The symmetry is spontaneously broken! As we can see, in this reaction the electric charge is still conserved. This is the remaining symmetry $U(1)_{\text{EM}}$.

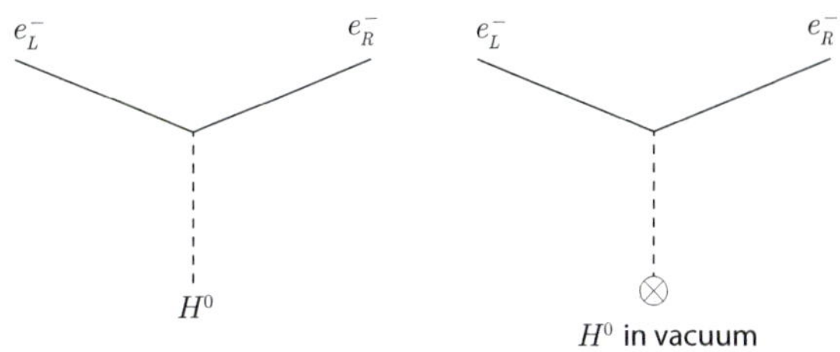

Electron-Higgs coupling

Big Bang → p. 280
Electromagnetism → p. 176
Up, Down, Left, Right → p. 180

Spontaneous Symmetry Breaking II

The Origin of the Photon

As described in the previous article (↓), in the early Universe at temperatures far above 246 GeV ($3 \cdot 10^{15}$ degrees Celsius), the Standard Model has the gauge symmetry $SU(2)_L \times U(1)_Y$. The $SU(2)_L$ part includes three gauge bosons, the charged W bosons: $W^{\pm}$, and a neutral W boson: W^0. The $U(1)_Y$ gauge symmetry, also called *hypercharge*, includes an electrically neutral gauge boson, denoted B^0. The symmetry predicts how the gauge bosons couple to the Higgs boson. For example, two charged W bosons couple to two neutral Higgs bosons, as shown on the left.

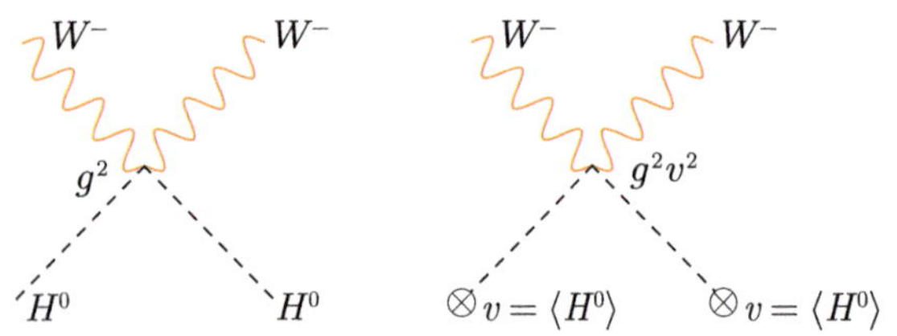

Coupling of charged *W* bosons to Higgs bosons

When the temperature in the early Universe dropped below 246 GeV, the symmetry was spontaneously broken. As a result, the Higgs boson has a vacuum expectation value, $v = \langle H^0 \rangle$, which is shown in the figure on the right as $\otimes$.

This leads to new interactions: The W bosons scatter off the background of Higgs bosons, like the electrons in the previous article. We do not see the Higgs bosons themselves, we have a pure W-W coupling. This is exactly what gives the $W^{\pm}$ boson a mass. This is the Higgs mechanism: through spontaneous symmetry breaking, the gauge bosons scatter on the Higgs background and acquire a mass.

The Feynman graph on the left in the figure is proportional to the $SU(2)_L$ gauge coupling squared: g^2. The graph on the right in the figure is then proportional to g^2v^2. One can show that the Higgs mechanism in the Standard Model predicts that the $W^{\pm}$ boson mass is given by $M_W^2 = ½\, g^2v^2$.

Now let's take an analogous look at the neutral gauge bosons, where things become a little more complicated. But it leads us to the somewhat unusual properties of the Z boson and to the massless photon of electromagnetism.

As mentioned, the $SU(2)_L$ gauge symmetry implies a third gauge boson, the W^0. This couples to the Higgs bosons as shown in the top left of the figure at the top of the next page: two W^0 interact with two H^0. The diagram is proportional to g^2, as in the previous figure. The gauge boson of the $U(1)_Y$ hypercharge, B^0, also couples to the Higgs boson, as shown at the top right of the figure, now two B^0 couple to two H^0. However, this diagram is proportional to g'^2, where g' is the coupling constant of the hypercharge interaction $U(1)_Y$.

There is another coupling for the neutral gauge bosons, shown at the bottom center of the figure. Here, a B^0 *and* a W^0 couple together to the two Higgs bosons. The diagram is proportional to gg'.

Spontaneous Symmetry Breaking I → p. 184

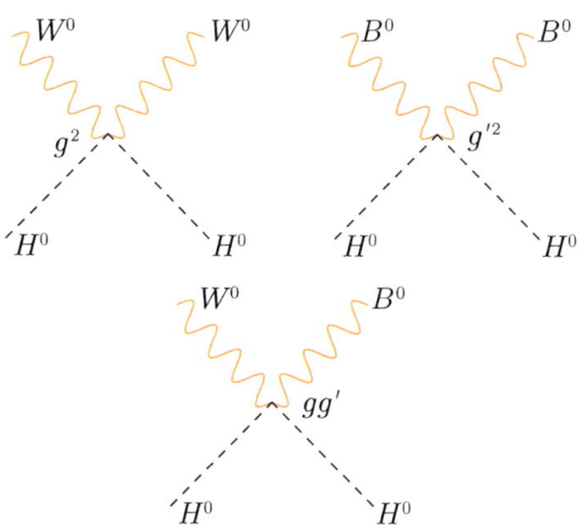

Coupling of the uncharged gauge bosons B^0 and W^0 to the Higgs

In the final step, we again replace the Higgs bosons with a coupling to the vacuum, $\otimes$, proportional to v, as shown in the figure below.

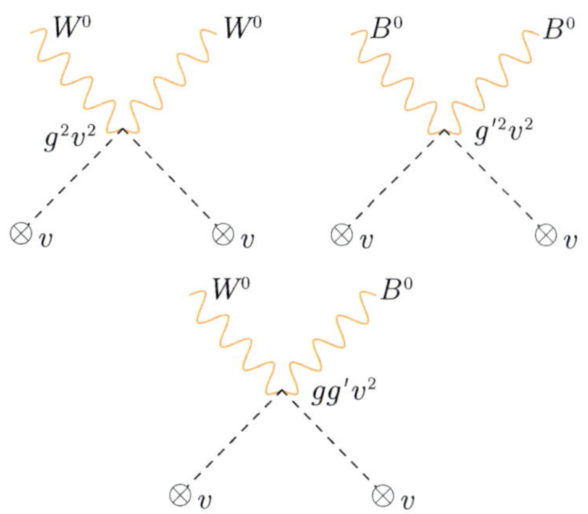

Coupling of the uncharged gauge bosons in vacuum

With the three couplings, we now have couplings of the form W^0W^0, B^0B^0, as well as a mixed term: W^0B^0. These mass terms are respectively proportional to v^2g^2, $v^2g'^2$, and v^2gg'. The last term contains both gauge couplings.

We have two apparently pure mass terms: W^0W^0 and B^0B^0, but also a mixed mass term: W^0B^0. To determine our gauge bosons and their masses, we have to rewrite the W^0 and the B^0 and express them by new bosons. This is illustrated in the figure below. It corresponds to a rotation of two vectors in the plane. The new fields are the photon γ and the Z boson. The angle of rotation is called the *electroweak mixing angle* θ_W.

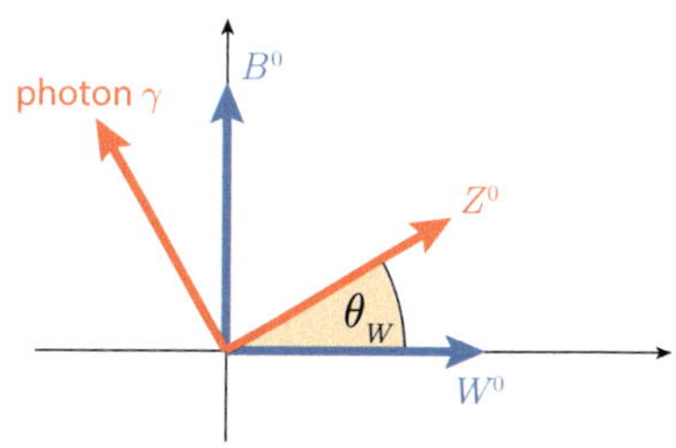

Mass mixing of the gauge bosons

The resulting particle is exactly the photon of electromagnetism with the corresponding couplings to the electric charge. However, these can be expressed by the other couplings, here, e.g., $e = g/\sin\theta_W$. The coupling of the Z^0 to the e_L^- is different from that to the e_R^- and also depends on $\sin\theta_W$. Another important prediction of this theory is the ratio of the W and Z boson masses $M_W/M_Z = \cos\theta_W = \sqrt{1-\sin^2\theta_W} = \sqrt{e^2-g^2}/e$, which has been confirmed experimentally.

Spontaneous Symmetry Breaking III

Relating the Bottom Quark Mass to the Higgs Boson

In the Standard Model, fermions couple directly to the Higgs boson. In the figure below we show as an example the couplings to an electron and to a bottom quark. It is always one left-handed and one right-handed fermion interacting (↓). The coupling strength for the electron is denoted h_e and for the bottom quark h_b, analogously for the other fermions—except for the neutrinos. Since there is no right-handed neutrino in the Standard Model, the neutrinos do *not* couple to the Higgs boson and are therefore massless (↓) in the Standard Model. The coupling parameters h_e, h_b, ... are free parameters, which need to be fixed by experiment.

At temperatures above 246 GeV in the early Universe, the Higgs field sits at the minimum of the potential at the origin and has no vacuum expectation value (↓). In this phase, the electron is massless, as it can only interact with free Higgs bosons. If the temperature in the Universe falls significantly below 246 GeV, the Higgs potential deforms and the Higgs boson gets a vacuum expectation value. Then there is a background of Higgs bosons on which electrons and bottom quarks can scatter. This is denoted in the figure by a ⊗ together with a v,

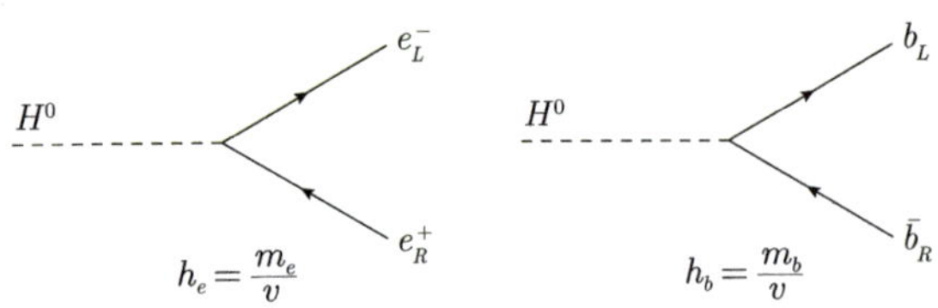

Coupling of an electron and a bottom quark to a Higgs boson, respectively

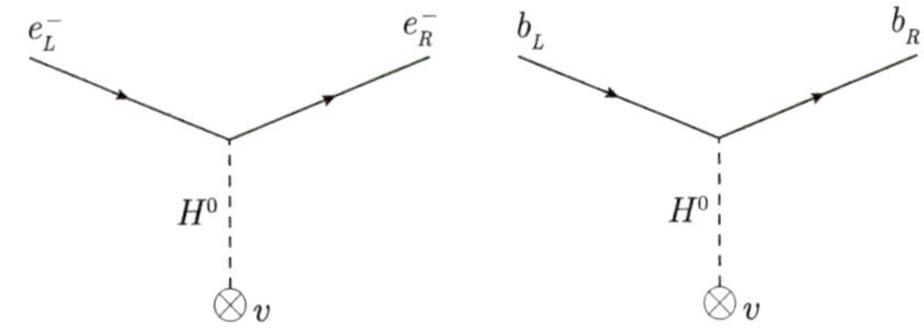

Electron and bottom quark mass terms, coupling to the Higgs vacuum expectation value

for the vacuum expectation value. This background of Higgs bosons is identical everywhere (homogeneous) and independent of the direction (isotropic). The electron acquires a universal property through scattering on this universal background: the *electron mass*. It is the same for the other fermions: μ, τ and the quarks u, d, s, c, b, t.

This Higgs mechanism determines the mass through the coupling to the Higgs field multiplied by the value v:

$$m_e = h_e \cdot v \ , \quad m_b = h_b \cdot v \ .$$

This also applies analogously to the other fermions except the neutrinos. It turns out that all particles in the Standard Model obtain their mass through the Higgs mechanism: the leptons and quarks, but also the W and Z bosons. The photon and the gluon remain massless, as they do not couple directly to the Higgs field (↓).

This is the *origin* of the mass of the elementary particles. Our own mass, however, comes primarily from the mass of the protons and neutrons in our body,

Up, Down, Left, Right → p. 180
Neutrino Masses → p. 272
Spontaneous Symmetry Breaking I → p. 184
Spontaneous Symmetry Breaking II → p. 186

which have their origin in the strong interaction (↓). The electrons contribute less than 0.1 % to the atomic mass: mass(electron)/mass(proton) = 0.054 %.

Interestingly, the atomic radius, i.e., the size of all atoms, is inversely proportional to the electron mass. If the electron would couple twice as strongly to the Higgs boson, it would be twice as heavy, but all atoms would only be half the size.

The masses of the fermions have been measured experimentally with great precision. We can reverse the above equations and write for the electron and the bottom quark:

$$h_e = \frac{m_e}{v} \ , \qquad h_b = \frac{m_b}{v} \ .$$

This means that we determine the Higgs coupling parameters by measuring the mass, as the vacuum expectation value is known. In the Higgs mechanism, the Higgs boson couples to the fermions in proportion to the fermion mass! In particular, the coupling to bottom quarks is larger than that to electrons by a factor $m_b/m_e \approx 8.000$. The gauge bosons $W^{\pm}$ and Z^0 also couple to the Higgs boson in proportion to their mass.

A large number of Higgs bosons (↓) have already been produced at the LHC. The probability that a Higgs boson decays into a specific fermion pair is directly proportional to the square of the coupling, as the mathematical expression of the Feynman diagram (↓) must be squared. The decay probabilities are:

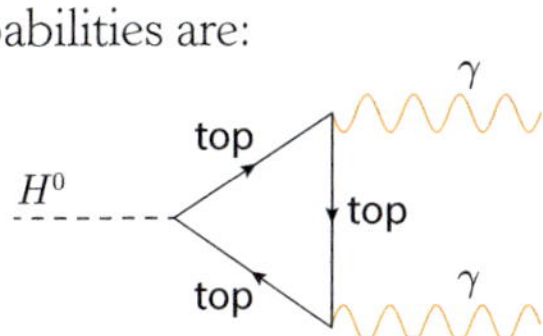

Decay of a Higgs boson to two photons via a virtual top quark loop

$$\text{probability:} \begin{cases} H^0 \to b\bar{b} & = 57.7\,\% \\ H^0 \to \tau^-\tau^+ & = 6.3\,\% \\ H^0 \to c\bar{c} & = 3.2\,\% \\ H^0 \to \mu^-\mu^+ & = 0.02\,\% \\ H^0 \to e^-e^+ & = 5{\cdot}10^{-9} \\ H^0 \to \gamma\gamma & = 0.04\,\% \\ H^0 \to W^+(W^-)^* & = 21.7\,\% \\ H^0 \to Z^0(Z^0)^* & = 2.6\,\% \end{cases}$$

The decay into bottom quarks is 100 million times more likely than into electrons, as the bottom quark is much heavier than the electron. The Higgs decay to two photons is an exception. The decays marked in blue have already been observed with the predicted frequency, within the experimental errors.

The Higgs boson was first observed via the relatively rare decay into two photons. The corresponding Feynman diagram is shown at the bottom left, with virtual top quarks as intermediate states. The observation of the photons was experimentally much easier than bottom quarks.

The Higgs boson cannot decay directly into two W or Z bosons because together they are heavier than the Higgs boson. One W or Z boson as a decay product must therefore be virtual, which is indicated by the *.

Decay of a Higgs boson into a real and a virtual W boson.

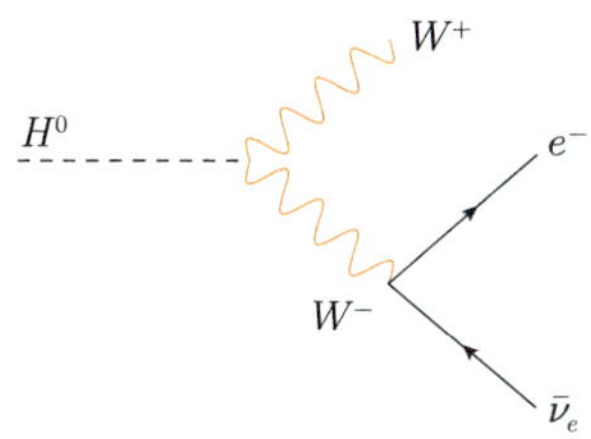

Where Does Our Mass Come From? → p. 220
Higgs boson: Is It Really the Higgs Boson? → p. 248
The Discovery of the Higgs Boson → p. 246
Feynman Diagrams → p. 48
On the Ranges of Interactions → p. 58

Discrete Symmetries: Violating *P* and *CP*

The Mirrored World Is Different

Beyond gauge symmetries (↓), so-called *discrete symmetries* are also very important in particle physics. These include the point reflection in space, called parity P, the time reversal T, and the charge conjugation C. In the figure below, the blue triangle is reflected at the origin O into the red triangle. We can see, for example, that the vector $\overrightarrow{OB}$ turns into $\overrightarrow{OB'} = -\overrightarrow{OB}$. Similarly time reversal can be imagined as a movie running backwards. Charge conjugation converts particles into antiparticles and vice versa.

Let's assume we film the collision between two balls on a billiard table (without friction). Then we show the original film, a spatially mirrored one, and a third film running backwards. Without having seen the original collision, you cannot say which is which. Originally, it was expected that these symmetries are universally valid, in particular also for the most elementary interactions between particles. And indeed, we have found that experimentally the electromagnetic (QED) and the strong interactions (QCD) always conserve C, P, and T with very high accuracy (Strong *CP* Problem ↓). However: the weak interactions do *not*!

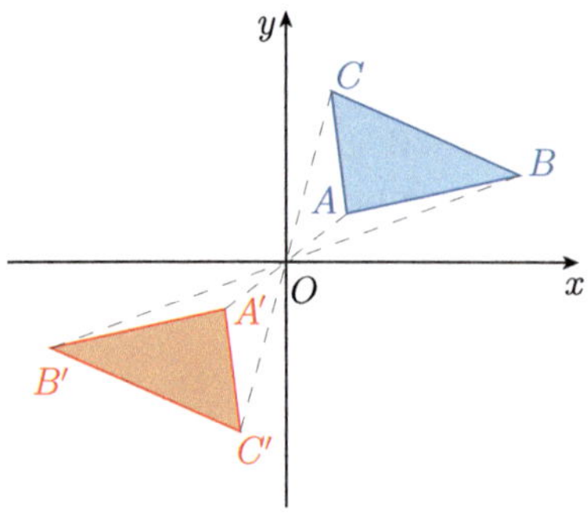

Point reflection of a triangle at the origin

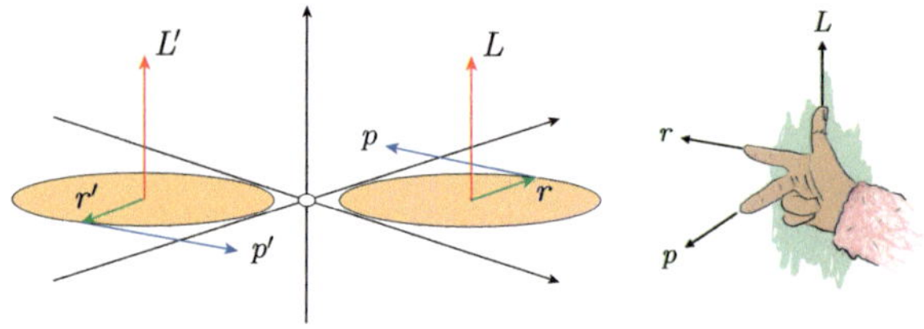

Reflection of the angular momentum: The right-hand rule still applies, the angular momentum is unchanged.

For the characterization of an elementary particle it is important to know how the *helicity* (↓) behaves under reflections (P). The helicity is the component of the spin (↓) in the direction of the particle's own momentum.

The figure at the top shows how the angular momentum behaves (Kinematics ↓) under point reflection. It is the product of the position vector and the momentum vector, where the three together follow the right-hand rule (see figure). Under the point reflection, the position vector $\vec{r}$ and the momentum vector $\vec{p}$ are exactly reversed and we get $\vec{r}\,' = -\vec{r}$ and $\vec{p}\,' = -\vec{p}$. If we apply the right-hand rule, we see that $\vec{L}\,' = \vec{L}$ is unchanged. The same applies to the spin. Under P the spin does *not* change.

Gauge Symmetries → p. 98
The Strong *CP* Problem → p. 290
Helicity: Up, Down, Left, Right → p. 180
Spin → p. 44
Kinematics → p. 40

Now let's take a look at the decay of a pion into a muon and a neutrino. Since the momentum reverses under parity, but the spin does not, the helicity changes sign under P as does the chirality. Thus right-handed particles change into left-handed and vice versa. This applies to the muon. However, this does not apply to neutrinos (↓), as there are no right-handed neutrinos. This is also different for the pion. It does not change under parity as it has no spin. Schematically:

$$P\colon (\pi^+,\ \mu_R^+,\ \nu_L) \Longleftrightarrow (\pi^+,\ \mu_L^+,\ \cancel{\nu_R}\ )\ .$$

However, this means that when a charged pion decays, the decay probabilities, D_P, connected by parity, are *not* equal:[1]

$$99\,\% = D_\mathrm{P}(\pi^+ \to \mu_R^+\ \nu_L) \neq D_\mathrm{P}(\pi^+ \to \mu_L^+\ \nu_R) = 0\ .$$

Parity is therefore maximally violated in the weak interaction decay of the pion.

Under charge conjugation C, particles change into antiparticles and vice versa, without change of helicity. This is possible for the pion and muon, but again not for the neutrino because there is no $\bar{\nu}_L$:

$$C\colon (\pi^+,\ \mu_R^+,\ \nu_L) \Longleftrightarrow (\pi^-,\ \mu_R^-,\ \cancel{\bar{\nu}_L}\ )\ .$$

We have for the decay probabilities:

$$99\,\% = D_\mathrm{P}(\pi^+ \to \mu_R^+\ \nu_L) \neq D_\mathrm{P}(\pi^- \to \mu_R^-\ \bar{\nu}_L) = 0\ ,$$

so C is also maximally violated in weak interaction processes.

Now we can combine the two transformations into one:

$$CP\colon (\pi^+,\ \mu_R^+,\ \nu_L) \Longleftrightarrow (\pi^-,\ \mu_L^-,\ \bar{\nu}_R)\ ,$$

which is viable and indeed we find for the decay probabilities

$$D_\mathrm{P}(\pi^+ \to \mu_R^+\ \nu_L) = D_\mathrm{P}(\pi^- \to \mu_L^-\ \bar{\nu}_R)$$

in agreement with observation.

CP seems to be a fundamental symmetry in Nature, and all weak interaction reactions preserve CP. Really all of them? No, not quite. In the Standard Model there is still a small violation of CP symmetry. Small means that CP-violating reactions occur very rarely.

The quarks d, s, b can mix in a similar way to neutrinos (↓). Via the Feynman diagram shown, the K^0 and the $\bar{K}^0$ meson mix with each other. And just like neutrinos they can convert into each other through oscillations. This violates CP and has in fact been experimentally observed in 1964. The analogous effect for B^0-$\bar{B}^0$ and D^0-$\bar{D}^0$ oscillations is described in the article Shapeshifters Among Themselves (↓).

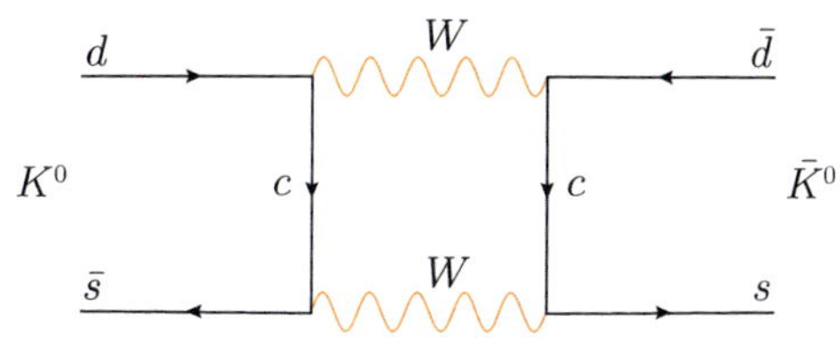

Mixing of kaons

A fundamental theorem of quantum field theory states that the combined symmetry CPT must *always* be preserved. Thus, a violation of CP is accompanied by a corresponding violation of the time reversal T. A CPT violation would be present, e.g., if the mass or the lifetime of a particle and of its antiparticle were different. However, these differences have never been observed to-date (Particle Traps ↓).

[1] In 1 % of the cases a pion decays to an electron and a neutrino.

Neutrinos → p. 196
The Path to Six Quarks in Three Families → p. 194 and Measurement of the Neutrino Masses → p. 256
Shapeshifters Among Themselves → p. 262
Particle Traps → p. 140

The Weak Decay of Pions

Left, Right, Faster, Slower: It All Matters!

Is it just a matter of definition what we call right and what we call left? Or does physics differentiate fundamentally between left and right? It turns out that one fundamental force, the weak interaction (↓), indeed does notice the difference. This is incredibly intriguing as the mirror symmetry (↓) P (or "parity") was long thought to be a fundamental symmetry of our Universe. Gravity and electromagnetism are invariant under an exchange of left and right. And even in the strong force, we see currently no evidence that parity as a symmetry is violated.

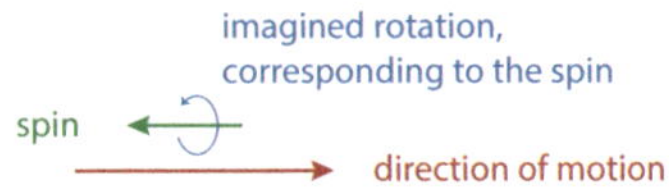

A left-handed particle with negative helicity $H = -1/2$: In the direction of motion, the imagined rotation corresponding to the spin is to the left (clockwise).

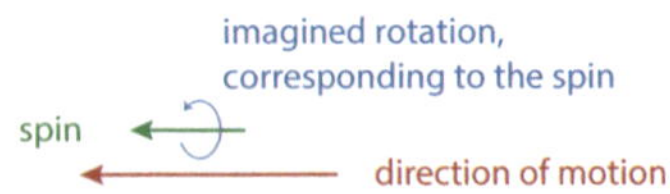

A right-handed particle with positive helicity $H = +1/2$: In the direction of motion, the imagined rotation corresponding to the spin is to the right (counterclockwise).

To build some intuition, we can use a simple example: Our right and left hands are nearly perfect mirror images of each other. However, this doesn't mean that left and right hands are identical: You for instance cannot wear the right-hand glove on the left hand, no matter how symmetrical your hands are! Amazingly, physics can assist you in the *definitive* determination what is right and what is left. For instance, spin (↓) is an intrinsic angular momentum of a particle. Helicity is the projection of spin onto the direction of motion. A left-handed particle is one whose spin corresponds to a leftward rotation (counterclockwise) when viewed along the direction of motion. Conversely, the angular momentum of the spin of a right-handed particle "turns" right (clockwise).

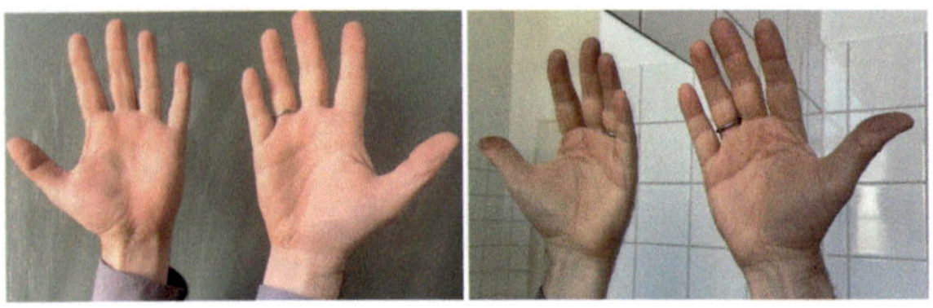

Two versions of right and left hands. But only one version is a reflection in a conventional mirror! Can you find the symmetry violation in one of the two images that allows you to clearly distinguish the mirrored image from the one with two different hands?

There is another property every particle has: *chirality*. While for nearly massless particles like neutrinos, chirality and helicity do coincide, once particles possess mass and cannot propagate with the speed of light, these properties start to differ. With a probability of $(1 - v/c)^2$, one finds a right-chiral particle as having left-handed helicity or vice versa (Up, Down, Left, Right ↓). Both properties are thus tied to the speed at which particles can travel. Particles whose energy is much larger than their rest mass travel nearly at the speed of light $v = c$. Therefore, for nearly massless particles, the probability of a difference between chirality and helicity is $(1 - c/c)^2 = 0$.

The Weak Interaction → p. 178
Symmetries → p. 60
Spin → p. 44
Up, Down, Left, Right → p. 180

The allowed decay of a π^+ with spin 0 into a left-handed (and therefore also left-chiral) neutrino ν_L

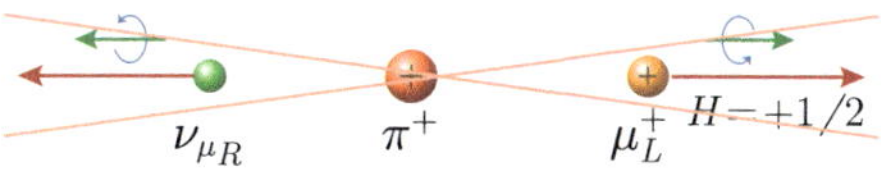

The forbidden decay of a π^+ with spin 0 into a non-existent right-handed (and therefore also right-chiral) neutrino ν_R

To define right and left in physics unequivocally, we need to measure a process or reaction that violates mirror symmetry. The decay of the charged pion $\pi^\pm$ has precisely this property. The pion possesses spin 0, thus its angular momentum is zero. Since the total angular momentum is conserved in any decay (Noether's Theorem ↓), and the spin also contributes to the angular momentum, the spins of the decay products in a decay into two particles must point in opposite directions to precisely add up to zero. Otherwise angular momentum would not be conserved. The π^+ decays with a probability of $(99.98770 \pm 0.00004)\,\%$ into a pair of antimuon μ^+ and neutrino $\nu_{\mu L}$, and only in $(0.0001230 \pm 0.0000004)\,\%$ of all cases into a positron e^+ and its associated neutrino ν_{eL}. This may initially seem very surprising: the weak interaction couples equally to all families, and therefore, the probability for both decays should be in the same order of magnitude (Rare Decays of Beauty ↓, Gauge Symmetries ↓).

The solution to the puzzle lies in parity violation: The gauge boson W^+, responsible for the π^+ decay, couples only to right-chiral antileptons $\ell^+_R = e^+_R$ or μ^+_R. However, from the right-chiral particle ℓ^+_R, a left-handed ℓ^+_L with negative helicity must emerge, for the spins of the left-handed neutrino and the lepton flying in the opposite direction to add up to the π^+ spin of zero. This requirement is dictated by angular momentum conservation. However, helicity and chirality can only disagree with a probability of $(1 - v/c)^2$ (CP and P violation ↓). The μ^+ is about 200 times heavier than the e^+, so, at a similar momentum (Kinematics ↓), it moves with a much lower velocity v_μ than the e^+ with $v_e \approx c$. The probability of encountering a left-handed muon, $(1 - v_\mu/c)^2$, is much greater than that of encountering a left-handed electron, $(1 - v_e/c)^2 \approx (1 - c/c)^2 = 0$. Therefore, it is more likely that the right-chiral μ^+_R corresponds to a particle with negative helicity than the e^+_R, which moves very close to the speed of light.

Hence, the decay rate into an antimuon dominates over the decay rate into a positron. This fact alone, without directly measuring the spin orientation, indicates that the weak interaction does not conserve parity P. Otherwise, there would always be enough ν_R, ℓ_L pairs, and electrons and muons would be produced with similar frequency.

We can thus define unequivocally: "left-handed" is the direction in which the spin of the antimuons from the decay of the positively charged pion rotates. Unfortunately, we are not quite there yet: We have assumed from the beginning that μ^- is a particle and μ^+ is its antiparticle. Establishing this *unique* distinction between matter and antimatter is achieved through CP violation—a further crucial property of the weak interaction.

Decay of a π^+, consisting of a u and $\bar{d}$ quark, into a left-chiral particle and a right-chiral antiparticle

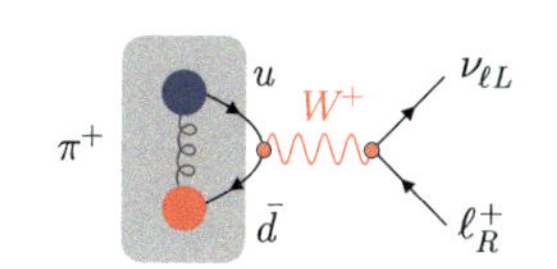

Noether's Theorem → p. 100
Rare Decays of Beauty → p. 260, Gauge Symmetries → p. 98
Discrete Symmetries: Violating P and CP → p. 190
Kinematics → p. 40
CP Violation in Beauty → p. 258

The Path to Six Quarks in Three Families

Who's Who in Interactions

With the discovery of kaons in 1947 (mesons K^0 with quark content $\bar{s}d$ and K^+ with $\bar{s}u$), particle physics suddenly encountered many peculiar properties (Quark Model ↓). For instance, the charged kaon predominantly decays into an antimuon and a muon neutrino $K^+ \to \mu^+ \nu_\mu$, but the decay of the neutral kaon into a muon and an antimuon, $K^0 \to \mu^+\mu^-$, although looked for, could not be observed at that time. The hard-earned certainty that the weak interaction universally couples to all particles, and thus the latter decay should occur (Gauge Symmetries ↓), seemed to be violated again, leaving physicists puzzled.

The first realization was that, in Nature, there exist no neutral currents with a Z^0 exchange that couple different quark flavors to each other: For example, there is the process $Z^0 \to d\bar{d}$, but not $Z^0 \to d\bar{s}$ (see illustration a). Therefore, the $W^\pm$ plays the special role of being the only interaction particle that transforms different quark flavors and quark families into each other (illustration b). No other interaction has this property. With the quark flavors known at that time (u, d, s), one could then calculate the probability of the transition $K^0 \to \mu^+\mu^-$ shown in illustration c). However, the predicted decay probability was much higher than measured in the experiment. The resolution of the mystery finally came through the idea of Sheldon Lee Glashow, John Iliopoulos, and Luciano Maiani in 1970: If there were another quark, the charm quark c, there would be another diagram d), which could destructively interfere with the decay diagram c). From this assumption the mass of the c quark was predicted correctly by a precise calculation by Mary K. Gaillard and Ben Lee.

Feynman diagrams for the GIM mechanism (left) and allowed and forbidden kaon decays (right)

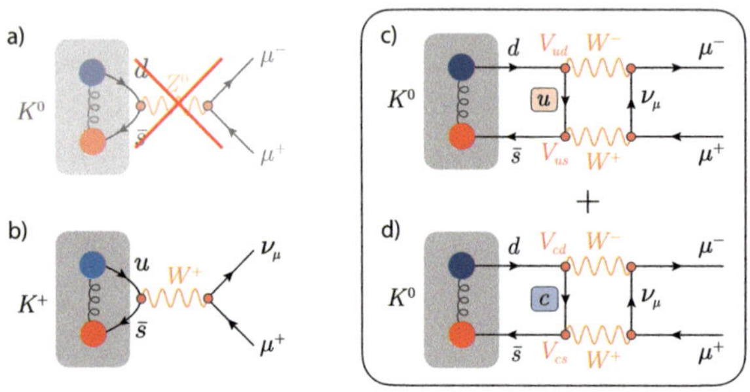

The figure of wave patterns at the next page illustrates how destructive interference in the sum can produce a vanishingly small total amplitude: The valleys and peaks of two wave functions are precisely shifted with respect to each other so that they cancel each other in the sum. In contrast, in constructive interference, valleys align with valleys, and peaks align with peaks, reinforcing each other. For a not quite perfect relative shift, this effect leads to the observed tiny decay probability. Four years later, the charm quark was indeed discovered (↓), and the explanation entered the history of physics as the *GIM mechanism*. Mathematically, destructive interference arises from the change in sign of the coupling: Feynman diagram c) with the up quark provides almost the same contribution as Feynman diagram d) with the charm quark, but with different signs. What is the underlying mechanism causing this? Physicists did not know at the time.

The Quark Model: A First Look → p. 20
Gauge Symmetries → p. 98
charm quark: The Accidental Discovery of the J/ψ → p. 238

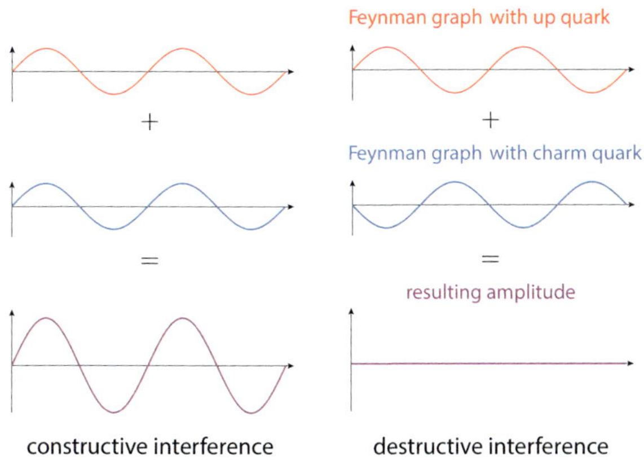

Illustration of constructive and destructive interference

The resolution to this question came from three scientists: Nicola Cabibbo, Makoto Kobayashi, and Toshihide Maskawa. (Kobayashi and Maskawa received the Nobel Prize in Physics in 2008 for their contributions.) Today, we know that there are not only two families $[(u, d), (s, c)]$ but three families with six quarks $[(u, d), (s, c), (t, b)]$ that couple to the $W^{\pm}$ boson. The quarks that the Higgs boson 'sees' (and thereby imparts mass to) are a mixture (↓) of the quarks that couple to the $W^{\pm}$ boson and interact weakly.

This mixture can be expressed as a rotation, elegantly represented in the form of the so-called *CKM matrix* (named after Cabibbo, Kobayashi, and Maskawa). Physicists usually write this matrix using the notation

$$V_{\mathrm{CKM}} = \begin{pmatrix} V_{ud} & V_{us} & V_{ub} \\ V_{cd} & V_{cs} & V_{cb} \\ V_{td} & V_{ts} & V_{tb} \end{pmatrix}.$$

The individual elements of the matrix parameterize how strongly individual quarks couple to each other via the $W^{\pm}$. The measured interaction strengths V_{ud}, etc., are shown in the illustration to the right as the thickness of the lines.

The coupling between up and down quark (V_{ud}) as well as charm and strange quark ($V_{cs} \approx V_{ud}$) are nearly equal. The coupling between up and strange quark (V_{us}) in magnitude is almost as large as the coupling between down and charm quark ($V_{us} \approx -V_{cd}$), but with the opposite sign. This is because the matrix between mass states and interaction states causes a rotation. When you rotate an object, one side lowers (negatively) just as much as the other side lifts (positively). We observed exactly this combination of effects in the GIM mechanism! The different signs lead to the destructive interference of both contributions.

It can be shown that the entire CKM matrix can be described with only four parameters, namely, three rotation angles and an additional parameter, the *KM phase* (named after Kobayashi and Maskawa) connected to *CP* violation. These angles and the phase cannot be predicted by the Standard Model and must be determined experimentally.

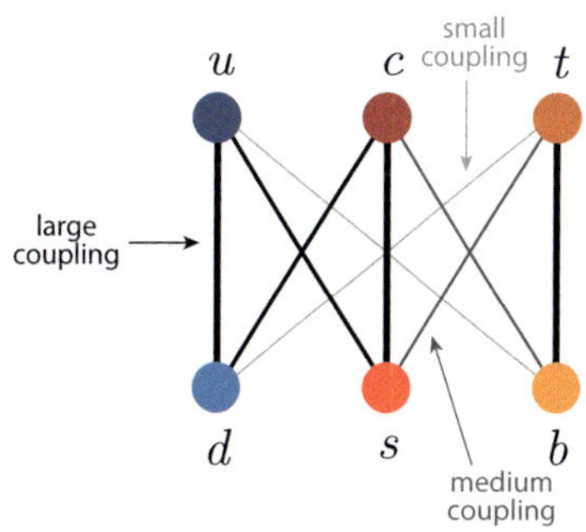

Strength of the effective coupling to the three families of quarks

Spontaneous Symmetry Breaking II → p. 186

Neutrinos

The Heavy Lightweights Among the Elementary Particles

Neutrinos are special in several respects. Their mass is orders of magnitude smaller than that of all the other fermions, and they are the only elementary fermions without an electric or strong charge. This is why they only participate in the weak interaction.

The existence of neutrinos was proposed by Wolfgang Pauli in 1930 to explain the apparent violation of energy conservation in beta decay. In this process, a neutron transforms into a proton, whereby an electron and an antineutrino are emitted. Initially, only the emission of an electron was observed with the technology available at the time. The energy conservation seemed to be violated, as the electron sometimes lacked more or less of the expected energy. The proton only receives a negligible amount of energy due to its comparatively much larger mass.

Should we question the conservation of energy? Pauli's proposal was that, in addition to the electron, another uncharged and therefore (for the available detectors) invisible particle (the neutrino) is emitted, which shares the available energy with the electron. Fortunately, neutrinos have turned out to be not completely invisible. It took a long time, but in 1956 neutrinos coming from a nuclear reactor were detected.

On average only one out of 10^{16} neutrinos from the nuclear reactor that passed through the detector interacted with the detector material. These small reaction rates make it very difficult to study them, as even very large detectors (↓)—for example, Super-Kamiokande, shown in the picture on the next page—rarely record signals from neutrinos.

All elementary fermions such as quarks, electrons and neutrinos have spin 1/2 (↓). The spin points either in the direction or in the opposite direction to the direction of propagation of the particle. As the spin has a sense of rotation we speak of right-handed or left-handed particles, analogous to screws with right-handed or left-handed threads. Normally, we observe all fermions with equal frequency: right-handed or left-handed. Not so neutrinos! As far as we know, they occur in Nature exclusively left-handed as neutrinos and exclusively right-handed as antineutrinos. The reason for this is that the weak interaction (↓) only couples to left-handed particles or right-handed antiparticles (Symmetries ↓ and *P* and *CP* violation ↓), and neutrinos can only be produced or detected in this way. The question arises: Do right-handed neutrinos even exist? Related to this is the question: What is the difference between neutrinos and antineutrinos?

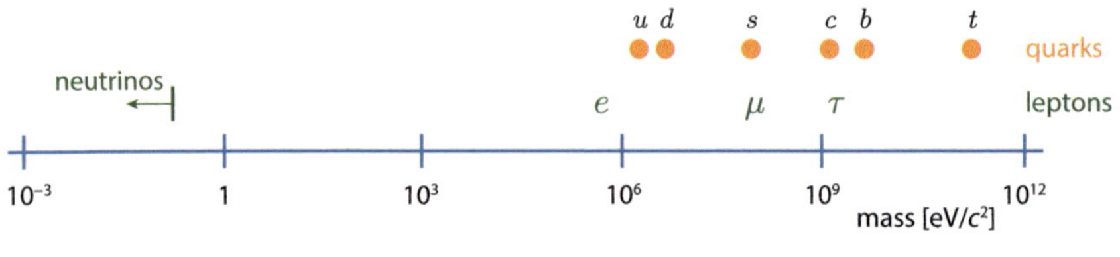

Masses of the elementary fermions. In orange the quarks, in green the leptons. Neutrino masses are unusually light.

Large Detectors Deep in a Mountain → p. 132
Spin → p. 44
The Weak Interaction → p. 178
Symmetries → p. 60
Discrete Symmetries: Violating *P* and *CP* → p. 190

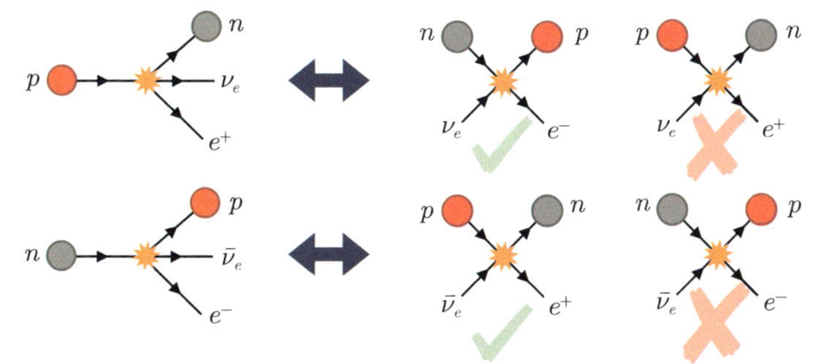

There is a difference between the neutrinos produced during the conversion of protons to neutrons and the neutrinos produced during the reverse conversion. This can be recognized by examining the possibilities for reverse reactions.

In the conversion of protons into neutrons through the weak interaction, such as in the Sun, neutrinos are produced, whereas the reverse conversion of neutrons to protons, e.g., in nuclear reactors, produces antineutrinos. Thus the first neutrinos observed in reactors in 1956 were antineutrinos! With neutrinos from the Sun, it is possible to convert neutrons back into protons, but not protons into neutrons. With the antineutrinos from reactors, it is exactly the opposite, which is why neutrinos and antineutrinos must be different.

Since neutrinos are uncharged, it could in principle be possible that they are their own antiparticles and that the difference in the reactions lies solely in the fact that the neutrinos from the Sun are left-handed, while those from the reactors are right-handed. The right-handed neutrinos would be those particles which we had previously interpreted as antineutrinos. This possibility seems very attractive, as it could elegantly explain some other open questions. For example, it is not yet clear why there is much more matter than antimatter in the Universe (Baryongenesis ↓). So far, no reactions of elementary particles have been observed in which the otherwise prevailing balance in the production of matter and antimatter would be disturbed to a sufficient extent. The neutrino as its own antiparticle would per se be a violation of the matter-antimatter balance and thus a possible contribution to understanding the oversupply of matter in the Universe. This also opens up possibilities to explain the unusually small neutrino mass (↓). Researchers are trying to answer the question of whether neutrinos are actually their own antiparticles with the search for neutrinoless double beta decay (↓). Neutrinos are therefore particularly interesting particles for studying the elementary structure and reactions of matter.

The Super-Kamiokande detector in Japan. A large tank filled with water in which the particles produced by neutrinos generate Cherenkov light, which is read out with photomultipliers. The picture was taken while the tank was being filled with water.

Baryogenesis → p. 328
Neutrino Masses → p. 272
Neutrinoless double beta decay: Searching for a New Experiment → p. 296

Neutrino Oscillations

Elusive Neutrinos

The *flavors* of the charged leptons are electron, muon and tau. The flavor of a charged lepton can be seen from its mass. The Standard Model of particle physics was developed under the assumption that neutrinos are massless (↓). But if neutrinos are massless, how do you recognize their flavor?

To determine the neutrino flavor, a neutrino must be converted into a charged lepton through an interaction. The leptons's flavor (i.e., its mass) determines the flavor of the initial neutrino. We know that only electron neutrinos can be produced in the solar interior (↓). Nevertheless, muon and tau neutrinos were observed coming from the Sun with the SNO experiment (↓). How can this be? The answer lies in the astonishing phenomenon of *neutrino oscillations*: Neutrinos can change their flavor in free flight which is only possible if neutrinos have mass after all. Neutrino mixing is another prerequisite.

Although there are three different masses for the three neutrinos, we cannot clearly assign one of the masses to, for example, the electron neutrino. Rather, an electron neutrino occurs in a mixture of the three masses. In the theoretical description, the wave function of an electron neutrino is a superposition of 68 % of the wave function of the lightest of the three masses and 32 % of the two heavier ones. If the mass of neutrinos could be measured directly, the mass of electron neutrinos would show up as the lightest mass with a probability of 68 % and as the two heavier masses with a probability of 32 %.

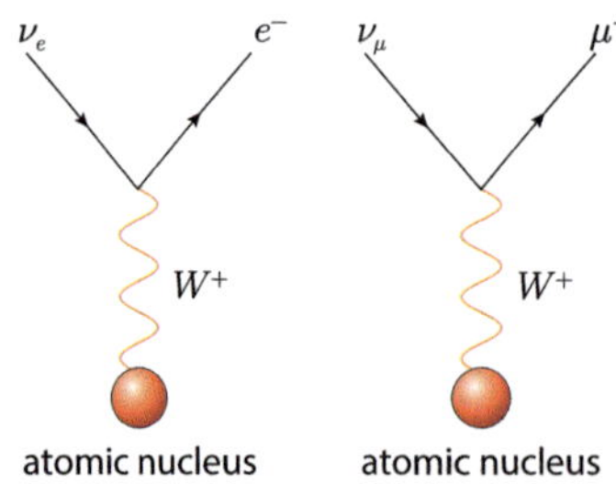

The flavor of a neutrino is measured by determining the mass and thus the flavor of the lepton produced in a charged current interaction.

To understand the phenomenon of neutrino oscillations, let us consider an analogy: Suppose there are two schools, the electron school and the muon school, both of which are attended by rabbits and hedgehogs. The schools classify rabbits and hedgehogs differently. At electron school, there are mixed bunny and hedgehog classes, while at muon school, rabbits and hedgehogs are in separate classes. If you see a pure hedgehog or a pure rabbit group on the way to school, you will assume that they belong to the muon school. If, on the other hand, you see mixed groups of rabbits and hedgehogs, you will assign them to the electron school.

In a running competition, one after the other the mixed classes of the electron school start. The first referee along the track (I) sees the mixed classes running past and notes: 'electron school'. After one lap, the hedgehogs fall back. Groups of rabbits only form in front, followed by groups of hedgehogs only. The next referee (II) sees these separate groups of rabbits and hedgehogs and assigns them to the muon school. If the rabbits of the following classes catch up with the hedgehogs of

Neutrinos → p. 196, Neutrino Masses → p. 272
Neutrinos From the Sun → p. 344
SNO-Experiment: Large Detectors Deep in a Mountain → p. 132

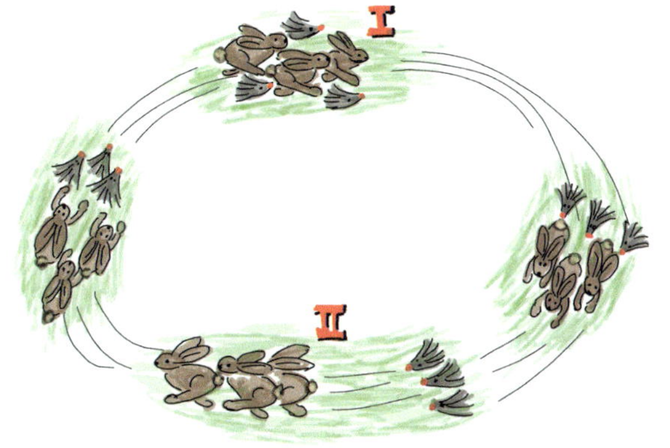

Initially, the rabbits and hedgehogs start out together, but after a while they separate into separate groups of rabbits and hedgehogs. Similarly, different mass fractions of the neutrinos diverge.

the preceding classes in the next round, mixed groups are formed again, which the referee consequently interprets as classes of the electron school. The origin of the classes changes for the referees between the electron and muon school, i.e., it oscillates. The same happens when the separate classes of the muon school start. After some time, the groups of hedgehogs and rabbits mix, and the classes previously appearing as from the muon school look like coming from the electron school. Decisive for this effect are the differently mixed classes in the schools and the different speeds of rabbits and hedgehogs. Analogous to the rabbits and hedgehogs, the masses mixed in the electron neutrino move at different speeds and diverge as the distance increases. The composition of the original electron neutrino changes so that it resembles more closely the mass mixture of a muon neutrino.

Such oscillations were first observed with neutrinos, which are produced by cosmic rays in particle collisions in the Earth's high atmosphere. This produces both muon and electron neutrinos. As neutrinos penetrate the Earth almost unhindered, it is possible to observe neutrinos from all directions after very different paths through the Earth, as shown in the figure below. With the Super-Kamiokande experiment in Japan, a deficit of muon neutrinos was found from below, while electron neutrinos were observed from all directions as expected. The explanation is that there are oscillations of the muon neutrinos on their way through the Earth. They transform into tau neutrinos, which are not visible in the detector for energetic reasons. Neutrino oscillation was then accepted as well to explain the solar neutrino puzzle detected much earlier by Ray Davis.

Since one of the prerequisites for these oscillations are different masses, their observation shows that at least two of the three neutrino masses are not vanishing. Due to the great significance of this experimental evidence of physics beyond the Standard Model, Arthur McDonald, head of the SNO experiment, and Takaaki Kajita, head of Super-Kamiokande, were awarded the Nobel Prize in 2015. Already in 2002, Ray Davis received the Nobel Prize for the detection of solar neutrinos. Further particle oscillations are described in "Shapeshifters among Themselves" (↓).

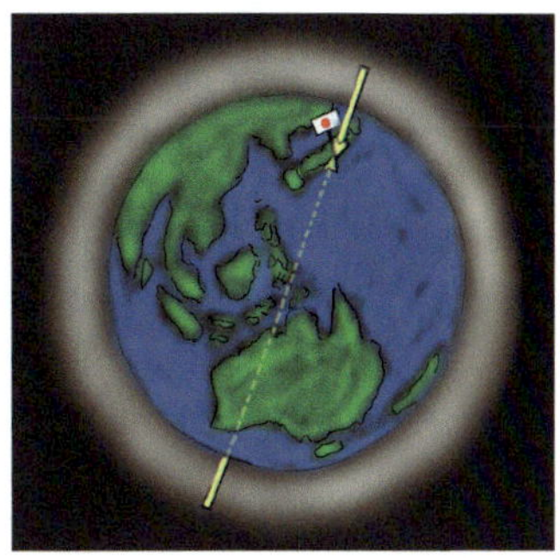

Neutrinos are produced in the Earth's upper atmosphere by cosmic radiation. In a detector, the neutrinos can be observed from all directions and thus after flying different distances.

Shapeshifters Among Themselves → p. 262

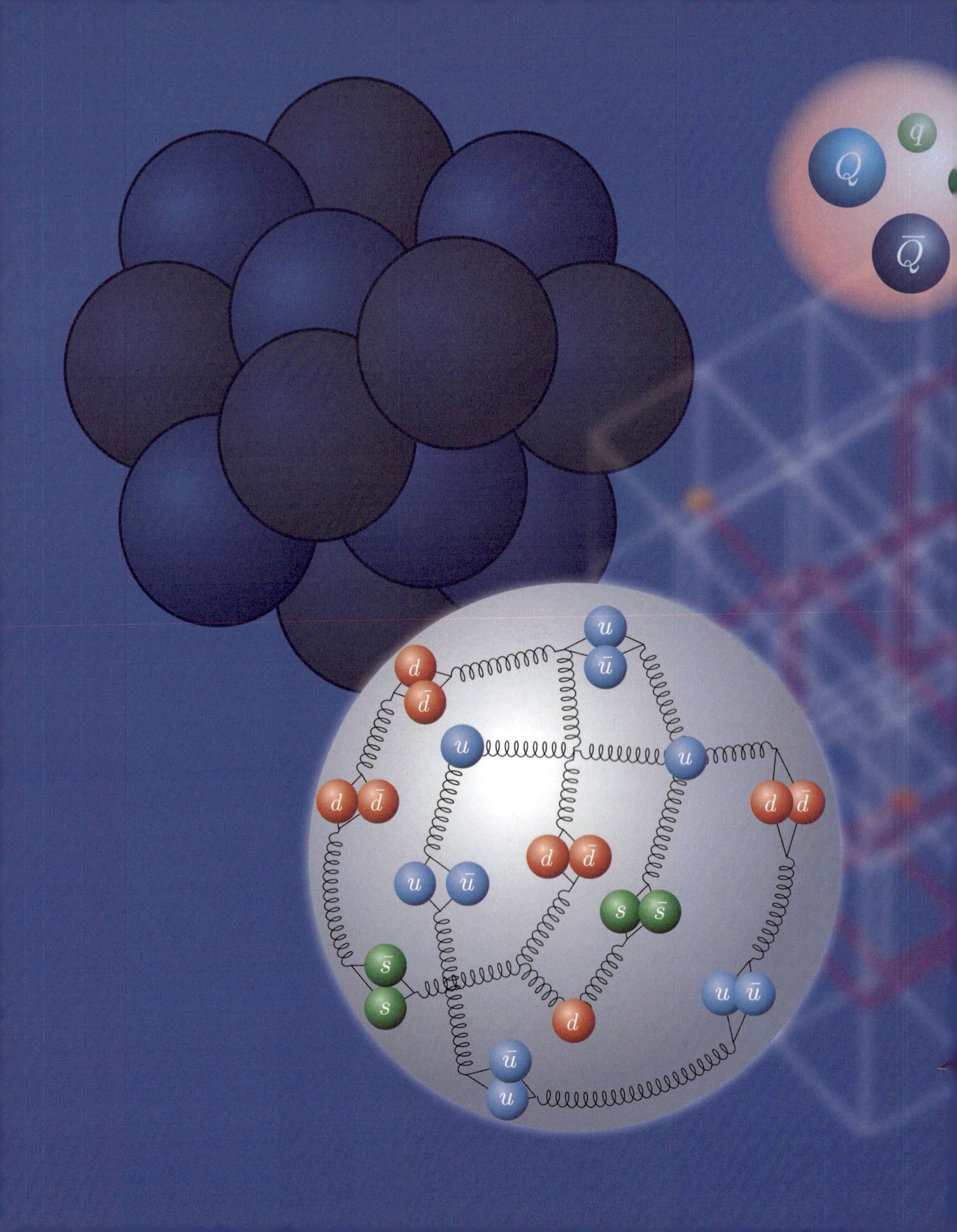

Q
q
$\overline{Q}$
u
$\bar{u}$
d
$\bar{d}$
u
u
d
$\bar{d}$
d
$\bar{d}$
d
$\bar{d}$
u
$\bar{u}$
s
$\bar{s}$
$\bar{s}$
s
d
u
$\bar{u}$
$\bar{u}$
u

8 The Peculiar Strong Interaction

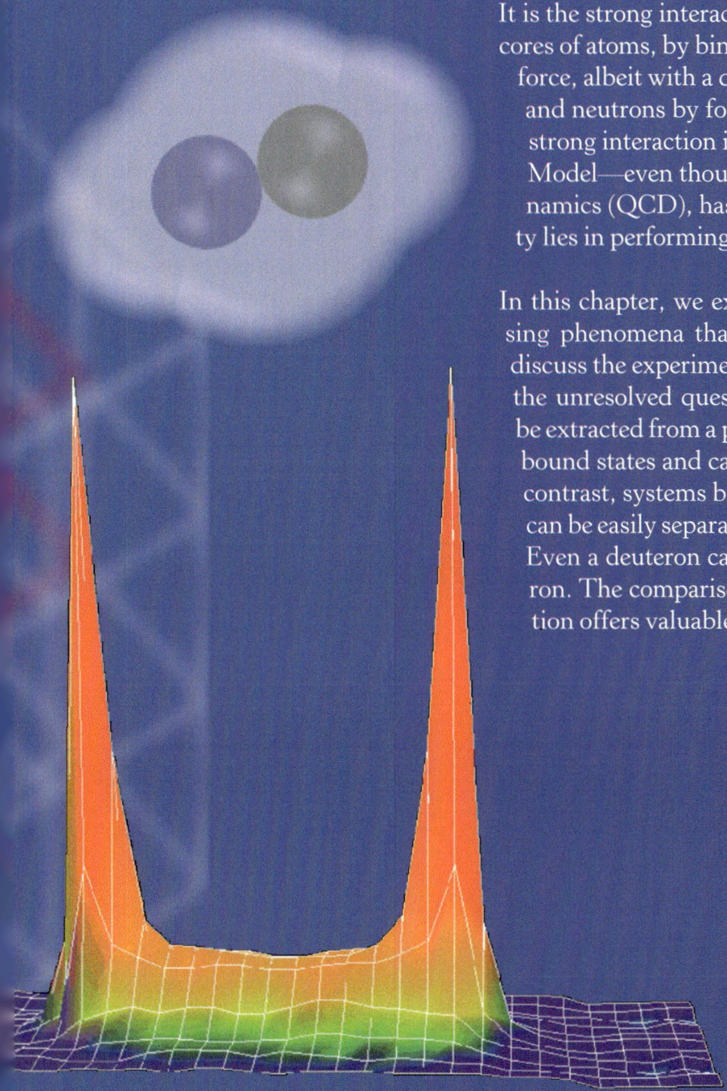

It is the strong interaction that is responsible for forming nuclei, the cores of atoms, by binding protons and neutrons together. The same force, albeit with a quite different mechanism, also creates protons and neutrons by forming bound states of quarks and gluons. The strong interaction is still the least understood part of the Standard Model—even though the underlying theory, quantum chromodynamics (QCD), has been known for several decades. The difficulty lies in performing calculations with this theory.

In this chapter, we explain why this is the case, explore the surprising phenomena that arise from quantum chromodynamics, and discuss the experimental and theoretical methods we use to address the unresolved questions. For example, individual quarks cannot be extracted from a proton or neutron—they are always confined in bound states and can only occur in very specific combinations. In contrast, systems bound by electromagnetic forces, such as atoms, can be easily separated—an electron can be removed from an atom. Even a deuteron can be split into its constituent proton and neutron. The comparison of electromagnetism and the strong interaction offers valuable insights into these distinct behaviors.

The Quark Model Revisited

How Quarks Tamed the Particle Chaos

In Chapter 1 the quark model (↓) was introduced as a simple way to reduce the number of fundamental building blocks. Here we go deeper into the concepts underlying the quark model. How did we actually come to the idea that things like quarks might exist? In the 1950s, accelerator technology had improved to a point where particle energies of 100 MeV to 1 GeV could be produced. At this point, to the surprise of the physicists, a large number of new particles were discovered. In addition to the three already known pions (π^+, π^0, π^- with spin 0) and the nuclear building blocks protons and neutrons (both with spin 1/2), 30 further particles were found—15 mesons (with integer spin, zero or one) and 15 baryons (with half-integer spin, 1/2 or 3/2). No one had expected this flood of particles. Subsequently, in the early 1960s Murray Gell-Mann and Yuval Ne'emann on the one hand and George Zweig on the other independently brought order into this particle zoo. They postulated that the discovered particles contain different combinations of only three distinct building blocks. Gell-Mann and Ne'emann named them *quarks*, Zweig called them *aces*. Today these constituents are called up (u), down (d) and strange quark (s), the corresponding antiparticles anti-up ($\bar{u}$), anti-down ($\bar{d}$) and anti-strange ($\bar{s}$), respectively.

Furthermore they postulated that the new force holding the quarks together, the strong force, does not distinguish between the types of quarks. As we will see in the article on where our mass comes from (↓), an important part of the total mass of systems composed of quarks, i.e., the mesons and baryons we are discussing here, comes from the field of the strong interaction, the gluons. Therefore, well-defined groups of particles composed of quarks (so-called ***multiplets***) should necessarily have almost the same mass. The small mass differences within the multiplets are due to differences in the quark properties, i.e., their mass and charge. All properties of all mesons and baryons known at that time could be explained by assigning the same mass to the up and down quark and

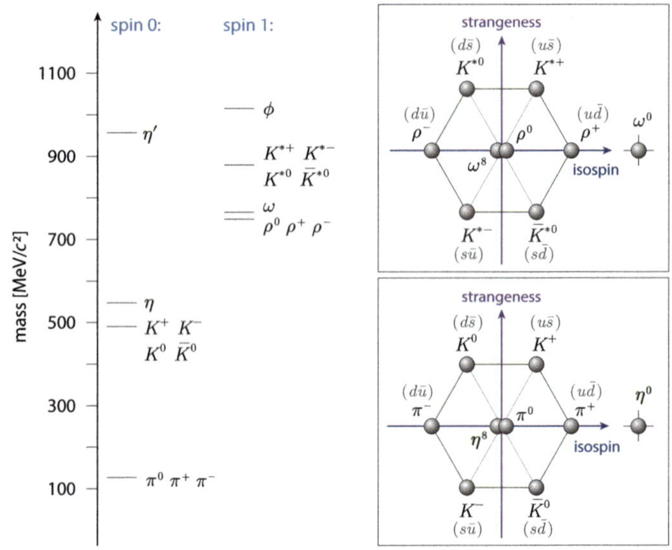

Mass spectrum and the associated multiplets of the simplest mesons in the quark model

The Quark Model: A First Look → p. 20
Where Does Our Mass Come From? → p. 220
The Quark Mass Effect → p. 224

a mass of about 200 MeV/c^2 higher to the strange quark (precise measurements reveal that the electromagnetic interaction also contributes to the mass differences, see Quark Mass Effect ↓). But the approach was capable of even more: It predicted the existence of another particle, the Ω^-, composed of three strange quarks, as well as its charge and mass. Its subsequent experimental discovery not only helped the quark model gain general recognition, even though a free quark has never been observed! It also earned Gell-Mann the Nobel Prize in 1969.

In the quark model, the mesons consist of a quark and an antiquark, and the baryons are composed of three quarks. Why exactly these quark combinations occur is explained in the article on confinement (↓). The above-mentioned multiplets can also be very well represented graphically. For this, the ***strangeness***, i.e., the number of strange quarks, is plotted on the y axis (according to convention, anti-strange quarks are counted positively and strange quarks negatively), and on the x axis the ***isospin***, which counts the number of up and down quarks, with up and anti-down counted positively and down and anti-up counted negatively (see figures). The left figure shows, for example, that pions occur in three charge states: The π^+ contains $u\bar{d}$, the π^0 a mixture of $u\bar{u}$ and $d\bar{d}$ and the π^- finally $d\bar{u}$. In the article "Seeing Quarks and Gluons?" (↓), we show that quarks are more than theoretical constructions to classify particles. In the following years, more particles were discovered and on the whole the quark model held up well, although it already had to be expanded with the discovery of the charm and bottom quarks.

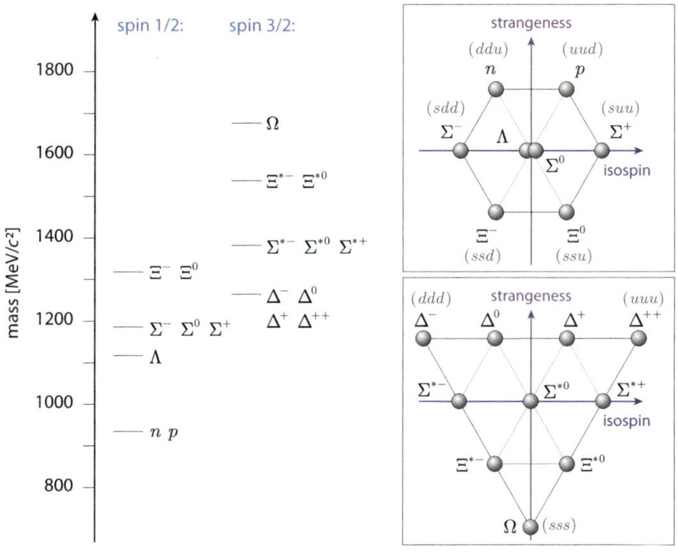

Mass spectrum and the associated multiplets of the simplest baryons in the quark model

By now we know that there are a total of six quarks (↓), with the heaviest among them, the top quark, having such a short lifetime (Particles' Births and Deaths) that it does not exist long enough to form composite objects. Charm and bottom quark, on the other hand, form bound states and until 2002 these could also be completely explained with the simplest quark model introduced above. But in 2003 the situation changed: Since then a large number of particles have been found, which do not fit into the scheme outlined above, calling for an extension of the quark model. We now know that there are mesons with at least four quarks and baryons with at least five—but that is a different story (Exotic States ↓).

Confinement → p. 206
Seeing Quarks and Gluons? → p. 216
The Path to Six Quarks in Three Families → p. 194
Particles' Births and Deaths → p. 54
Exotic States → p. 230

Quantum Chromodynamics

From One to Three: a Small Difference With Dramatic Implications

The theory of electromagnetism (↓) is called quantum electrodynamics (QED). The theory of the strong interaction is denoted *quantum chromodynamics* (QCD). Both theories are described by the same mathematics, with the one fundamental difference: In quantum electrodynamics there is just one charge (−) and its anticharge (+); in quantum chromodynamics there are three charges and three anticharges. The latter are named for better illustration after the three primary colors red, green, and blue, even though they have nothing to do with the "colors" we see with our eyes. Therefore, we also speak of *color charges* and hence quantum "chromo"-dynamics, as *chroma* is Greek for color.

At first, the difference between QED and QCD seems tiny—three instead of one charge, so what? But the implications of this difference are dramatic! In atoms, the electrons are electromagnetically bound to a nucleus. Typically the number of negatively charged electrons in the shell matches the positive charge of the nucleus and thus the sum of all charges of the atom vanishes—atoms are usually electrically neutral. However, it is no problem to remove electrons from the shell and thus create ions. In contrast, QCD *forces* all particles that carry a color charge to connect with others in such a way that the composite object does not carry any overall color charge. In particular, it proves impossible to extract a building block that carries a color charge from this compound. This phenomenon is known as color confinement or *confinement* (↓).

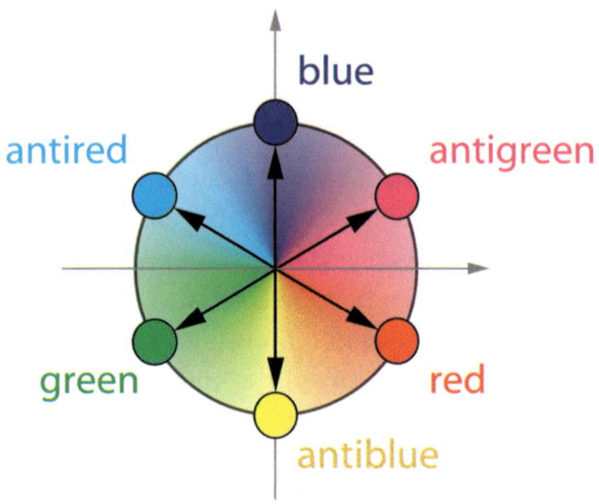

The color charges of the strong interaction

The mathematical structure of a theory also determines how many exchange particles there are and what properties they have. Thus, the exchange particle of QED, the photon, has no electric charge itself. However, QCD has eight exchange particles, and these each carry different combinations of color charges. Since they glue the quarks together, so to speak, they are collectively called *gluons*. These color charges of the gluons are the origin of the very special properties of QCD. Here, in addition to the aforementioned confinement, the coupling strength (↓) of QCD is noteworthy, which changes in a very peculiar way as the energy is varied. The charges of the exchange particles also have an influence on the range of the strong interaction (↓).

But how do we know that there are three colors, when quarks are always trapped in hadrons? The simplest quantity that directly measures the number of colors

Electromagnetism → p. 176
Confinement → p. 206
The Running Coupling → p. 210
On the Ranges of Interactions → p. 58

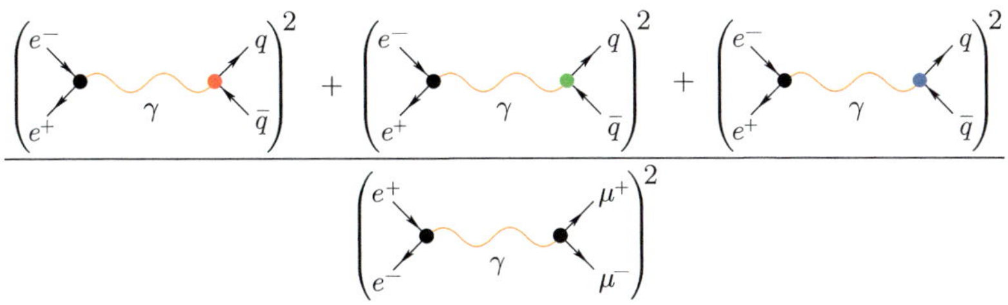

$$= 3 \; \frac{(\text{electric quark charge})^2}{(\text{electric muon charge})^2}$$

The contribution of a single quark type to the R ratio—the 3 counts the number of color charges.

is the so-called R ratio. For this, one compares how frequently a muon-antimuon pair and how frequently strongly interacting particles are produced, when electrons and positrons collide. The creation of strongly interacting particles is based on the creation of a quark-antiquark pair. The contribution of each quark type to the R ratio is shown in the figure above.

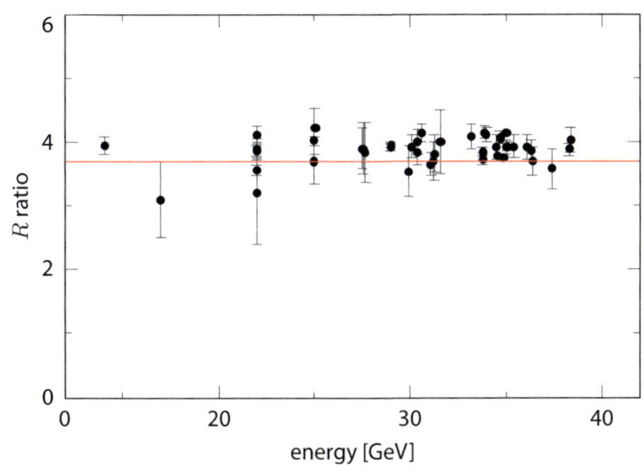

The R ratio: Comparison of the measurements with the calculation (the red line) for three colors

Now let us consider R in an energy range where there are no resonances (↓), for example between 15 and 40 GeV. These energies are high enough to generate the lightest five types of quarks (two with charge $(2/3)\,e$ and three with charge $-(1/3)\,e$; for comparison: a muon carries -1 unit of e).

If we now count the contributions of the individual types of quarks, we get a prediction for the R ratio (recall: the rate is proportional to the amplitude—depicted as a Feynman diagram—squared)

$$3\left(3\left(\frac{1}{3}\right)^2 + 2\left(\frac{2}{3}\right)^2\right) = \frac{11}{3} \approx 3.7\,,$$

in good agreement with the experiment, as can be seen in the left figure. An even better description of the data is obtained if the corrections dictated by field theory (QFT ↓) are also taken into account. In contrast, the results for the R ratio for two or four colors deviate significantly from the data.

Resonances: Particles' Births and Deaths → p. 54
Quantum Field Theory → p. 104

Confinement

Why There Are no Isolated, Free Quarks

The term *confinement* describes the fact that objects (composite or elementary) which carry a strong charge, like individual quarks or gluons, cannot be separated from their partners in a bound state, not even by a very large amount of energy—they remain confined. But why is this so?

The theory of strong interaction, QCD (↓), has three charges (the so-called *color charges*) and their anticharges. The mathematical structure of the theory (Gauge Symmetries ↓) then dictates that, unlike the exchange particles of the QED (the photons), the exchange particles of QCD (the gluons) are not neutral, but carry a color charge themselves and thus can attract each other. This leads to drastic effects that can be understood qualitatively by a simple geometric consideration.

For that, let us compare the force of QED with that of QCD: To understand the strength of a field, it is helpful to consider the *field lines*. Their local density or area density, i.e., the number of field lines penetrating a certain area, is a measure of the force acting there. If one considers a single, isolated electromagnetic charge, the field lines run radially outward (see Acceleration and Radiation ↓) Accordingly, their area density decreases as the surface of a sphere, at the center of which the charge sits, increases, i.e., with the square of the radius. This corresponds to Coulomb's law. In the presence of two oppositely charged electric charges, the spherical symmetry is broken, but the field lines still fill the entire space—this is indicated in the left part of the figure below.

However, if the field lines attract each other, as is the case with the gluons, which carry their own color charges, they do not fill the entire space, but form a kind of tube that connects the color charges. This is shown in the right part of the figure below. In this case, however, the thickness of the tube and thus the force between the charges is independent of their distance. If one tries to separate these color charges, i.e., to extend their distance to infinity, an infinite amount of energy is needed—the quarks therefore inevitably remain trapped. There we have it: confinement.

Unfortunately, the effect just described of the formation of gluonic flux tubes cannot be directly measured experimentally. In experiments, the confinement manifests itself in the fact that free

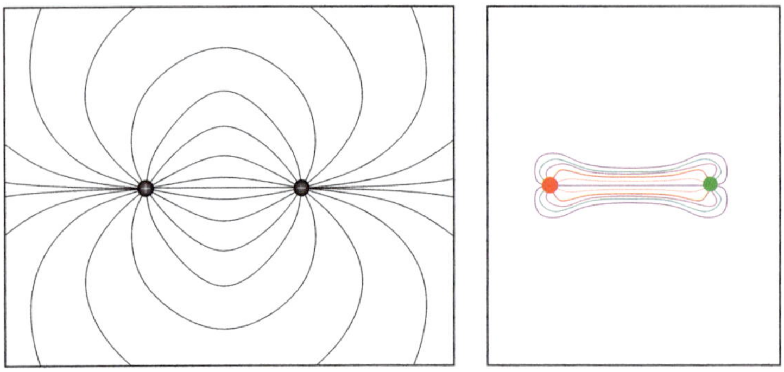

Comparison of the field lines of QED (left) and QCD (right)

Quantum Chromodynamics → p. 204
Gauge Symmetries → p. 98
Acceleration and Radiation → p. 42

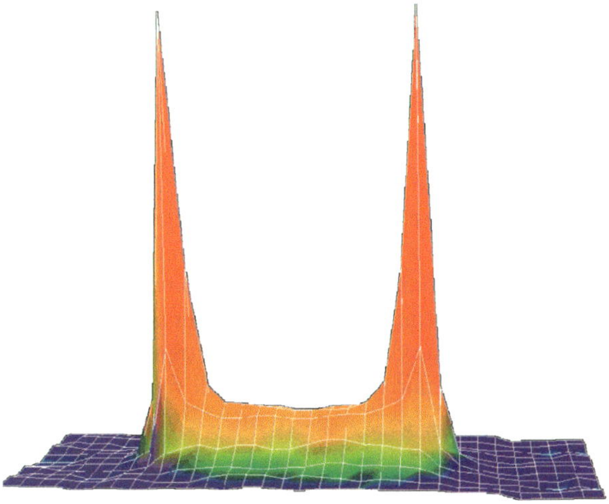

Calculation of the energy in the gluon field between two point charges.

color-charged objects have never been measured. By the way, it is quite unlikely that the corresponding signal was overlooked: Since free quarks have a fractional charge, they would have certainly been noticed in the many measurements that particle physicists have performed over a huge energy range (see also Search for free Quarks ↓).

However, it can be shown that QCD actually leads to the formation of flux tubes, using numerical simulations. One of the following articles introduces lattice QCD (↓), with which QCD interactions can be simulated on a computer. A very nice result of this endeavor is shown in the picture above: Here, a quark and an antiquark were placed at the positions that show the sharp peaks. Then the energy density in the surrounding space was computed in the numerical simulation. In the picture, it is clearly visible that a tube with constant energy per length has formed between the two quarks. It is important to note that besides the two fixed quarks in this simulation there were no further quarks, so that the flux tube could not tear. This so-called *string breaking* would lead to the generation of new hadrons, as described in the article Seeing Quarks and Gluons (↓).

Before closing we briefly address the question of which combinations of color charges are permissible, since confinement demands color neutrality. Interestingly, the color analogy also applies here: As a rainbow shows, the combination of red, green, and blue results in white. Accordingly, three quarks with this color combination form a permissible state; these are the *baryons* of the quark model (↓).

In addition to the color charges, there are also the corresponding anticolor charges, which appear as color charges of the antiquarks—just as the antiparticle of the electron, the positron, carries the corresponding anticharge, i.e., it is positively charged. Charge and anticharge neutralize each other; so, for example, a red quark with an anti-red antiquark can form a permissible state. These are the *mesons* of the quark model.

More complex combinations are also formally possible, such as states which consist only of gluons—however, these so-called *glueballs* have not yet been identified experimentally, although candidates exist. What does not exist as a physical state, is, for example, the combination of two quarks; from only two primary colors, one cannot mix white: This combination is fundamentally color-charged and therefore does not appear isolated in Nature.

Image: Gunnar Bali, http://www.physik.uni-regensburg.de/forschung/bali/pics.html

The Search for Free Quarks → p. 208
Lattice QCD → p. 222
Seeing Quarks and Gluons? → p. 216
The Quark Model: A First Look → p. 20

The Search for Free Quarks

Don't Be Fooled by Statistics

Quarks were introduced in the early 1960s to explain the structure of the multitude of hadrons from a few building blocks. This concept was initially met with great skepticism. No detector had ever registered a quark. Today we know that the attempt to detect a free quark is in vain. Due to confinement (↓) of quantum chromodynamics (↓), quarks never appear as free particles in the detector. However, before this was fully understood, free quarks were searched for experimentally.

The special thing about quarks is that they only carry a fraction of an elementary charge, either $\pm 1/3$ or $\pm 2/3$. The idea was to exploit this in their direct detection. When a charged particle passes through a cloud chamber, liquid droplets form along the particle's path and make it visible. The probability of droplet formation is proportional to the square of the charge. For a quark, which carries only 1/3 of an elementary charge, for example, the probability is therefore a factor of $1/9 \approx 0.11$ smaller than for an electron or proton, which each carry a full elementary charge.

In an experiment, particles of cosmic radiation (↓) were examined in the hope of detecting a single such quark. Some 10 000 particles passed through a cloud chamber and on average 229 droplets were counted per particle track. For comparison: If you count raindrops that hit a certain area within a certain time interval, you also get an average value dependent on the strength of the rain. However, the number of drops will fluctuate around this value with each measurement. In such counting experiments, this fluctuation is generally described by a Poisson distribution (Statistics ↓). This was initially also assumed in the experiment for quark detection.

In the experiment, a track was also found where only 110 droplets were counted. As can be seen from the left figure at the next page, this is very unlikely. The probability of 110 droplets or even fewer occurring when an average of 229 is expected is only $1.6 \cdot 10^{-18}$. That is, on average, one expects only a track that generates 110 or fewer droplets for every $1/(1.6 \cdot 10^{-18}) \approx 6.3 \cdot 10^{17}$ tracks. However, since only 55 000 tracks were analyzed in the experiment, it was concluded that this one track must be a particle that carries less than one elementary charge, such as a quark, for example.

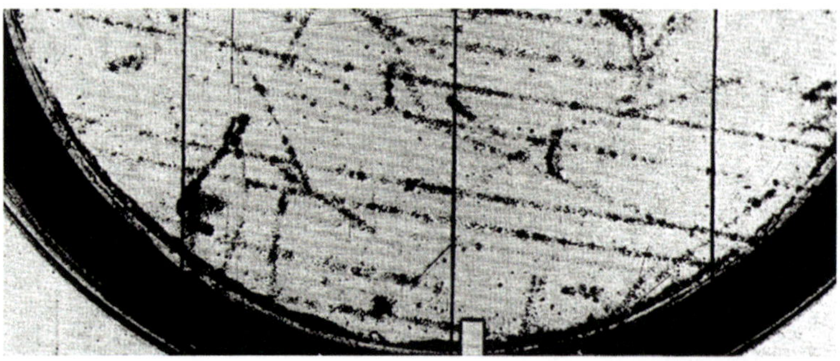

Cloud chamber image from the original publication[1]

Reprint of the cloud chamber image with permission from C. B. A. Mc Cusker, I. Cairns, Physical Review Letters, 23, 658 (1969), Copyright 1969 by the American Physical Society

Confinement → p. 206
Quantum Chromodynamics → p. 204
Cosmic Rays → p. 336

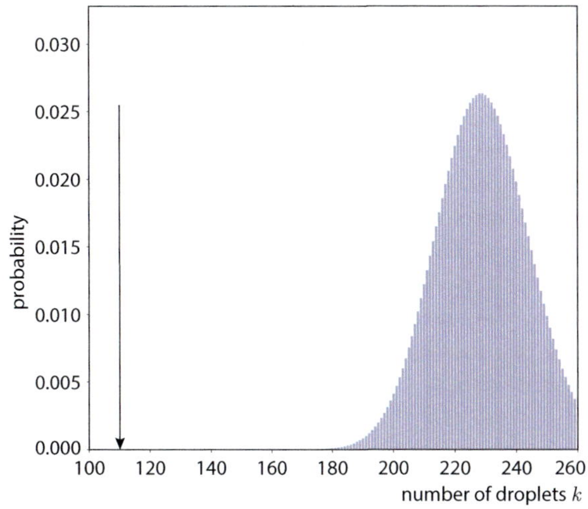

Probabilities for a Poisson distribution with a mean value of 229. The figure shows, for example, that the probability of 220 droplets occurring is about 0.022 or 2.2 %. The probability of only 110 droplets occurring is so small that it cannot be seen given the linear scale on the vertical axis.

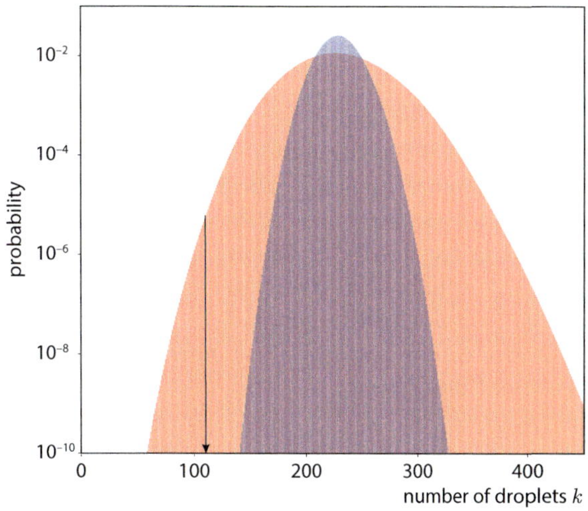

Probabilities for the occurrence of k events in a Poisson distribution (blue, identical to the left figure) and a two-stage Poisson distribution (red) with the same average of 229. Here, a logarithmic vertical scale was chosen so that one can still read the value of the red distribution.

Later, however, it was found[2], that the process of droplet formation occurs in two stages. A particle carrying an elementary charge, on average, causes about 57 collision processes (first stage) with molecules in the cloud chamber. On average, 4 droplets are formed per collision (second stage). This does not change the average ($4 \cdot 57 \approx 229$). However, since the process now occurs in two stages, the fluctuations in the number of droplets from track to track are much larger and follow the red distribution in the right figure, which is now plotted on a logarithmic scale on the y axis.

Now there is a probability of $4.2 \cdot 10^{-5}$ that 110 droplets or less are observed. This is still a very small number. The fact that an event has a very small probability does not mean that it does not occur. One just has to try often enough. Since 55 000 tracks were analyzed, one expects $55\,000 \cdot 4.2 \cdot 10^{-5} = 2.2$ tracks with 110 droplets or less. The result of the measurement is therefore compatible with the hypothesis that the track with 110 droplets was also generated by a particle carrying exactly one elementary charge. As mentioned at the beginning, we know today, that confinement is a result of quantum chromodynamics and explains why we cannot observe isolated quarks.

[1] C. B. A. McCusker, I. Cairns 1969, *Evidence of Quarks in Air-Shower Cores*, Physical Review Letters, 23, 658, doi:10.1103/PhysRevLett.23.658
[2] R. Adair, H. Kasha 1969, *Analysis of Some Results of Quark Searches*, Physical Review Letters, 23, 1355, doi:10.1103/PhysRevLett.23.1355

Statistics → p. 84

The Running Coupling

A Pseudo-Giant

A fundamental principle of quantum mechanics (↓) is that the spatial resolution of a measurement depends on the energy of the probe. Therefore, it is not surprising that the relevant energy scale also plays an important role in the actual calculation for every observable. To illustrate this, let us take a closer look at the coupling of a gauge boson (photon or gluon) to a matter particle (lepton or quark). Formally, this is referred to as *charge renormalization* (↓).

How can one imagine this? While a photon couples to a charged matter particle, it can, for example, generate electron-positron pairs (see last image in the figure for QED). These, however, shield the charge of the matter particle. Thus, with increasing energy (= increasing resolution), the effective coupling to the matter particle will increase as one gets closer to the actual charge of the matter particle, and penetrates the cloud of electron-positron pairs.

Central to the process just described is that the photon itself carries no electric charge and thus does not directly interact with itself. In contrast, gluons carry strong charges and therefore not only couple to quark-antiquark pairs, but also interact with each other and with themselves (see figure). As a result, the strong charge is larger at low energies, corresponding to large wavelengths or poor resolution, than at high energies and thus high resolution. Thus the strength of the strong interaction behaves opposite to what one is used to: It is large from far away but gets smaller as one approaches it—following a famous German childrens-novel one may call it a pseudo-giant.

Vertex functions with their leading field-theoretical corrections for QED (above) and QCD (below). Solid lines show fermion fields (quarks or leptons), wavy lines the photon and curly lines the gluons.

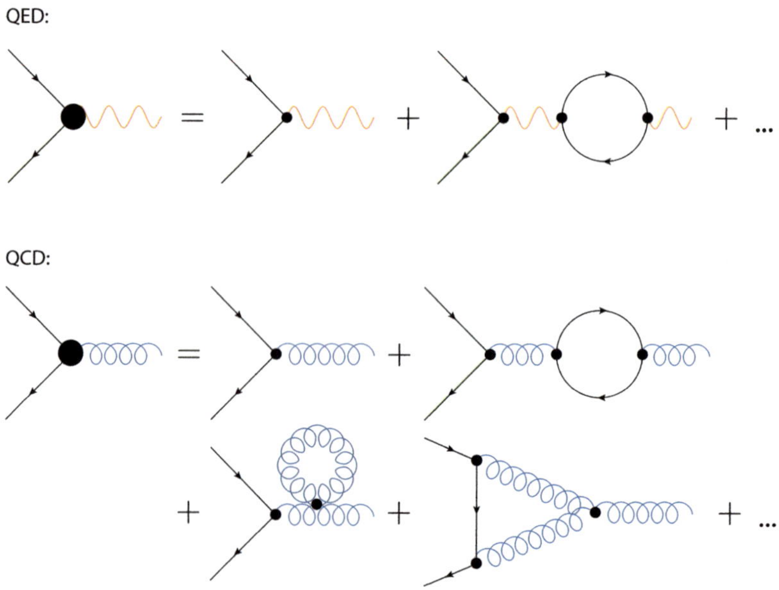

Concepts of Quantum Mechanics → p. 50
Gauge Symmetries → p. 98
Renormalization → p. 108

What has been described here in words can be directly calculated using the so-called *renormalization group equations* of quantum field theory. In the case of QED, the diagrams of the first line are included in the calculation, for QCD it is the diagrams of the lower two lines. The result for the coupling strength of QCD, α_{QCD}, is shown in the right figure and compared for energy scales $Q > 1$ GeV with experimental results. The increasing spread of the lines for smaller values of Q indicates that one can no longer perform precise calculations of the coupling strength here. Below the energy scale denoted Λ_{QCD} the calculation breaks down. A method to extract α_{QCD} from data for *jets*, i.e., bundled hadron beams, will be introduced in the article section on seeing quarks and gluons (↓). In the figure, there are also other reactions listed, some of which require considerable mathematical effort to get access to α_{QCD}.

Accordingly, quarks can move almost freely at high energies, as the coupling to the gluon field is small. This phenomenon is called *asymptotic freedom*. At low energies, however, the coupling becomes large. It is this area of large coupling, in which the quarks combine into bound states, the hadrons. As shown in the figure, the scale that separates the areas between large and small couplings is called Λ_{QCD} (green area). It has a value of 200 to 300 MeV.

In contrast to the behavior of α_{QCD}, the renormalization group equations for the QED coupling yield values that are significantly smaller and slowly increase with an increasing energy scale Q: For $Q < 1$ MeV, one finds $\alpha_{\text{QED}} = 1/137$, for $Q \approx 100$ GeV its value has grown to 1/128. And of course, the coupling of the weak interaction also depends on the energy. We will return to the different energy dependencies of the Standard Model couplings in the article GUT (↓).

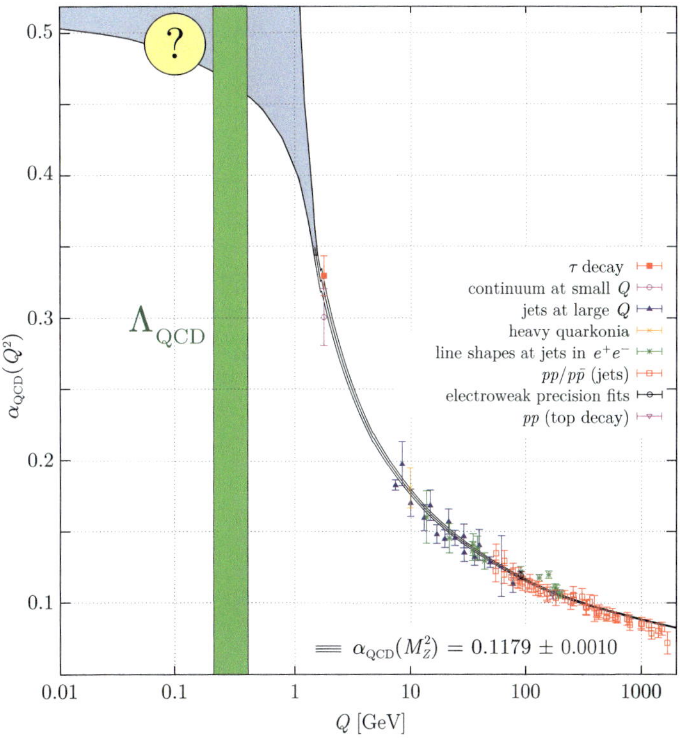

Dependence of the strong coupling on the energy scale Q

Seeing Quarks and Gluons? → p. 216
GUT → p. 276

The Structure of the Proton

A Rich Inner Life

By the early 1930s, our current understanding of the structure of atoms had been established. They consist of a nucleus, which in turn is composed of positively charged protons and electrically neutral neutrons, surrounded by negatively charged electrons.

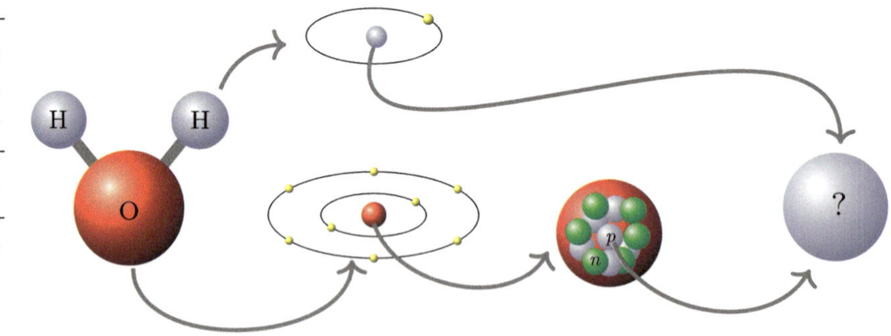

From a water molecule to the proton

The question then arose whether these particles are elementary or whether they in turn consist of further components. The electron is, according to current knowledge, truly elementary and has no spatial extension. However, for protons and neutrons the situation is different—in fact, a rich inner life was discovered. As is often the case in particle physics, scattering experiments (↓) played a crucial role in uncovering this.

By bombarding protons with an electron beam, it was first shown that the proton has a spatial extension of about 10^{-15} m, five orders of magnitude smaller than an atom. By the end of the 1960s, the achievable electron energy was so high that the inner life of the proton could be examined more closely.

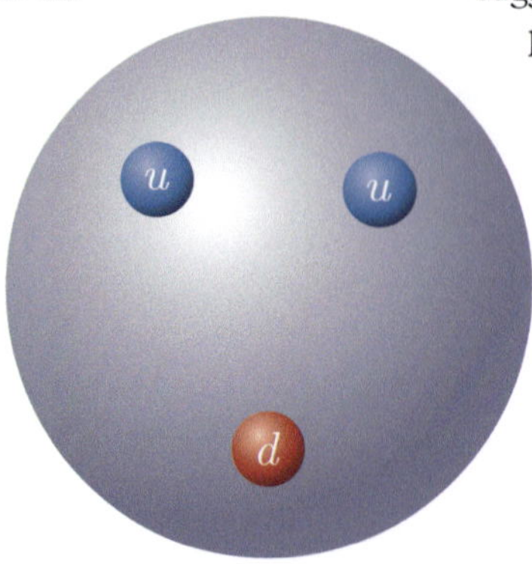

Proton composed of three quarks

The quark model (↓), developed in the early 1960s, was initially an elegant mathematical model to bring order to the multitude of observed particles. By bombarding protons with very high-energy electrons it was possible to look deeper and deeper into the proton. The observed deflection of the scattered electrons suggested that protons consist internally of point-like particles, which were eventually identified as quarks. Further experiments with neutrino beams (↓) finally confirmed the fractional charge of quarks predicted in the quark model. This showed that quarks are real and more than just a mathematical concept (Seeing Quarks ↓).

Scattering Experiments → p. 46
The Quark Model Revisited → p. 202
The Discovery of Neutral Current Interactions → p. 236
Seeing Quarks and Gluons? → p. 216

Very simply put, the proton can be described as a particle consisting of two up quarks, each carrying 2/3 of the elementary charge, and a down quark, which carries −1/3 of the elementary charge. In total, this results in a positive elementary charge of the proton. A neutron consists accordingly of two down quarks and one up quark, making it electrically neutral. This simple picture of the proton and neutron proves to be very useful for calculating certain properties, such as the charge.

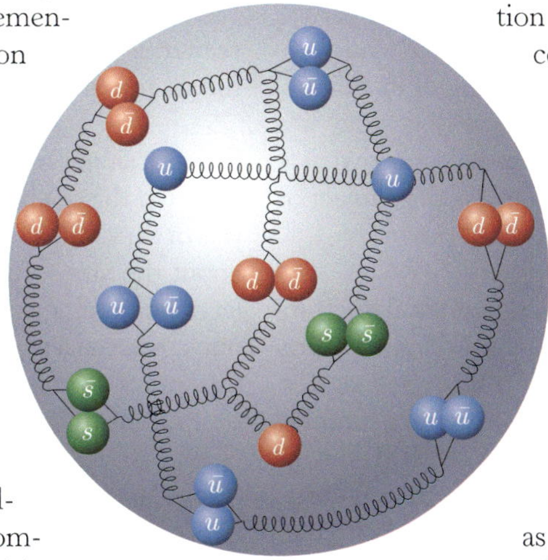

Proton composed of quarks and gluons

Further experiments however, from which the theory of strong interaction, quantum chromodynamics (QCD ↓), has developed, provide a much more complex picture, as outlined in the right figure. In addition to the quarks, the proton contains *gluons*. They are the exchange particles of the strong interaction and glue the quarks together. Gluons can in turn form quark-antiquark pairs. Thus, the proton can be imagined as a dynamic system, in which gluons and quark-antiquark pairs are constantly being created and destroyed.

We therefore see that the statement that a proton consists of two up quarks and one down quark is very simplistic. Even the statement that a proton consists of a certain number of up quarks, down quarks and gluons is not possible. This can be clarified by the following consideration. If you want to spatially resolve the proton at ever finer scales, you have to do this with particle beams of ever higher energy. However, according to Einstein's relation $E = mc^2$, this energy can also be used to create new particles. The question of whether these particles are components of the proton or have been created by the probing interaction cannot be easily answered. A decomposition of the proton into its building blocks, as we know it for everyday objects, is therefore not possible. The number of particles you encounter in the proton depends, so to speak, on how closely you look. The higher the energy, the finer the structures you can resolve.

Nevertheless, it has been possible with measurements, such as those carried out at the HERA accelerator at DESY in Hamburg, and QCD to describe the quark and gluon content in the proton very accurately. Only with the help of these so-called quark and gluon distributions (↓), is it possible to understand the proton-proton collisions at the LHC (↓) accurately.

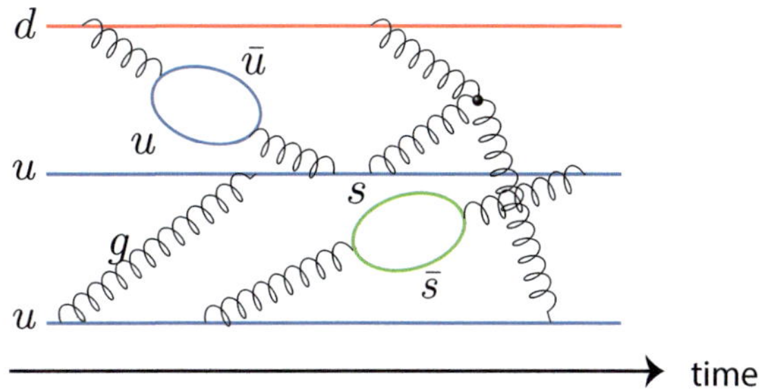

Proton as a dynamic system of quarks and gluons

Quantum Chromodynamics → p. 204
Quark Distributions Inside the Proton → p. 214
The LHC Accelerator Complex at CERN → p. 116

Quark Distributions Inside the Proton

More Than Just Three Quarks

In the article about the inner life of the proton (↓) we have seen that a proton is made up of quarks and gluons. But we have also seen that a listing of its components, as we are used to from objects of everyday life, is not possible. How one can nevertheless arrive at a quantitative description of the proton is the subject of this article.

For this purpose, it is useful to imagine the flying proton as a beam of quarks and gluons, which are summarized under the term *partons* in this context (see figure). Each of the partons carries a certain momentum fraction x (Kinematics ↓) of the proton. The sum of these fractions must of course be one, which corresponds to the total momentum of the proton. With this concept one can assign so-called *parton distributions* to each type of quark and also to the gluons. These indicate how many partons of a type carry a certain momentum fraction x in the proton.

How do we know this? In scattering experiments (↓), where protons are bombarded with electrons, the electron and proton exchange a photon. If the energy of the photon is high enough and thus its wavelength is small enough, it can resolve structures within the proton (↓). The scattering process then does not involve the entire proton, but randomly one of the quarks. In the figure, an up quark is hit by the photon. The probability of hitting a certain quark is higher, the more quarks of the corresponding type are present in the proton. Interestingly, the momentum fraction of the struck quark can be determined solely from the momentum of the electron before and after this collision process. Such a scattering process is referred to as *deep-inelastic*, as the proton breaks apart in this case.

From a single scattering event, one cannot find out which of the quarks was struck. Even through the analysis of a multitude of events, one only learns something about the sum of the contributions of all quarks. To learn something about the individual types of quarks, one again makes use of symmetry considerations. For this purpose, scattering experiments are carried out on protons, in which electrons are shot at a hydrogen target, as well as on heavy hydrogen, i.e., deuterium (From Nucleus to Nuclei ↓). The nucleus of deuterium, denoted deuteron, consists of a neutron and a proton. The symmetry exploited here is *isospin symmetry*. It states that the distribution of up quarks in the proton looks the same as the distribution of down quarks in the neutron. In the quark model, proton and neutron

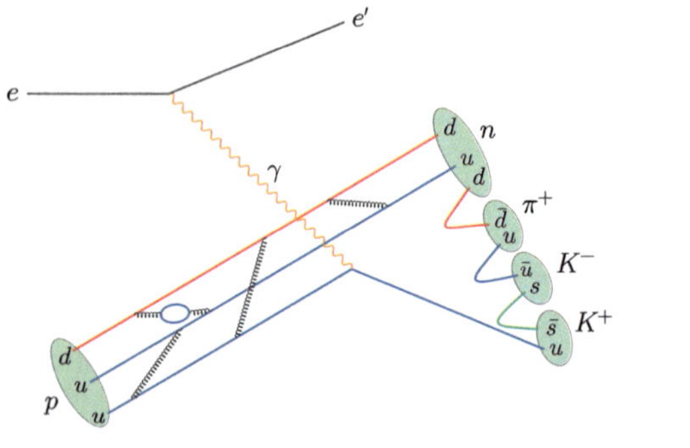

Deep inelastic scattering process $e + p \to e' + n + \pi^+ + K^- + K^+$ in the quark model

The Structure of the Proton → p. 212
Kinematics → p. 40
Scattering Experiments → p. 46
Seeing Quarks and Gluons? → p. 216, Confinement → p. 206
From Nucleus to Nuclei → p. 226

only differ by swapping these two types of quarks. If one also uses results from deep-inelastic scattering with neutrino beams, one can extract momentum distribution functions of all types of quarks individually from the experimental data.

What do these parton momentum distributions tell us? From the figure below, one can read how many quarks of a given type are present in the proton at a certain momentum fraction. We have seen that the proton in the simple quark model (↓) consists of two up and one down quark. This is also reflected in the parton distributions. If one sums or, mathematically more precisely formulated, integrates over all momentum contributions of the up quarks and subtracts the contributions of the anti-up quarks, one obtains:

$$\int_0^1 (u(x) - \bar{u}(x))\,\mathrm{d}x = 2 \ .$$

The corresponding integral over the down quarks yields the value one. This shows that the net quark content of the proton is indeed two up and one down quark.

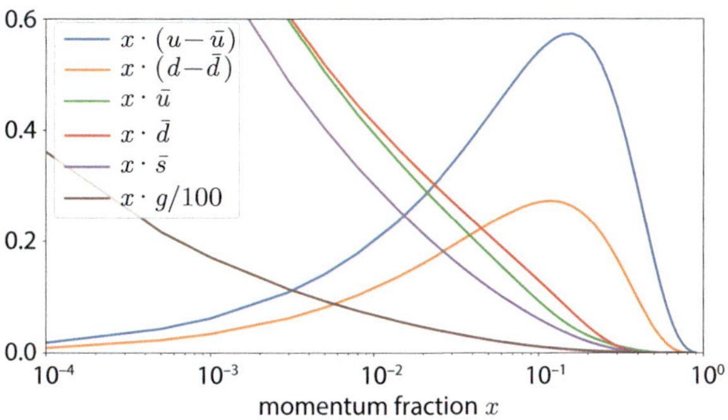

The quark and gluon distributions multiplied by the momentum fraction x

These quark distributions are not static. It turns out: The deeper one looks into the proton, i.e., the greater the momentum transferred by the photon and thus its resolution, the finer structures one can resolve. Thus these distributions depend not only on the momentum fraction x, but also on how large the energy of the photon is with which one examines the proton. This dependence can then also be used to indirectly determine the gluon distribution, although the electrically neutral gluons do not directly interact with the photons. Here we come back to the statement made at the beginning. One cannot simply say that the proton consists of so and so many quarks. It depends on how closely one looks. With the parton distributions, it has been possible to describe the proton very accurately in quantitative terms. They are, for example, indispensable for understanding and analyzing proton-proton collisions at the LHC (↓)

The HERA accelerator at DESY in Hamburg, where fundamental experiments on the structure of the proton were conducted

The Quark Model: A First Look → p. 20
The LHC Accelerator Complex at CERN → p. 116

Seeing Quarks and Gluons?

Making the Trapped Visible

Originally, quarks were introduced to bring order to the particle zoo (↓). But are these objects real? Or do they merely serve to classify particles? In this article we will see that quarks and gluons are indeed real. They can be "seen" in experiments, even more directly than described in the article on the structure of the proton (↓)—but of course not as free particles (↓), as this is prohibited by confinement (↓).

If electrons and positrons are shot at each other with equal energies and a quark-antiquark pair is created, energy and momentum conservation demand that they fly apart in opposite directions. But quark and antiquark are color-charged. So a *flux tube of gluons* forms between them. If enough energy is stored in this tube, then this energy can also convert into one or more quark-antiquark pairs, which in turn can group together into formations that carry no color charge. This is sketched in the figure below. So from a quark-antiquark pair with a lot of energy in the binding tube, a whole bunch of hadrons is created. These move with a large amount of energy. The process just described is called *hadronization*, since a pair of quarks created from the photon gets converted into bunches of hadrons (= bound states of quarks). The resulting bundled beams of many particles are called *jets*. Since the initial quarks flew exactly in opposite directions, in the situation just described, two jets consisting of many hadrons fly out of the reaction area in opposite directions. Clearly, for all this to happen the energy of the whole system must be high enough (at least a few GeV) to produce the required sizeable number of hadrons. On the left side in the figure on the next page, you can see the visualization of such a two-jet event, as it was measured with the JADE experiment at the PETRA accelerator at DESY (how such an experiment works is described in chapter 3 ↓).

What is very exciting is that one can "see" gluons in basically the same way: With a certain probability, one of the quarks emits a gluon shortly after its creation. At this moment we have not only two (the quark-antiquark pair, which directly arises from the photon), but

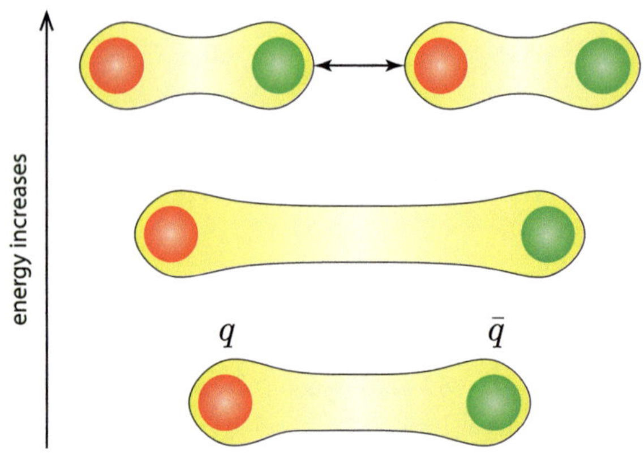

Illustration of particle production by tearing the flux tube

The Quark Model: A First Look → p. 20
The Structure of the Proton → p. 212
The Search for Free Quarks → p. 208
Confinement → p. 206
Chapter 3 Experimental Basics → p. 65

even three color-charged particles—after all, a gluon also carries a color charge. So hadronization occurs again, only this time not two, but three jets are created. This process was first observed at the PETRA accelerator at DESY, Hamburg, in 1979—a corresponding event is shown on the right side of the figure below. This is considered the experimental discovery of the gluon.

Interestingly, the ratio of the number of three-jet to two-jet events is a direct measure of the probability that a quark emits a gluon and thus a measure of the strong coupling α_{QCD}. Therefore, by counting the corresponding events, the size of the strong coupling constant can be extracted at a given total energy. Some of the values of α_{QCD} in the figure from the article "The Running Coupling" have in fact been determined in this way.

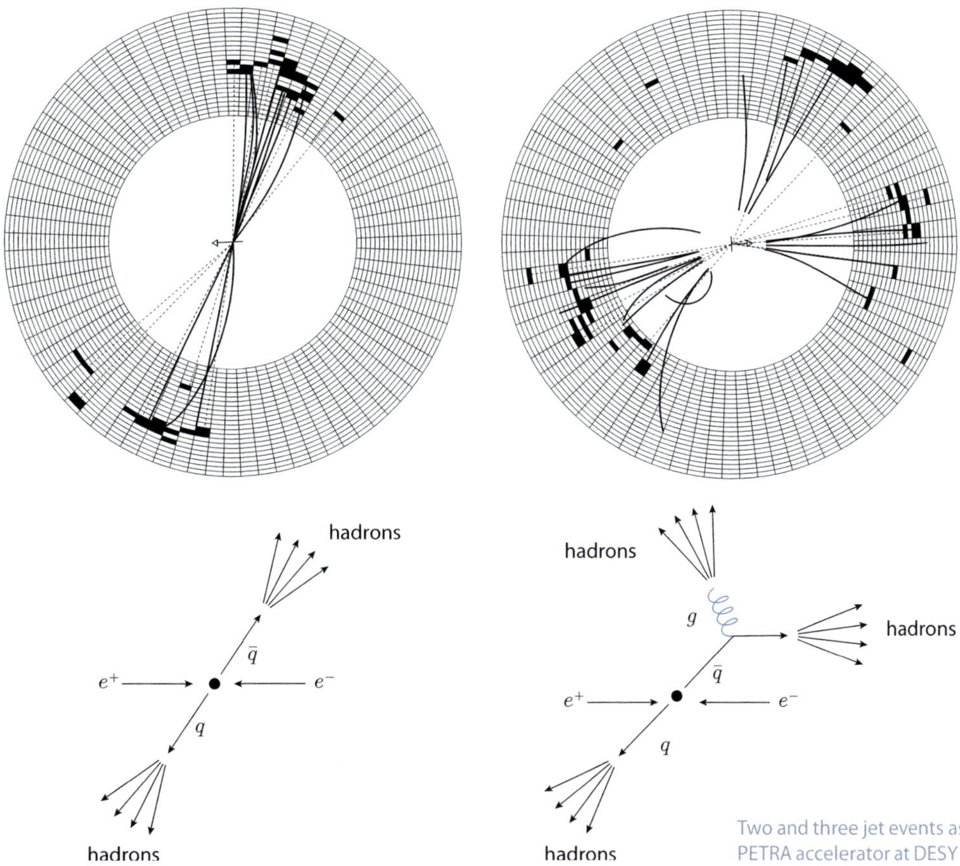

Two and three jet events as seen at the PETRA accelerator at DESY in Hamburg

The Running Coupling → p. 210

Quark-Gluon Plasma

Hot and Dense

In several articles (↓) we came across a fascinating state of matter known as plasma, where, e.g., electrons are no longer bound to individual atomic nuclei, but instead move freely throughout the material. This raises an intriguing question: Could something similar happen at a deeper level? Could quarks and gluons also exist in a free-floating state, no longer confined within hadrons like protons and neutrons?

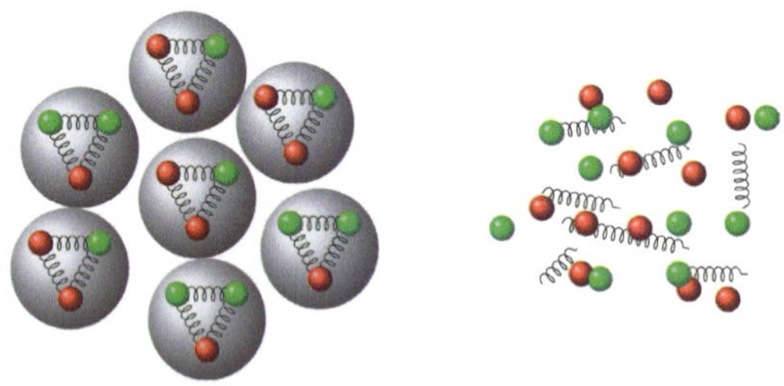

Left: Quarks and gluons confined in hadrons, right: quarks and gluons build a quark-gluon plasma.

It is already difficult to create the kind of plasma mentioned above because of the high densities and/or temperatures required. However, because of the confinement (↓) in strong interactions, it is even more difficult to create the corresponding state for quarks and gluons. Estimates show that temperatures of the order of 10^{12} Kelvin are necessary to form such a so-called *quark-gluon plasma* (QGP), where the boundaries between individual hadrons disappear. Conditions like these existed about one microsecond after the Big Bang, and may also play an important role during the formation of neutron stars. An understanding of the evolution and the properties of the quark-gluon plasma is thus mandatory for a deeper understanding of the development of our Universe.

While it is impossible to create a macroscopic volume of quark-gluon plasma in a laboratory, such conditions can be briefly recreated in high-energy heavy-ion collisions, for example by using lead or gold nuclei. The Relativistic Heavy Ion collider RHIC at Brookhaven National Laboratory on Long Island in the United States was—as the name suggests—built to study these types of interactions. One of the big LHC experiments, ALICE (A Large Ion Collider Experiment) is also dedicated to heavy-ion physics.

Free quarks and gluons in a QGP cannot be observed directly in a detector due to confinement. Instead, physicists look for indirect signatures—for example, changes in the spectra and abundances of hadrons produced in heavy-ion collisions, compared to those from proton-proton collisions, where a QGP is not expected to form.

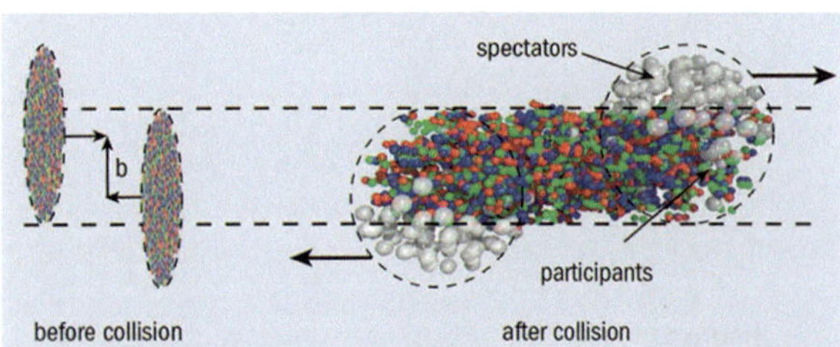

Left: Two nuclei before they collide, right: in the central region the nucleons inside the nuclei interact and can form a quark-gluon plasma.

Plasma Acceleration → p. 120, Spontaneous Symmetry Breaking I → p. 184, Open Questions → p. 270
Confinement → p. 206
The Running Coupling → p. 210

A heavy ion collision observed by the CMS experiment at CERN. The lower left jet has a much higher energy compared to the jet in the upper right. Note the huge amount of additional particles produced moving in all directions.

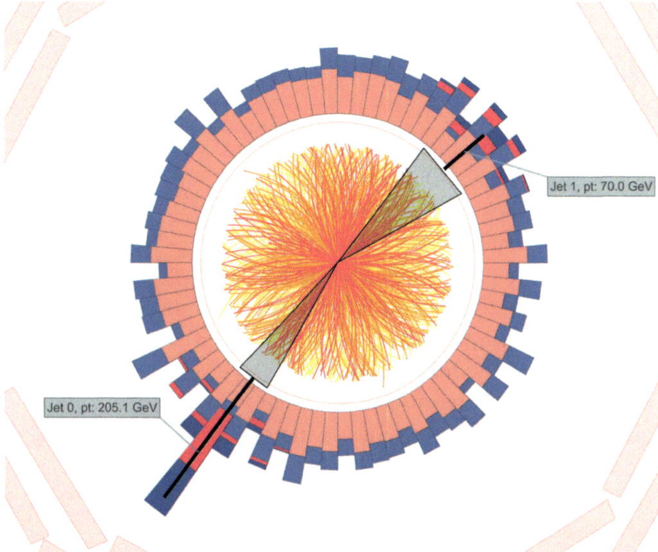

One example is called *jet quenching*. In ordinary high-energy collisions (↓), two jets of particles—produced by back-to-back quarks—carry roughly equal energy. However, if one of the quarks loses more energy in the quark-gluon plasma than the other, this results in asymmetrical jets. This energy loss is visible in detectors (see figure) and is a strong indicator of QGP formation.

First evidence for the formation was reported by studying lead-lead collisions at the Super Proton Synchrotron at CERN at the beginning of the millennium. This was later confirmed by measurements at the Relativistic Heavy Ion Collider RHIC colliding two beams of gold nuclei. Ongoing research at the LHC continues to explore the formation and properties of this exotic state of matter.

The experimental results show that heavy ion collisions can be described with variables like temperature, pressure, chemical potential and viscosity–quantities usually not used by particle physicists studying single particle collisions. But it turns out that the description in terms of these quantities borrowed from thermodynamics and fluid dynamics is appropriate here.

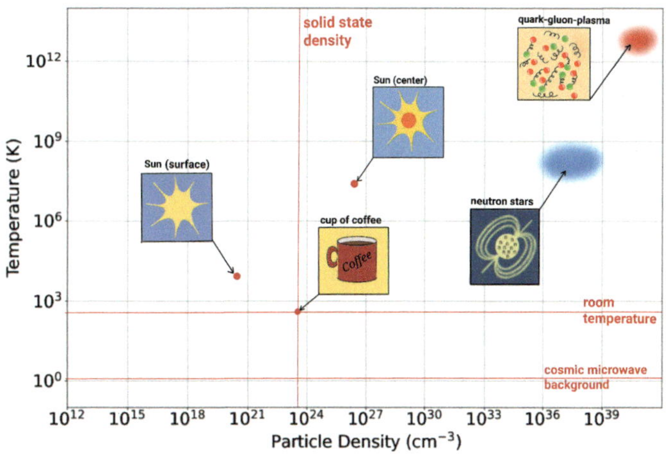

Temperature versus particle density diagram of various objects.

Seeing Quarks and Gluons? → p. 216

Where Does Our Mass Come From?

It's Not Primarily the Higgs Boson!

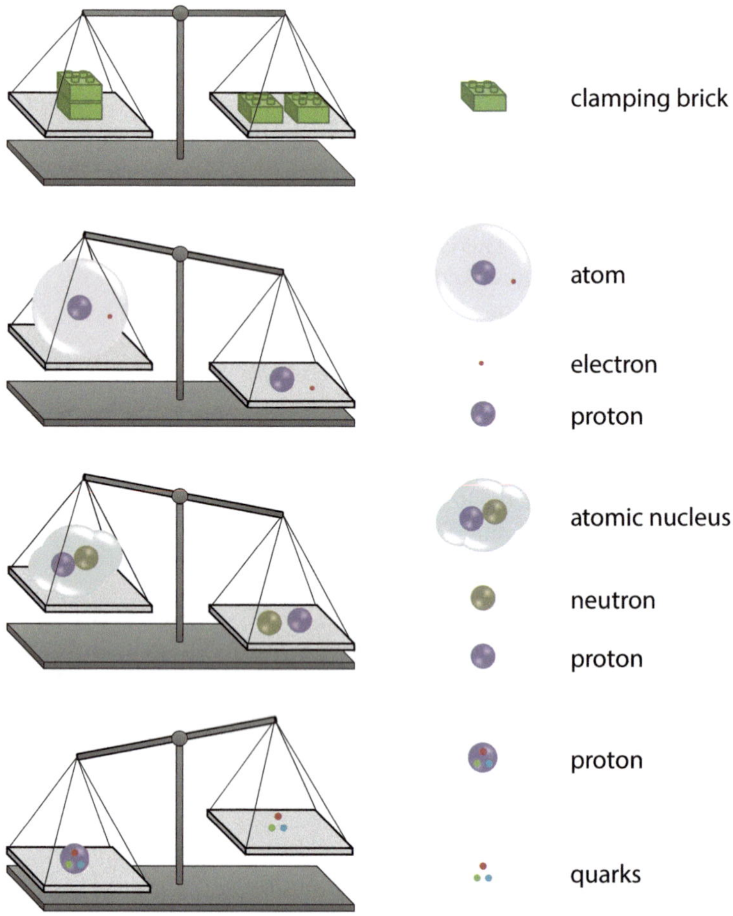

Weight comparisons between objects and their building blocks

At first glance, the answer to this question seems simple: Our mass comes from the masses of the various building blocks that make up our body, namely the atoms. Their masses come primarily from their nuclei. These are made up of protons and neutrons, which are thus the main source of our mass. So far, so good. The exciting question for particle physics, which we want to discuss in the following, is not only how accurate that statement is, but also where protons and neutrons get their masses from. In doing so, we will encounter some surprises.

To illustrate this, let us look at a few composite objects. If you, for example, stack two building blocks on top of each other, the total mass of the compound is naturally equal to the sum of the masses of the blocks. At the atomic level, however, this is no longer true: When an electron is bound to a proton, energy is released, the so-called *binding energy*. As a result, the atom is lighter than the sum of the electron and proton mass—exactly by the binding energy divided by c^2 due to $E = mc^2$. This is a very small but measurable effect in atoms: The weight of the atom is smaller by one part in a hundred million.

Something analogous happens when a proton and neutron combine to form a deuteron. Only that here, since the strong interaction is at work, the binding energy and thus the mass loss is much larger. The mass gets reduced by one per mille.

And where do protons and neutrons get their masses from? In the first article on the quark model (↓) we saw that their building blocks are three quarks each. Following up on what we just discussed one would thus expect that the masses of protons and neutrons is slightly less than the sum of masses of the pertinent quarks. However, this is not the case: The quark masses only make up about one percent of the proton and neutron masses. The vast majority of their mass comes from the energy of the gluon field that holds the quarks together. The masses of atoms and atomic nuclei are slightly smaller than the sum of those of their constituents. In contrast the masses of the nucleons are much larger than the sum of their constituents, the quarks.

The fact that the field energy of the gluons is so large is closely related to the phenomenon of confinement (↓); after all, the gluons have to ensure that the quarks always stay together. We can actually "calculate" that this picture is correct—how this is done is discussed in the article on lattice QCD (↓). The figure on the left page compares the masses of the composite objects just discussed with those of their building blocks—although the last line is a cheat: After all, there are no isolated quarks; here the illustration reaches its limits.

Interestingly, the picture is somewhat different for hadrons that contain heavy quarks like the charm or the bottom quark: Here, the mass of the composite system is actually by far dominated by the mass of the heavy quark. However, here too, the mass of the composite systems is larger than the sum of the masses of the building blocks. The reason for this drastic difference is that the strong coupling constant at such large energy scales as provided by the masses of the heavy quarks is much smaller than at low energies (The Running Coupling ↓), so that the field energy is correspondingly reduced. However, the mechanism of confinement also applies here: Just like the light quarks, the heavy ones cannot be isolated from a hadron.

How to Estimate the Masses of the Light Quarks

Due to confinement, quarks are not observable as isolated particles, which makes it difficult to determine their masses. But with the knowledge from the article on the quark model (↓), we can at least make an estimate. As mentioned, the mass difference, for example, between the proton (uud) and the Σ^+ particle (uus), mainly comes from the mass difference between down and strange quarks, which we therefore estimate to be approximately 200 MeV/c^2. This value is also an estimate of the mass of the strange quark, as the down quark mass is very small.

Furthermore, pions (π) and kaons (and η mesons) are closely linked to the spontaneous symmetry breaking (↓) of the strong interaction. In particular, $M_\pi^2 = B(m_u + m_d)$, $M_{K^+}^2 = B(m_u + m_s)$ and $M_{K^0}^2 = B(m_d + m_s)$ apply. From this, one can take 20 as the ratio of strange quark mass to down quark mass and thus obtain an estimate for the down quark mass of 10 MeV/c^2; approximately one percent of the proton mass.

The Quark Model: A First Look → p. 20
Confinement → p. 206
Lattice QCD → p. 222
The Running Coupling → p. 210
Spontaneous Symmetry Breaking → p. 62

Lattice QCD

Brought to You ... by Supercomputers

The physics of the strong interaction at low energies is very complicated. For example, since the probability that quarks radiate gluons is so high, there is at all times a large varying number of gluons present in a proton. Furthermore, the gluons also interact directly with each other, so that their dynamics as a whole is very complicated (Structure of the Proton ↓). For the curious readers, we now describe how, despite this, predictions can be made from QCD, for example for the masses or the scattering properties of hadrons, using numerical simulations (↓). The basic idea is the same as outlined in the earlier article. The main difference is that in addition to time, now also space must be segmented. This makes it possible to represent dynamical processes on the computer. A lattice is configured in space and time, on it quarks and gluons can "move" (see figure). Therefore, the method is called *lattice QCD*.

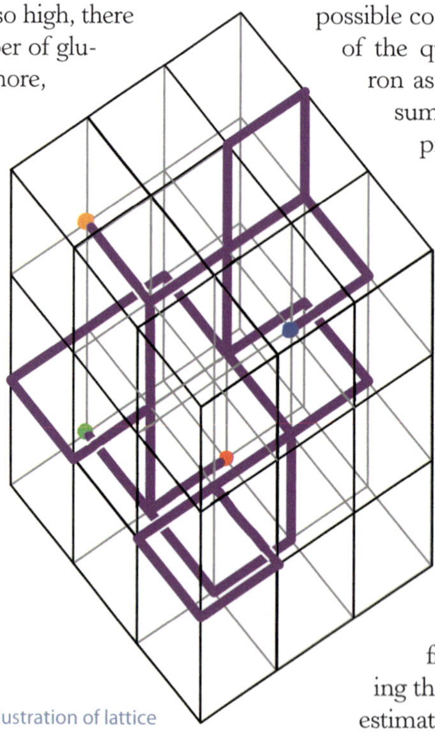

Illustration of lattice QCD—the colorful balls are the quarks, the purple lines are the gluons.

What actually needs to be simulated? In the article on quantum mechanics (↓) we discussed that the *action* is a central quantity of quantum mechanics. It is also essential for quantum field theory. For example, to calculate how a hadron moves from A to B, all possible configurations, i.e., all possible paths of the quarks and gluons within the hadron as well as their interactions must be summed up. From the total sum the properties of the hadron, e.g., the mass can be extracted. In the calculation each configuration is given a weighting factor, which is evaluated from the action. This means that unlikely paths contribute less to the sum than likely ones.

Of course, not all paths can be calculated—there are infinitely many—and the world is not a finite lattice. But with appropriate effort, one can approximate the correct result as closely as desired by making a finer grained lattice and/or increasing the computing power. The art lies in estimating, for example, what effect the finite spacing of the lattice has on the final result. To do this, calculations are carried out with different lattice sizes in order to extrapolate the behavior for an infinitely large lattice from the results. The same procedure is used with the finite lattice spacing, as well as the fact that only a finite number of paths can be simulated.

The Structure of the Proton → p. 212
Simulations in Theoretical Physics → p. 112
Concepts of Quantum Mechanics → p. 50

Furthermore, the parameters of the simulation must be adjusted so that they represent Nature. This is to be demonstrated using the example of a calculation of the masses of some hadrons consisting of the three lightest quarks (up, down, strange) in various combinations. In the simulation, the results of which are shown in the figure below, the two lightest quark types (up and down) were assumed to be of equal mass (their mass differences are relatively small—but important, as described in the article on the quark mass effect (↓)).

Thus, for the hadrons studied there are in total three parameters: The up/down quark mass, the strange quark mass, and the strength of the strong coupling (↓). The masses of the lightest quarks are set so that the mass of the pion corresponds to the physical value. The mass of the strange quark is determined by the mass of the kaons (↓) and finally the strength of the strong interaction is determined by the mass of the Ξ baryon. With all parameters fixed, additional hadron masses can be predicted.

The results of the calculation are represented in the figure by the red symbols; the blue circles show the masses that were used to determine the parameters. The black lines show the experimental values of the particle masses, the gray boxes their widths (Resonances ↓). As you can see, the match between experiment and theory is perfect. We therefore understand very well how the interaction of quarks with gluons and of gluons among themselves results in the masses of, for example, protons and neutrons (summarized in the figure as N (nucleon))—and thus also our mass. After all, this comes almost entirely from the masses of the atomic nuclei and thus to a large extent from those of protons and neutrons.

By the way, the above-mentioned calculation of the hadron spectrum required 45 million core hours of computing time at a bandwidth of 13.6 GB/s per core. If you want to do such a calculation on a laptop with 8 cores each with about 6 GB/s, it would take over 1400 years. Supercomputers are used to obtain the results in a reasonable time span. They have many more and faster cores which can work in parallel. On the Jülich computer JUGENE[1], which had 295 000 cores, a few weeks of pure computing time at full load would have been sufficient. In reality, various supercomputers were used utilizing only parts of the computing power. In addition, as described above, the parameters of the simulation had to be fixed, so that the whole project took almost five years in total. But the effort was worth it!

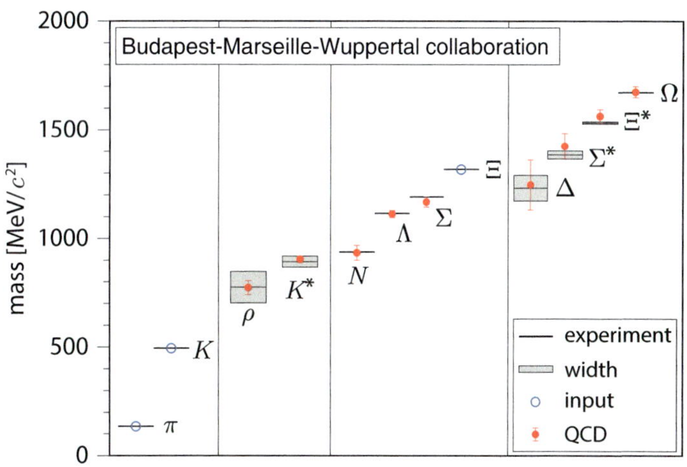

Masses of the lightest hadrons calculated with lattice QCD

[1] An image of the supercomputer JUGENE can be found at page 112.

The Quark Mass Effect → p. 224
The Running Coupling → p. 210
Where Does Our Mass Come From? → p. 220
Resonances → p. 56

The Quark Mass Effect

We Only Exist Because of the Quark Mass Difference

Quarks are the building blocks of protons and neutrons, which in turn are the building blocks of atomic nuclei (↓), which are about 10 000 times smaller than atoms. Can something hidden under so many layers have a direct influence on our existence? Yes, it can—in this article we argue that quark properties are in fact a key feature making our existence possible.

The proton (p) is slightly lighter than the neutron (n), just enough so that beta decay (Particles' Births and Deaths ↓) $n \to p + e^- + \bar{\nu}_e$, i.e., the decay of a neutron into a proton, an electron (e^-) and an electron antineutrino ($\bar{\nu}_e$) (Weak Interaction ↓), is energetically allowed. Thus, free neutrons decay, whereas free protons are stable. However, if the latter were so much heavier than the neutron that the reverse beta decay, $p \to n + e^+ + \bar{\nu}_e$, were energetically allowed, then the neutron would be stable and not the proton. But since the neutron is electrically neutral, it cannot bind electrons and thus cannot form atoms. In a world where the proton is heavier than the neutron there would be no elementary hydrogen, no water and ultimately no life as we know it.

But why is the proton lighter than the neutron? The central building blocks of the proton are two up and one down quark, while the neutron consists of two down and one up quark (Structure of the Proton ↓). The strong interaction does not distinguish between the types of quarks. Therefore, the mass difference between proton and neutron must be due to differences in the quark properties.

beta decay of free neutron occurs, because::	beta decay of free proton does *not* occur, because:
$M_p + M_e + M_\nu < M_n$	$M_n + M_e + M_\nu > M_p$

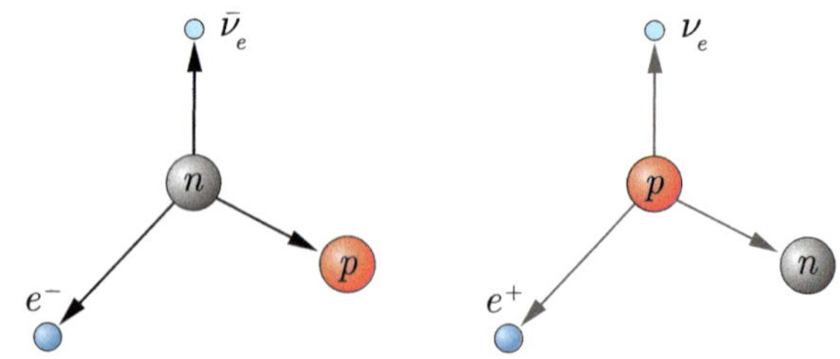

Comparison of the beta decay of neutron and proton

An up quark has the electric charge $+2/3\,e$ and a down quark $-1/3\,e$, where $-e$ denotes the charge of the electron. If you add these charges together you get, as expected, a total charge of $1\,e$ for a proton and zero for a neutron. In the proton, therefore, more electric charge is "trapped" in a small space than in the neutron. But equal charges repel each other and therefore it takes energy to keep these charges close together. Thus the electromagnetic interaction gives a proton a slightly higher energy content than a neutron!

With the help of $E = Mc^2$, this energy difference can be translated into a mass difference, so that from the point of view of the quark charges and their interactions a proton should be heavier than a neutron. A simple estimate (see info box) that can be confirmed by detailed studies shows that the electromagnetic in-

From Nucleus to Nuclei → p. 226
Particles' Births and Deaths → p. 54
The Weak Interaction → p. 178
The Structure of the Proton → p. 212

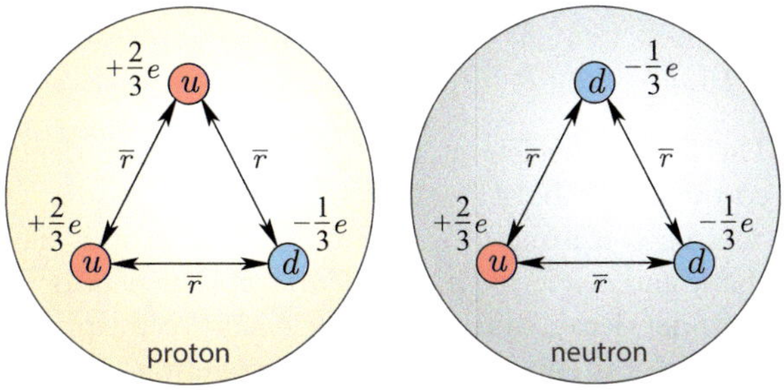

Very simple model of nucleons.

teraction between the quarks makes the proton about $0.5\ \text{MeV}/c^2$ heavier than the neutron. If this were the only source of the proton-neutron mass difference, the proton could decay into a neutron, with the previously described fatal consequences for our existence.

It is the quark mass difference, which is responsible for the fact that the proton is *lighter* than the neutron. Since the up quark is lighter than the down quark, there is a second contribution to the proton-neutron mass difference. And this is slightly larger than the electromagnetic effect, giving an overall *lighter* proton. We therefore owe our existence to the mass difference of the light quarks (see also chance ↓).

By the way, it is generally true that charged particles are heavier than neutral ones, if no other effects contribute. Since charged and neutral pions (Quark Model ↓) each contain equal proportions of up and down quarks or antiquarks (for π^+, π^0 and π^-, the quark content is $u\bar{d}$, $u\bar{u} - d\bar{d}$ and $d\bar{u}$)), quark mass effects are strongly suppressed. The charged pion is therefore heavier than the neutral one. In contrast, the negatively charged kaon (a $\bar{u}s$ state) is lighter than the neutral one (a $\bar{d}s$ state), as here the quark mass effects not only contribute, but are larger than the electromagnetic effect.

Mathematical Supplement: Electromagnetic Mass Difference

The order of magnitude of the electromagnetic contribution to the proton-neutron mass difference can be estimated using a classical consideration. The mechanism of confinement (↓) ensures that the average distance between the quarks, $\bar{r}$, remains below $10^{-15}\ \text{m} = 1\ \text{fm}$. We thus obtain an estimate of the energy difference generated by electromagnetism between the proton and neutron by considering the quarks as static electric charges with relative distance $\bar{r}$ (see image). For the electromagnetic (e.m.) energy between two charges q_1 and q_2 the Coulomb law gives

$$E_{\text{e.m.}} = \frac{1}{(4\pi\varepsilon_0)\,\bar{r}}\,(q_1 q_2)\,.$$

This results in the e.m. energy difference between proton and neutron

$$\frac{1}{(4\pi\varepsilon_0)\,\bar{r}}\,\big(q_u^2 + 2q_u q_d - (q_d^2 + 2q_u q_d)\big) = \frac{\alpha\hbar c}{3\bar{r}}\,,$$

where the electromagnetic fine structure constant $\alpha = e^2/(4\pi\epsilon_0\hbar c) \approx 1/137$ and $\hbar c \approx 197\ \text{MeV fm}$ denotes the so-called conversion factor. For $\bar{r} = 1\ \text{fm}$, this results in a $0.5\ \text{MeV}$ higher energy content of the proton. This value is very close to the $(0.7 \pm 0.3)\ \text{MeV}$ resulting from a complex theoretical analysis.

Did Our Life Emerge by Chance? → p. 292
The Quark Model: A First Look → p. 20
Confinement → p. 206

From Nucleus to Nuclei

Astonishing Complexity From Just Two Building Blocks

In the article on quantum mechanics (↓), we saw that the Coulomb interaction binds electrons with well-defined energies to atomic nuclei to form atoms. In this process, so-called *shells* are formed, which determine the chemical properties of the atoms. Within the shells, the differences between the various allowed binding energies are small, between the shells they are relatively large. Analogously, the so-called *nuclear force*, driven by the strong interaction, binds protons and neutrons together to form atomic nuclei. Since the nuclear force is much stronger than the Coulomb force and also much shorter in range (↓), atomic nuclei are much smaller than the atomic cloud that surrounds it. The allowed energy levels within nuclei can also be sorted into shells—these are responsible for the nuclear structure. Nuclei whose shells are filled are particularly stable.

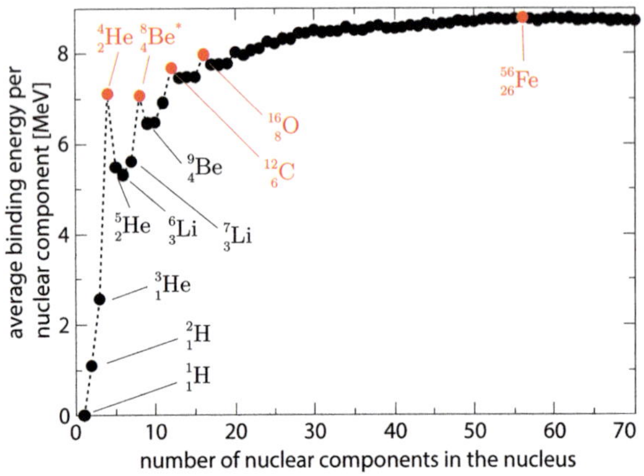

The lightest nucleus is the hydrogen nucleus, denoted by the symbol H, consisting of a single proton.[1] A nucleus consisting of a proton and a neutron is also stable and is referred to as a *deuteron*. The neutron in it cannot decay, since its binding energy to the proton (2.3 MeV) is larger than the proton-neutron mass difference (1.3 MeV). Accordingly, the possible final state after such a decay, namely $ppe^-\bar{\nu}_e$, is heavier than the deuteron and the decay is blocked energetically (Particles' Births and Deaths ↓). By the way, all nuclei with the same number of protons have the same name, regardless of the number of neutrons. After all, they form atoms with chemically identical properties, as only the total nuclear charge matters for the structure of the electron shell. Nuclei that differ only in the number of neutrons are called *isotopes*. Often, atomic nuclei are denoted with the symbol

$$^{\text{Number of nuclear components}}_{\text{Number of protons}}\text{Name}\,,$$

although the information content of "Name" and "Number of protons" is identical. So the symbol for the deuteron is $^{2}_{1}$H.

Two neutrons do not form a bound state, as the attraction between them is slightly weaker than that between proton and neutron—why this is so is not known. Also two protons do not form a bound system. However, the triton, $^{3}_{1}$H, consisting of a proton and two neutrons, can form, but decays with a mean lifetime of just over twelve years through beta decay analogously to the neutron according to $^{3}_{1}\text{H} \rightarrow {}^{3}_{2}\text{He} + e^- + \bar{\nu}_e$ into a helium-3 nucleus, which consists of two protons and one

[1] H is also used as symbol for the atom.

Concepts of Quantum Mechanics → p. 50
On the Ranges of Interactions → p. 58
Particles' Births and Deaths → p. 54

neutron. This is possible because the neutron in ${}^3_1$H is with 0.76 MeV only slightly more bound than the proton in ${}^3_2$He, which is insufficient to overcome the proton-neutron mass difference.

In the left figure, the average binding energies per nuclear building block are plotted against the number of buildings blocks. It is noticeable that some nuclei are significantly more strongly bound than their neighbors. These are marked in red in the figure. This happens exactly when both the number of protons and of neutrons fill the corresponding shells—these numbers, 2, 8, 20, 28, 50, 82, 126, ..., are referred to as *magic numbers*. Particularly striking here is the ${}^4_2$He nucleus, also known as an alpha particle, as it is much more strongly bound than its neighbors. By the way, the horizontal axis of the figure does not need to stop at 70. However, the diagram from there on is somewhat boring, as the average binding energy from the most strongly bound nucleus, iron (${}^{56}_{26}$Fe) with 8.79 MeV, only drops very slowly and smoothly to uranium 235, ${}^{235}_{92}$U, with an average binding energy of 7.59 MeV.

What seems boring as a diagram has drastic implications. Up to iron, a lot of energy is released in *nuclear fusion*, as the combined heavier nuclei are more strongly bound than the lighter ones. This is a fantastic source of energy, if the electrical repulsion between the nuclei is somehow overcome—and it works: This is exactly where stars, including our Sun, get their energy from (Birthplace of Nuclei ↓). Here it is gravitation that leads to such high densities in the centers of stars that the light nuclei get close enough for the weak and strong interaction to take over. For heavy nuclei, energy is released in *nuclear fission*, as discovered by Otto Hahn and Fritz Straßmann at the end of the 1930s and theoretically described by Lise Meitner and Otto Frisch. While the energy release in this process is much less per nuclear building block than in fusion, it is still enough to provide the energy source of nuclear power plants. The left figure mostly lists stable nuclei—only ${}^8_4$Be is unstable and therefore marked with an asterisk *. It decays with a mean lifetime of 10^{-17} seconds into two ${}^4_2$He nuclei. This nucleus is included because it is one of the particularly strongly bound nuclei (only that the mass of two ${}^4_2$He nuclei is even smaller) and plays an important role in the formation of elements in stars, with exciting implications (Did Life emerge by Chance? ↓).

${}^1_1$H	proton (p)	stable
	neutron (n)	unstable
	two-neutron	not binding
${}^2_1$H	deuteron	stable
${}^3_1$H	triton	unstable
${}^3_2$He	helium-3	stable
${}^4_2$He	helium-4 (α)	very stable

The lightest nuclei

Besides the structure of the light nuclei, in this chapter we also met some of the most important nuclear decays: In addition to beta decay, discussed frequently in this book, there is alpha decay, in which a helium nucleus is emitted. In addition to these, a nucleus can be formed in an excited state and emit photons as gamma rays when transitioning to the ground state—analogous to the atomic transitions (Chance, Interference and Predictions ↓). The difference is that the energies of the photons emitted in nuclear transitions are much higher, so that γ radiation can cause great damage.

The Birthplace of Nuclei → p. 342
Did Our Life Emerge by Chance? → p. 292
Chance, Interference and Predictions → p. 52

The Karlsruhe Nuclide Chart

About the Valley of Stability

An overview of the known atomic nuclei is provided by the Karlsruhe Nuclide Chart, which is schematically represented in the figure. Horizontally from left to right, N, the number of neutrons increases. Vertically from bottom to top, Z, the number of protons increases. Each little box represents a nucleus. The small black boxes denote the stable nuclei. The figure shows that for stable light nuclei, the number of protons is close to the number of neutrons. However, as the number of protons increases, the electrical repulsion in the nucleus grows, so that an increasing neutron surplus is needed for increasingly heavier nuclei—from a neutron number of 21 up, all stable nuclei are below the line $N = Z$ marked in red. Already in calcium $^{48}_{20}\mathrm{Ca}$ there is a neutron surplus of 8, which for lead $^{208}_{82}\mathrm{Pb}$ has grown to 44. The figure also shows, in addition to the stable nuclei, unstable ones that have a sufficiently long lifetime to be observed in experiment. The color code shows their dominant decay modes.

The so-called *valley of stable nuclei* is clearly visible as the sequence of black boxes. For a larger number of protons, the nuclei mostly decay by the transition of a bound proton into a bound neutron (marked in red), either undergoing a β^+ decay ($p \rightarrow n + e^+ + \nu_e$) or capturing a shell electron($p + e^- \rightarrow n + \nu_e$, marked as ε in the figure). In some cases, the emission of a proton (light brown) or an alpha particle (yellow) may also occur. If there is a neutron surplus relative to the valley of stability, then either beta decay occurs ($n \rightarrow p + e^- + \bar{\nu}_e$—dark blue) or a neutron is emitted (light blue). Very heavy nuclei can simply fall apart, which, if the decay products are of similar size, is referred to as *spontaneous fission* (sf), otherwise as *cluster emission* (CE). Finally, there are long-lived excited states which transition to their ground states. These are referred to as *isomeric transitions* (IT).

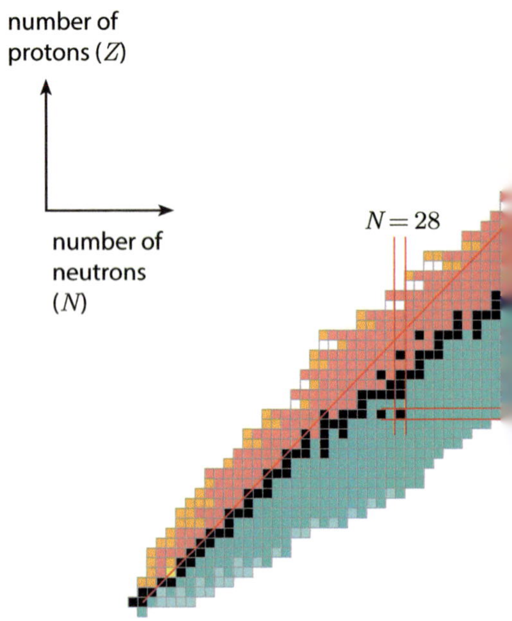

For the unstable nuclei, the nuclide chart also contains information about the half-life, i.e., the time in which, on average, half of a given type of nuclei has decayed. Also shown are other decay characteristics, as can be seen in the inlay. A detailed description can be found, for example, in the article given below[1].

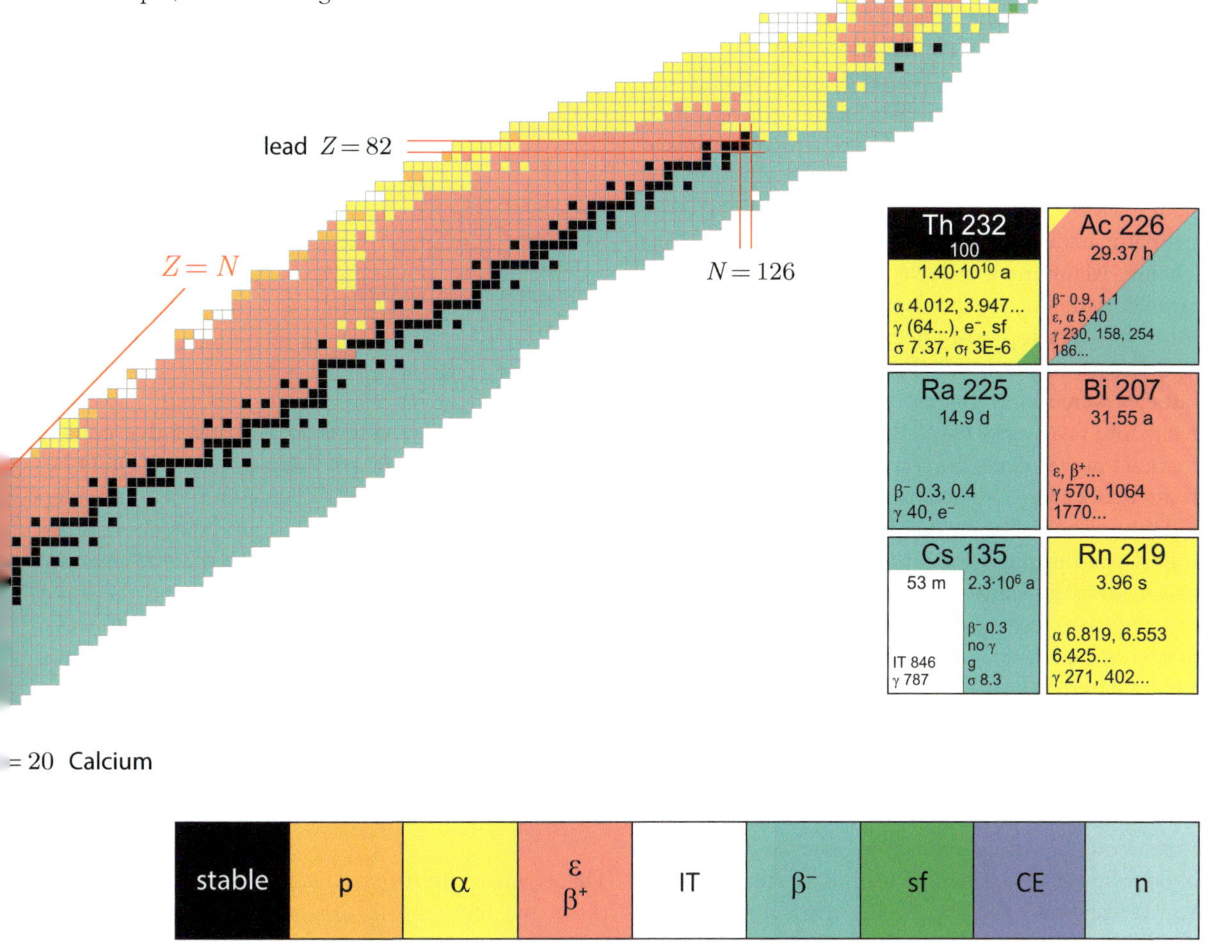

[1] Z. Sóti, J. Magill, R. Dreher 2019, *Karlsruhe Nuclide Chart—New 10th edition 2018*, EPJ Nuclear Science Technology, 5, 6, doi:10.1051/epjn/2019004

Exotic States

A New Particle Zoo

As described in the articles on the quark model (↓), for a long time the simplest quark model successfully explained the properties of particles that experience the strong interaction. In particular, those that have an integer spin were described as bound states of a quark and an antiquark and those with half-integer spin as bound states of three quarks. These are the simplest states allowed by the rules of QCD (Confinement ↓). But this picture was shown to be incomplete, when in 2003 the first particle was observed that did not fit at all into this scheme. Many more followed, especially among the particles with a heavy quark and its antiquark (charm or bottom—the top quark does not form bound states, as it decays too quickly). For example, quite a few particles were found that carry an electric charge and decay into final states with a heavy quark and its antiquark. However, since the combination of a quark and its antiquark always leads to an electrically neutral state and the creation of heavy quarks in decays is extremely unlikely (at this mass scale, the strong coupling (↓) is too small), these states must have a more complicated structure than just quark-antiquark. The charge must come from additional light quarks already present in the decaying state.

But what is the structure of these new states? Here, there is still a lot of room for speculation. Researchers active in this field are convinced that we will know a lot more about the strong interaction once we have understood why certain structures form and others do not. An overview over the approaches most intensely discussed at present is shown in the figure above.

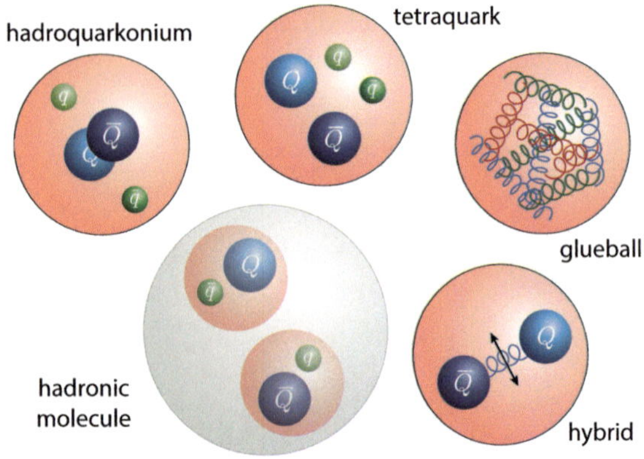

Popular interpretations of mesons with two heavy quarks, which do not fit into the traditional quark model

Glueballs are referred to as bound states that consist only of gluons. In principle this is possible in QCD, because gluons carry a strong charge and therefore interact with each other. In contrast, bound states consisting purely of photons are impossible, as photons can only interact very indirectly with each other and the force they then exert on each other is also extremely weak. *Hybrids* are states in which, in addition to the quarks, the gluons also play an active role, e.g., contribute to their quantum numbers. For the above-mentioned electrically charged heavy particles, however, we need a different explanation, as gluons carry a strong, but no electric charge.

This is where *multiquark states* come into play. For these, the literature focuses mainly on three groups, compact tetraquarks, hadroquarkonia and hadronic

The Quark Model: A First Look → p. 20, The Quark Model Revisited → p. 202
Confinement → p. 206
The Running Coupling → p. 210
From Nucleus to Nuclei → p. 226

molecules, which all differ in how the quarks group together within the particles. Compact *tetraquarks* are a fairly obvious extension of the simplest quark states: The underlying assumption is that in these exotic structures two quarks (diquark) and two antiquarks (antidiquark) group together in pairs to form diquark-antidiquark bound states. The interaction between the building blocks is assumed in analogy to the established, most simple quark model.

In *hadroquarkonia*, it is assumed that the heavy quark and its antiquark form the compact core of the exotic particle, which is surrounded by a cloud of light quarks. *Hadronic molecules* are close relatives of atomic nuclei (↓), except that their building blocks are not protons and neutrons, but other hadrons. At first glance, this structure seems very similar to that of tetraquarks. However, what appears to be a simple rearrangement dramatically changes the properties of the particles, so it is justified to speak of a new structure. This is mainly because diquarks and anti-diquarks carry a color charge, heavy mesons do not. Therefore, the size of compact tetraquarks is limited to the confinement radius of less than one fermi. This does not apply to hadronic molecules that can in turn extend over much larger scales. Moreover, molecules are most likely to decay into their building blocks—other decay channels are suppressed.

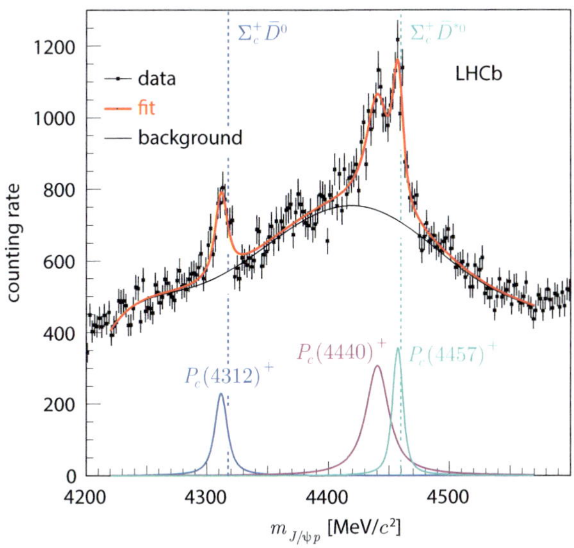

Signals of pentaquark states

Also baryons with at least five quarks (four quarks and one antiquark), the so-called *pentaquarks*, were found experimentally and the concepts just described were transferred to these. The figure shows the corresponding experimental signals found in the decay $\Lambda_b \to K J/\psi p$. On the x axis, the invariant mass (↓) of J/ψ and proton is plotted. Three pentaquarks show up as resonances (↓), as there are clearly visible narrow structures at three energies. Also, the energies at which the named particles can be produced are marked in the figure. The proximity of the masses of the pentaquarks to these so-called thresholds is seen as an indication that the pentaquarks are hadronic molecules made of $\Sigma_c\bar{D}$ or $\Sigma_c\bar{D}^*$.

The attempt to decipher the large number of exotic states is currently a very active field of particle physics, promising some exciting insights in the coming years and decades.

R. Aaij et al. (LHCb Collaboration) 2019, *Observation of a Narrow Pentaquark State, $P_c(4312)^+$, and of the Two-Peak Structure of the $P_c(4450)^+$*, Physical Review Letters, 122, 222001, doi:10.1103/PhysRevLett.122.222001

Identifying the Particle Type via the Mass → p. 156
Resonances → p. 56

Radiocarbon Dating With ^{14}C

Using Nuclear Physics for Archaeology

Cosmic radiation (↓) contains, among other particles, also high-energy neutrons. When such a neutron hits the nucleus of a nitrogen atom in the atmosphere, in some collisions an isotope of carbon is created, which consists of eight neutrons and six protons, thus containing a total of 14 nuclear building blocks (↓) and is therefore referred to as carbon-14, or ^{14}C.

	n	+	^{14}N	$\rightarrow$	^{14}C	+	p
neutrons	1	+	7	=	8		
protons			7	=	6	+	1

In contrast to the most common stable carbon isotope ^{12}C, which is made up of six protons and six neutrons, ^{14}C is radioactive. It decays through a beta decay (Particles' Births and Deaths ↓) back into nitrogen, an electron and an electron antineutrino:

$$^{14}C \rightarrow {}^{14}N + e^- + \bar{\nu}_e$$

with a half-life of 5730 years. The proportion of ^{14}C in the atmosphere is only about 10^{-12}. So for every 1000 billion C atoms, there is only one ^{14}C atom. Since chemically there is no difference between the isotope ^{14}C and the stable isotopes ^{12}C and ^{13}C, the latter of which makes up about 1 % of carbon, all three contribute to the formation of carbon dioxide CO_2.

Plants, animals and humans take up carbon, for example in the form of carbon dioxide (CO_2), including of course the small proportion of ^{14}C mentioned above. When an organism dies, it no longer produces or absorbs CO_2. This can be used for age determination.

At the time of death, the ^{14}C proportion in the organism is the same as in the atmosphere. After that, no more ^{14}C enters the organism and the proportion only decreases due to radioactive decay. Roughly speaking, this means that if you know that initially the ^{14}C proportion was 10^{-12} and only half is detected in the measurement, the object under investigation is 5730 years old, which is exactly the half-life.

A proportion of 10^{-12} seems very small. However, considering that a gram of carbon consists of about $5 \cdot 10^{22}$ carbon atoms, there are already $5 \cdot 10^{10}$ ^{14}C nuclei contained in it. Of these, an average of 17 nuclei decay per minute.

Cosmic Rays → p. 336
From Nucleus to Nuclei → p. 226
Particles' Births and Deaths → p. 54

To measure the ^{14}C proportion in a sample whose age is to be determined, various methods are used. Either the decays of the ^{14}C can be detected, or the ^{14}C proportion is determined directly by mass spectrometry. The latter method is more accurate, but also more complex.

As is so often the case, the basic principle is easily explained, but the devil is in the details. In the above calculation, it was assumed that the ^{14}C proportion of 10^{-12} in the atmosphere was constant over millennia. However, this is not the case. The strength of cosmic radiation has changed over time. In addition, the carbon cycle is influenced by climate changes. The Earth's magnetic field is not constant over time, which also leads to variations in the ^{14}C production.

Calibration curve: The so-called radiocarbon age, determined under the assumption of a constant ^{14}C proportion, is plotted on the vertical axis. The true age can be read off the horizontal axis via the curve.

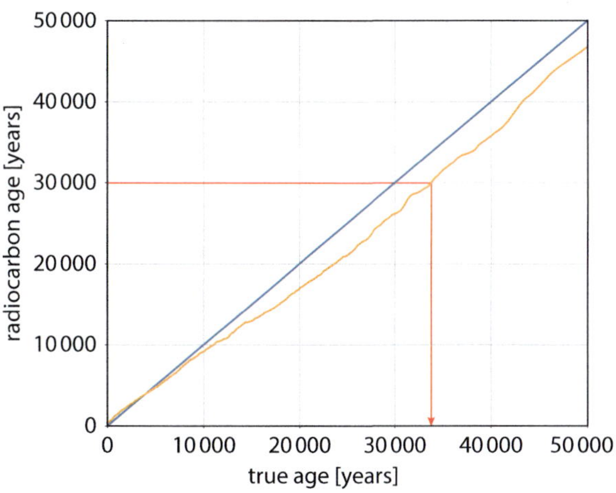

Tree rings are used to determine the ^{14}C proportion in the atmosphere in different years. These results then flow into the calibration curve.

Furthermore, humans are increasingly influencing it. For example, the CO_2 from the combustion of fossil fuels contains almost no ^{14}C, as oil and coal were formed millions of years ago. Nuclear weapons tests released neutrons, which temporarily led to a doubling of the ^{14}C content in the atmosphere.

To determine the real age from the age determined under the assumption of a constant ^{14}C proportion, a calibration curve (see figure) must be used, which, among other things, relies on the analysis of tree rings. Here, the ^{14}C content can practically be measured year by year. Further information is provided by Lake Suigetsu in Japan. Its sediments contain annual layers, which can be counted and analyzed similarly to the annual rings of a tree.

With the ^{14}C method, age determinations up to about 50 000 years are possible. First, as described above, the age is determined under the assumption of a constant ^{14}C proportion in the atmosphere. From this value on the vertical axis of the figure, which is referred to as the radiocarbon age, a horizontal line is then drawn to the curve. From there, a vertical line leads to the true age, which is plotted on the horizontal axis.

9 Incredible Success and Yet Possible Failure of the Standard Model

It has long been a goal of particle physics to find a single unified mathematical description of all fundamental measurable phenomena. The previous chapters have outlined the concepts and ideas behind such a comprehensive theory, namely the Standard Model of particle physics. It describes three of the four known fundamental forces of Nature in a fully consistent model: the electroweak interaction, which unifies electromagnetism and the weak force, and the strong force. Together with the independent description of gravity by general relativity, this allows us to describe "in principle" all known phenomena that can be measured in precision experiments on Earth. This represents a historically unique situation and a major cultural achievement. This chapter presents the most outstanding experimental steps that established the Standard Model as *the* description of the forces in the microcosm.

Despite the great successes of the Standard Model, physicists have still far from completed their work. They know that the Standard Model cannot be the "Theory of Everything". The ever more precise experiments in cosmology demand an extension or a complete replacement of the Standard Model. They require an explanation for dark matter through a new particle, as well as a mechanism to account for the observed difference in the amount of matter and antimatter in the Universe. Additionally, precision measurements testing the Standard Model's own predictions, including anomalous magnetic moments or rare meson decay observations, show deviations in some areas that might be random but could also indicate a path beyond the Standard Model. Researchers are therefore diligently searching for the precise point where the Standard Model no longer accurately describes the nature of the microcosm. They are in pursuit of a future groundbreaking discovery.

The Discovery of Neutral Current Interactions

The First Breakthrough of the Standard Model

The history of great scientific discoveries can be divided into three types: predicted discoveries, accidental discoveries of phenomena that were in principle already expected, and truly surprising measurements. In the first case, there is a theory that convincingly explains why a particular experiment should make a discovery that would not necessarily have been conceived without the theory.

In 1973 the first such discovery as predicted by the Standard Model was observed: the *neutral currents* of the weak interaction (↓). Since 1934, when Enrico Fermi developed an effective quantum field theory (↓) of the weak interactions, the concept of the *charged current* had been known. This refers to a weak interaction that changes the electric charge q of the participating fermions by $\Delta q = \pm 1$. In 1967, with the completion of the electroweak Standard Model theory by Sheldon Glashow, Abdus Salam, and Steven Weinberg, a new, unexpected interaction was postulated: an interaction similar to the charged current but without a change in charge ($\Delta q = 0$): the (electrically) neutral current.

charged current neutral current

In the Standard Model, the charged current is mediated by the charged $W^\pm$ boson, and the neutral current is mediated by the neutral Z^0 boson. Both must have a very high mass compared to the other elementary particles known at the time in order for the theory's predictions to match the experimental measurements. However, the value of this mass was unknown. Only the ratio of the coupling (between the new bosons and the Standard Model fermions) to the masses of the gauge bosons was predicted exactly.

As of 1973, neither the $W^\pm$ boson nor the Z^0 boson had been directly observed as real particles (Identification via Mass ↓). The idea that they must exist was based solely on the mathematical structure of the Standard Model. With these particles and the neutral currents the Standard Model was simply much more symmetrical, and thus more "beautiful", and was more precisely calculable than all other ideas about possible fundamental interactions known at the time (↓). This was reason enough to undertake a large-scale effort at CERN, using the *Gargamelle* bubble chamber (↓) to search for neutral currents.

There is only one problem: The only known way to have a weak interaction while turning off electromagnetism is by using a neutrino beam (↓). Thus, a quasi-invisible beam enters the experiment, in this case a ν_μ beam. A weak interaction occurs only very rarely. In some cases,

The Weak Interaction → p. 178
Effective field theory: Accurate Calculation of the Unsolvable → p. 110
Identifying the Particle Type via the Mass → p. 156
External and Internal Symmetries → p. 96 and A Deep Dive Into the Scientific Method → p. 30
Bubble chamber: Liquified Noble Gases as Detectors → p. 138

The reconstruction of neutral current interactions using the Gargamelle bubble chamber

through a charged current, a muon neutrino ν_μ is converted into a muon μ^-. The muon penetrates large parts of the detector with observable electromagnetic interactions (Signatures of Particles ↓). It can be easily used to detect the interaction and take a picture of the event. In the even rarer predicted neutral current events, however, the muon neutrino remains a neutrino and leaves the detector unnoticed after a possible scattering. Only the particles off which it scatters, such as an electron e^- from an atomic shell in the liquid of the bubble chamber, are accelerated enough to become visible as a track. A high-energy electron transfers much more energy to the surrounding material over the same distance than a muon and therefore comes to a stop within the bubble chamber. Since all generated particle tracks start and end in the bubble chamber, nothing visible enters, and nothing visible exits the detector. Thus, it requires motivation from a convincing and elegant theory to keep taking pictures of possible events, even if most of the photos show nothing due to very rare neutrino interactions. In 1973, the first of these unique type of events was finally captured by chance,[1] where nothing seemed to enter the detector, nothing seemed to exit, yet it was possible to take a picture of something happening in between!

This detection of a neutral current event was the first of many great victories of the Standard Model and convinced the broader physics community that the Standard Model is the (so far) best theory of elementary particles! However, the Z^0 boson (and the $W^\pm$) had not yet been discovered as a particle. Every major discovery is a reason for further investigations: in this case, the search for the real Z^0 particle (↓).

[1] F. J. Hasert et al. 1973, *Observation of neutrino-like interactions without muon or electron in the gargamelle neutrino experiment,* Physics Letters B, 46, 1, 138, doi:10.1016/0370-2693(73)90499-1

Neutrinos → p. 196
Signatures of Particles → p. 80
The Discovery of the W and Z Bosons → p. 240

The Accidental Discovery of the J/ψ

The Second Breakthrough of the Standard Model

Even in the large, long-term planned experiments of particle physics there are surprises. Sometimes a great deal of luck plays a role in a discovery, and the interplay of different types of experiments is also significant.

One such discovery is the first observation of a particle with the unusual name J/ψ. It is a meson composed of two quarks—a charm quark c and its antiparticle $\bar{c}$. At the beginning of 1974, only the three light quarks u, d, and s were known. The Standard Model was a promising theory, which had just experienced its first experimental breakthrough a few months earlier (↓). If the Standard Model were to be a good description of elementary particles, then according to the *GIM mechanism* (↓) there would have to be a fourth quark, the charm c, which is necessary to explain the observed properties of kaon K decays (Up, Down, Left, Right ↓).

Although it was known that another quark might exist, there was no sufficiently precise prediction on where to find it. Mary Gaillard and Ben Lee had predicted the mass range in which it should be found, but still various experimental approaches were necessary to explore previously unexamined interactions and energy ranges—much like an expedition into unknown territories. One approach involved using an accelerator to generate high-energy protons and directing them at a block—the target—made of different materials. Behind this target was a detector capable of distinguishing between different types of particles (↓). If, in the interaction between the proton and the nuclei in the target, an unknown particle with a given mass and lifetime decayed into two clearly identifiable particles, one could determine the mass of the particle from the energies and momenta of the decay products (↓).

Sam Ting and his team conducted such an experiment at the Brookhaven National Laboratory in the eastern United States in 1974[1]—and a narrow, previously unknown signal appeared at a mass of 3.1 GeV. Since identifying the decay products—here, an electron and a positron—was not straightforward, the group was not entirely certain that they had discovered something new.

At the same time, Burt Richter and his colleagues on the West Coast of the United States were conducting experiments at the SPEAR storage ring at the Stanford National Accelerator Laboratory (SLAC) with the collision of electrons and positrons.[2] The detector, for instance, measured how frequently an interaction occurred. The experimenters asked the accelerator team to gradually increase the collision energy—like an experimental journey through an unknown landscape. E.g., after reaching a collision energy of 3 GeV, the interaction rate was measured, and then 3.1 GeV would follow, then 3.2 GeV, and so on.

Fortunately, the operational team was not too precise with the numbers—3.105 is almost the same as 3.1, right? But suddenly, everything changed; the collision rate had increased by a factor of 100—like going from a leisurely walk up a 88-meter hill to a step up to the summit of Mount Everest!

The Discovery of Neutral Current Interactions → p. 236
GIM mechanism: The Path to Six Quarks in Three Families → p. 194
Up, Down, Left, Right → p. 180
Signatures of Particles → p. 80
Identifying the Particle Type via the Mass → p. 156

A new particle, which showed up as a resonance (↓) at this energy, allows for new interactions and hence a much higher rate: the J/ψ. At the SPEAR accelerator, it was produced in a manner that was exactly the "reverse" of how it decays in Sam Ting's experiment—a beautiful symbol of the power of time-reversal symmetry in particle physics.

Production of a $J/\psi\,(c\bar{c})$ at SPEAR

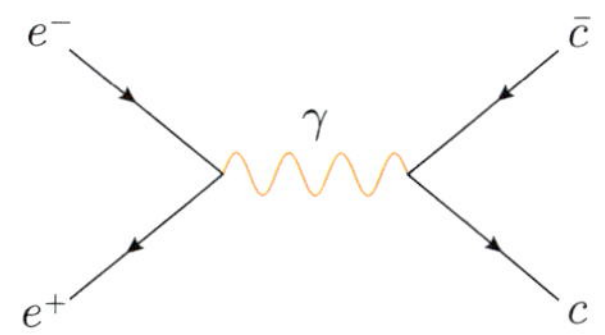

Decay of a $J/\psi\,(c\bar{c})$ into an electron-positron pair

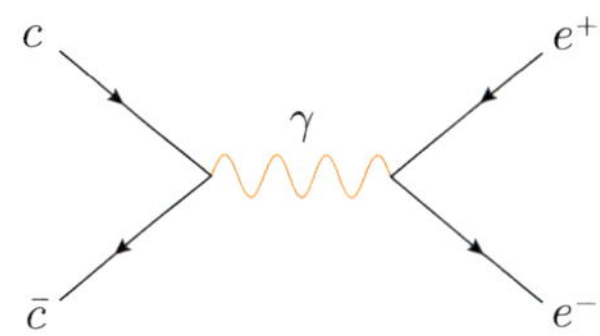

Now that two independent experiments had revealed the same new particle, any remaining doubts about the Brookhaven result dissolved, and particle physicists agreed: something new had been discovered! Further measurements showed that it was likely a meson composed of the charm quarks predicted by the GIM mechanism—another triumph for the Standard Model, as this quark behaves in its interactions exactly like the other quarks of the Standard Model already known at the time, however, with a significantly higher mass. Also important is the methodological independence and complementarity of experiments with electrons in the initial state on the one hand and hadrons on the other—a key principle that continues to be important in modern particle physics.

But why the unusual name? It honors the two independent experiments: the J was chosen by Sam Ting, and the ψ by Burt Richter.

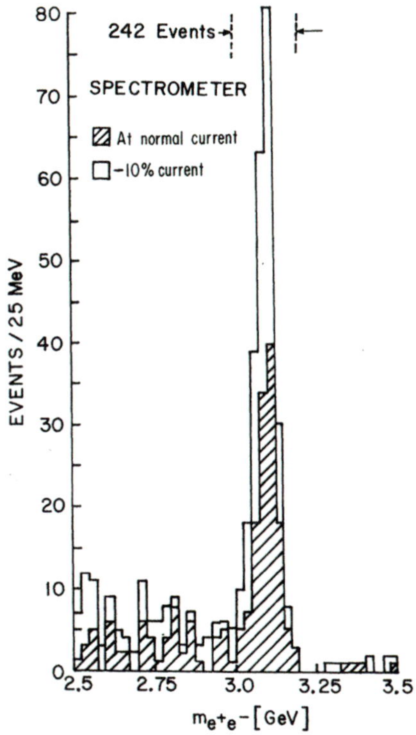

The observed decay of $J/\psi \to e^+e^-$

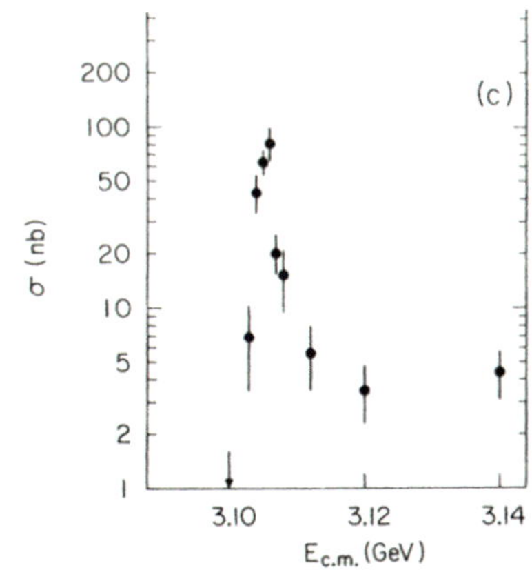

The observed production of $e^+e^- \to J/\psi$

[1] J. J. Aubert et al.1974, *Experimental Observation of a Heavy Particle J,* Physical Review Letters, 33, 23, 1404, doi:10.1103/PhysRevLett.33.1404

[2] J.-E. Austin et al. 1974, *Discovery of a Narrow Resonance in e^+e^- Annihilation,* Phyiscal Review Letters, 33, 23, 1406, doi:10.1103/PhysRevLett.33.1406

Resonances → p. 56 and The Invisible Width of the Z Boson → p. 182

The Discovery of the *W* and *Z* Bosons

From the Observation of a Current to the Discovery of a Particle

The electromagnetic force is mediated by the exchange of photons. Similarly, the weak force in the Standard Model is mediated by the exchange of other particles. If the Standard Model is correct, then not only all the interactions it predicts must exist (Discovery Neutral Currents ↓), but it must also be possible to produce the exchange particles of these interactions as *real* particles in a resonance (↓). "Real" means that—seen in the rest frame of the produced particle—the interaction in which the particle is produced transfers roughly as much energy into the particle as its own mass. In quantum mechanics (↓), it is also possible to produce *virtual* particles in which more or less energy is transferred from the interaction than corresponds to the particle's mass (EFT ↓).

When a particle is not produced as a real particle, its properties can only be inferred indirectly from the observed interaction—a situation that is unsatisfactory from an experimental perspective. Since 1974, the Standard Model had achieved great successes, including the successful discovery of the predicted neutral currents—but a direct observation of the $W^{\pm}$ and Z^0 bosons, which the Standard Model posits as mediators of the weak force, was still pending.

By the end of the 1970s, however, something very important had been achieved: By analyzing the interaction rates of neutral and charged currents in neutrino scattering experiments, it was possible to measure the *electroweak mixing angle*, θ_W. With this new experimental information, the Standard Model allowed for a prediction of the masses of the real $W^{\pm}$ and Z^0 bosons in the range of about 80 and 90 GeV/c^2, respectively. Thus, the concrete experimental search for these particles could begin: The combination of the "beautiful" and "simple" theory of the Standard Model with its high predictive power, due to a relatively small number of free parameters on the one hand, and the revolutionary measurements of neutral currents on the other hand, enabled the exact prediction of all properties of the gauge bosons of the weak interaction $W^{\pm}$ and Z^0 before these particles were discovered.

There were now two possibilities: Either these particles existed as predicted by the Standard Model—that would be a tremendous triumph for the Standard Model—or they did not exist or had different properties than predicted—which would likely mean the end of the Standard Model and the beginning of a search for a new, alternative fundamental theory. In either case, it would be a scientific sensation! However, this guaranteed sensation was not easy to achieve: A mass of

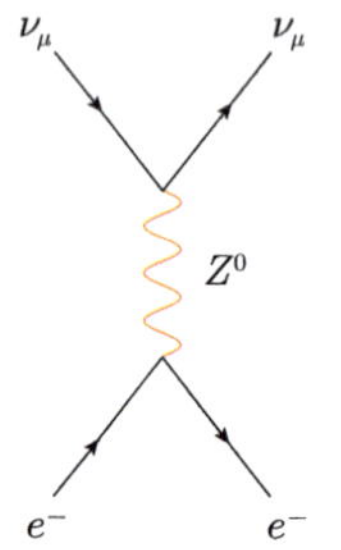

neutral current

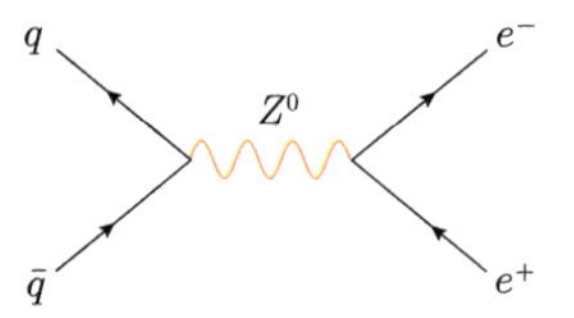

Z production and decay at Sp$\bar{p}$S

The Discovery of Neutral Current Interactions → p. 236
Resonances → p. 56
Concepts of Quantum Mechanics → p. 50
EFT: Accurate Calculation of the Unsolvable → p. 110
The Weak Interaction → p. 178

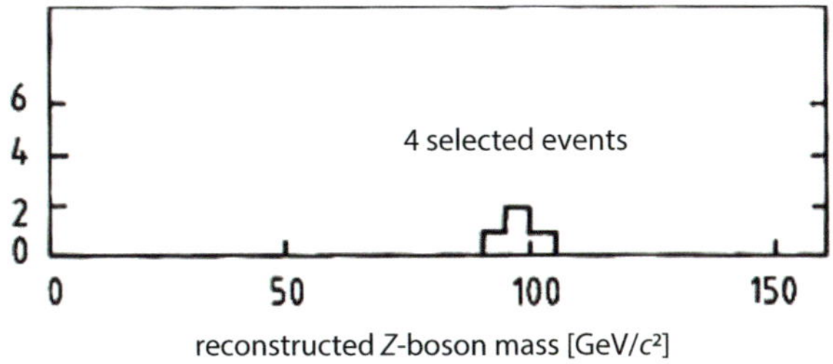

The first four observed Z bosons in the UA1 experiment

Carlo Rubbia and Simon van der Meer at the announcement of the Nobel Prize

80 GeV/c^2 is about 20 times heavier than the heaviest known elementary particle at that time, the b quark. Significant experimental advances were required to realize this measurement.

In the discovery of the $W^\pm$ and Z^0 bosons as real particles, two physicists played a special role: Simon van der Meer and Carlo Rubbia. The former was an accelerator physicist who found a way to direct an antiproton beam with *sufficient intensity* onto another oncoming proton beam at the Sp$\bar{p}$S Collider—achieving a center-of-mass energy (Kinematics ↓) of 450 GeV, which was unprecedented at that time. Carlo Rubbia, on the other hand, led one of the experimental collaborations and had significantly shaped the entire project at the Sp$\bar{p}$S. The UA1 and UA2 experiments were of a complexity previously unknown. Many of their features were pioneering for later large detectors at the LEP, Tevatron, and LHC accelerators.

Crucial to the measurement principle is the reconstruction of a particle's mass from the momentum and energy of its decay products, as the real gauge bosons are produced and decay in the vacuum of the accelerator beam pipe, before reaching the measurement devices of the detector (Identification of Particles ↓). They knew exactly what they were looking for, as the sought Z boson would be nothing more than a real neutral current—the Feynman diagrams are virtually identical between the previous and the new experiment, but logically "rotated" in space and time. In 1982, both experiments discovered the $W^\pm$ boson, and in 1983, they found the Z^0 boson with the precisely predicted respective masses at 80.4 GeV/c^2 and 91.2 GeV/c^2. In the histogram of pairs of simultaneously produced electrons and positrons, there was a statistically significant (↓) peak at the predicted mass.[1] This was a triumph, and Carlo Rubbia and Simon van der Meer celebrated their Nobel Prize on a gallery in the experimental hall in 1984.

[1] L. Di Lella and C. Rubbia 2015, *The Discovery of the W and Z Particles,* Advanced Series on Directions in High Energy Physics, 137, doi:10.1142/9789814644150_0006

Kinematics → p. 40
Identifying the Particle Type via the Mass → p. 156
Drawing Conclusions From Statistics → p. 160

Discovery and Properties of the Top Quark

The Heaviest Elementary Particle Known to Mankind

The development of increasingly higher-energy accelerators (↓) has allowed physicists to discover particles with ever greater mass. The Standard Model of particle physics (↓) was firmly established as the best-known fundamental theory of Nature after the predicted discovery of the W and Z bosons (↓). There were only two remaining candidates for particles that could be heavier than the W and Z: the Higgs boson (↓) and the top quark t. According to theory, the top quark had to exist, required by the mathematical consistency of the theory and by subtle indirect hints in the data of the highest-energy e^+e^- accelerators PETRA, TRISTAN and LEP at the time. Data from the LEP accelerator (1989–2000) (↓) was precise enough to even allow to predict the mass range in which the top quark was to be found.

In the 1980s and 1990s, a new accelerator was built at the American particle physics laboratory Fermilab, near Chicago, capable of accelerating protons and antiprotons to a combined center-of-mass energy of 1.96 TeV—more than four times the energy of the Sp$\bar{\text{p}}$S accelerator, where the W and Z bosons were discovered. This represented by far the highest center-of-mass energy of any accelerator worldwide in the 1990s. However, even under these conditions, the discovery and measurement of the top quark is not easy. Since it is so heavy, it behaves very differently from all other quarks.

The five other quarks each form a hadron with one or more other quarks before they possibly decay. For example, the second-heaviest quark, the b quark, forms a meson together with another quark. E.g. a B^- meson contains a b and a $\bar{u}$ quark. The meson lives long enough to be observed as its own particle in the detector before it decays in a characteristic manner into lighter hadrons and possibly leptons. However, the top quark is so heavy and thus has so much energy available for the decay into a wide variety of lighter particles, that it decays faster than the strong interaction would take to bind the top quark with another quark in a hadron. It is created and decays within less than 10^{-24} s.

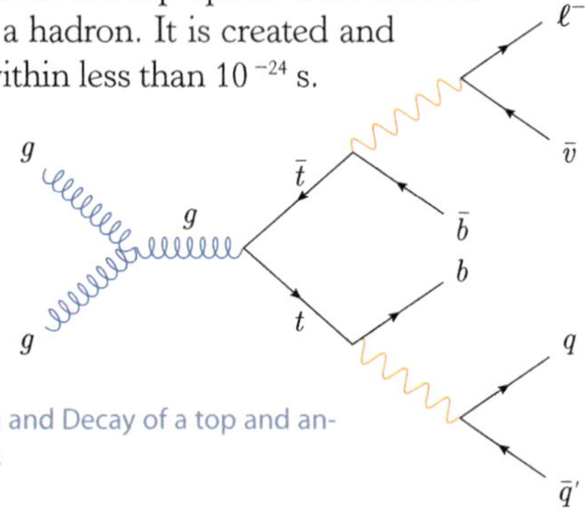

Production and Decay of a top and antitop quark

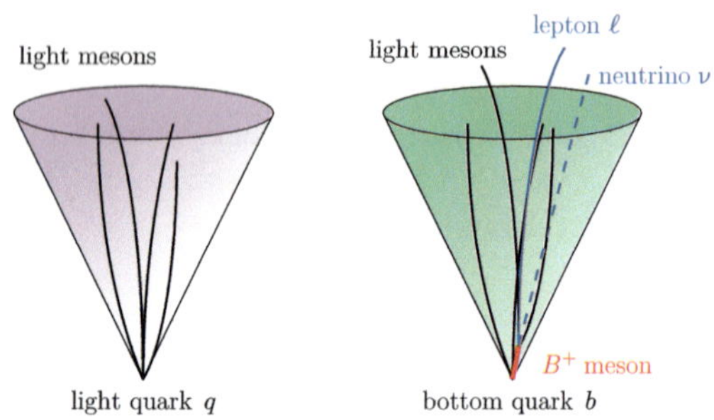

The difference between jets originating from light quarks (left) and from heavy quarks (right)

How Does a Particle Accelerator Work? → p. 66 and The LHC Accelerator Complex at CERN → p. 116
Chapter 7 The Electroweak Standard Model of Particle Physics → p. 175
The Discovery of the W and Z Bosons → p. 240
The Discovery of the Higgs Boson → p. 246
LEP accelerator: Why the Standard Model? → p. 244

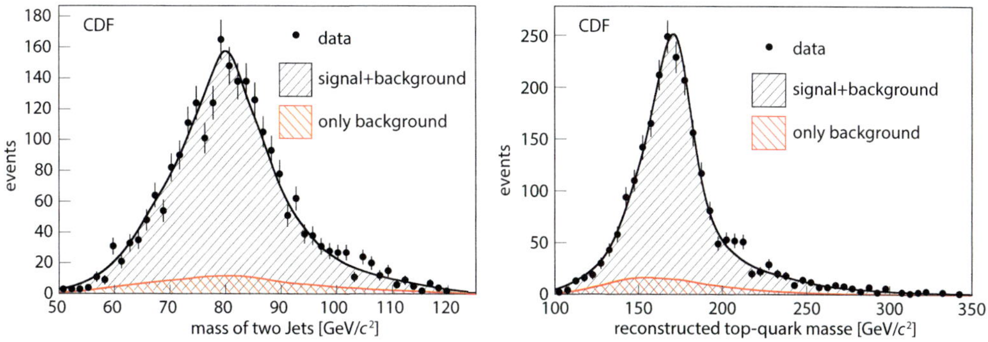

Left: Mass of the W boson candidates, which together with a b-jet form the mass distribution of the top quark candidates shown on the right

This has two important consequences: The good news for particle physicists is that the top quark is the only quark that can be studied in a "free" state—all other quarks can only be studied when bound in hadrons. However, the downside is that the top quark decays into many particles—the detector is filled with numerous decay products, from which the properties of the top quark must be reconstructed based on their combined characteristics.

In such a complex situation, two principles helped in discovering the top quark: First, light quarks form bundles of hadrons in the detector, the so-called *jets* (↓). These bundles collectively carry roughly the momentum and energy of the original quark. Second, the top quark in almost all occurences decays into a bottom quark, which can be identified through the resulting B meson. An event in which a top quark and an anti-top quark are produced must therefore be reconstructed from many individual elements in the final state. If reconstructed correctly, the mass of the top quark can be determined from the momenta and energies of its decay products (↓). The measurement by the CDF experiment at the Tevatron shows a pronounced peak at a mass of about 175 GeV/c^2—the properties of the events in this peak match exactly with a top quark of such a mass. This makes it the heaviest known elementary particle. It was discovered in 1995 and since then, its properties have been measured with increasing precision at the Tevatron and the LHC. These properties have so far all matched the predictions of the Standard Model. All the same more and more properties are being measured with increasing precision—in the hope of finding a concrete deviation from the Standard Model.

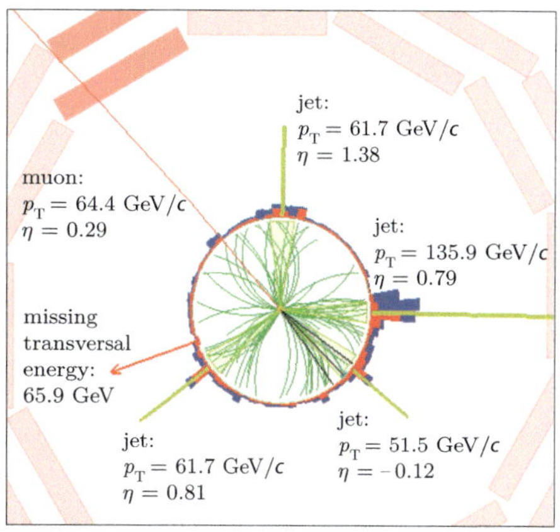

A $t\bar{t}$-collision event in the CMS detector

Jets: Seeing Quarks and Gluons? → p. 216
Identifying the Particle Type via the Mass → p. 156

Why the Standard Model?

Precision Physics at LEP and Prediction of the Higgs Boson Mass

The Standard Model of particle physics (↓) is highly successful and, at the time of this book's publication, the undisputed benchmark against which all theories about the fundamental structure of Nature must be measured. But why is that so? Is it just because it happens to contain six quarks and six leptons? These particles have of course all been discovered, and one can use them like building blocks to assemble many other known particles such as protons and neutrons, and from them nuclei. The history of particle physics sometimes suggests that predicting and discovering the "right" particles is the main content of particle physics.

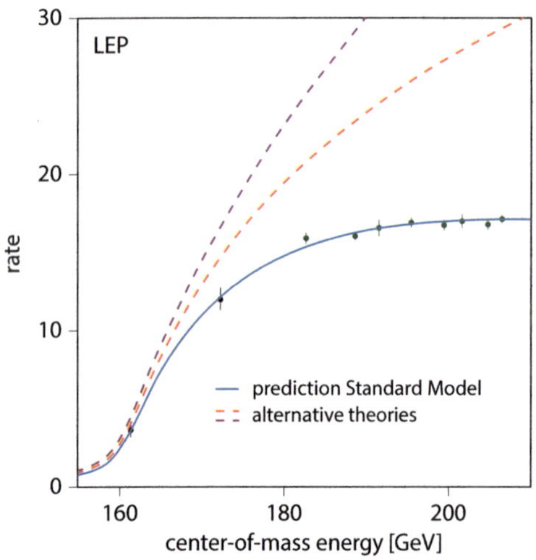

The measured rate of the $e^+e^- \rightarrow W^+W^-$ process as a function of the center-of-mass energy matches exactly with the Standard Model, but not with alternative theories.

However, the majority of the work in particle physics lies in the very precise measurement and calculation of the properties of particles and their interactions. A good example is the LEP accelerator at CERN, which operated from 1990 to 2000 with the four experiments ALEPH, DELPHI, L3 and OPAL. With a circumference of 27 kilometers, it was the largest accelerator at the time. Unlike the proton-antiproton accelerators Sp$\bar{p}$S and Tevatron, where the W and Z bosons, as well as the top quark were discovered, LEP accelerated electrons and their antiparticles, the positrons, and brought them into collision. Although lower energies are reached—"only" 90 to 209 GeV compared to the 1960 GeV of the Tevatron—*elementary* particles are directly brought into collision. Thus, it is not two bags of gluons and quarks colliding, as in the case of protons and antiprotons, but two point-like objects with precisely known properties.

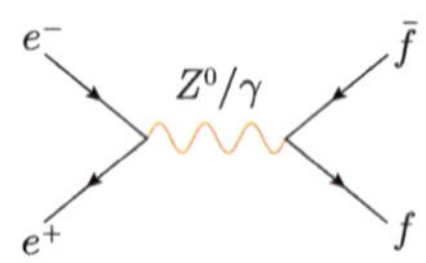

Z production and direct decay

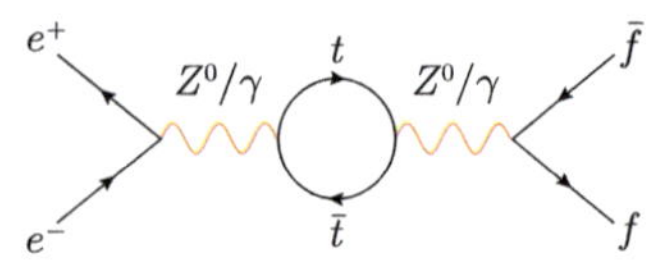

Z production and decay with contributions from a virtual top quark

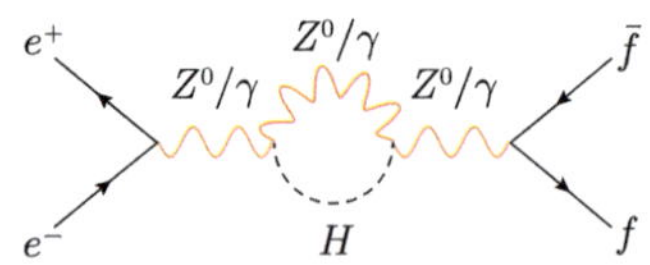

Z production and decay with contributions from a virtual Higgs boson

Chapter 7 The Electroweak Standard Model of Particle Physics → p. 175

This has many advantages. Above all, many measurements become much more precise. This is illustrated by the measurement of the frequency of W boson pair production as a function of the precisely known energy of the colliding electrons and positrons in the left figure. The measured values of the frequency (black and green points with error bars) match exactly the prediction of the Standard Model (blue line), which is a sum of many contributions. If we drop any one of these contributions, the measurement and prediction disagrees wildly (dashed lines). Any theory that aims to replace the Standard Model as the fundamental description of Nature must match the data in this and many other measurements at least as well as the Standard Model—and that is not easy!

This principle can be applied not only to particles directly produced at LEP, such as the W boson, but also to particles predicted by the Standard Model, which remained undiscovered during the LEP operation, as they are too heavy. The three Feynman diagrams (left) illustrate the production of a fermion-antifermion pair at LEP (QCD ↓). This can occur *directly*, or other particles may appear in *loops* (↓) in "virtual" intermediate states, even though they have too much mass to be produced in the final state. Nonetheless, according to the Standard Model, they influence the precise prediction of the properties of the fermion-antifermion pairs in the final state. Because these properties were measured with unprecedented accuracy at LEP, and since all measurements agree with the Standard Model (see Width of Z Boson ↓), the measurement principle can be reversed: The measurements can only be explained if there is a particle that behaves exactly like the Higgs boson of the Standard Model. And this is even though the Higgs boson was not *directly* observed at LEP!

The precise measurements in conjunction with the predictive power of the theory even allowed to significantly constrain the Higgs boson mass before its discovery. Consider the inside of the blue contour, allowed by the LEP measurements of fermion-antifermion pair production in the figure below. It overlaps with the green area of the top quark mass measurement from the Tevatron and the yellow area from direct Higgs searches. This corresponds to a prediction of the Higgs boson mass within the Standard Model between 114.4 and about 280 GeV/c^2. It seems again to be about a discovery. However, this was only possible due to an unprecedented consistency of all measurements within a single highly predictive theory. It was a tremendous triumph that the Higgs boson, whose existence was first predicted in 1964, was indeed found at the LHC in 2012 within exactly this mass range!

The overlap of the blue ellipse, the green line, and the yellow area is the prediction of the Higgs boson's mass from precision measurements before it was discovered.

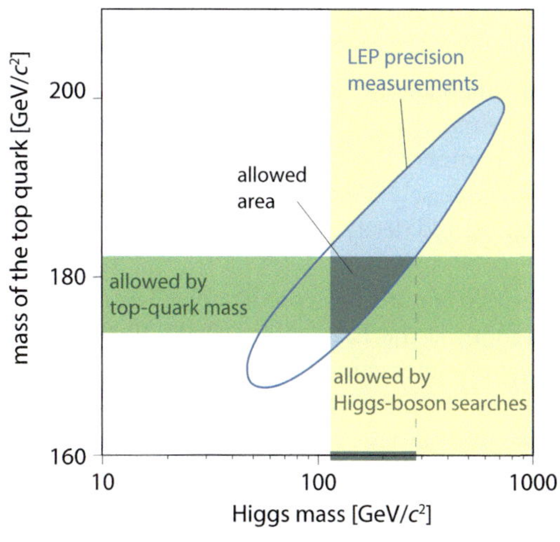

QCD: Quantum Chromodynamics → p. 204
Loops → p. 106
The Invisible Width of the Z Boson → p. 182

The Discovery of the Higgs Boson

The Ultimate Victory of the Standard Model

The Higgs boson of the Standard Model of particle physics triggers many of the major open questions in modern particle physics: How are electromagnetism and the weak force (↓) related? Why are there gauge bosons with large masses like the W and Z bosons (↓) and others with no mass at all, like the photon and the gluon? And why do all quarks and charged leptons have mass, while the neutrinos have barely any mass? All these (and more) questions are related to the Higgs mechanism and spontaneous symmetry breaking (↓) in the Standard Model.

From the measurements at the LEP and Tevatron accelerators (↓), described in the previous pages, physicists knew by around the year 2000 that the mass of the Standard Model Higgs boson had to be somewhere between 114.4 and 280 GeV/c^2. The Higgs boson had to be found within this narrow range if the Standard Model was correct. In addition, given a Higgs boson mass, the Standard Model has no additional free, i.e. undetermined and thus adjustable, parameters. Then all properties of the Higgs boson in the Standard Model were precisely known, in particular how it would be produced and decay.

But would Nature be so predictable? The experimental conditions at the LHC allow nearly 1000 times more energy per interaction than in the experiments of the 1970s on which the Standard Model was based. Never before had humans observed Nature under the conditions of the LHC. Would the Standard Model prevail? Or would the LHC perhaps present a major surprise?

Such a condition is ideal at the start of a new experiment. In particle physics, it is called a *no lose theorem*: Previous experiments and a well established theory together have set the stage to allow a convincing prediction for a new experiment which then either underscores the depth of the understanding of Nature encapsulated in the theory if the prediction is precisely fulfilled, or which brings upon a revolution in our understanding of reality if the prediction is not fulfilled. Then, the exact modes of failure of the prediction often are crucial for the development of an even better theory.

Since 2009, the analysis of data (↓) from the ATLAS and CMS experiments at the Large Hadron Collider had been in full swing. The two primary signals used to search for and identify the Standard Model Higgs boson (Discovery of a Particle ↓) were the decay into two photons or the decay into four light charged leptons: a pair of electrons and positrons and/or muons and antimuons.

On July 4, 2012, the results from both experiments were published.[1,2] They had analyzed their data independently and both found a small excess at the same mass (↓) of about 125 GeV/c^2, in both decay channels! The statistical analysis showed that in each experiment the probability of a random excess of background events at a mass of 125 GeV/c^2 was less than 1 in 1 000 000. The threshold of five standard deviations of significance was reached, and the Higgs boson was discovered.

The Weak Interaction → p. 178
Typical Detectors at Accelerators → p. 128
The Discovery of the W and Z Bosons → p. 240
Spontaneous Symmetry Breaking I → p. 184
Data Analysis → p. 152 and Neural Networks and Machine Learning → p. 170

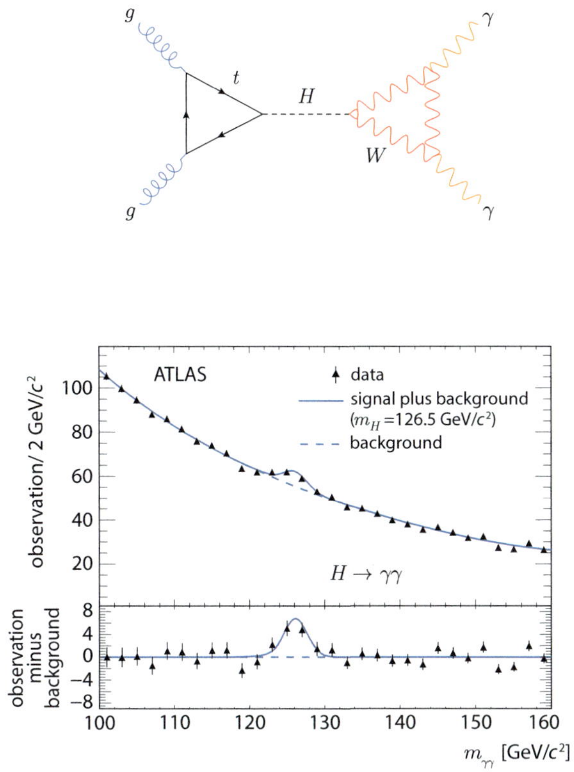

The production of a Higgs boson H at the LHC and its decay into two photons γ

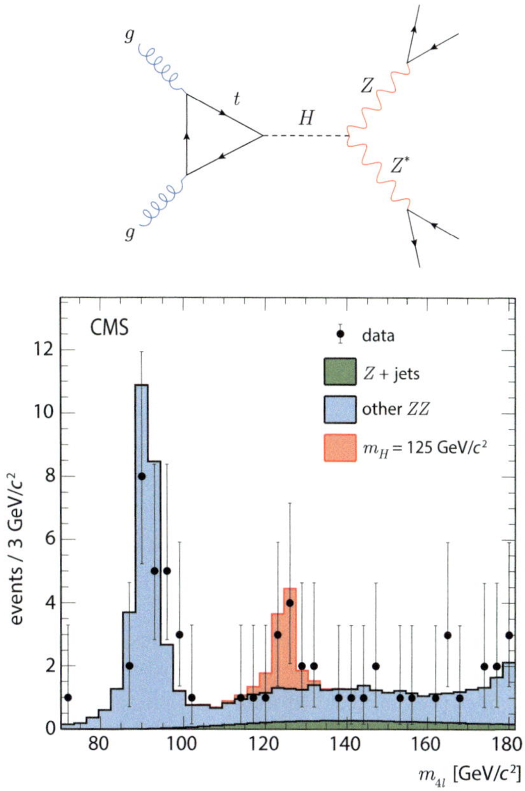

Production of a Higgs boson H and decay into two Z bosons and their subsequent decay into four leptons

This discovery was an indescribable triumph for Peter Higgs and his colleagues, who had laid the foundations of this prediction from 1964 to 1973. Or could it perhaps be an unexpected particle accidentially mistaken for the Standard Model Higgs boson? In fact, it was determined that the data cannot be explained without the existence of a new particle that is at least *similar* to the Standard Model Higgs boson. As with the Z boson, after the discovery, the next step is to measure all properties of the new particle as precisely as possible. Then one can see how it fits into the theoretical framework of the particle world.

[1] **ATLAS Collaboration 2012,** Science, 338, 6114, 1576, doi:10.1126/science.123200
[2] **CMS Collaboration 2013,** Journal of High Energy Physics, 081, doi:10.1007/JHEP06(2013)081

The Discovery of a Particle → p. 164
Identifying the Particle Type via the Mass → p. 156

Is It Really the Higgs Boson?

An Experimental Effigy of a Particle

In 2012, with the discovery of the Higgs boson (↓), the last missing particle of the Standard Model was found. If the history of particle physics were only about the discovery of new fundamental particles, then particle physicists would not have been very successful since then—at least up to the publication of this text. But the history of particle physics is primarily about the precise measurement—and the improvement of the theoretical understanding—of the *properties* of the particles and their interactions.

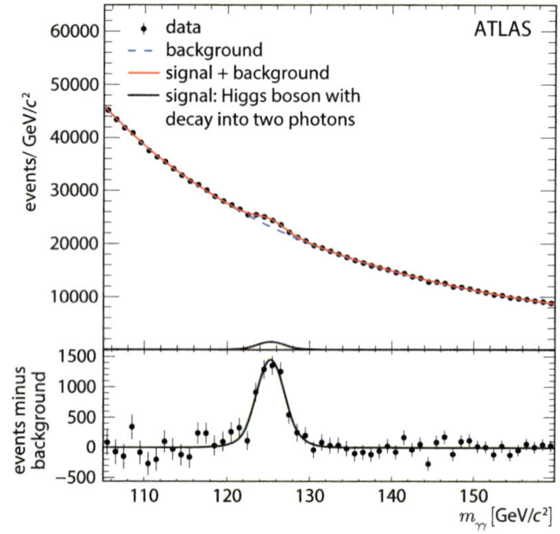

This can mean a lot of detailed and seemingly tedious work. The Higgs boson and the experimental status of its understanding up until the time frame of 2022 to 2024—around the ten-year anniversary of its discovery—shall serve as an example here. Because at the time of the discovery an important question remained unanswered: What was actually discovered? The Higgs boson of the Standard Model? Or a Higgs boson from an extended model (Supersymmetry ↓)? Or something entirely different that just happened to look like a Higgs boson at first glance?

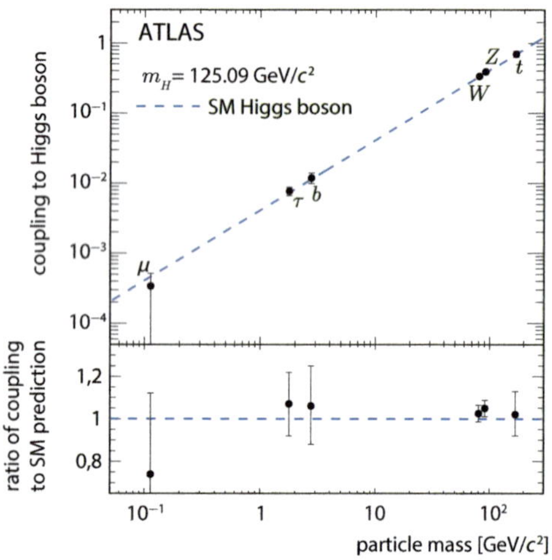

The masses of the particles correspond exactly to the expectation from their measured couplings to the Higgs boson.

Although the discovered signal might initially have appeared somewhat feeble, that definitely changed with the data recorded up until 2018. There was then no doubt that there is a signal at a mass of 125.1 GeV/c^2 in the decay of the new particle into two photons. However, other decay channels were also observed. This included the decays to W and Z boson pairs, as well as to pairs of b quarks, and τ, and μ leptons. In all cases all possible properties such as angular distributions

The Discovery of the Higgs Boson → p. 246
Supersymmetry → p. 278

and energies of decay products and other particles in the event were measured. From these measurements a measure of the coupling between the new particle and the known particles can also be derived.

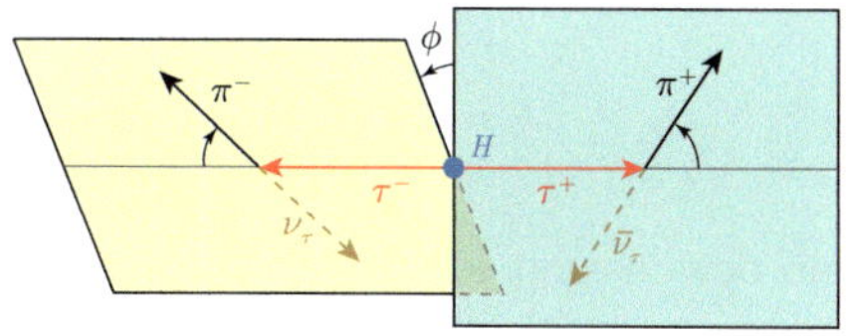

The angle ϕ that spans the decay planes of both τ leptons is measured.

If the new particle is indeed the Higgs boson of the Standard Model, then it must generate the mass of the other particles (↓). This means that the mass of the particles and their measured coupling to the new particle should be in a fixed relationship—and lo and behold, the relationship in the graph on the left page is unmistakable! This is not proof that the particle with a mass of $125.10 \pm 0.14\ \text{GeV}/c^2$ is *exactly* the Higgs boson of the Standard Model. Therefore, physicists are still searching for deviations in ever more precise measurements. However, the resemblance to the Higgs boson predicted over 58 years ago is very convincing. For this reason, physicists have started referring to it as *the* Standard Model Higgs boson.

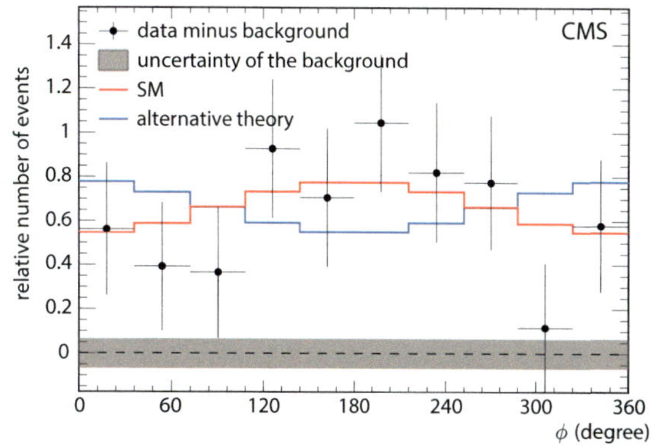

Measuring the *CP* symmetry properties of the Higgs boson

Not only the relationship between mass and coupling characterizes the Higgs boson. The Standard Model also predicts that it has spin 0 (↓) and does not cause any violation of *CP* symmetry (↓). Models that deviate from the Standard Model suggest that a violation of *CP* symmetry in Higgs couplings could be a possible explanation for the observed asymmetry between particles and antiparticles in the Universe (↓). The physicists at the ATLAS and CMS experiments at the LHC first count the rate of Higgs bosons produced and deduce the Higgs boson couplings from these measurements. Moreover, they also look in detail at the decay of the Higgs boson into a pair of τ leptons. Here, for each Higgs boson that decays in this manner, they try to reconstruct the angle between the decay planes of the particles into which the τ leptons decay.

CP violation in the couplings of the Higgs boson to other particles would manifest itself in the angular distribution of the particles produced in the τ decays. In the distribution of the angle between the decay planes ϕ shown in red, *CP* symmetry is exactly conserved, whereas in the blue distribution, it is always violated. The red distribution is predicted by the Standard Model, and it fits the data better. It is actually very unfortunate that this measurement seems to confirm the conservation of the *CP* symmetry in the Higgs sector and thus does not offer a solution to the cosmological problem of baryogenesis. Even in this complex measurement, Nature has so far denied physicists a surprising sensation. On the other hand, it can be regarded as utterly sensational that the now more than 50-year old prediction of the Higgs boson prevails to date.

Particle Masses: Spontaneous Symmetry Breaking III → p. 188
Spin → p. 44
Symmetries → p. 60 and Discrete Symmetries: Violating *P* and *CP* → p. 190
Asymmetry antiparticles: Baryogenesis → p. 328

Precision Experiments

0.001 159 652 ...

How do we gain new insights in physics? Ideally, one has a theory and compares it with measurements. This either leads to confirmation or to the rejection or extension of the theory.

Thus, Newton's theory of gravity explains why the planets orbit the Sun in elliptical paths. As we know, this is very well confirmed by observations. When small deviations were later found, these could be explained by extending Newton's laws within the framework of the theory of general relativity.

With the Standard Model, particle physics has a theoretical construct available that allows comparisons between theoretical prediction and experiment with an unparalleled accuracy. The prime example of such a comparison is the so-called *anomalous magnetic moment* of the electron, for which the variable a_e is used. This quantity describes how an electron behaves in a magnetic field. The measurement is carried out in particle traps, which are discussed in the corresponding article (↓).

In the language of Feynman diagrams (↓), the interaction of an electron with a magnetic field is expressed by the following diagram:

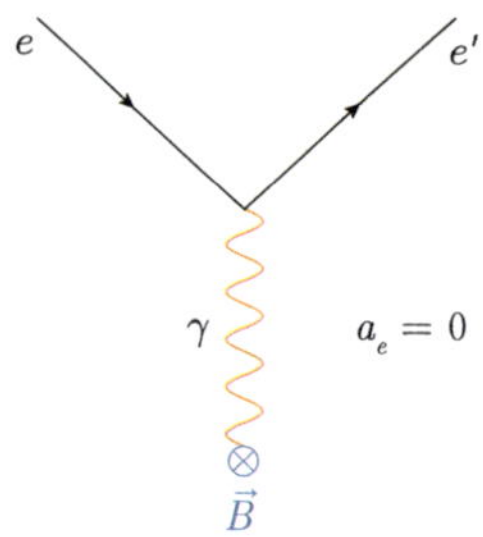

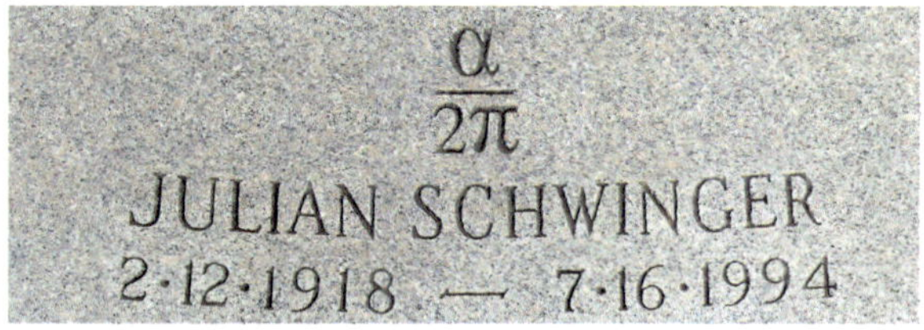

The gravestone of Julian Schwinger at Mt. Auburn Cemetery in Cambridge, USA

The electron (e) reacts with a photon (γ) that represents the magnetic field ($\vec{B}$). With only this diagram the electron couples to the magnetic field as naively expected. It would not be anomalous and $a_e = 0$, which also agreed with the first observations. However in a quantum theory also loop diagrams are important. These give a non-zero contributions to a_e, which were also confirmed in more precise measurements

Julian Schwinger was able to show that due to quantum electrodynamics (↓) a value of $a_e \approx \alpha_{\text{QED}}/(2\pi) \approx 0.001$ is expected. Since this is an effect of quantum electrodynamics, it is not surprising that the fine structure constant (↓) α_{QED} appears here. Thus, theory and experiment could be reconciled again. In the corresponding Feynman diagram, a photon is exchanged between the two electrons.

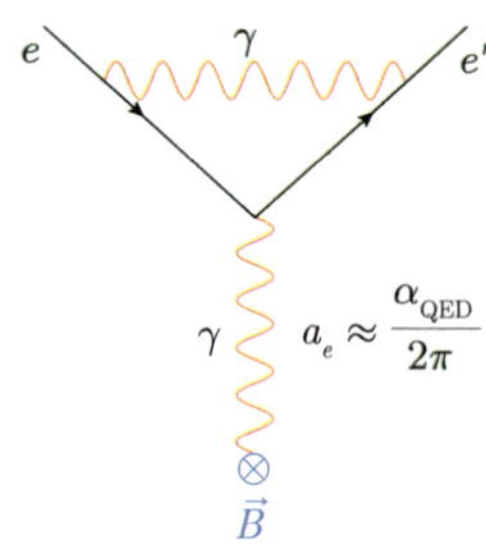

Particle Traps → p. 140
Feynman Diagrams → p. 48
Quantum electrodynamics: Electromagnetism → p. 176
Fine structure constant: Concepts of Quantum Mechanics → p. 50

The story does not end here. The figure below shows how the measurement accuracy has been improved by many orders of magnitude over the last 50 years.

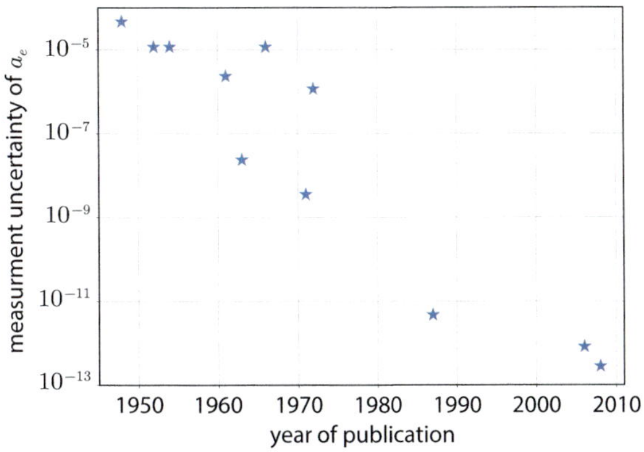

The development of measurement accuracy on the anomalous magnetic moment of the electron a_e

Some of the Feynman graphs that contribute to the determination of the theoretical value

The theoretical calculation has also become more precise over time. By now, contributions from over 12 000 Feynman diagrams, some of which are shown in the figure at the bottom left, have been calculated.

As of 2023, a measured value

$$a_e^{\text{exp}} = 0.001\,159\,652\,180\,62(13)$$

and a theoretical calculation

$$a_e^{\text{SM}} = 0.001\,159\,652\,180\,41(08)$$

are opposed. In the first twelve decimal places, both values agree. Only then differences arise, which can be expected given the uncertainties (↓) indicated in brackets. It should be noted that in order to calculate the Standard Model value one needs the fine structure constant as input. Currently, there are slightly contradicting measurements. Here we took the average to evaluate a_e^{SM}. The measurement of the electron's sister particle, the muon (↓), is somewhat more complex and thus the measurement uncertainty is greater. But this does not make the measurement any less interesting.

The agreement of the theoretical with the experimental value within the measurement errors is one of the great triumphs of the Standard Model. There is probably no other scientific discipline in which theory and experiment can be tested with such accuracy.

The experimental and the theoretically calculated value of the anomalous magnetic moment of the electron

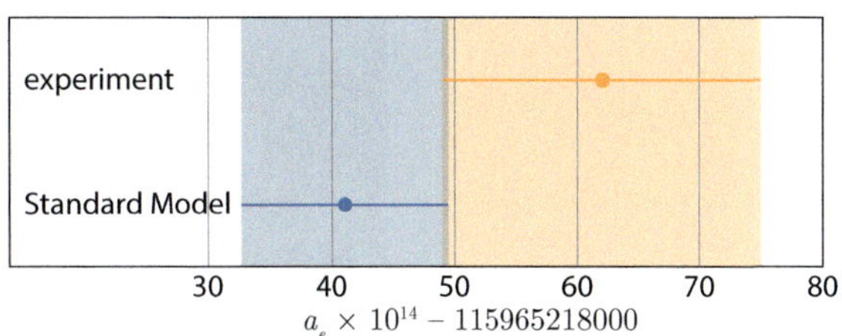

Measurement Uncertainties → p. 86
The Muon → p. 252

The Muon

Big Sister of the Electron

When the muon was discovered in 1936, it came as a big surprise. Since the early 1930s, the structure of atoms made up of protons, neutrons, and electrons had been explained. Further particles were not really necessary. The Nobel laureate Isidor Rabi therefore commented on the discovery of the muon: "Who ordered that?" (↓) The muon has many similarities with the electron, such as electric charge and spin. The main difference is that it is about 200 times heavier than the electron. This property makes the muon particularly interesting for studies in particle physics.

In the previous article, the anomalous magnetic moment of the electron was discussed (↓). The magnetic moment (Spin ↓) determines how quickly the spin of a particle rotates in a magnetic field. The magnetic moment is described by the so-called *g factor*. For leptons such as the electron or the muon, the Dirac equation, the relativistic variant of the Schrödinger equation, predicts a value of exactly $g = 2$. Due to quantum field theoretical corrections of higher order, deviations are expected. These can be calculated very precisely within the framework of the Standard Model on the one hand, and measured very precisely on the other hand. The relative deviation of the g factor from the value 2, $a_\mu = (g-2)/2$ is referred to as the *anomalous magnetic moment*.

Since muons must first be produced in sufficient numbers using accelerators and they decay again after a few microseconds, it is not surprising that measurements on muons are significantly more difficult than on electrons. One consequence is that the anomalous magnetic moment of the muon can only be measured with an accuracy that is a factor of 1000 worse than for the electron. One might now ask whether the effort is still worthwhile. The answer is a clear "Yes". The aim of such measurements is always to put the Standard Model to the test. It therefore not only depends on how accurately a quantity can be measured, but also on what deviations are expected if physics beyond the Standard Model contributes. Since this physics is not yet known, this is of course difficult to estimate. Nevertheless, as a rule of thumb: The heavier a particle is, the more likely influences of still unknown particles and interactions become noticeable.

The magnet of the experiment was used in a previous experiment at the Brookhaven National Laboratory on Long Island. The picture shows the magnet during transport to the Fermi National Accelerator Laboratory near Chicago, where the magnet was used again in a follow-up experiment.

Origins of Particle Physics → p. 10
Precision Experiments → p. 250
Spin → p. 44

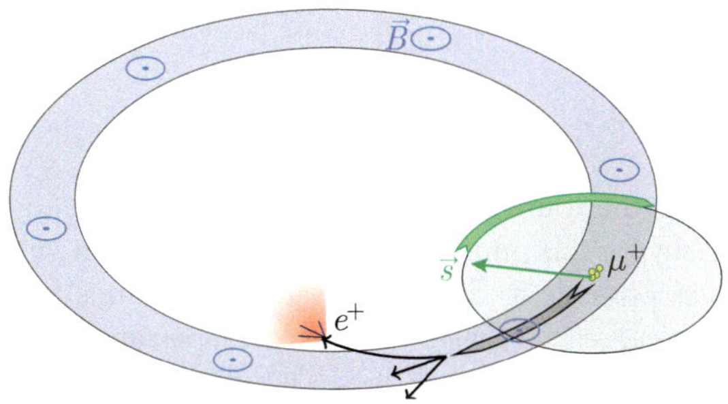

Basic setup of the experiment

Therefore, despite the lower measurement accuracy, significant deviations from the Standard Model value are even expected for the muon.

To measure the anomalous magnetic moment of the muon, first muons are generated by directing a proton beam onto a target. In the collision process pions are created, which in turn decay into muons. This is a decay of the weak interaction, which results in the spins of the muons being aligned parallel to each other. The muons then orbit in a ring-shaped superconducting magnet. Their spin also rotates in this magnetic field

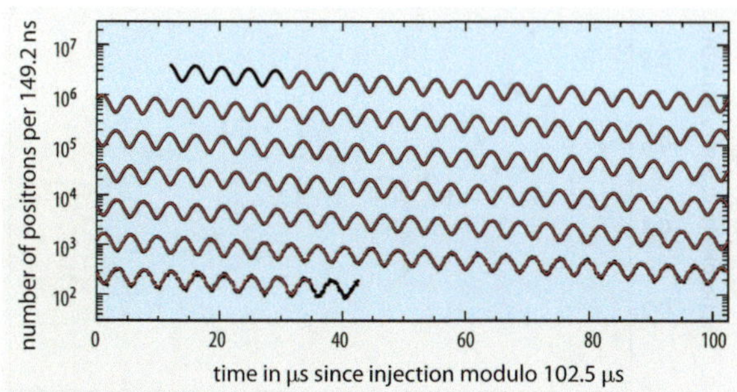

Time spectrum of the decay positrons from the muon decay. The observed modulation is directly proportional to the sought-after anomalous magnetic moment *a*.

at a frequency which directly provides information about the sought-after anomalous magnetic moment a_μ. Within a millisecond, almost all muons decay into a positron and two neutrinos:

$$\mu^+ \to e^+ + \nu_e + \bar{\nu}_\mu \ .$$

The positron is then detected. The decay, induced by the weak interaction, has the nice property that the positron direction provides information about the current direction of the muon spin. Therefore, from the oscillations observed in the time spectrum of the positrons (see figure at bottom left) the quantity a_μ can be directly determined.

The figure below shows a comparison of the measured value with the value calculated from the Standard Model. In previous years, there was a large difference between these values, leading to many debates and suggestions for new theories. With the newest results, given the uncertainties (↓) on the experimental and the theoretical value, both quantities now agree with each other. In this case, the deviation was resolved through improving the Standard Model prediction, mainly due to progress in lattice QCD calculations (↓). This agreement between theory and experiment puts stringent constraints on possible extensions of the Standard Model.

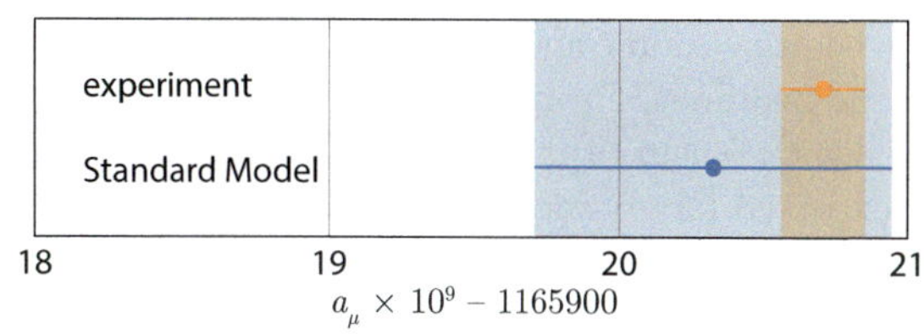

Comparison of the value of the muon's anomalous magnetic moment calculated in the Standard Model and the measured value.

Determining Measurement Uncertainties → p. 158
Lattice QCD → p. 222

Verifiying Neutrino Oscillations

Impressive Confirmation of a Deviation

One of the special features of neutrinos (↓) is the ability of electron, muon and tau neutrinos to transform into each other. The phenomenon was discovered with electron neutrinos from the Sun (↓) and with muon neutrinos produced by cosmic rays in the atmosphere. These neutrino oscillations (↓) were not expected within the Standard Model. In both cases, the neutrinos originate from natural astrophysical sources. Can the phenomena be reproduced with neutrinos from man-made sources? In this article, we take a look at how this effect was investigated.

While the electron neutrinos from the Sun have energies in the MeV range, the muon neutrinos produced in the atmosphere have energies of a few GeV. The beta decays of the fission products in nuclear reactors generate a high flux of electron antineutrinos with suitable energies in the MeV range. Neutrinos from nuclear reactors are therefore suitable for testing the oscillation of solar neutrinos. In contrast, muon neutrinos with energies in the GeV range to test the oscillation of atmospheric neutrinos have to be specially produced using particle accelerators. A decisive parameter for the oscillation is the quotient L/E of the distance L between the neutrino source and the detector and the energy E of the neutrinos. From the theoretical description, an oscillation length of thirty to one hundred kilometers is expected for an energy of a few MeV. This means that at half the oscillation length from a nuclear reactor, the proportion of muon antineutrinos produced by oscillations should be maximum. After a full oscillation length it should be minimum again, and so on.

With the KamLAND detector, which was set up especially for this purpose in the Kamioka mine in Japan, it was possible to detect electron antineutrinos from all nuclear reactors distributed across Japan and to a small extent as well from Korea. The plot of the measured rate of electron antineutrinos as a function of L/E in the figure nicely confirms the ups and downs of the oscillations. L/E is given by the known distance L to the reactors and the measured energy E.

At 1000 km and more, the oscillation length for atmospheric muon neutrinos with energies of a few GeV is significantly longer. In order to reproduce these oscillations with terrestrial sources, special accelerator facil-

Ratio measured by KamLAND between the rate of electron antineutrinos and the rate expected without oscillations as a function of L/E. You can see very nicely how the rate of electron antineutrinos fluctuates up and down due to the oscillation.

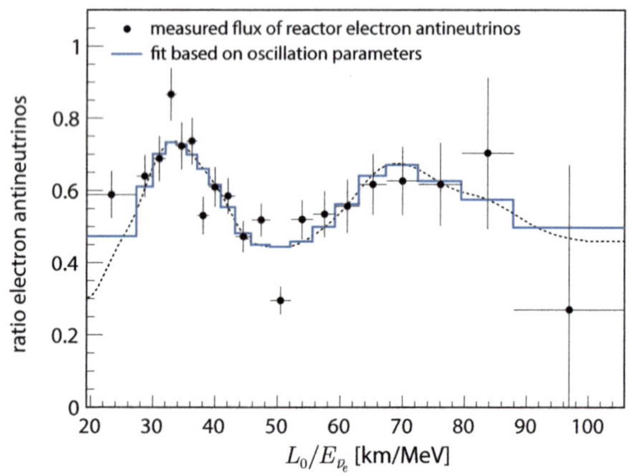

Neutrinos → p. 196
Neutrinos From the Sun → p. 344
Neutrino Oscillations → p. 198

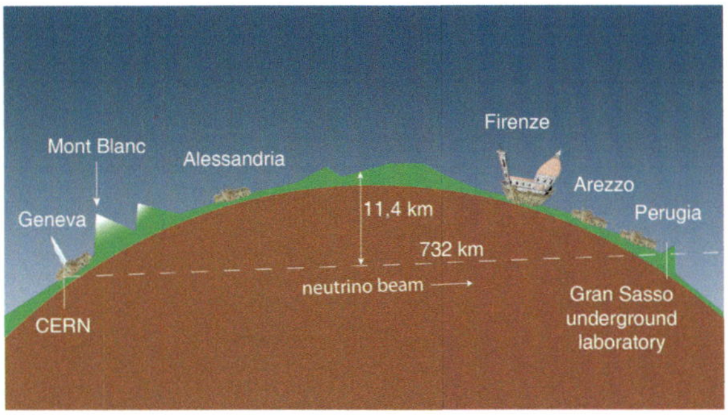

Path of the neutrino beam from CERN to the Gran Sasso underground laboratory in Abruzzo, east of Rome

ities in Japan, the USA and at CERN in Europe were used to generate beams of muon neutrinos, which were aimed at distant large neutrino detectors (↓).

The muon neutrinos generated at CERN in Geneva were directed to the Gran Sasso underground laboratory east of Rome, 732 km away, where a special detector—OPERA—was set up. This was used to investigate whether tau neutrinos are produced by oscillations in the muon neutrino beam from CERN, as would be expected from neutrino oscillations. Detecting tau neutrinos is a major challenge. To do this, the tracks of the particles generated in the detector must be measured with millimeter precision. Not easy in a detector that at the same time has to be as big as a skyscraper. This was achieved by stacking millions of very thin photo plates, that were later developed like photographic film, and in which particle tracks were searched for. After four years of measurements, five tau neutrinos were detected. That doesn't sound like much, but it meets expectations. And the mere existence of tau neutrinos in an otherwise pure muon neutrino beam proves that neutrinos oscillate.

In the USA, the neutrinos were generated at Fermilab near Chicago and sent on a journey of 735 km to the MINOS detector in a mine in northern Minnesota. This detector was not sensitive to tau neutrinos. Instead, the oscillations of the muon neutrinos show up in an energy dependent reduction of the muon neutrino flux.

Not expected in the Standard Model, the confirmation of neutrino oscillations by such experiments was a great success for particle physicists who together developed these challenging experiments.

The energy spectrum measured by MINOS shows clear deviations from the expectation shown in red without oscillations. The data can be described very well with the expectation shown in green with neutrino oscillations. The relative deviation of the data from the expectation without oscillations is shown below.

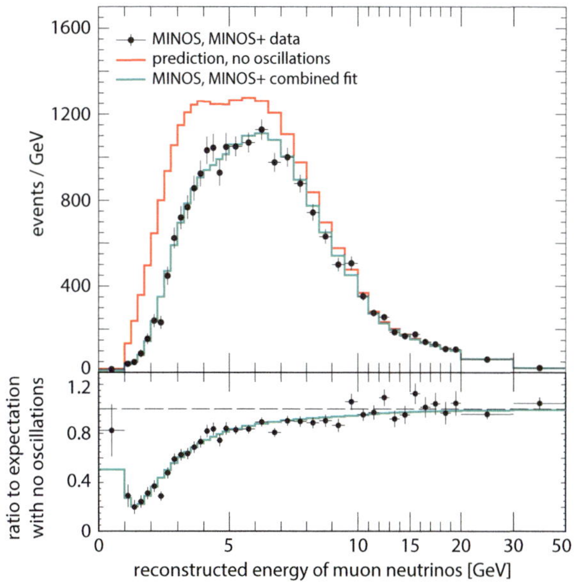

Large Detectors Deep in a Mountain → p. 132

Measurement of the Neutrino Masses

Small Mass—Enormous Scale

The mass of elementary particles is usually determined by exerting electromagnetic forces on them and studying their reaction, e.g., via the curvature of their trajectory. Unfortunately, it is not possible to exert forces on neutrinos, as they carry neither an electromagnetic nor a strong charge. Other methods are therefore needed to determine their mass. The equivalence of energy and mass comes in handy here. If a neutrino is produced in a reaction, at least the energy equivalent to the neutrino mass is required. This energy is lacking in the other reaction partners, whose energy is lower than expected. By measuring the energy of electrons that are emitted together with neutrinos in beta decays, information on the neutrino mass can be inferred. In particular, the beta decay of tritium is investigated for this purpose.

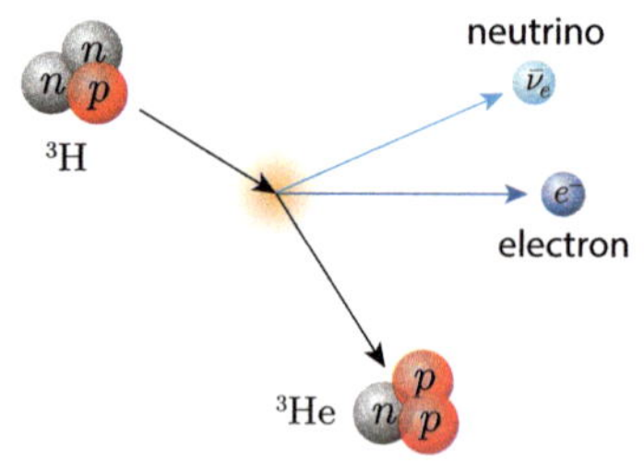

Beta decay of tritium to He-3. This produces an electron and a neutrino. The maximum possible energy of the electron is lower than expected by the value equivalent to the neutrino mass.

In the Standard Model, the neutrino mass is zero. But with the detection of neutrino oscillations (↓), we know neutrinos have masses, but we do not know their values. Precise knowledge of the neutrino mass is an important but still missing building block in our understanding of the elementary structure of matter. Neutrinos also play a major role cosmologically, as they more or less influence the formation of structure in the Universe depending on their mass. The determination of neutrino mass has therefore been high on the agenda of global particle physics for decades.

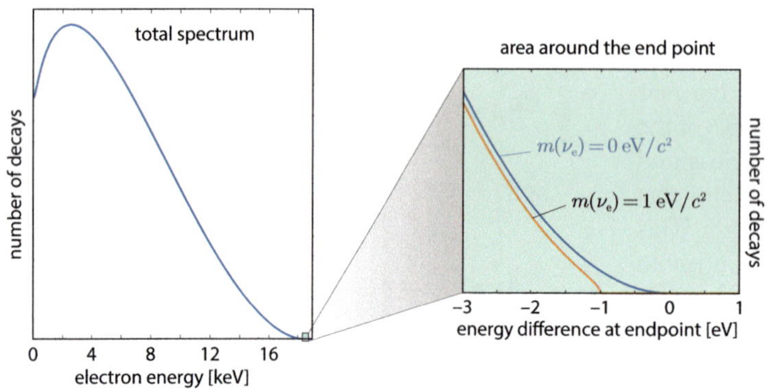

The shape of the beta spectrum at the endpoint is influenced by the value of the neutrino mass. The maximum electron energy is reduced by the neutrino mass. One difficulty for the measurement is the extremely small proportion of beta decays of interest for the mass determination.

The world's most sensitive experiment is the "Karlsruhe Tritium Neutrino Experiment"—KATRIN. In a collaboration of university groups and institutes from all over the world, a unique spectrometer for

Neutrino Oscillations → p. 198

the precise measurement of the electron energy in tritium decay has been realized. To measure the energy of the electrons, they are allowed to run against an electrical voltage. Only electrons whose kinetic energy is sufficient to overcome the voltage reach a detector, which counts them. This is similar to kicking a ball up a ramp.

It only reaches the top of the ramp, if it is kicked hard enough. By varying the voltage, which corresponds to the height of the ramp, the distribution of the electron energy, the so-called beta spectrum, can be measured step by step. In order to achieve the highest possible accuracy, the electrons are guided by a magnetic field, which requires the spectrometer to be as large as possible. The spectrometer tank was built in Deggendorf, just 350 km from Karlsruhe. It was so large that it could not be transported via normal highways. Instead, it had to be transported by ship on an 8600 km journey via the Danube, Black Sea, Mediterranean Sea, Atlantic Ocean, English Channel, North Sea and finally the Rhine River to Karlsruhe. The most spectacular part of the journey was the last 7 km by heavy transport from the Rhine River to the research center.

Transport route of the KATRIN spectrometer

KATRIN will take data for several years and increase the sensitivity to such an extent that a mass greater than 0.2 eV/c^2 should become visible. This value is two and a half million times smaller than the mass of the electron. As it is difficult to further increase the size of a spectrometer, development projects are underway to perhaps investigate beta decays in future using other methods, e.g., low-temperature detectors (↓). So far, the best way to narrow down the neutrino mass is through cosmological observations. However, we have to rely on assumptions in our cosmological models. Therefore, conversely, KATRIN can also make a contribution to cosmology with a model-independent value for the neutrino mass.

The last few kilometers of the transport were the most difficult.

Cryogenic Detectors → p. 136

CP Violation in Beauty

Which Side of the Mirror Are We On?

Symmetries (↓) play a fundamental role in particle physics—they are essentially the foundation of the Standard Model. Among the many discrete (↓) and continuous symmetries in particle physics, *CP* symmetry—the combination of charge symmetry *C* and spatial reflection *P*—plays a special role: An application of *CP* reflection transforms a left- or right-handed particle into a right- or left-handed antiparticle, and vice versa (as in the picture below).

If *CP* symmetry were always an exact symmetry of Nature, then particles and antiparticles would be perfect mirror images of each other. However, this does not mean that particles and antiparticles are identical: the positive charge distinguishes the positron from the electron, which has a negative charge. But *what* we call positive and *what* we call negative charge would be independent of all other properties of the particles that are identical for particles and antiparticles. Ultimately, it would be a matter of definition what we call a particle or an antiparticle (Pion Decay ↓).

In the decay of *B* mesons, which are bound states of a bottom quark (↓) with an up or down antiquark, a deviation and thus a violation of *CP* symmetry has been observed (Spontaneous Symmetry Breaking ↓). For this purpose, particle physicists have built a series of specialized experiments at e^+e^- accelerators such as the Belle and BaBar experiments, and also at the LHC (the LHCb experiment). One of the studied decays of the B^0 meson into $K^+\pi^-$ or its antiparticle $\bar{B}^0$ into $K^-\pi^+$ is shown in the Feynman diagrams below. Both decays appear to proceed exactly the same way—only in the second diagram each particle has been replaced by its antiparticle compared to the first. If *CP* symmetry were perfectly satisfied, meaning particles and antiparticles are exact mirror images of each other, then both decays would occur with exactly the same frequency.

The measured result of the Belle experiment proves: particles and antiparticles are *not* exact mirror images of each other, as $K^+\pi^-$ final states are more frequent than $K^-\pi^+$. This is very clear in the next graphs: the

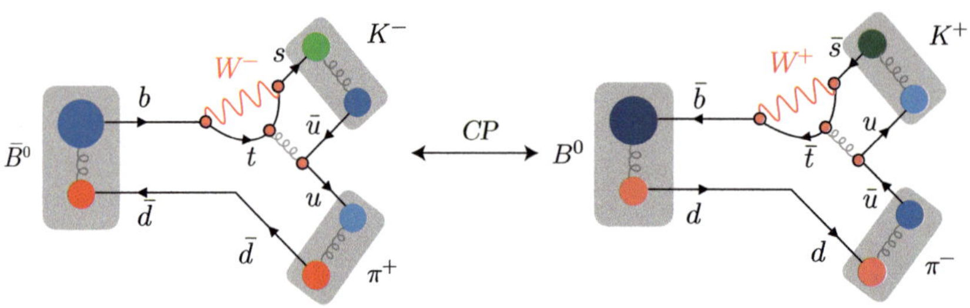

The decays of the $\bar{B}^0$ meson and its antiparticle B^0 into a kaon K and a pion π: a direct *CP*-inversion

Symmetries → p. 60
Discrete Symmetries: Violating *P* and *CP* → p. 190
The Weak Decay of Pions → p. 192
The Quark Model: A First Look → p. 20
Spontaneous Symmetry Breaking I → p. 184

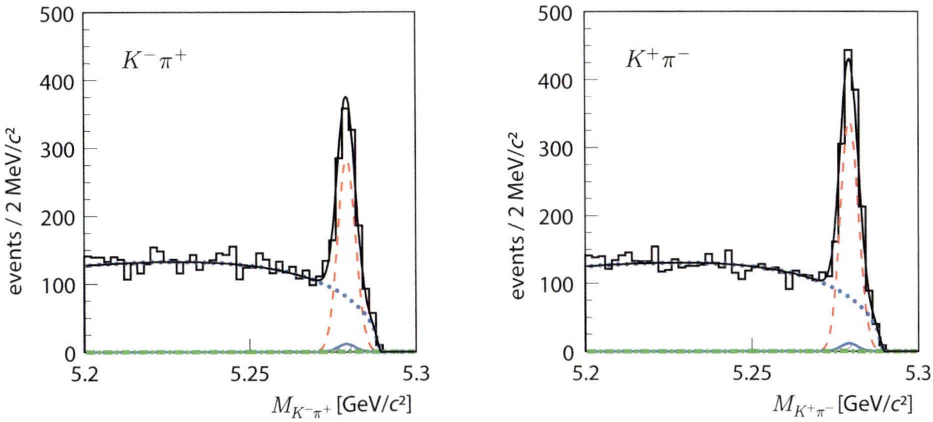

The $\bar{B}^0$ decays less frequently into $K^-\pi^+$ than its antiparticle B^0 decays into $K^+\pi^-$: The mirror symmetry is not exact!

broad background level is the same for both decays, but the sharp peak of the observed $K^+\pi^-$ decays is higher than the peak of the $K^-\pi^+$ decays! *CP* violation appears in the direct *CP* reflection of a decay diagram. Therefore, this type is also called *direct CP violation*.

But how can a *CP* reflection change a decay rate, i.e., the frequency of a decay? This is illustrated by the drawing on the right: Suppose the length of arrows in a two-dimensional plane represents the decay rate. The *CP* reflection, applied to the CKM matrix (↓), involves a reflection about the horizontal axis in this graph. The contribution for particles (red, thin arrow) points upwards, while the *CP*-reflected part for antiparticles (blue, thin arrow) points downwards. Both arrows are of equal length due to the reflection. However, the contribution from the CKM matrix is not the only contribution to the decay: there are also other effects from different parts of the Standard Model (gray), whose contributions are independent of the *CP* reflection. Their contributions for particles and antiparticles are added to the contributions from the CKM phase in the image. In this way, the lengths of the summed arrows for particles and antiparticles (thick orange and blue arrows) suddenly differ, and thus a *CP* reflection causes a change in the counting rate!

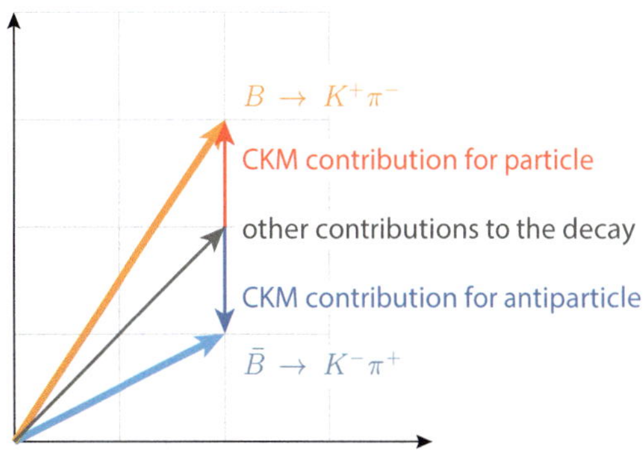

Illustration of the differing predictions for the rates (represented by the length of the thick arrows) of $B^0 \to K^+\pi^-$ and $\bar{B}^0 \to K^-\pi^+$ in the Standard Model

CKM matrix: The Path to Six Quarks in Three Families → p. 194

Rare Decays of Beauty

Marginal Hints of the Big Scoop?

There are good theoretical and cosmological reasons why the Standard Model cannot be the "final or ultimate answer" to all questions (↓) about the structure of matter and its interactions. Therefore, intense searches are conducted for deviations between measurements and predictions of the Standard Model. An interesting group of statistically marginally significant (↓) deviations appears in so-called *rare decays* of B mesons. Rare decays refer to situations where only very few of the particles in the initial state decay into the studied final state. Typically, there are specific mechanisms that suppress such decays in the Standard Model (see Up, Down, Left, Right ↓, and The Path to Six Quarks ↓). Another simple example is the decay of a charged pion $\pi^+ \rightarrow e^+\nu_e$, which could be quite common but is "suppressed" by the structure of the weak interaction and the resulting parity violation, making it very rare (Pion Decay ↓).

However, if the Standard Model is not the fundamental theory of Nature, then even the tiniest contribution from "Physics Beyond the Standard Model" (↓), such as from yet unknown particles that are much heavier than the top quark, could represent a significant *relative* change compared to the tiny contribution from the Standard Model.

One of the most interesting measurements of this kind as of the publication of this book is the ratio of the number of B mesons decaying into an excited D meson (hereafter referred to as D^*) and a pair of τ lepton and τ neutrino, relative to the number of B mesons decaying into a D^* and—instead of the τ and ν_τ—into their sibling particles: an electron and an electron neutrino, or a muon and a muon neutrino. These two processes are shown in the Feynman diagrams below, including the subsequent decay of the tau lepton.

Behind the choice of this ratio as a measurable quantity lies another assumption: For various reasons, it is often a good guess that unknown physics is more likely to occur together with heavier Standard Model particles such as the τ than with lighter ones like the μ. When measuring a deviation from an expectation, it is always of highest importance to reduce the uncertainties of the prediction for the expectation. Therefore, the experimental investigation at the BaBar, Belle, Belle II, and LHCb experiments uses a trick to reduce systematic modeling

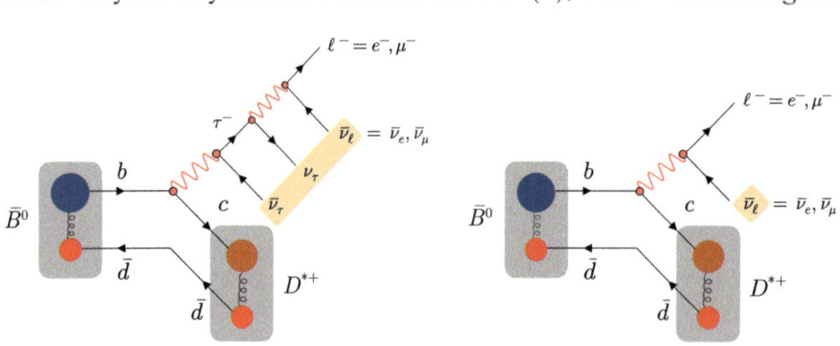

The ratio of the rate of both processes deviate from the expectation from the Standard Model of Particle Physics

Open Questions → p. 270
Significance, confidence: Drawing Conclusions From Statistics → p. 160
Up, Down, Left, Right → p. 180, The Path to Six Quarks in Three Families → p. 194
The Weak Decay of Pions → p. 192
Searching for New Particles at the LHC I → p. 302

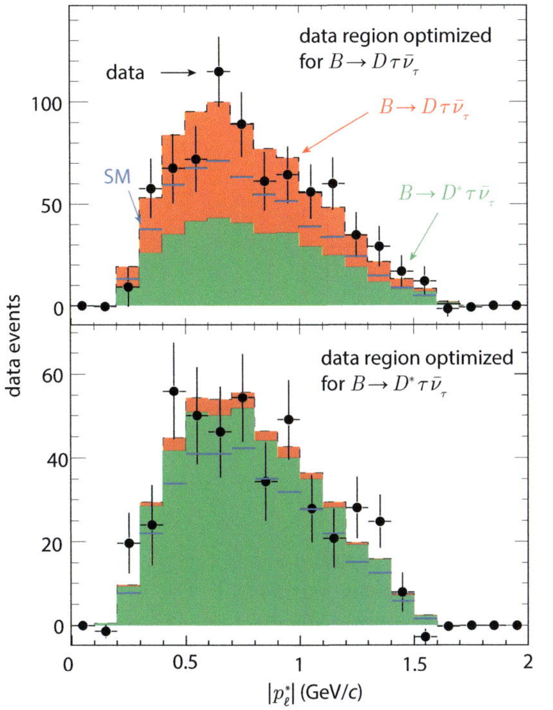

The rate of $B \to D\tau\bar{\nu}_\tau$ and $B \to D^*\tau\bar{\nu}_\tau$ followed by the decay $\tau \to \ell\,\bar{\nu}_\ell\nu_\tau$ measured by the BaBar experiment as a function of the lepton momentum

uncertainties (↓) on the Standard Model expectation: only the decay of the τ lepton into an electron or muon are considered. This way, the final state of the two decays differs only by the amount of energy carried away by invisible neutrinos. While this reduces the τ rate, as not all decay channels are measured, it ensures that almost all modeling errors occur equally in both the numerator and denominator of the ratio of the two decay rates! This significantly reduces the systematic error.

In the left figure, the measurement with electrons and muons in the final state compares $B \to D\tau\bar{\nu}_\tau$ and $B \to D^*\tau\bar{\nu}_\tau$ with the Standard Model expectation. The momentum of the observed electrons and muons is shown for two data regions, optimized to select either the D or D^* final states. The selection is not perfect, so that $B \to D^*\tau\bar{\nu}_\tau$ events (shown as a green histogram) also show up in the D optimized region and vice versa. The Standard Model expectation for each $|p^*_\ell|$ is always shown as a blue line. It is significantly lower than the observed rates.

In combination with other measurements of this type, the excess over the Standard Model expectation is 3.3 standard deviations. If this were created by chance due to a statistical fluctuation around the Standard Model value, such an outcome would be expected in one of a thousand cases. Is this evidence of specific *new physics*? Or even the discovery of a new particle such as the *leptoquark LQ*? (GUT ↓) The statistical significance is not yet sufficient for this as to claim a discovery: for that, particle physicists require a chance of one in ten million; more precise measurements will bring interesting further developments—either for or against the Standard Model.

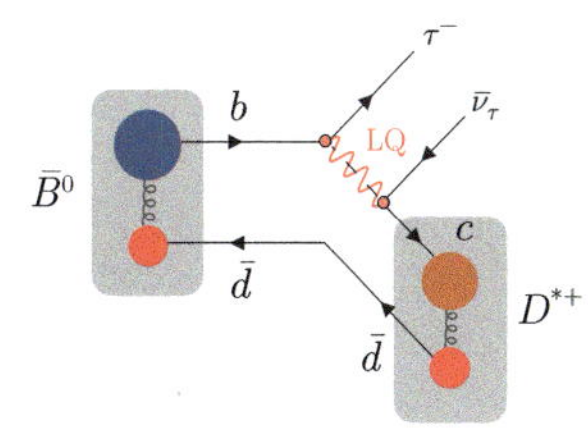

A possible explanation for the deviation: the presence of a new particle, the so-called leptoquark LQ.

Determining Measurement Uncertainties → p. 158
GUT → p. 276

Shapeshifters Among Themselves

A Story of Charm, Beauty[1] and Strangeness

There is another group of particles that, like neutrinos, are true shapeshifters: neutral mesons. Mesons are states composed of a quark and an antiquark. However, the transformation of neutral mesons differs fundamentally from neutrino oscillations (↓): while neutrinos can transform into each other, some neutral mesons can transform into their antiparticles! An additional effect comes into play: the mesons under consideration decay into other particles after a short lifetime. This makes experimental detection difficult. The first such system investigated was neutral kaons, with which CP violation (↓) was first observed in 1964.

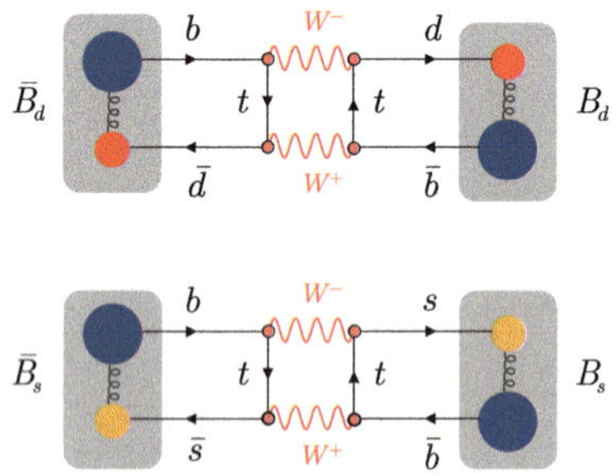

The transformation of B mesons into their antiparticles and back

A particularly beautiful example of such a transformation occurs in systems with bottom quarks: Here, a b quark can transform into its antiparticle $\bar{b}$ if in parallel an antiquark, e.g., a $\bar{d}$, is transformed into a d quark. This happens when an anti-B meson ($\bar{B}_d$, composed of a bottom quark b and a down antiquark $\bar{d}$) transforms into its antiparticle (B_d, composed of $\bar{b}$ and d). In the illustration, this is achieved with the help of a loop of top quarks and $W^{\pm}$ bosons. The B_d meson and its antiparticle have a lifetime of only $1.5 \cdot 10^{-12}$ s. However, the average period of an oscillation is about $1.2 \cdot 10^{-11}$ s. That is, only a few B mesons live long enough to oscillate into their antiparticle. For this process to be possible, quarks of different families must couple via the W boson and the CKM matrix (↓) to each other, here, e.g., the top quark to a down quark. If the transformations from matter → antimatter and antimatter → matter were exactly equally probable, then there would be no observable CP violation in the oscillation.

This rare transformation was first observed in 1987 in the ARGUS experiment[2] at the Deutsches Elektronen-Synchrotron DESY, the German laboratory in Hamburg. The first event observed is shown in the image on the next page: electrons and positrons are collided, producing simultaneously a B meson and its antiparticle, a $\bar{B}$ meson. The image displays a cross-section through the detector, with the e^+e^- beam perpendicular to the surface of the page. By analyzing the final state particles, one can reconstruct the decay cascade and, using the charge and kinematic properties, assign the four charged pions and the two kaons to the two initial mesons (indicated in red and blue, also see Neutral Currents ↓). Although a B meson and its antiparticle were actually produced, the image shows the decay products of two B mesons, meaning the anti-B meson has transformed into a B meson via an oscillation. Fantastic!

[1] The bottom quark sometimes is also called "beauty" quark. That's why there's beauty in the title..
[2] see ARGUS Collaboration et al. 1987, *Observation of B^0-B^0 mixing*, Physics Letters B, 192, 1–2, 245, doi:10.1016/0370-2693(87)91177-4

Verifiying Neutrino Oscillations → p. 254
Discrete Symmetries: Violating P and CP → p. 190 and CP Violation in Beauty → p. 258

Beautiful oscillations also occur in other systems. It becomes particularly spectacular when the down quark is replaced by a strange quark. Then we can study the oscillation of so-called B_s mesons. These oscillations occur incredibly quickly. With an oscillation frequency of $2.9 \cdot 10^{12}$ Hz and a lifetime of $1.5 \cdot 10^{-12}$ s, the B_s meson oscillates on average four times before it decays. This phenomenon was first measured at the Tevatron accelerator at Fermilab and has been confirmed with incredible accuracy by the LHCb experiment at CERN[3].

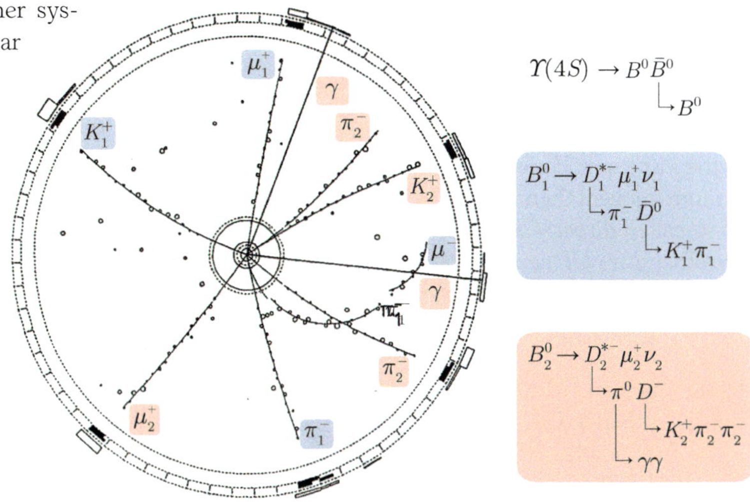

The discovery of B meson oscillations: A recorded event from the ARGUS experiment observing two B_d and no $\bar{B}_d$ meson.

The figure bellow shows the oscillations of B_s into $\bar{B}_s$ and vice versa: The blue and red curves represent the measured number of B_s and $\bar{B}_s$ mesons at different times after the production.

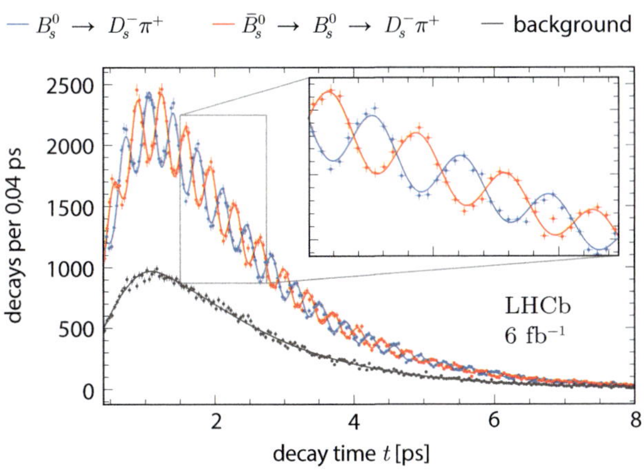

The experimental observation of B_s oscillations

Recently, LHCb has also detected oscillations in the fourth and final neutral meson system where such oscillations can occur: neutral D mesons, which are composed of a charm quark and an up antiquark. An intriguing aspect is that the rate of conversion from matter to antimatter is not exactly equal to the rate of conversion from antimatter to matter (↓).

[3] LHCb Collaboration 2022, *Precise determination of the frequency*, Nature Physics, 18, 1, doi:10.1038/s41567-021-01394-x

CKM-Matrix: The Path to Six Quarks in Three Families → p. 194
The Discovery of Neutral Current Interactions → p. 236
What Is Antimatter? → p. 22 and The Search for Electric Dipole Moments → p. 318

Measuring the CKM Triangle

Where Everything Comes Together

As we have seen, matter and antimatter do not behave in the same way: For example, the decay of $B^0 \to K^+\pi^-$ is more frequent than $\bar{B}^0 \to K^-\pi^+$, which is an example of so-called *direct CP violation* (↓). However, there are other ways *CP* can be violated in Nature. For those interested in more details about this exciting aspect, further reading is available in this article.

In the Standard Model of particle physics, *CP* violation is embedded in the CKM matrix (↓). It has already been explained how the CKM matrix can generate *CP* violation through the reflection of its phase in the decay of *B* mesons into a kaon, *K*, and a pion, π. A spectacular type of *CP* violation is the different average lifetimes of B^0 and $\bar{B}^0$ in decays where the final state particles cannot be distinguished. This occurs in the decay of $B^0 \to J/\psi\, \bar{K}^0$ with $\bar{K}^0 \to \pi^+\pi^-$, which we discuss here.

This process can occur in two ways: The B^0 can directly decay into a J/ψ and a $\bar{K}^0$, or it can first transform into its antiparticle $\bar{B}^0$ and then decay into $J/\psi\,\pi^+\pi^-$, as illustrated in the figure. Analogously for a produced $\bar{B}^0$, it can either decay directly or first transform into a B^0 (meson oscillation ↓). Using a trick, particle physicists can determine the originally produced state B^0 or $\bar{B}^0$ at e^+e^- colliders: *B* mesons are always produced in pairs and are entangled. If one determines the state of the *other B* meson at the time of its decay into a final state which allows to determine the original *B* state (e.g. B^0), one knows the state of the *studied B* meson at that exact moment (then, e.g., $\bar{B}^0$). Due to quantum entanglement, the decay of the first *B* meson determines the state (B^0 or $\bar{B}^0$) of the second *B* meson ($\bar{B}^0$ or B^0) at a specific time. With this information, one can now measure the time dependence of *CP* violation. This is shown in the figure: In this specific decay, the B^0 mesons (red line) decay on average later than their antiparticles (blue line).

These and other measurements can be combined to measure *CP* violation in quarks. All measurements can be graphically summarized in the *unitarity triangle*, shown in black in the center of the left figure. The col-

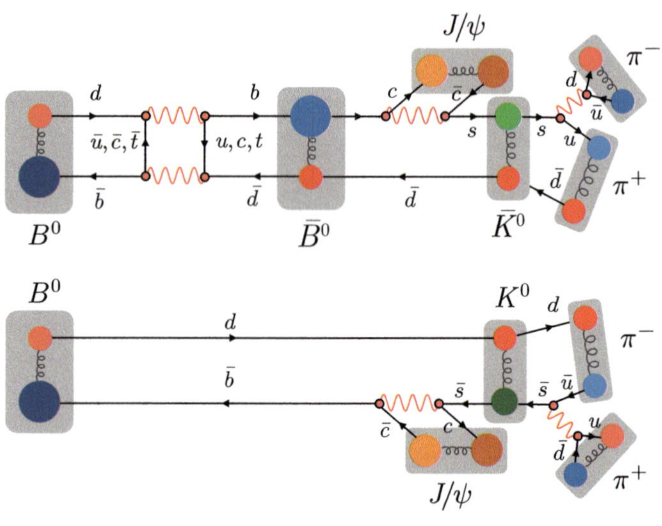

The decay of B^0 and $\bar{B}^0$ into the same *CP* eigenstate

CP Violation in Beauty → p. 258 and Discrete Symmetries: Violating *P* and *CP* → p. 190
CKM-Matrix: The Path to Six Quarks in Three Families → p. 194
Meson oscillation: Shapeshifters Among Themselves → p. 262

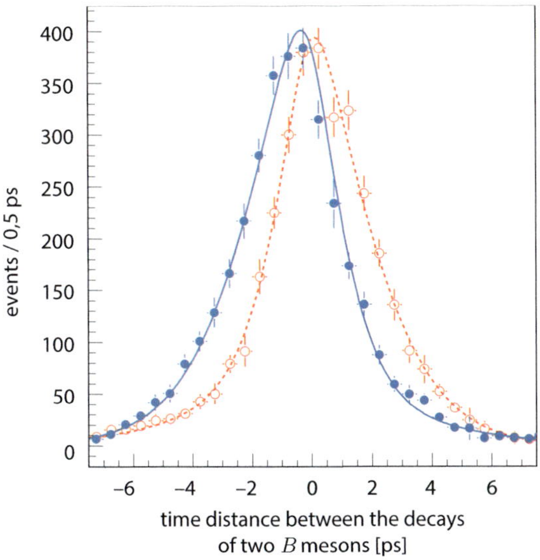

The B mesons produced as $\bar{B}^0$ decay, on average, later (red line), whereas those produced as B^0 decay earlier (blue line) into the same final state. This indicates a violation of T symmetry and CP symmetry.

ored areas show the regions allowed for the top point associated with the angle α by different measurements in the space of two of the four parameters of the CKM matrix, here denoted as $\bar{\rho}$ and $\bar{\eta}$. For example, the area of the green ring is the allowed region due to experimental constraints on the CKM matrix from the measured B meson decay $B \to \pi \ell \nu$, which involves the quark transition $b \to u$. Similarily, each other measurement allows another colored region. If the Standard Model can explain the observed CP violations, the allowed regions of all measurements must overlap at least at *one* common point, otherwise there would be no unified explanation of all measured phenomena of quark mixing through the CKM matrix, i.e., there would be no triangle consistent with all measurements. But there is: In a tiny red-bordered point at the top of the triangle all areas intersect, and thus the CKM matrix of the Standard Model provides a consistent picture.

Once again the Standard Model triumphs, leading to the Nobel Prize in 2008 for Kobayashi and Maskawa. Since in the Standard Model the combination of CP and T symmetry (time reversal, see Antimatter ↓) is conserved, CP violation also implies the violation of T symmetry. The conservation of CPT symmetry in the Standard Model also implies that the overall lifetimes of B^0 and $\bar{B}^0$, taking all decays and not only specific ones into account, are indistinguishable.

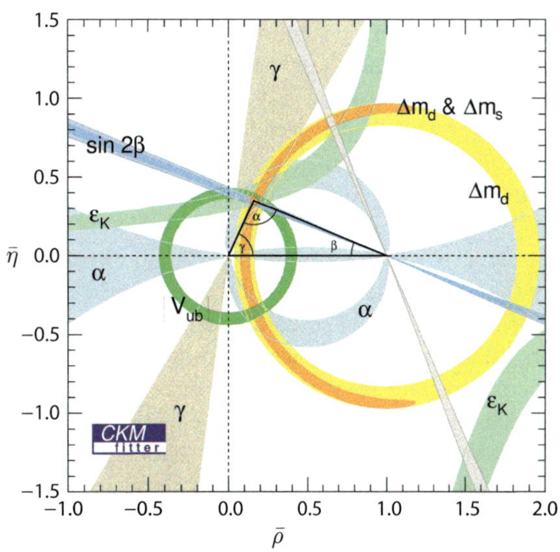

The colored areas illustrate the allowed regions for the parameters of the unitarity triangle from various measurements. They all intersect at exactly one point (small red circle at the top of the triangle): The CKM matrix can describe the mixing of quark families and CP violation.

What Is Antimatter? → p. 22

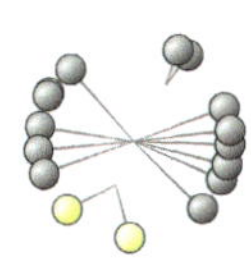

The Hurdles of Success

What New Theories Must Overcome

A crucial part of theoretical physics is to develop new ideas for describing Nature. Major breakthroughs in human cultural history are associated with such developments. For example, the heliocentric worldview of Copernicus, Kepler, and Galileo; Newtonian mechanics; or the development of modern physics with quantum mechanics and relativity theory, which underpin our modern information society through quantum states in computer chips or through enabling global positioning systems.

Experiencing such a revolution would be a remarkable event. However, all these previous revolutions were not only based on a few very brilliant minds creating new theories but also on decades of precise elaboration of theoretical predictions, as well as thorough experiments and accurate measurements.

The same is true today: We particle physicists hope that the beginning of the end of the Standard Model of particle physics as the best and most complete description of the microcosm is near. We hope that improved experiments will provide more concrete clues about the forthcoming revolution. However, a very important characteristic of these revolutions is that they must also explain all previous experiments—after all, the experiments were not wrong. Any new theory must therefore account for everything almost exactly as the Standard

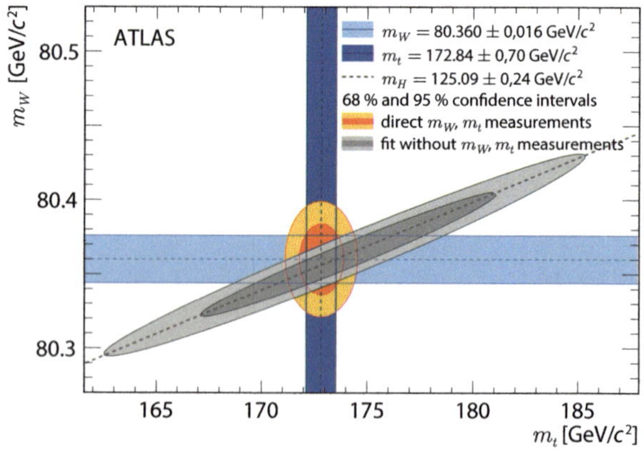

The measured masses of the top quark t, the W boson, and the Higgs boson H are consistent with each other in the Standard Model: all measurements converge at a single point.

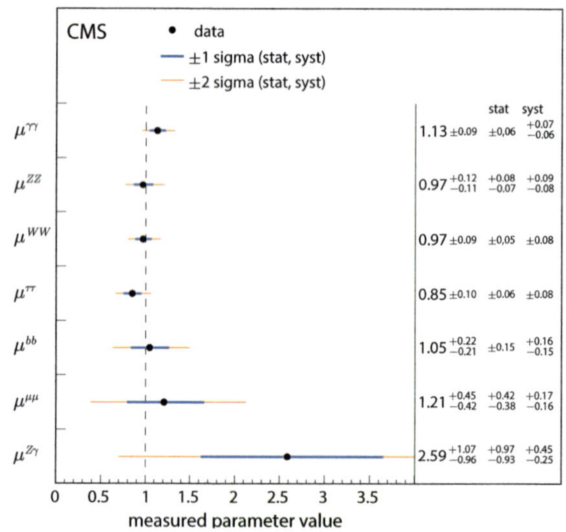

The measured properties of the Higgs boson, in terms of the measured couplings in seven different final states, are in agreement with the Standard Model expectation of a value of one.

Model does (within some tiny uncertainties, see Precision Experiments ↓). Researchers say that the better, more fundamental, simpler, more beautiful, and more comprehensive new theory must encompass the old one. This has been the case in the past.

It is not sufficient, for example, to claim: "We are actually living in a computer simulation, where, for example, cellular automata are the fundamental unit" (Imagining Reality ↓). This may be an interesting idea—but to be taken seriously as a physical theory, this hypothesis of cellular automata would need to explain at least everything that the Standard Model successfully explains within the measurement uncertainties. Of course alternative theories to the Standard Model exist. They contain the Standard Model and extend it and will be covered in the next chapter. However, a fully successful theory built on entirely independent foundations from the Standard Model does not yet exist.

The growing challenge for theorists: The amount of precise measurements from the history of particle physics (illustrated here with selected examples) is enormous. These measurements are all consistent with the Standard Model within their experimental uncertainties (Statistics ↓). This entire legacy of precise measurements in particle physics must also be accounted for by any future more complete theory, along with many new phenomena: a monumental task.

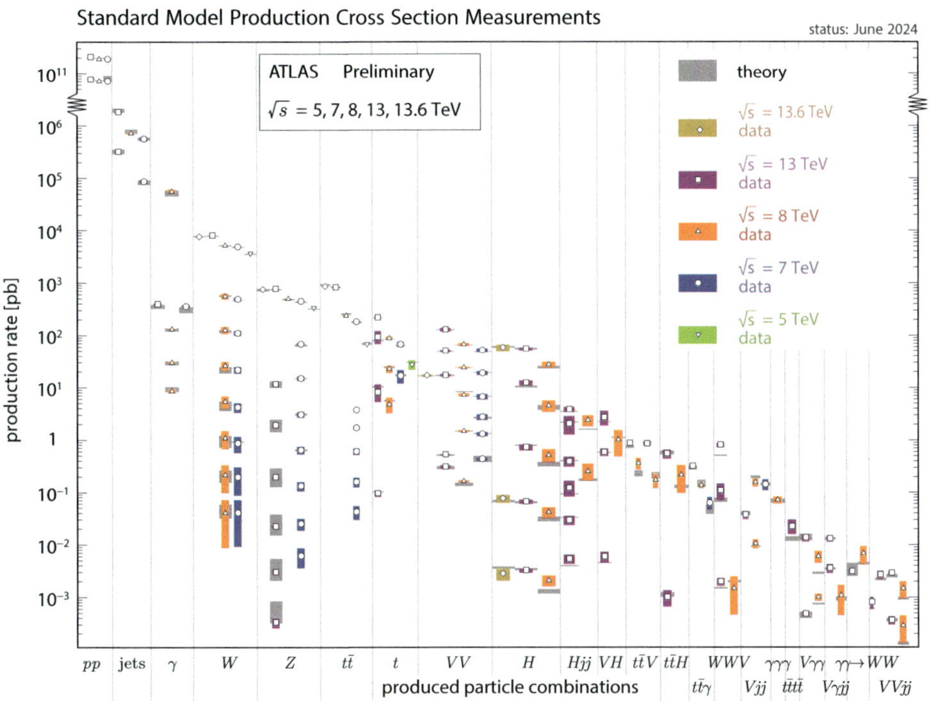

The predictions of production rates for various processes at the LHC agree with the Standard Model to within 14 orders of magnitude.

Precision Experiments → p. 250
Imagining Reality → p. 32
Drawing Conclusions From Statistics → p. 160

10 Limitations of the Standard Model

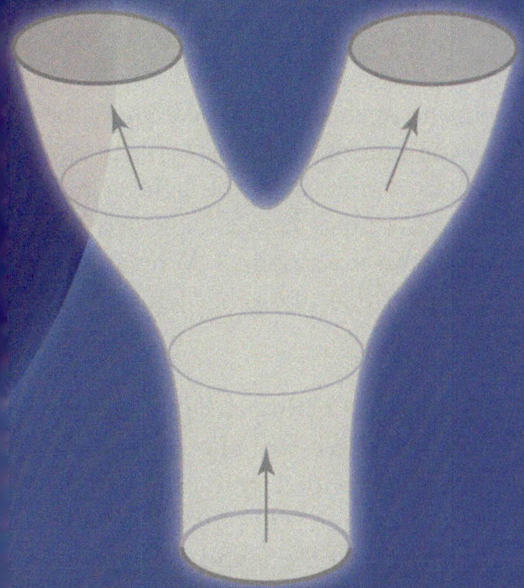

In this chapter, we look out into the unknown world beyond the Standard Model. In the previous chapters we have seen that matter consists of a relatively small set of building blocks, which appear to be elementary. Or could they be made of even smaller substructures? This does not seem to be the case. Such models with further substructures have been constructed, but they have all been refuted by the very precise data from the LEP collider. However, other questions remain unanswered within the Standard Model: Where do the neutrino masses come from? Why are the electron's and proton's electric charges exactly equal and opposite even though one is a lepton and the other made of quarks? How can we explain the dark matter indirectly observed in the Universe? And what is dark energy actually all about?

In this chapter, we take a closer look at proposed scenarios beyond the Standard Model. Electricity and magnetism are unified into electromagnetism. They are different sides of the same coin. Perhaps all the forces of the Standard Model originate from one large unified force? Such a theory is called a Grand Unified Theory or GUT. Perhaps there are new fundamental symmetries or even a supersymmetry? Perhaps we are also wrong in assuming the elementary particles as point-like and they are actually very very small extended strings. Or are we even living in a world with more than four dimensions? In exploring these questions we can maybe get a glimpse of the world beyond the Standard Model.

Open Questions

Why We Keep Searching

Since the discovery of the Higgs boson in 2012 the Standard Model of particle physics is complete. As such it is a tremendous success; it describes all laboratory experiments to date, in some cases with sensational precision (↓). The experiments span more than 13 orders of magnitude in energy, from less than 1 eV in atomic physics up to 13 TeV at the LHC.

With the help of the Standard Model, our understanding reaches far out into the Universe (Big Bang ↓). We understand in detail how the Sun works, including the stream of neutrinos (↓) that originate deep inside. In the interplay of the Standard Model and the general theory of relativity, we understand the scattering of neutrinos in a supernova shortly after its explosion. This scattering causes their arrival on Earth to be distributed in time, as observed. Also directly related to general relativity, we understand how the light elements hydrogen, helium and lithium were produced in the first three minutes of the early Universe and in what abundance (Nucleosynthesis ↓). We also largely understand the production of the heavier elements in stars (The Birthplace of Nuclei ↓).

At this point we might want to lean back and enjoy what we have achieved. And yet our gaze continues to wander. New questions arise that encourage us to look further. Here we present a list of the most important *open* questions in the Standard Model. The answers should lie in future new experiments and in deeper theoretical structures.

1. The different masses of the more than 100 different elements in the periodic table are explained by our understanding of the atomic structure with a nucleus (↓) made of protons and neutrons and a shell of electrons. In contrast, in the Standard Model we have twelve different fermions, with masses ranging from 173 GeV/c^2 for the top quark to a fraction of an electron volt per c^2 for the neutrinos. To date we only describe them with free parameters. Why are the masses of the fermions so different?

2. What are the neutrino masses exactly? We know the difference between the neutrino masses (↓), but not the absolute values. In particular: is the lightest neutrino massless? Do the neutrinos have a Dirac or a Majorana mass? Following on from point 1: Why are the neutrinos so much lighter than the other fermions in the Standard Model?

3. Why is the charge of the hydrogen atom exactly zero? Experimentally, we know that the sum of the proton and electron charge divided by the value of the electron charge satisfies $(q_p + q_e)/e < 10^{-21}$! In other words: Why are the charges of the u and d quarks exactly $+2/3\ e$ and $-1/3\ e$, respectively? This is known as the ***problem of charge quantization***. In the unified theories (GUTs) there is an answer. But are the forces unified? Do they have a common origin?

4. The Universe contains a large amount of *dark matter*. This can probably be explained by one or more new(!) elementary particles. This is not possible with-

Precision Experiments → p. 250
Big Bang → p. 280
Neutrinos From the Sun → p. 344
Nucleosynthesis After the Big Bang → p. 330
The Birthplace of Nuclei → p. 342

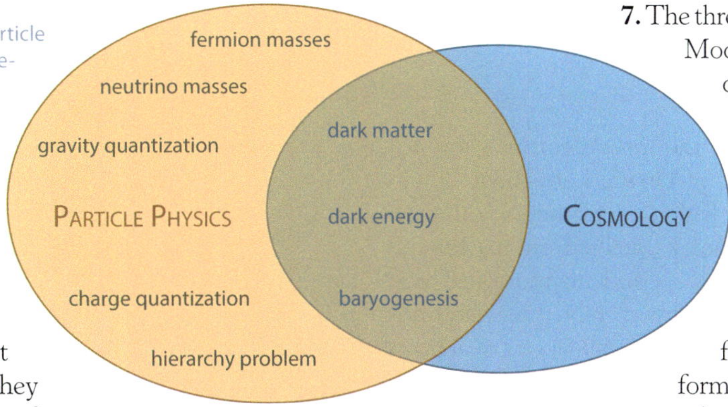

Open questions in particle physics, including those related to cosmology. Open questions related only to cosmology are not included

in the Standard Model. What are these new particles? How can we detect them? And how do they fit into the overall theory?

5. After the Big Bang, the Universe began as an extremely hot plasma. Thermodynamics tells us that in such a hot state there must be equal numbers of particles and antiparticles. Thus the baryon number, the number of baryons minus the number of antibaryons, should have been exactly zero. Today we observe almost only baryons in the entire Universe, and no antibaryons. Thus the baryon number today is not zero. How could the Universe have evolved from its original matter-antimatter symmetric state to today's asymmetry, even though baryon number is conserved to very high accuracy in the Standard Model? This is the problem of *baryogenesis* (↓).

6. Today we observe an accelerated expansion of the Universe, which can be explained by a small vacuum energy, also known as *dark energy* (↓). What is the origin of this vacuum energy? And how is the much larger vacuum energy in the Standard Model seemingly compensated by a yet unknown mechanism, such that it does not contribute to the expansion rate of the Universe?

7. The three forces of the Standard Model are all excellently described by quantum field theories. What about gravity? Can it also be quantized? Is there a *graviton* as an exchange particle? We do not know. String theory offers one possibility. It is formulated at the Planck energy scale (10^{19} GeV). There, gravity should be of the same strength as the other forces. But how can we show that there is a string theory? All predictions to date have been energetically far beyond the reach of today's experiments.

8. The Higgs boson has a mass of around 125 GeV/c^2. At very small distances, 10^{-35} m, the Planck length (or very high energies, 10^{19} GeV), gravity should be comparably strong to the other forces. Quantum effects should occur. These should connect directly or indirectly with the Higgs mass and make it much larger than observed. This is the so-called *hierarchy problem*. Supersymmetry (↓) is a possible solution. It protects the Higgs mass from the Planck scale corrections. However, there is (still) no experimental proof of supersymmetry.

The Standard Model is a self-contained, complete theory. However, as we have described, there remain many open questions. In addition, we can now look so far back and so precisely into the history of the Universe that this also raises many cosmological questions unanswered by the Standard Model.

From Nucleus to Nuclei → p. 226
Neutrino Masses → p. 272
Baryogenesis → p. 328
Dark Energy → p. 284
Supersymmetry → p. 278

Neutrino Masses

Small but Oh So Nice

In the Standard Model, the three neutrinos ν_{eL}, $\nu_{\mu L}$, $\nu_{\tau L}$ are all left-handed (Up, Down, Left, Right ↓) and massless. From the oscillation experiments with solar and atmospheric neutrinos, as well as from reactor and accelerator experiments (↓), and cosmological constraints, we know that at least two of the three neutrinos are massive, with masses of about $0.01\ \mathrm{eV}/c^2$. This is very small but non-vanishing. How can neutrinos have mass? And why is their mass so much smaller than that of the charged leptons and quarks?

In the Higgs mechanism, for example, a left-handed anti-top quark ($\bar{t}_L$) and a right-handed top quark (t_R) couple to the Higgs boson (H^0) with coupling strength h_{top}: $h_{\text{top}} H^0 \bar{t}_L t_R$. *After* spontaneous symmetry breaking (↓) the Higgs boson receives a vacuum expectation value $H^0 \longrightarrow v$ and the coupling leads to the top quark mass: $m_{\text{top}} \bar{t}_L t_R$, , with $m_{\text{top}} = h_{\text{top}} v$.

Before spontaneous symmetry breaking, the top quark is massless, since the Higgs boson has no vacuum expectation value. A mass that is based on such a left-right coupling $\bar{t}_L t_R$ is called a *Dirac mass*. All quarks and all charged leptons are massless before spontaneous symmetry breaking. After spontaneous symmetry breaking they have a Dirac mass, i.e., a mass term with a left-handed and a right-handed particle.

Since there are no right-handed neutrinos in the Standard Model, no Dirac mass can be generated. The neutrinos are massless, even after spontaneous symmetry breaking. This contradicts experimental observation.

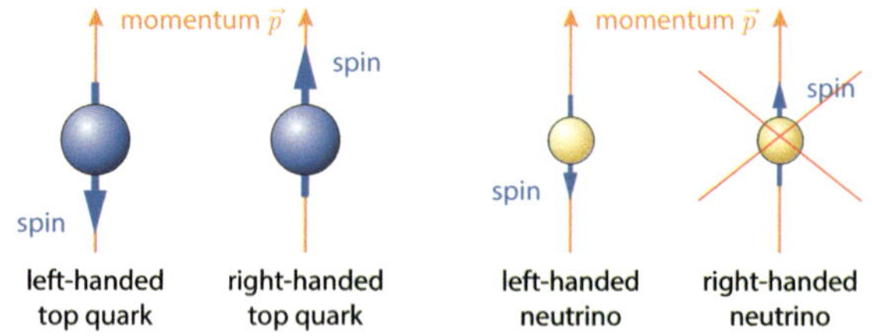

In the Standard Model there are no right-handed neutrinos.

However, we can extend the Standard Model by three right-handed neutrinos: ν_{eR}, $\nu_{\mu R}$ and $\nu_{\tau R}$. Analogously to the charged leptons and quarks, this leads, via the Higgs mechanism, to three Dirac neutrino masses, m_D. Problem solved! However the coupling: $h_{\text{neutrino}} \approx 10^{-14}$ is unnaturally small, compared to $h_\mu \approx 4 \cdot 10^{-4}$ or even $h_{\text{top}} \approx 1$. Thus one can ask: why are the neutrino masses so very much lighter than the other fermions?

For the sake of simplicity, let's consider *one* left-handed, ν_L, and *one* right-handed neutrino, ν_R. Both are electrically neutral. The ν_L couples to the $W^\pm$, Z^0 bosons. However, the ν_R has no charge under any Standard Model force. It does not couple to any gauge boson. We know from quantum field theory that such a spin-1/2 particle (fermion) without a gauge charge can be its own antiparticle. It is then called a *Majorana particle*.

For a Majorana particle it is possible to form a gauge invariant mass term, i.e., one which satisfies all the symmetry requirements of the Standard Model, with the right-handed neutrinos alone: $M_R \nu_R \nu_R$. The resulting mass is also called a *Majorana mass*. This Majorana

Up, Down, Left, Right → p. 180
Verifiying Neutrino Oscillations → p. 254
Spontaneous Symmetry Breaking III → p. 188 and Spontaneous Symmetry Breaking I → p. 184

mass is *independent* of the Higgs mechanism and therefore not bound to the energy scale of the Higgs vacuum expectation value $v = 246$ GeV. The special thing about right-handed neutrinos is that they are particles without charge quantum numbers. Therefore only they can have a mass independent of the Higgs mechanism. This mass is arbitrary and can be lighter or much heavier than for example the top quark mass.

So now we have both for our neutrinos: a Dirac mass (left-right, $\bar{\nu}_L \nu_R$) as for the other particles, but additionally a Majorana mass (right-right, $\nu_R \nu_R$). This means that the left-handed and right-handed neutrinos mix, a peculiarity of quantum mechanics, which also applies to neutrino oscillations (↓). It turns out that there are then two mixed states with fixed mass: a light one with mass m_L and a heavy one with mass $M_H \approx M_R$. Each is a different mixture of ν_L and ν_R. The product of these two masses in this model, however, must be constant, denoted $m_D{}^2$, independent of the heavy Majorana mass M_R:

$$m_L \cdot M_R = m_D^2 = \text{constant} \; .$$

Here m_D is the Dirac mass from the Higgs mechanism given above. In a model where the unknown mass M_R is very large, then the other mass m_L must become very small, and vice versa: If M_R is very small, then m_L becomes very large. It's like a seesaw on the playground: If one side goes down, the other must go up. This is why this model is called the *seesaw mechanism*.

As an example, if we take $m_D = m_{\text{top}}$ and $M_R = M_{\text{GUT}} = 10^{15}\,\text{GeV}/c^2$, the large grand unification scale (GUT ↓), we obtain $m_L = 0.01\,\text{eV}/c^2$. Incidentally, the light state is now an almost pure ν_L state and also a Majorana neutrino. In this model, these are the neutrinos which we observe in experiments. So if indeed Nature has a ν_L and ν_R and the seesaw mechanism, we should soon observe neutrinoless double beta decay (↓) experimentally. This would be a very strong indication of the seesaw mechanism.

Can the seesaw mechanism explain the very light neutrino masses?

Neutrino Oscillations → p. 198
GUT → p. 276
Neutrinoless double beta decay: Searching for a New Experiment → p. 296

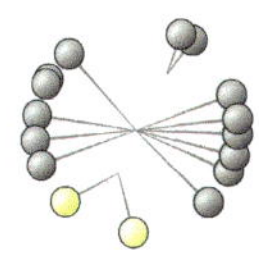

Coupling Unification

Can the Standard Model Forces Have a Common Origin?

In this and the next article we want to investigate whether the three forces of the Standard Model might have a common origin. The resulting hypothetical structure is denoted a *Grand Unified Theory*, or GUT (↓). Let us however first consider the measured values of the respective coupling constants

$$\alpha(M_Z) = 0.0078167 \pm 0.0000005$$
$$\alpha_{\text{weak}}(M_Z) = 0.034989 \pm 0.000002$$
$$\alpha_s(M_Z) = 0.1187 \pm 0.0017$$

Here $\alpha(M_Z)$ denotes the electromagnetic fine structure constant $[\alpha = e^2/(4\pi)]$ as measured at the energy scale of the Z^0 boson mass: $E = M_Z c^2 = 91.19$ GeV. The parameter " $-e$" is the value of the electric charge of the electron. Similarly $\alpha_{\text{weak}}(M_Z)$ and $\alpha_s(M_Z)$ denote the weak and strong coupling constant as measured at $M_Z c^2$, respectively. The three couplings have quite different values. However, in the article on renormalization (↓) we have seen that due to quantum field theory corrections, their respective strengths change with the energy scale. We denote this as *running couplings* (↓).

Here we use the running of the couplings to extrapolate their respective values to higher energies. It is important to note that the three couplings do *not* run in the same manner. This is because different Feynman diagrams contribute to the respective running. As shown in the right figure we first see two diagrams for the electromagnetic coupling (left side). In the lower diagram we show the contribution from an electron in the loop. Completely analogously all charged fermions can contribute: leptons and quarks, and via a related diagram the charged W bosons. At the right side of the figure we see two analogous diagrams for the strong coupling. In the second of these diagrams, in the loop, only quarks can contribute, but no leptons. However, we also have a third class of diagrams, with no counterpart in electromagnetism, with gluons also in the loop, as they interact with each other.[1] We have a separate set of diagrams for the weak interactions, not shown here. Thus for each coupling different diagrams contribute and they evolve completely separately and differently.

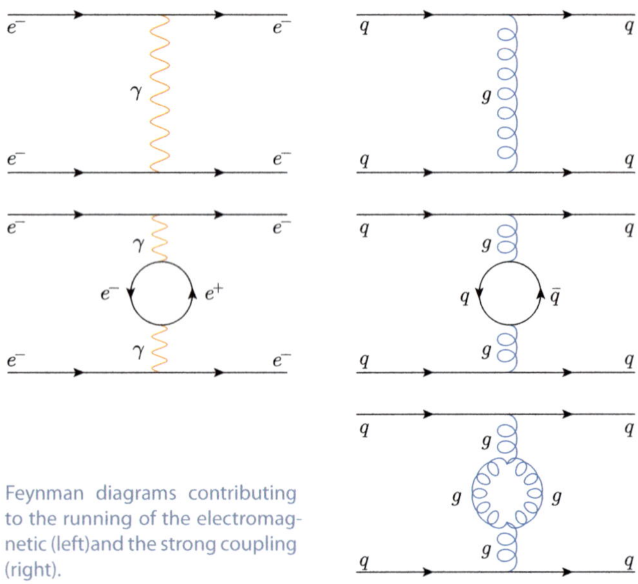

Feynman diagrams contributing to the running of the electromagnetic (left)and the strong coupling (right).

[1] For a more extensive set of diagrams see Loops → p. 106 and The Running Coupling → p. 210

GUT → p. 276
Renormalization → p. 108
The Running Coupling → p. 210

In the figure below, we have, starting from the above measured values, extrapolated all three couplings to very very high energies. Note, instead of plotting the couplings themselves we plot their inverse: $1/\alpha(E)$, $1/\alpha_{\text{weak}}(E)$, and $1/\alpha_s(E)$ as a function of energy, as this leads to straight lines. It is important to note that a particle contributes to the running via the loops in the Feynman diagrams only for energies larger than its mass: $E > mc^2$. Thus for E below 1 GeV, only the following particles contribute to the running of the electromagnetic coupling via the loop: e^-, μ^-, u, d, s. With increasing energy as we successively cross each mass threshold, we must include further particles: c, τ^-, b, $W^\pm$, Z^0, H^0, t.

The left part of the figure shows the overall result for the Standard Model above $E = m_{\text{top}}c^2$. The (small) widths of the curves reflect the small experimental errors at low energies where the couplings are measured.

We see the couplings do indeed converge, however they do not meet in a point, and instead form a larger triangle at an energy around 10^{15} GeV. There seems to be no common origin for the forces.

However, if we extend the Standard Model by supersymmetry (↓) the spectrum is roughly doubled and we must alter the running of the couplings above the respective thresholds. Assuming the new particles all have a mass of about 10^4 GeV we show the new extrapolation in the third figure. Note the kink at $E = 10^4$ GeV, where the supersymmetric particles change the running. The couplings do now meet at one point at the very high energy of 10^{16} GeV. Beyond this energy there could be a Grand Unified Theory with just one coupling constant, indicated by the black line in the right part of the figure. This motivates the search for grand unification and supersymmetry.

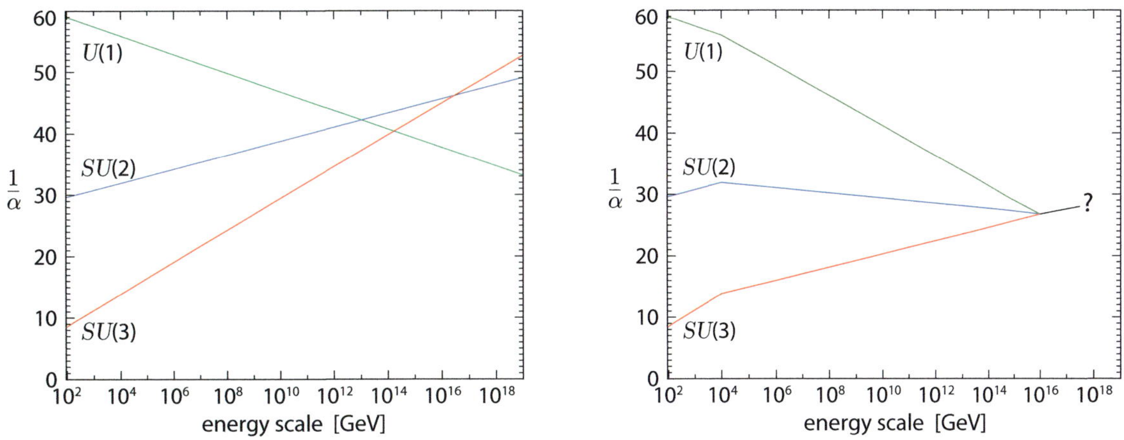

Running of the inverse couplings as a function of energy. Extrapolation of the Standard Model (left) and assuming supersymmetry (right).

Supersymmetry → p. 278

GUT

The Grand Unified Theory

Newton realized that the force holding the Moon on its orbit around the Earth is the same as the force which causes an apple to fall from a tree. Universal gravity was the first unification in physics. Suddenly the dynamics of the heavens became open to investigation. Today we compute the orbits of satellites which fly out to the edge of the solar system. Physicists next realized that the electric and the magnetic force have a common origin. Magnetic fields are caused by moving electric charges, and changing magnetic fields induce electric fields. Overall there is only one relevant coupling $\alpha = e^2/(4\pi)$ (↓). All electromagnetic phenomena are encoded in four coupled equations, the Maxwell equations.

In the Standard Model we have seen that the electromagnetic and the weak force are combined into the electroweak interaction above the vacuum expectation value of the Higgs boson: $v = 246$ GeV, see the article Spontaneous Symmetry Breaking II (↓). In the figure we see that the cross sections for the neutral current and the charged current merge at high energy $Q^2 > (100\ \text{GeV})^2$, corresponding to equal strength interactions, as expected.

In the previous article we saw that, when including supersymmetry, the three couplings of the Standard Model run and merge at about 10^{16} GeV. It is thus imaginable that a new unified theory exists, formulated at this scale, which combines these three forces into one universal force. This we would call a *Grand Unified Theory* (GUT). What could this GUT look like?

For this we look a bit closer at the symmetries of the Standard Model. Each of the three forces is described by a gauge symmetry (↓). These are *internal* symmetries based on a rotation in an abstract space. For electromagnetism this symmetry is denoted $U(1)_{EM}$ and is a one-dimensional rotation. For the weak interaction it is a 3-parameter rotation denoted $SU(2)_L$. And for the strong interaction it is an 8-parameter rotation denoted $SU(3)_C$. "L" and "C" stand for left and color, respectively. Each of the 12 parameters corresponds to an in-

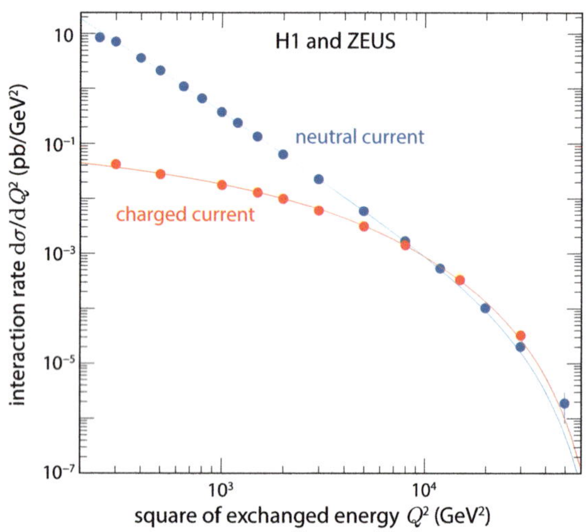

Electroweak unification observed at HERA. The blue curve denotes the neutral current interaction rate as a function of the exchanged energy squared. Here the electron and the proton can exchange a photon or a Z boson. The red curve denotes the charged current interaction rate via a W boson. At high energy these unify to the electroweak interaction, with equal strength.

The Running Coupling → p. 210
Spontaneous Symmetry Breaking II → p. 186
Gauge Symmetries → p. 98

dependent gauge boson: the photon, the W^+, W^-, and Z^0, as well as the 8 gluons.

If we want to construct a combined unified symmetry, by continuity it must *at least* contain the Standard Model symmetry, and its interactions. It must therefore be at least a 12-parameter rotation symmetry. Furthermore, there is something called the *rank* of such a symmetry. It corresponds to the number of different (generalized) neutral currents, which do not change charge, flavor or color. These are the photon and the Z^0 boson, of the electroweak sector, as well as two of the eight gluons of the strong interaction. Thus we need at least a rank-4 symmetry. All of these internal symmetries have been classified by Sophus Lie, a Norwegian mathematician. Thus we just have to look in a mathematics table. There we find the minimal symmetry with rank ≥ 4 and ≥ 12 rotation parameters to be $SU(5)$. It contains $U(1)_{\mathrm{EM}}$, $SU(2)_L$, and $SU(3)_C$ as "subsymmetries".

$SU(5)$ has rank 4 and 24 independent phases corresponding to 24 gauge bosons: the 12 gauge bosons of the Standard Model plus 12 new ones, which we call $X_a^{\pm 4/3}$, $Y_a^{\pm 1/3}$. Here $a = r, b, g$ is the color charge and $\pm 4/3$ and $\pm 1/3$ are the electric charges. The spontaneous symmetry breaking (↓) through an extended Higgs sector results in

$$SU(5) \longrightarrow SU(3)_C \times SU(2)_L \times U(1)_Y\,,$$

giving the SM again at low energy. Through the symmetry breaking the new $X^{\pm 4/3}$- and $Y^{\pm 1/3}$-gauge bosons get a mass of about 10^{16} GeV/c^2.

For $SU(2)_L$ we had, e.g., the doublets $(\nu_e, e^-)_L$ and $(u, d)_L$ which couple to a $W^\pm$ boson, changing into each other (Up, Down, Left, Right ↓). Similarly, in $SU(3)_C$ the three different colors of quarks are combined into a triplet, and six of the eight gluons each couple to two different colors. In $SU(5)$, a family of the Standard Model is organized in a quintuplet (5) *and* in a decuplet (10):

$$\mathbf{5} = (d_R^r,\ d_R^g,\ d_R^b,\ e_L^-,\ \nu_{eL})$$
$$\mathbf{10} = (u_R^r,\ u_R^g,\ u_R^b,\ u_L^r,\ u_L^g,\ u_L^b,\ d_L^r,\ d_L^g,\ d_L^b,\ e_R^-)$$

where, e.g., u_R^r is the right-handed (R) red (r) up quark. The $W^\pm$ boson and the gluons each couple to different elements of the quintuplet (5) or decuplet (10), as before. The new gauge bosons $X^{\pm 4/3}$ and $Y^{\pm 1/3}$ mediate new interactions between the elements of the 5- and 10-plets respectively. Examples are the decays:

$$X_r^{+4/3} \longrightarrow u^g + u^b\ , \qquad X_r^{+4/3} \longrightarrow e^+ + \bar{d}_R^r\ ,$$
$$Y_r^{+1/3} \longrightarrow \bar{\nu}_e + \bar{d}_R^r\ , \qquad Y_r^{+1/3} \longrightarrow u_L^b + d_L^g\ .$$

Since the $X^{\pm 4/3}$ and $Y^{\pm 1/3}$ couple to leptons and quarks at the same time, they are also called *leptoquarks*. These decays violate baryon (B) and lepton number (L), but conserve the combination $B - L$. This enables proton decay (↓) $p \longrightarrow \pi^0 + e^+$, as shown in the Feynman diagram below. This is the most important experimental prediction of grand unification and is therefore intensively searched for. In very large tanks of water the spontaneous disintegration of a proton would lead to an energetic positron, which produces Cherenkov radiation, as well as an annihilation signature with an electron: two energetic gamma rays. The pion decay also gives two further gamma rays $\pi^0 \longrightarrow \gamma\gamma$.

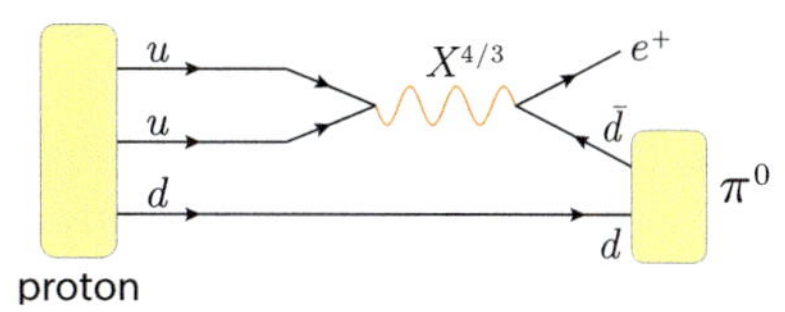

Proton decay

Spontaneous Symmetry Breaking I → p. 184
Up, Down, Left, Right → p. 180
Proton Decay → p. 320

Supersymmetry

The Maximal Possible Symmetry

Supersymmetry is an exciting possible extension of the Standard Model. Here, the external symmetries (space and time translation, rotation, and the boosts of special relativity, see External and Interal Symmetries ↓) are supplemented by an additional transformation, namely between particles with different spin (↓):

Standard Model		**New Particles**
spin 1	⟷	spin $\frac{1}{2}$
spin $\frac{1}{2}$	⟷	spin 0
spin 0	⟷	spin $\frac{1}{2}$

If the theory is invariant under these transformations, then it is an additional external symmetry, since the spin, as the intrinsic angular momentum of the particles, is an external property in space and time. What is important here is that the transformation changes *only* the spin, all other quantum numbers, such as mass, electric charge, and color charge remain the same.

Supersymmetry postulates that for every spin-1/2 particle, e.g., the electron, there is a new particle that is the same as the electron in every respect except with spin 0. Since spin-0 particles are also called *scalars*, the new particles are called: scalar electrons or selectrons for short. We extend the particle symbol by a tilde: $\tilde{e}$, $\tilde{\mu}$, $\tilde{\tau}$; $\tilde{\nu}_e$, $\tilde{\nu}_\mu$, $\tilde{\nu}_\tau$; $\tilde{u}$, $\tilde{c}$, $\tilde{t}$; $\tilde{d}$, $\tilde{s}$, $\tilde{b}$ (see the figure at the right page). For the gauge bosons (spin 1), new spin-1/2 partners are postulated. These new particles are named with the suffix "-ino". Thus, the partner of the photon is called the *photino* and is also given a tilde: $\tilde{\gamma}$. The partner of the gluon is called the *gluino* and is written $\tilde{g}$. The partners of the $W^\pm$ and Z^0 are called *Wino* and *Zino* and are written as $\widetilde{W}^\pm$ and $\widetilde{Z}^0$, respectively.

For the mathematical consistency of the theory, the Higgs sector must be expanded. Beyond the Higgs boson of the Standard Model (H^0) there are three further spin-0, i.e., scalar, Higgs bosons: $\bar{H}^0$, A, $H^\pm$. The corresponding supersymmetric spin-1/2 partners are called *Higgsinos* and are summarized as $\widetilde{H}_u$, $\widetilde{H}_d$. Our simple Standard Model has become much more complicated.

Supersymmetry requires a spin-0 particle with the same mass and charge as the electron. This should have been observed long ago—but it hasn't. There is no selectron with mass 511 keV. Therefore, the supersymmetry transformation cannot be exact. It must be a broken symmetry, if it actually exists. The $SU(2)_L$ symmetry of the Standard Model is spontaneously broken, leading to massive $W^\pm$ and Z^0 bosons, as well as leptons and quarks. So maybe supersymmetry is also a broken symmetry. The energy scale of supersymmetry breaking leads to correspondingly heavy selectrons, which have not yet been observed. The same applies to all other supersymmetric partners. So far, after an intensive search, the LHC has set lower bounds of around 1 TeV/c^2 for most supersymmetric masses (↓). Supersymmetry has therefore not yet been confirmed experimentally.

Why even bother to introduce this already broken symmetry? There are two reasons: one for mathematical and aesthetic reasons, and another because of a tech-

External and Internal Symmetries → p. 96
Spin → p. 44
Searching for New Particles at the LHC I → p. 302

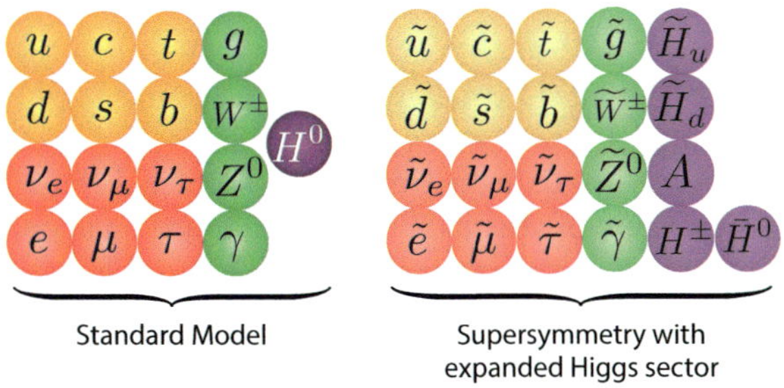

nical problem in the Standard Model, the so-called *hierarchy problem*. Both are briefly described below.

Without any symmetry restrictions, a three-dimensional object can have any shape. An object which is rotationally symmetric around one axis looks like a turned piece of wood. There are still many possible shapes, but already significantly fewer than without symmetry. If the object is completely rotationally symmetric, i.e., around three axes, only one shape remains: the perfect sphere.

Similarly, there are many symmetries in particle physics, and the more symmetries you require, the more restrictive are the possible forms, for example, of the scattering distributions, such as $e^+e^- \to \mu^+\mu^-$, in accelerator experiments. There is now an extraordinary mathematical statement, which we would like to explain briefly.

Consider a quantum field theory with the external symmetries of the Standard Model as listed above. If we now extend this theory to include supersymmetry, then the theorem states that *this is absolutely the maximum external symmetry allowed*. Why? If you further extend the external symmetries of the Standard Model, all scatterings become so simple as to vanish. There is *no more scattering*. Particles can no longer interact, they fly right by each other. A world in which we do not want to, let alone can live in!

Thus, supersymmetry is the last possible extension of the *external* symmetries of the Standard Model; the theory with the highest possible symmetry. The question arises whether Nature has fully exploited this potential or not, i.e., whether supersymmetry actually exists. This is what we are trying to find out at the LHC.

The second point, the hierarchy problem, is even more technical. It is about corrections to the Higgs mass in the quantum field theory. Assume there is a new higher energy scale *beyond* the Standard Model, as in the GUT models (↓). Then the quantum corrections to the Higgs mass originating from this new theory are such that the Higgs mass must be as large as this new scale. But the Higgs boson weighs 125 GeV/c^2 and not 10^{15} GeV/c^2.

The only way out within the Standard Model would be to extremely fine tune the parameters of the unified theory (Did Our Life Emerge by Chance? ↓), namely to one part in 10^{30}!!! This seems extremely unnatural. Supersymmetry offers a solution. In unbroken supersymmetry, these quantum corrections are exactly zero. The contributions from the bosons (spin 0 and spin 1) and the fermions (spin 1/2) exactly cancel each other, as guaranteed by the symmetry. An amazing result! If the supersymmetry is spontaneously broken, the corrections remain small as long as the masses of the new supersymmetric particles are smaller than about 10 TeV/c^2. For this mechanism to work, supersymmetric particles should be found at the LHC.

Special Relativity → p. 38
GUT → p. 276
Did Our Life Emerge by Chance? → p. 292

Big Bang

The First Particle Laboratory

We know from observations that the Universe is expanding and cooling. There must therefore have been a *Big Bang* at the beginning, an extremely dense and hot initial state. Higher temperatures mean higher energy to produce particles. This is why all kinds of particles were created in this phase and why their properties had a major influence on the development of the Universe. There is a very strong connection between the properties of the elementary particles and the current structure of the Universe. It is therefore worth taking a closer look at the early phases of the Universe after the Big Bang, the first laboratory of particle physics.

Even in separate, very distant regions of the Universe, we observe the same conditions, such as temperature or structure sizes. It is assumed that this homogeneity was created by a brief phase of sudden expansion of the Universe by many orders of magnitude, known as *inflation*, in the early phase of the Universe. Previously causally linked regions have become so far removed from each other in this rapid expansion that they now appear unconnected. Inflation can be understood through a mathematical description of the Universe, but a field—the so-called *inflaton*, which caused the inflation—would have to exist. Its existence has not yet been proven experimentally.

We assume that shortly after the Big Bang (really shortly, about 10^{-35} s) the strong, weak and electromagnetic forces were combined into one "unified force" (↓) until the Universe had cooled down to a temperature of about 10^{15} GeV.[1] Afterwards, a spontaneous symmetry breaking (↓) made it possible to distinguish the strength of the strong interaction from the strength of the other two forces.

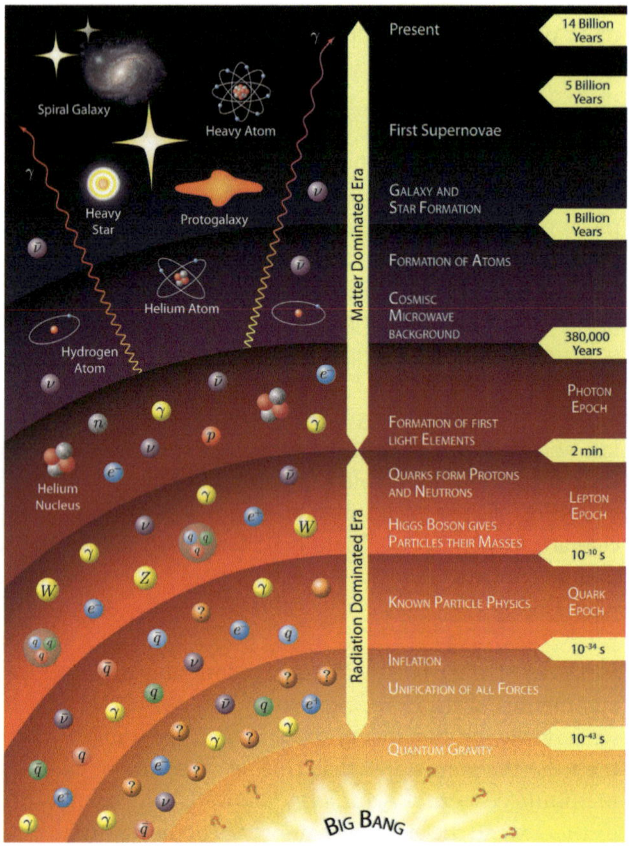

[1] Since the temperature is a measure of the average energy of the particles, it is practical to express it in energy units.

GUT → p. 276
Spontaneous Symmetry Breaking → p. 62

Only after cooling to about 10^3 GeV can we be comparatively sure of what has happened, since these energies are reached in accelerators such as the LHC and we can measure the processes. Only about one picosecond ($1\ \text{ps} = 10^{-12}\ \text{s}$) had passed since the beginning of the Big Bang down to this temperature. All the particles we know from the Standard Model are still highly abundant. If the temperature and thus the average energy falls below the mass mc^2 of a particle during further cooling, these particles annihilate with their antiparticles and form radiation. They no longer participate in the dynamics of the Universe. Thus, first the heavy, then the lighter particles gradually disappeared until only the light u and d quarks, electrons, muons and neutrinos with their respective antiparticles were left. In this process, a further symmetry breaking occurred, as a result of which the Higgs particle gave mass to the bosons of the weak interaction. After this, the electromagnetic and weak forces were also distinguishable.

At a temperature of around 200 MeV, the first hadrons were formed from the quarks (↓). The quarks and antiquarks were mainly bound to form pions (e.g. $\pi^+ = u\bar{d}$). Only a tiny excess of quarks compared to antiquarks of about one in a billion formed protons (uud) and neutrons (udd). Why this tiny excess of matter over antimatter exists is not understood in the Standard Model (Baryogenesis ↓).

After about one second had passed, at about 2 MeV, the density of neutrinos and electrons had decreased to such an extent that the time between two reactions of a neutrino was longer than the age of the Universe. Neutrinos therefore no longer reacted and their annihilation was halted. The neutrinos created in the Big Bang still exist today and, together with photons, are the most common of all known particles. For a long time, they were therefore thought to be the perfect explanation for dark matter (↓), but this turned out to be wrong. Since they are so light and thus travel close to the speed of light, they would lead to the wrong distribution of galactic structures. In the course of the events described so far, other particles still unknown to us must therefore have been created as an explanation for dark matter.

After just two minutes, at a temperature of around 0.1 MeV, protons and neutrons from the tiny excess of matter over antimatter formed the first light elements (hydrogen, helium, lithium) whose abundance can be measured today (Nucleosynthesis ↓). This makes the formation of the light elements the earliest stage of the Big Bang documented by measurements. It took 10 000 years to reach the next milestone. Then, at a temperature corresponding to 1 eV, the energy of the photons had fallen so far that, despite their immense abundance, their energy density fell below that of matter particles. Until then, the Universe had been dominated by radiation, matter now predominated. Only then the formation of the structures from which galaxies later emerged could begin.

380 000 years after the Big Bang, with the temperature corresponding to about 0.3 eV—still a temperature 1000 times higher than today—electrons and atomic nuclei were able to bind together to form neutral atoms. Whereas photons were previously scattered by the charged atomic nuclei and electrons, they were now able to move almost freely through the Universe and are now measured as the ubiquitous cosmic microwave background (CMB, ↓) radiation. The discovery of the CMB in 1964 is therefore considered the first successful test of the Big Bang theory.

The Quark Model: A First Look → p. 20
Baryogenesis → p. 328
Dark Matter → p. 282
Nucleosynthesis After the Big Bang → p. 330
Cosmic Microwave Background → p. 332

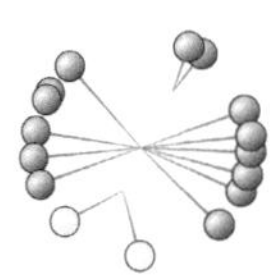

Dark Matter

A Challenge for Elementary Particle Physics

Ever since Johannes Kepler and Isaac Newton established the laws of planetary motion and much later Albert Einstein significantly improved our understanding of gravity, we have been able to calculate the orbits and velocities of all objects in the solar system very accurately. So accurately that we can send probes to other planets to precisely predetermined locations. It is therefore all the more astonishing that the known laws of gravity do not seem to apply when we look at larger scales, such as galaxies or galaxy clusters. At great distances, objects orbit each other at a higher speed than should be the case according to the law of gravity. Jan Hendrik Oort and Fritz Zwicky observed this in galaxy clusters in the 1930s. In the 1960s, this was confirmed by Vera Rubin with systematic studies of galaxies. Clusters of galaxies should actually fly apart due to the high centrifugal forces. However, this does not happen, which is why the gravitational force must be stronger than we assume.

There are two possible solutions to this problem: Either we have to modify the laws of gravity, as we know them from the solar system, at greater distances, or there is additional matter in the Universe that is invisible to us, which causes a greater mass attraction. It has not yet been possible to formulate a contradiction-free modification of the laws of gravity that reconciles all observations, which is why we assume that there is additional matter. As this matter is still invisible, it is called *dark matter*. This is not a small correction, but rather, at around 85 %, the vast majority of matter in the Universe consists of dark matter. The fact that the nature of most of our Universe is unknown is a little scary, but also an exciting question for researchers in cosmology and particle physics.

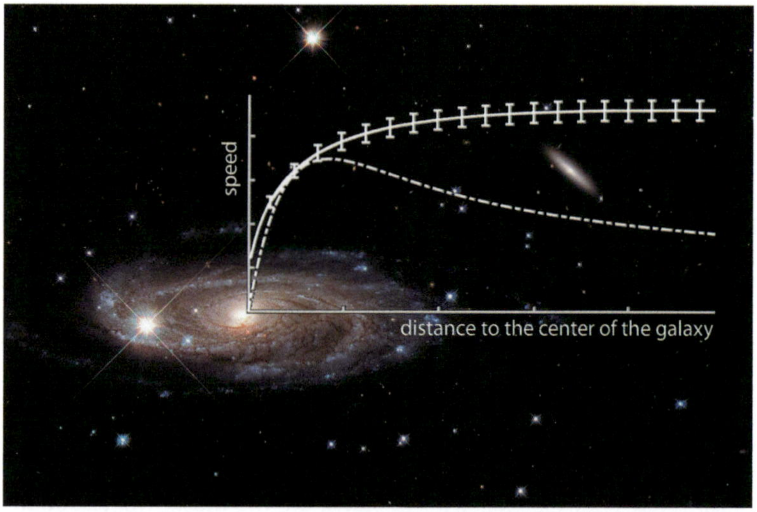

If objects orbit around a galaxy, there is a balance between mass attraction and centrifugal force. However, the measured speed at large distances is significantly higher than would be expected by the gravitational force of visible matter alone.

Open Questions → p. 270

As an explanation for the constituents of dark matter, all of the known elementary particles of the Standard Model can be ruled out. In fact, the astrophysical observations of dark matter are the clearest proof to date that the Standard Model of particle physics needs to be extended (Open Questions ↓).

Before introducing new elementary particles as an explanation, it may be more obvious to assume that non-luminous objects made of normal matter, such as black holes, are behind dark matter. They are referred to as *MACHOs*, which stands for *Massive Compact Halo Objects*. However, all attempts with elaborate observation campaigns to detect MACHOs in sufficient numbers have been unsuccessful. They can be ruled out as an explanation for dark matter.

Precise studies of the cosmic microwave background (↓) still show us today the pattern of oscillations and density fluctuations of electrons and atomic nuclei in the early phase of the Universe. Astonishingly, the density fluctuations observed are nowhere nearly sufficient to later form the structure observed today in the Universe with galaxies, galaxy clusters, etc. In addition to electrons and atomic nuclei, other forms of matter must therefore exist that had already begun to form structures long before the formation of the microwave background. These can only be uncharged particles, as they would otherwise have visibly influenced the measurable density fluctuations of the microwave background. The only known uncharged stable elementary mass particles are neutrinos, but they are far too light and were therefore far too fast in the early phase of the Universe to form structures (The Largest and the Smalles ↓).

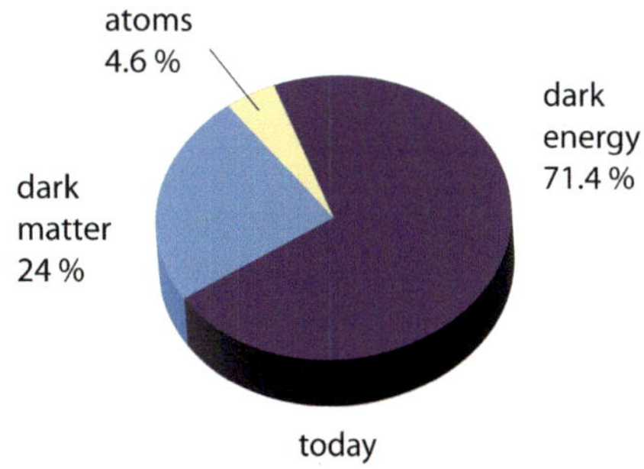

Energy composition of the Universe. 71.4 % consists of an unknown form of energy, so-called dark energy, 24 % of dark matter. Only 4.6 % consists of the matter we know.

In order to form structures early on, the particles must be slow—i.e., either be heavy particles or be generated at rest for other reasons—and, apart from the gravitational interaction they should participate at most in the weak interaction. Physicists call these hypothetical heavy particles *WIMPs*—*weakly interacting massive particles*. This term succinctly summarizes the necessary properties that a particle must have to explain dark matter. However, there is another variant in which super-light particles, so-called *axions* (↓), could also explain dark matter, if they are produced as slow particles.

There are many efforts worldwide to detect WIMPs or axions, be it to detect them directly as part of the dark matter (↓) surrounding us, to search for indirect astrophysical signals generated by possible decay products, or to produce them at high-energy accelerators. So far, no trace of them has been found. The astonishing realization remains that the predominant components of the Universe remain unknown to us and that the knowledge we have gained about the elementary structure of matter is incomplete.

Cosmic Microwave Background → p. 332
The Largest and the Smallest → p. 334
The Search for Axions → p. 322
The Mystery of Dark Matter → p. 308, Direct Search for Dark Matter → p. 310

Dark Energy

Is the Expansion of the Universe Accelerating?

The entire Universe is expanding, much like a balloon that you inflate. In this analogy, the three-dimensional Universe is represented by the two-dimensional surface of the balloon. If you attach paper cutouts of galaxies to the balloon, as shown in the illustration below, and then inflate it, the distance between all the galaxies increases because the balloon's surface stretches. However, the paper cutouts themselves—the galaxies—do not expand, as in the real Universe.

How do we observe the expansion of our three-dimensional Universe? All atoms emit light when they are excited, producing characteristic wavelengths that form their atomic spectra (↓). These spectra can be measured precisely in the laboratory. Similarly, stars in galaxies are hot and contain atoms that radiate light. We can collect this light with telescopes on Earth and analyze the spectra. However, the spectra of light from distant galaxies are all shifted toward longer wavelengths—into the red part of the spectrum—compared to the laboratory measurements on Earth, as shown in the illustration. This *redshift* occurs because the light waves are stretched during their long journey through the expanding intermediate space. Furthermore, the farther away a galaxy, the greater its redshift. This is similar to the situation on the balloon: pieces of paper that are farther apart from each other separate more quickly as the balloon inflates.

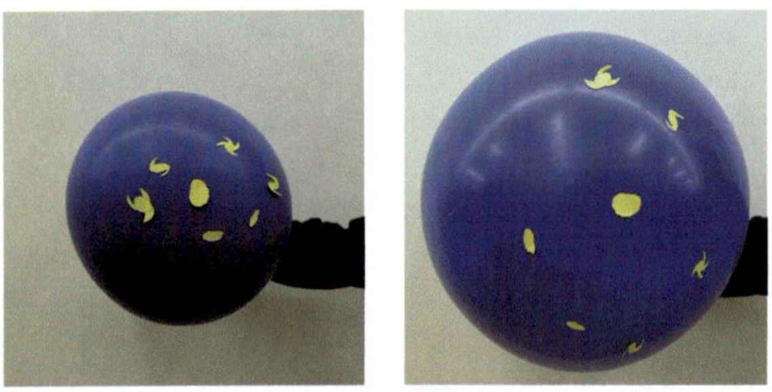

The Universe expands, similar to the two-dimensional surface of an inflated balloon with taped paper galaxies.

The expanding Universe can be understood in the context of an initial cosmological event, the Big Bang (↓). According to the general theory of relativity, the entire matter in the Universe—comprising all the particles' mass—as well as the electromagnetic radiation should cause a gradual slowing of the expansion. This is similar to gravity on Earth: if you throw a ball straight up, it rises for a while, slowing down until it eventually stops and then falls back to the ground.

More recently, in addition to measurements of galaxy spectra, the redshift of a large number of supernovae in far-away galaxies has been studied in detail. Thermonuclear supernovae (type Ia) are exploding stars that are well understood from a physical standpoint. These stars accumulate mass from their surroundings and explode when they reach a specific mass threshold. Thus always the same mass energy is converted into light (and also neutrinos), giving type Ia supernovae a known absolute luminosity. By measuring their luminosity on Earth, we can compare the observed brightness to the expected brightness, which allows us to determine the distance to the supernova.

Spectra: Chance, Interference and Predictions → p. 52
Big Bang → p. 280

When combined with the redshift measurements mentioned above, this technique has led to improved estimates of the expansion rate of the Universe. It was discovered that about 10 billion years ago, the expansion of the Universe began accelerating![1] This means that galaxies are moving apart faster and faster, rather than slowing down. This phenomenon cannot be explained by ordinary matter and electromagnetic radiation alone.

A very compact white dwarf star (left) sucks material from its companion. When its mass reaches a critical value, the dwarf star explodes in a supernova (type Ia).

In an expanding Universe, the total amount of matter remains constant, but the volume increases. As a result, the energy density of matter decreases. For radiation, the energy density decreases at an even faster rate, since the expansion of the Universe stretches the wavelengths of the radiation. In the general theory of relativity, in addition to matter and radiation, there is another form of energy, known as the *cosmological constant*. Unlike matter and radiation, whose energy density decrease as the Universe expands, the energy density of the cosmological constant remains *constant* as the Universe expands. However, it is not directly observable on cosmologically small scales, like on a planet or in a solar system. It is therefore referred to as *dark energy*. The equations of general relativity show that if this constant energy density dominates the Universe, it leads to an exponentially accelerating expansion. In this scenario, it seems as if different parts of the Universe are repelling each other! This behavior is consistent with current observations.

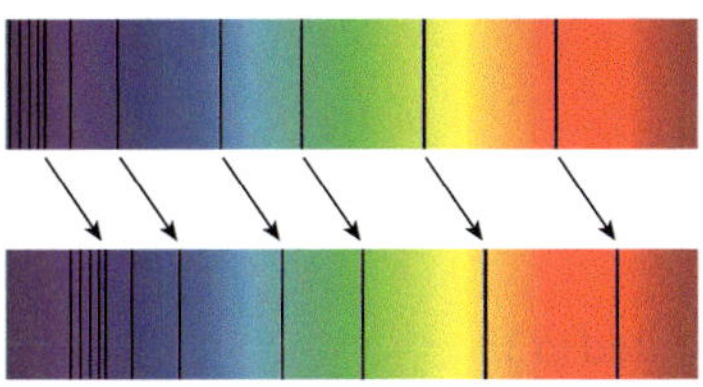

The Sun's absorption spectrum (left), and of a distant galaxy (right). The lines are shifted to lower frequency, i.e., towards the red.

Interestingly, the vacuum energy of the Higgs boson, following spontaneous symmetry breaking in the Standard Model, leads to a constant energy density in the Universe. However, its natural value is over 60 orders of magnitude too large! If this were the true value, it would rule out our existence, as the Universe would have already expanded so rapidly in its early stages that stars and planets could never have formed. This is known as the *cosmological constant problem*. Solving it requires a mechanism that compensates for the contribution of the Higgs boson to the vacuum energy density, potentially through a symmetry, such as supersymmetry. At the same time, another unknown scalar field must generate a much smaller contribution to the vacuum energy density that matches the observed acceleration. These issues remain unsolved.

It is important to note that the interpretation of supernova data is more complex than presented here. The absolute luminosity of supernovae is not always exactly the same, and there are still unresolved issues related to the scattering of the data. Although the discovery of the accelerated expansion of the Universe, based on supernova observations, earned scientists the Nobel Prize in Physics in 2012, the interpretation of these results remains a topic of ongoing debate.

[1] The Universe is now approximately 14 billion years old.

Extra Dimensions

Maybe Gravity is not as Weak as We Think

The Standard Model leaves some questions unanswered: For example, what is the origin and composition of dark matter (↓) and dark energy (↓)? Moreover, so far gravity is not part of the Standard Model. It is not even understood why it is so much weaker than all the other forces of Nature (↓). As discussed in detail in the article on the range of forces (↓), the strengths of the interactions can only be compared as long as one is within the interaction range of the other interactions—but then the differences are dramatic. For example, at atomic distances of 10^{-10} m, gravity is suppressed by over 30 orders of magnitude compared to electromagnetism. Such large differences demand some explanation.

One idea discussed in this context are the so-called *extra dimensions*. This hypothesis is described here, among other reasons because it demonstrates well how the scientific method is applied to such puzzles: Once a possible explanation for a particular phenomenon has been found, it is to be worked out in detail and observable consequences are to be identified against which it is tested. The idea here is that gravity could be so weak because the gravitational field can spread into more spatial dimensions than just the three spatial and one time dimension we know. These extra dimensions are not accessible to the other interactions and particles of the Standard Model. In such a world, the gravity we know would be so weak because we feel just a fraction of its full strength. This effect is indicated in the figure.

A possible analogy is that one cannot tell from the front side (two-dimensional) of a building how deep it is (third dimension), and thus one does not know, for example, how many apartments are in it. The strength of gravity that we perceive would be as unrepresentative of its true strength as the head side of a building is of its true size.

However, such additional dimensions, if they exist, would also affect our world. They would, for example, modify Newton's law of gravitation, because n extra dimensions would change it from $1/r^2$ to $1/r^{2+n}$. This is a straightforward generalization of the geometric derivation of Coulomb's law, which we discussed in the article on confinement (↓). But since Newton's force

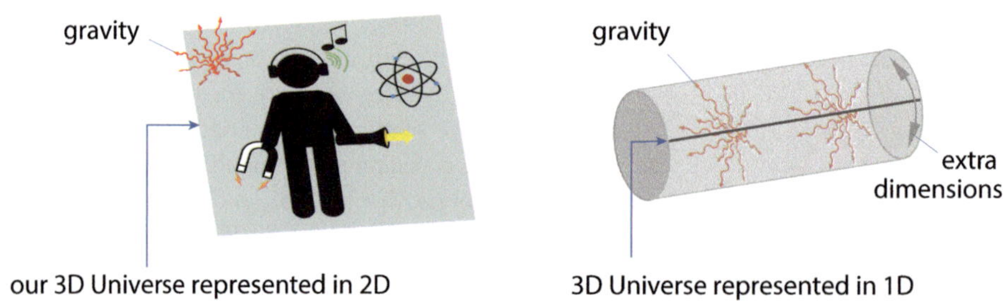

Dark Matter → p. 282
Dark Energy → p. 284
The Four Fundamental Forces of Nature → p. 16
On the Ranges of Interactions → p. 58
Confinement → p. 206

law is experimentally confirmed, the additional dimensions cannot extend infinitely. They must instead be *compact*. In this case gravity is modified only on length scales of the order of the size of the extra dimensions. This is also shown in the figure below.

To explain the weakness of gravity, a certain minimum size is required for the extra dimensions. On the other hand, their extension R must not be too large, because for distances smaller than R the above-mentioned modification of Newton's law would set in. It turns out that the number of extra dimensions also plays a role: If one wants to bring the strength of gravity to within an order of magnitude of the weak interaction with only one additional dimension, then it would have to be so large that even the planetary movements in our solar system would no longer be determined by the $1/r^2$ force. Clearly, this is not the case.[1] From two extra dimensions onwards, it gets exciting, as gravity would now change at distances of less than one millimeter.

How can this be verified? One possibility is direct tests. For this, one studies whether a pendulum really behaves exactly as the law of gravitation dictates. This sounds simple; but if one wants to make such a measurement at distances smaller than one millimeter, it requires a tremendous amount of technical effort.

With this torsion pendulum, it was possible to verify the validity of Newton's law of gravity down to micrometer distances. The diameter of the perforated disc is ten centimeters. The suspension is designed so that the upper part can rotate relative to the lower perforated disc around the axis of symmetry.

The photo shows the setup with which it was possible to confirm Newton's law of gravitation at distances of $55\,\mu\text{m} = 55 \cdot 10^{-6}\,\text{m}$. In the execution of the experiment, the upper perforated disc performs an oscillating rotating motion around the vertical axis. The temporal course of this rotation is measured—with a fascinating sensitivity. The apparatus even registers when raindrops on the trees outside the laboratory change the gravitational field.[2]

Further tests for such additional dimensions can be carried out at the LHC (↓). Moreover, it is even possible to employ the cooling rate of a neutron star after a supernova explosion in the search for extra dimensions. Its heat could also be carried by gravity into these additional dimensions, so that the star would cool faster than allowed by the Standard Model. This exciting idea is elaborated in the article on supernovae (↓).

[1] In fact, deviations from Newton's law of gravitation are observed, but these can be fully explained by general relativity.
[2] Experiment: E. G. Adelberger et al. 2009, Progress in Particle and Nuclear Physics, 62, 102

Additional Spatial Dimensions at the LHC → p. 306
Supernovae as Particle Physics Laboratories → p. 324

String Theory

It's All in a String?

String theory is a very promising but hypothetical theory for extending the Standard Model. However, it is regarded by many physicists as our best candidate for solving an ambitious fundamental quest: the unification of the three forces of the Standard Model with gravitation to form a *complete quantum theory*. The Standard Model is a quantum field theory in which the fundamental constituents, e.g., the electron and the photon, are point-like particles, each described by a separate quantum field. The particles are independent and have different quantum numbers, e.g., spin or electric charge.

String theory is instead based on the assumption that there is only a *single* kind of basic building block, namely a very very small (10^{-35} m) string. This string can be open or closed, see the figure below (top row). When these strings move, they sweep an open area or tube in spacetime. With a point particle it is always just a line.

The different particles, as we know them, are different fundamental vibrations of this one type of string. There are definitely many strings (just as there are many electrons), but the strings can be excited differently. If a string vibrates in a certain way, it is an electron; if it vibrates in a different way, it is a photon. This is an amazing unification: All particles are different manifestations of one and the same type of string. The equivalent of a Feynman diagram—here for a decay—is shown in the bottom row of the figure below for open and closed strings.

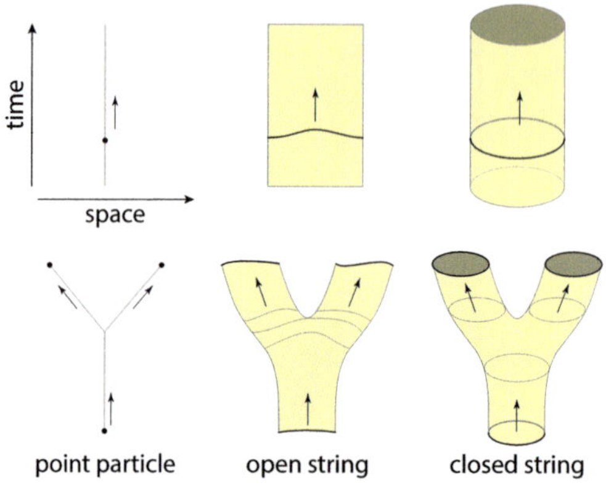

When formulating a quantized string theory some special features emerge, compared to a quantum field theory for point particles.

1. From the structure of the general theory of relativity we know that classical gravitational waves have a different polarization structure than electromagnetic waves. This polarization structure has recently been confirmed in the observation of gravitational waves by the LIGO experiment. In a quantum field theory this structure corresponds to a *spin-2* particle, not spin-1, like the gauge bosons in the Standard Model (see Multimessenger Astronomy ↓). This new exchange particle is called a *graviton*. Interestingly, the quantized string theory *always* automatically has exactly one massless spin-2 particle, which can be identified with the graviton. The graviton is a special excitation of the closed string. String theory is therefore also automatically a quantized theory of gravity. This makes it so extremely attractive! All other attempts to quantize gravity have so far failed.

Multimessenger Astronomy → p. 340

2. In string theory, there is only one energy or mass scale. This is the so-called *Planck energy scale* at around 10^{19} GeV. At this energy, quantum gravity effects become significant. This is an extremely high energy, very far beyond all current experiments. However, it has been proven that the general theory of relativity and quantum field theory are consistently obtained as low-energy limits of string theory. Just as general relativity is compatible with Newton's gravity in the limit of flat spacetime, string theory is compatible with quantum field theory and general relativity at low energies compared to the Planck scale. This is a high hurdle that new theories have to take (↓). As an analogy consider looking at a one meter long fluorescent tube. Up close it is a rod, from afar (about 2 km) it appears as a point-shaped light source.

According to quantum theory, the Planck energy scale corresponds to the Planck length scale of about 10^{-35} m. This is the length of the string. If two objects approach to within 10^{-35} m the gravitational force becomes as large as the strong force. Note also, a proton is 10^{20} times larger than the string length! The lower energy scales, such as the electroweak energy scale, are created by renormalizing the couplings (↓).

3. Quantized string theory must include supersymmetry, otherwise the theory does not include fermions. This means that there is a symmetry transformation on the string, which relates excitations with different spins. This is a further motivation for low-energy supersymmetry (↓).

4. The supersymmetric, quantized string theory can only be formulated consistently in 10 spacetime dimensions. At the energy scales that we use in everyday life and in the laboratory we only experience 4 spacetime dimensions. The remaining 6 dimensions must therefore be, as we say, *compactified* (Extra Dimensions ↓). As an example, you can look at the bridge cable of the Bonn North Bridge with ants on it, see the illustration. From a distance, the cable looks like a one-dimensional string. If you get very close, you can see an ant, that can move in *two* directions. One very long extended direction along the cable and a short one perpendicular to it. The latter forms a closed circle around the cable, so that the ant reaches its starting point again after a short time. This second direction is called *compactified*. It only has a finite extension. This must be done with 6 of the 10 dimensions in string theory.

For an ant, a bridge cable has two dimensions.

String theory is a fascinating conceptual extension of quantum field theory, which can unify all four forces in Nature. Unfortunately, it is not clear if and when we will be able to test it experimentally.

The Hurdles of Success → p. 266
Renormalization → p. 108
Spontaneous Symmetry Breaking → p. 62
Supersymmetry → p. 278
Extra Dimensions → p. 286

The Strong *CP* Problem

How Theorists Wash Away Problems

The point reflection P (parity transformation) and the charge transformation C, in which each particle is replaced by its antiparticle are important symmetries (↓) for particle physics. C, P and the combined symmetry CP are only violated in the Standard Model by the weak interaction (↓). This is surprising, because due to its complex structure, QCD should also violate the CP symmetry. The strength of the associated CP-violating interaction is parameterized by the variable Θ—the corresponding interaction is called the Θ-term. A non-vanishing Θ-term would necessarily lead to a non-vanishing electric dipole moment (↓) of the neutron and this has not been found, despite extremely precise measurements. Thus we know that the strong interaction violates the CP symmetry very weakly at most. The parameter Θ behaves like an angle and can therefore take any value between 0 and 360 degrees, but experimentally the value of Θ is determined to be less than 10^{-10} degrees. Thus, although the strong interaction is allowed to violate the CP symmetry, it apparently does not make use of this. However, we do not know why—this is referred to as the *strong CP problem*.

This problem could be solved by identifying a natural reason why the value of Θ is so small. A mechanism that accomplishes this is most likely to be found outside the Standard Model. A very popular solution is the *Peccei-Quinn mechanism*, proposed by Helen Quinn (Origins of Particle Physics ↓) and Roberto Peccei. It involves postulating an additional scalar field and an additional symmetry of QCD. The construction works as follows: The scalar field creates a potential in the shape of a sombrero. This leads to a spontaneous symmetry breaking (↓), which is illustrated by a ball rolling from the center of the Mexican hat to the edge and coming to rest there at some arbitrary location—as mentioned above, at any angle between 0 and 360 degrees. The procedure is associated with the appearance of a new, initially massless particle (similar to the mesons π, η and K in the strong interaction), which we shall call A. The whole thing is made so that this additional particle couples to gluons in the same way as the Θ-term—Θ and A always appear together.

So far, we have not gained anything: There can still be any amount of CP violation. However, Peccei and Quinn pointed out that the complex structure of QCD leads to the mentioned new symmetry being explicitly broken in addition to the spontaneous symmetry breaking. In the sombrero picture, this corresponds to a tilt of the hat. According to the construction, the direction of tilt is exactly so that the effect of Θ is always

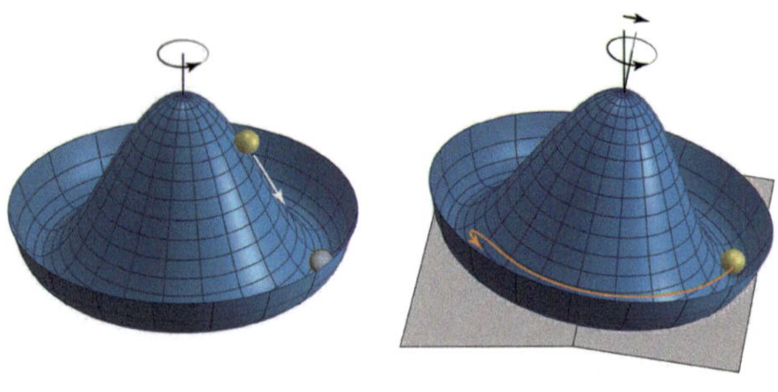

The Mexican hat potential

Symmetries → p. 60
Discrete Symmetries: Violating *P* and *CP* → p. 190 and *CP* Violation in Beauty → p. 258
The Search for Electric Dipole Moments → p. 318
Origins of Particle Physics → p. 10
Spontaneous Symmetry Breaking → p. 62

exactly compensated by the new particle A, so that there is no net *CP* violation. Thus, a constant part of the A field is fixed so that it exactly compensates the effect of Θ. But excitations of A are still possible compared to this ground state. These excitations correspond to a particle, which we call *axion*. The axion acquires a mass through the mentioned explicit symmetry breaking.

The ship swing always hangs straight, no matter how the ground is inclined.

One can imagine the mechanism as a ship swinging symmetrically around the suspension point. Without gravity, it could point in any direction—all angles are equal, there is a symmetry. This is no longer the case when we turn gravity on, which breaks this symmetry. Then the swing always hangs downwards. And this is true even when the ground is inclined. The inclined ground corresponds to $\Theta \neq 0$, which would result in strong *CP* violation. However what matters is the angle of the swing with respect to the vertical direction. This is always zero, due to gravity.

This would solve the strong *CP* problem! But the solution has its price in the form of a new particle, the axion. By the way, this can also be seen in our analogy, because the ship swing can oscillate around its rest position. Oscillations around the rest position correspond to the axion. If one of the two people in the swing moves, the swing will oscillate. The person has transferred some of their kinetic energy to the swing. In particle physics, one would say they have emitted an axion. The second person will therefore also wobble. Through the exchange of an axion—through the oscillation of the swing—the two people have exchanged energy.

We know that, for example, neutral pions can decay into two photons. Since axions (if they exist) owe their existence to a spontaneous symmetry breaking, like pions, one should expect axions to do the same. This is an experimentally verifiable consequence of the proposed solution to the strong *CP* problem.

The search for axions (↓) thus tries to clarify whether the *CP* conservation of the strong interaction has a deeper reason or is just a coincidence. The motivation for this search is considerably strengthened by the fact that the properties of axions make them a perfect candidate for dark matter (↓). Axions that were created in the early Universe could thus make up the majority of the matter in the Universe.

Being able to solve two independent problems at once makes axions very popular particles among physicists. Nobel laureate Frank Wilczek named the axions after a commercial detergent, as the particle can wash away two problems so nicely.

The Search for Axions → p. 322
Dark Matter → p. 282

Did Our Life Emerge by Chance?

Small Changes—Huge Effects

Based on Darwin's theory of evolution, it is often discussed whether the emergence of life through random mutations is realistic. After all, the formation of complex molecules necessary for this is already extremely unlikely—just like winning a lottery is unlikely. However, there are many planets in the Universe, and just by chance some of them might have won the "lottery of life", just as there are some lottery winners amongst a large number of players. So it is at least understandable that life could have arisen on Earth by chance (and perhaps not only on Earth—who knows?). But the probability referred to in the title of this article is a different one.

The chance mentioned there refers to the part of this question that touches on particle physics: How likely is the emergence of life or rather of conditions in which life can arise, assuming that the parameters of the Standard Model have random values? Or put differently: How *special* is our Standard Model, with its parameters taking exactly the measured values? This is indeed a very relevant question. If, for example, the masses of the light quarks were only a few percent different, there would be insufficient production of carbon and/or oxygen in stars, and life as we know it would be impossible in the Universe (Birthplace of Nuclei ↓). The Earth would then probably not be the wonderful place it is, but rather an unfriendly, inhospitable rock. Another example from this category is dark energy (↓)—the mechanism that drives the Universe apart that is not at all understood up to date. If this had dominated the dynamics in the Universe earlier, there might not have been enough time to form structures like galaxies and planetary systems.

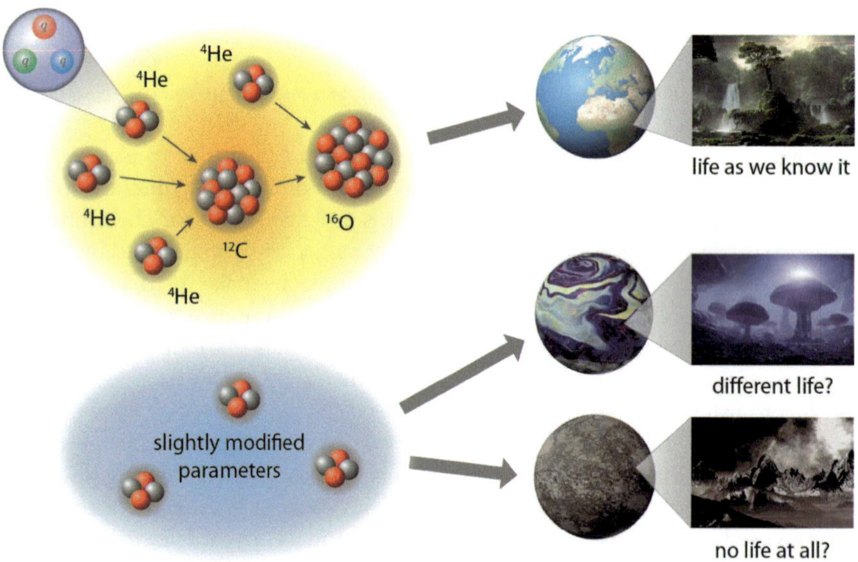

Illustration of the influence of carbon production in stars on life on Earth

The Birthplace of Nuclei → p. 342
Dark Energy → p. 284

The *anthropic principle* calls for a different view on the issues described above. It essentially states that the parameters of the fundamental interactions must be so that life can arise such that someone (us humans!) can inquire about the origin of life. From this, one can deduce that arguing with a too low probability for the emergence of life is not meaningful, since it clearly exists, for otherwise there would be no one to argue—independent of how low the probability is. One would also never tell a person, who won in a lottery, that chances for that are too low to ever occur.

This spiral galaxy is hundreds of millions of light years away from us and is just one of countless galaxies in the Universe. If the dark energy had been stronger in the early Universe, structures like this would probably not have been able to form. And we humans would not exist either.

An important research focus of modern particle physics is to find evidence for a more fundamental theory "hidden" behind the Standard Model. Ideally, this should solve at least some of the problems of the Standard Model (Open Questions ↓) and also for example provide insights as to why the quark masses take exactly the values they have and not slightly different ones.

In light of this discussion, the question that titles this article can be reformulated: Can we understand why the parameters of the Standard Model took exactly the values that allow for life?[1] This discussion takes on a special flavor in light of the admittedly rather weak experimental evidence[2] that the fundamental parameters of physics could be location- or time-dependent. Should this be confirmed, it would revolutionize the way we think about the Universe. The question about the values of the parameters would then no longer be "Why do the parameters have the values that they have?", but rather "Life originated in the zone and at the time in the Universe, where the fundamental parameters of the Standard Model were within the appropriate value ranges." This simultaneously raises the exciting question of what the other areas of the Universe might look like. An alternative, theoretically conceivable explanation that does not require fine-tuning of the parameters in our Universe would be *multiverses*. In this scenario, a large number of universes exist in parallel, shaped by different sets of fundamental parameters. Then we would simply live in the Universe that allows for our life. Unfortunately, it is not clear whether it is even possible to ever test this theory or rather hypothesis.

At the present time, we do not know the answer to the central question of this section—and also not whether we will ever find it. However, this should not prevent us from searching for it.

---

[1] For a readable treatise, see R. N. Cahn 1996, *The eighteen arbitrary parameters of the standard model in your everyday life*, Reviews of Modern Physics, 68, 951, doi:10.1103/RevModPhys.68.951

[2] e.g. J. K. Webb et al. 2011, *Indications of a spatial variation of the fine structure constant*, Physical Review Letters, 107, 191101, doi:10.1103/PhysRevLett.107.191101

Open Questions → p. 270

11 Searches for Physics Beyond the Standard Model

Although the Standard Model has been remarkably successful in enhancing our understanding of the elementary structure of matter and in accurately describing all observed particle physics processes to date, as noted in the previous chapter, there remain a number of unanswered questions. For example, the Standard Model does not explain why we observe only matter in the Universe but no or nearly no antimatter. Furthermore, the particles of the Standard Model are not sufficient to build up all the components of the Universe. To explain the nature of dark matter, we need at least one further particle beyond those of the Standard Model.

Extensions of the theory designed to address these unresolved questions must be experimentally testable. It is an exciting challenge to search for observable effects that could confirm these well-motivated theoretical ideas.

In this chapter, we explore several experimental searches for particles, decays or other effects that arise from extensions of the Standard Model. If any of these experiments were to detect the predicted effects, it would provide a significant indication of the direction in which the theory should be further developed.

Searching for a New Experiment

The Need for "Fictional Science"

In this book, we have so far discussed a wide variety of particle physics experimental facilities, covering more than 100 years of history, and including projects which range from fitting in a small laboratory room (Precision Experiments ↓), to the largest scientific facility of the world, the Large Hadron Collider, LHC (↓). Searching for new phenomena and particles beyond our current experimental knowledge requires the constant technological and scientific evolution of these facilities. Maintaining the technological and methodical variety of these projects is important, because it is their interplay—the way they tackle scientific questions from different views—which allows us to discover, constrain or exclude physical theories from all angles, and most importantly, which allows us to leave room for the unexpected.

Some of these projects can be funded by a single university group, but many require the collaboration of researchers from several universities or even countries. The fact that there is more than one on a similar topic at the same time ensures the required diversity of technologies, methods and sensitivities. The biggest facilities, such as the LHC, are truly international, where many thousands of scientists from all continents contribute to only one project at a time. In every case, it is important to take into account the scientific interest of all contributors based on a very well studied *physics case*: The *promised* groundbreaking physics results the project will deliver if the hitherto known physics prevails (in our current case: the Standard Model), and the *possible* physics results if yet unexcluded alternatives to the Standard Model were true.

As an example, consider a new facility designed to measure why matter and antimatter behave differently, so-called *CP* violation (↓). Such a facility requires a physics case that explains its experimental goals and capabilities. First, it must show how the facility can measure the Standard Model prediction for *CP* violation with much greater precision than ever before. Second, it should address why the Standard Model, perplexingly, does not appear to exhibit enough *CP* violation, contrary to cosmological expectations (Baryogenesis ↓). These goals represent the *promised* aspect of the physics case, as they are grounded entirely in prior ex-

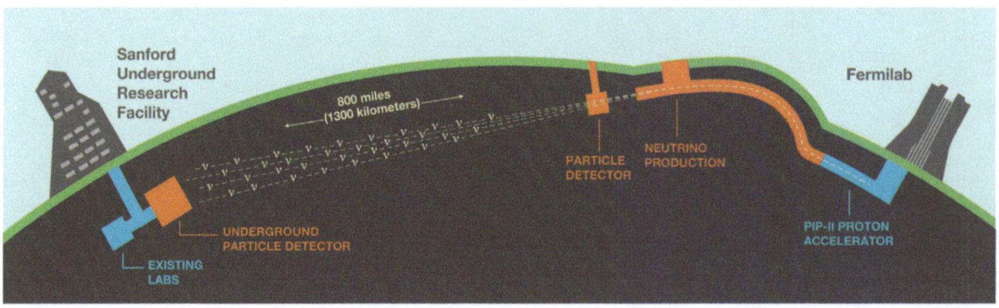

A sketch of the DUNE experiment. Its physics case has been carefully studied and we know how precisely it could discover possible *CP* violation in neutrino oscillations, if such *CP* violation exists.

Precision Experiments → p. 250
The LHC Accelerator Complex at CERN → p. 116
CP Violation in Beauty → p. 258
Baryogenesis → p. 328
Is It Really the Higgs Boson? → p. 248

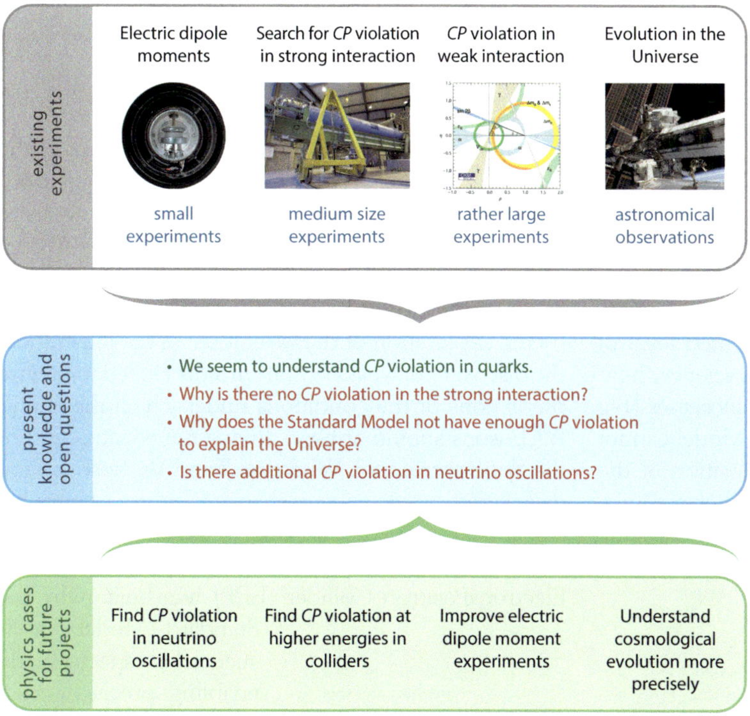

How small and large experiments with many different experimental approaches come together to understand *CP* violation—or rather, highlight what we do not understand yet about its origin and its role in the history of the Universe, and what follows from that for the next generation of experiments. All the experiments in the upper row are explained in separate articles in this book.

perimental knowledge. Finally, the physics case should explore whether additional, experimentally previously unknown sources of *CP* violation might exist, since there is a good theoretical motivation for *CP* violation beyond the Standard Model. This represents the *possible* aspect of the physics case, which extends beyond existing knowledge, and which can be described as a productive form of ***Fictional Science***: Science that we could do in the future if Nature behaves in ways we envisage (see, e.g., Is it really the Higgs Boson? ↓). This interplay of current and possible future experiments, and of theoretical expectations and measurements, is shown in the figure.

It is an interesting feature of the diverse combination of experiments shown here that small experiments can be as sensitive a probe of a given phenomenon as big experiments. For instance, in our example *CP* violation is very strongly constrained by the measurement of the electric dipole moment of particles, e.g., the neutron (↓). For many possible theories predicting more *CP* violation, as required by cosmology, the constraint from small experiments is as big of a hurdle as the measurements from big experiments (CKM Triangle ↓). However, when it comes to investigating the underlying mechanisms behind a given phenomenon, often it turns out that large experiments are necessary. Such accelerators often operate at high energies that allow for *direct* tests of the mechanisms of how fundamental particles interact, e.g., via new heavy particles (↓). Therefore, the particle physics community is currently discussing a future large new facility. It is thought to be indispensable for the progress of particle physics, even in a very diverse field of exceptionally creative smaller experiments. Both approaches complement each other and have the potential to discover yet unknown particles or phenomena.

The Search for Electric Dipole Moments → p. 318
Measuring the CKM Triangle → p. 264
The Case for a Future Higgs Factory → p. 298

The Case for a Future Higgs Factory

How to Leave no Loop Holes for the Standard Model

Over the last 50 years, particle physics has made extraordinary progress with the Standard Model. In principle, it describes all known properties of microscopic objects at all experimentally accessible energies—from the atomic binding energies underlying chemistry and biology to the highest collision energies at the LHC and in cosmic rays (↓). This unprecedented level of precision and apparent completeness in science, however, leaves us puzzled: despite its great successes (Big Bang Nucleosynthesis ↓), the Standard Model cannot fully account for the history and composition of the Universe and has significant conceptual shortcomings. Tackling these challenges requires new experiments, any of which might lead to a revolution by finally refuting the Standard Model (↓).

To strategize our best attempt at finding flaws in the Standard Model, we can draw valuable insights from particle physics' history. Progress often came from studying a specific property of the Standard Model, which pointed to where the next breakthrough could occur—either to confirm or refute it. This is exemplified by the exploration of the weak force in the illustration: the discovery of weak neutral currents (↓) led to precise predictions of the conditions under which the Z and $W^{\pm}$ bosons should appear, driving the construction of the $Sp\bar{p}S$ accelerator. When the Z and $W^{\pm}$ bosons were discovered (↓), the focus changed: either a Higgs boson and a top quark must exist or a major scientific upheaval was imminent. To confirm this, CERN's Large Electron-Positron Collider (LEP) was built (why the Standard Model ↓). It succeeded spectacularly, enabling precision measurements of the Z and $W^{\pm}$ bosons that revealed the predicted properties of these missing Standard Model components.

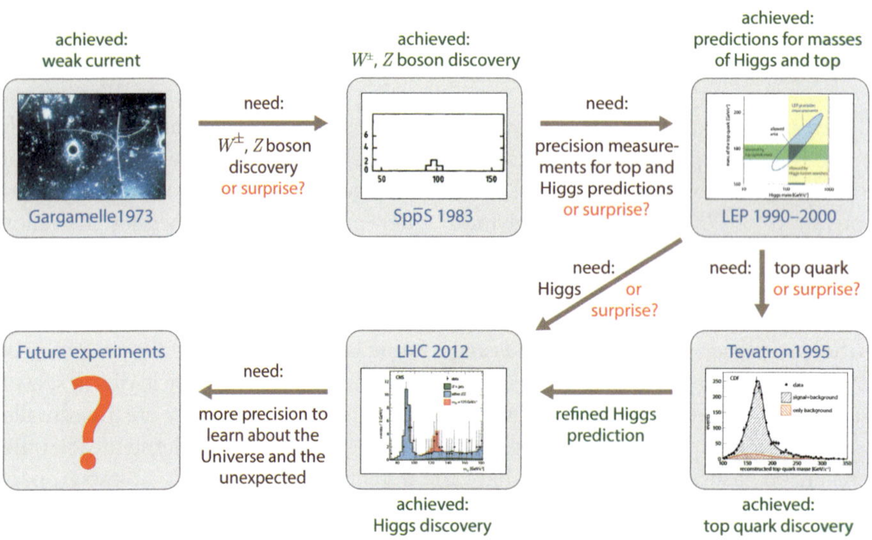

The progression of flagship projects of particle physics in the last 50 years, which cemented the electroweak Standard Model as the explanation of the microcosm. All the experiments are explained in separate articles in Chapter 9 of this book.

The Hurdles of Success → p. 266, Cosmic Rays → p. 336
Nucleosynthesis After the Big Bang → p. 330
Open Questions → p. 270
The Discovery of Neutral Current Interactions → p. 236
The Discovery of the W and Z Bosons → p. 240

With this, the stage was set to construct two new accelerators to discover the top quark (↓) and the Higgs boson (↓): the Tevatron and the LHC. After their respective discoveries, attention shifted to studying their properties. So far they have behaved as predicted.

To make progress beyond the current precision, the global particle physics community is now considering the next major step: a *Higgs factory*. This would involve electron-positron collisions at energies up to 350 GeV or even 500 GeV, providing a clean experimental environment for unprecedented precision studies. Concepts for its realization are, e.g., the Chinese Electron Positron Collider CEPC, the Future Circular Collider FCC at CERN, or the International Linear Collider ILC. Such a facility would not only probe the Higgs boson in much greater detail but also enable more precise measurements of the properties of other heavy Standard Model particles (Z, W, and top), going well beyond the precision of the LHC. The Higgs boson, with its central role in the Standard Model, holds the key to many unanswered questions, including the origins of masses, the flavor structure of particles, the matter-antimatter asymmetry (Baryogenesis ↓), disparate energy scales, and the cosmological constant.

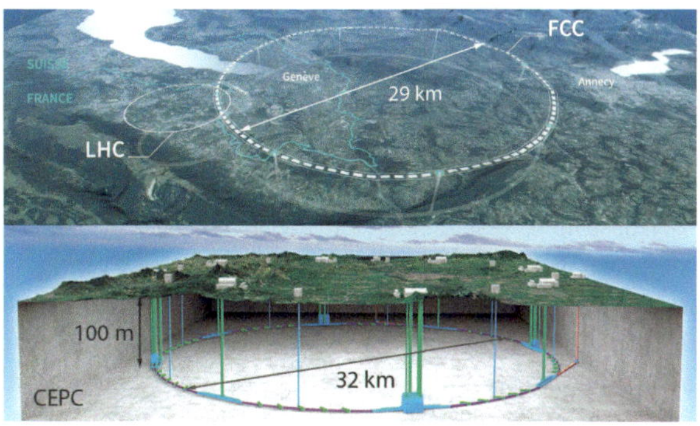

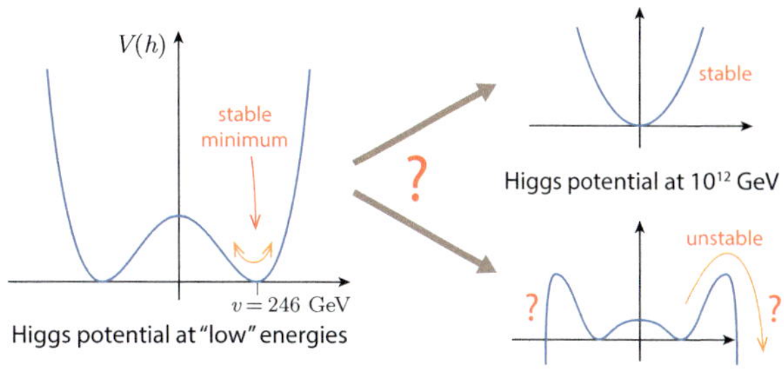

The two different ways how the Higgs potential could evolve when the energy is increased from those currently prevailing in the Universe (left) to much higher energies (right).

One particularly interesting open question for this machine revolves around the shape of the Higgs potential under extreme conditions. At the highest energies, the spontaneous symmetry breaking (↓) could simply disappear (following the upper path in the left illustration). In this case, baryogenesis—an essential process in cosmology—would not be possible within the Standard Model. Knowledge of the precise shape of the Higgs potential at our laboratory energies will provide insights into what happened in the early Universe at much higher energies when matter and antimatter were formed. Moreover, the precision of current experiments suggests that at energies above 10^{12} GeV, the Standard Model predicts the collapse of the Higgs potential, shown as the lower path in the graph. Whether this actually is a problem or not depends on the probability of the Universe falling out of the present state and spontaneously transforming into something else! We definitely would want to know that! In summary, the physics case of a dedicated e^+e^- Higgs factory is compelling: After all, we all want to understand how it is possible that we exist at all!

Two examples for a possible future accelerator project to create a "Higgs factory": the FCCee at CERN and the CEPC in China.

Discovery and Properties of the Top Quark → p. 242
The Discovery of the Higgs Boson → p. 246
Baryogenesis → p. 328
Spontaneous Symmetry Breaking I → p. 184

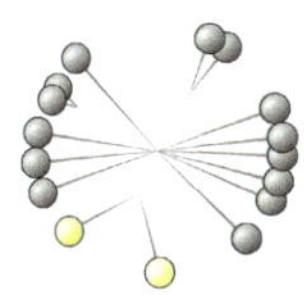

Is the Neutrino Its Own Antiparticle?

The Search for Neutrinoless Double Beta Decay

Why is there matter in the Universe, but hardly any antimatter (Baryogenesis ↓)? If the neutrino (↓) were its own antiparticle, it could contribute a lot to answering this question. One way to test this is to investigate double beta decays. These are two simultaneous beta decays of a single atomic nucleus: $2\,n \rightarrow 2\,p + 2\,e^- + 2\,\bar{\nu}_e$. The decay was predicted to exist by Maria Goeppert Mayer in 1935 and first detected in 1986.

The LEGEND experiment is surrounded by a water Cherenkov detector to detect penetrating cosmic rays. In the center, the germanium detectors are located inside a liquid argon tank.

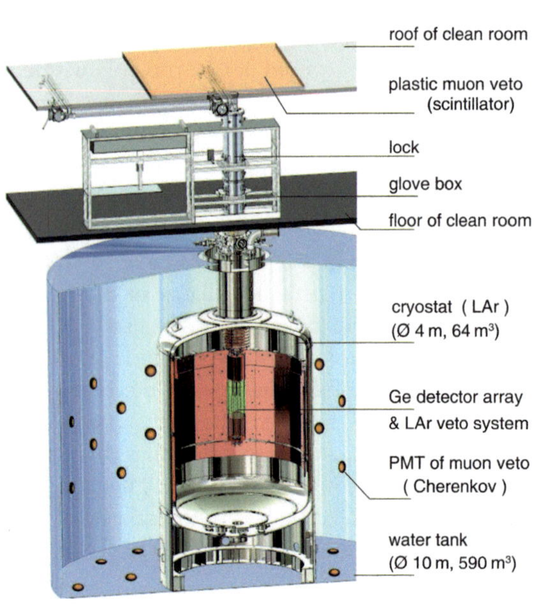

Double beta decays are extremely rare, but have all the same been observed in a whole series of nuclei despite the very long half-lives. Two electrons and two antineutrinos are emitted during the decay. If the neutrino were its own antiparticle, the antineutrino emitted in one of the decays could be absorbed as a neutrino in the other decay and thus lead to a neutrinoless double beta decay ($0\nu\beta\beta$—where (0ν) indicates that zero neutrinos occur and ($\beta\beta$) the emission of two β particles, i.e., electrons).

Worldwide there are many searches for $0\nu\beta\beta$ decay. Detectors have been constructed from materials that contain one of the following isotopes: $^{76}_{32}$Ge, $^{130}_{52}$Te, $^{136}_{54}$Xe, and $^{150}_{60}$Nd. The two electrons emitted during the decay deposit their kinetic energy in the detector. In the $\beta\beta$ decay *with neutrinos* the energy can be distributed over the electrons and neutrinos in various ways. In contrast, in the $0\nu\beta\beta$ decay the electrons are emitted at a very specific energy due to energy and momentum conservation.

Various techniques are used to meet the high demands of the search for such an extremely rare decay: semiconductors, various types of scintillators, gas counters and cryogenic calorimeters. So far, nothing has been found. However, astonishing progress has been made in increasing the sensitivity of the search. It is currently possible to test lifetimes up to around 10^{27} years. To observe even one decay event on average per year, you need 10^{27} atomic nuclei, which corresponds to about 100 kilograms of the corresponding isotope. As the isotopes are generally not very common, the material

Baryogenesis → p. 328
Neutrinos → p. 196

Low temperature detectors from TeO_2 of the CUORE experiment in the Gran Sasso underground laboratory.

View into the water Cherenkov detector of the LEGEND experiment. In the center the tank with the germanium detectors.

must be enriched. Natural germanium consists of five isotopes and only 7.8 % of it is $^{76}_{32}$Ge. In order to produce detectors to search for the $0\nu\beta\beta$ decay, this isotope is first enriched to 86 % abundance in a complex process. Once a sufficient amount of detector material is available, the background radioactivity from the surrounding material must be reduced to such an extent that a possible $0\nu\beta\beta$ event does not go undetected in a flood of disturbing signals from α, β or γ decays in the surroundings. In addition, cosmic radiation can also lead to interfering signals.

This is why all experiments are set up in underground laboratories to shield them from cosmic radiation. The LEGEND experiment is located in the Gran Sasso underground laboratory in Italy and is thus protected from cosmic radiation by 1.5 kilometers of rock. Nevertheless, cosmic radiation particles still reach the experiment too frequently. To detect and thus be able to reject these as signal events, the entire setup is located inside a large water Cherenkov detector (↓). The germanium detectors of the LEGEND experiment are immersed in liquid argon. The argon not only cools the detectors to the temperature of about 80 K (about minus 190 °C) required for operation, it is also used as a scintillation detector to detect radiation penetrating from outside, close to the $^{76}_{32}$Ge detectors. The Gran Sasso underground laboratory is also home to CUORE, the world's largest low-temperature detector system to date. For this purpose, 1000 detectors consisting of a total of one ton of tellurium oxide (TeO_2) were cooled down to 20 thousandths of a degree from absolute zero. Furthermore, the former detectors for measuring solar neutrinos, KamLAND in Japan and SNO in Canada, are filled with scintillators loaded with Xe or Te. The EXO experiment, currently being set up in an underground laboratory in New Mexico, USA, uses liquid xenon as a scintillator to search for $0\nu\beta\beta$ decays in $^{136}_{54}$Xe. If neutrinoless double beta decay exists, it should show up in several of these experiments.

For particle physics, this would be a great gain in several respects: it would show that the neutrino is its own antiparticle. This would be a violation of the matter-antimatter symmetry and thus a possible building block for explaining the dominance of matter over antimatter (↓). It might also explain the extremely small value of the neutrino mass (↓).

Cherenkov Radiation and Particle Identification → p. 82
Matter and Antimatter → p. 314
Neutrino Masses → p. 272

Searching for New Particles at the LHC I

Searching for the Specific

In the search for *New Physics* beyond the Standard Model, two interesting dilemmata arise: First, all resources are limited. It is not practically feasible to search for *everything* imaginable. Second, it is statistically not very sensitive to search for *arbitrary* deviations from known physics, because looking for deviations of the data from the background at any arbitrary point in an extremely large space of potential measurements usually only chases random fluctuations (Blind Analysis ↓).

Therefore, it may be sensible to approach the search for new particles or interactions with a certain "theoretical prejudice". Successful basic principles of constructing physical theories suggest simplicity, maximum symmetry, maximum predictive power in many physical areas, and the absence of "fine-tuning" (↓) as a good basis. Some researchers adhere to the "prejudice" that it would be reasonable to assume that Nature is *supersymmetric* (↓). Then there would be a number of precisely predictable *signatures*, i.e., properties of events that should be found at high-energy accelerators like the LHC. Figuratively, with such searches, particle physicists shine a spotlight on certain areas of Nature that have been in the dark so far. This makes the view sharper, the distraction by random deviations smaller, and the search more promising—if the theory is correct.

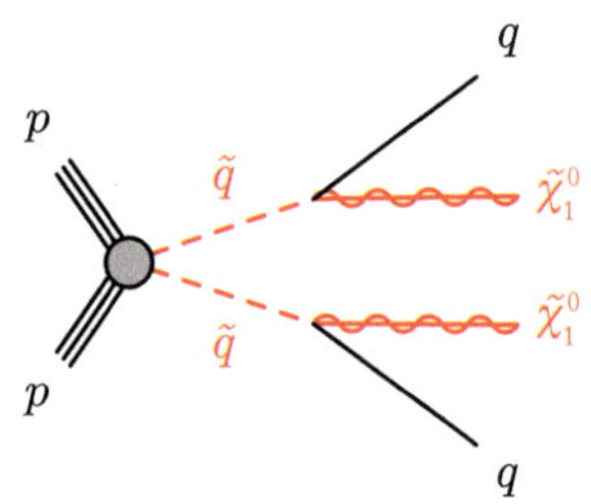

The possible production of supersymmetric particles. In this case, squarks $\tilde{q}$ and neutralinos $\tilde{\chi}_1^0$

An example of such a signature at the LHC is found in the Feynman diagram shown here.[1] The $\tilde{q}$ is a scalar quark (*squark*): the hypothetical supersymmetric partner of a quark with spin 0, with the same couplings as a Standard Model quark, but with significantly higher mass. The $\tilde{\chi}_1^0$ is the *neutralino*—the hypothetical supersymmetric partner particle of the photon and the Z boson with spin 1/2. It can be heavy, is electrically neutral, and is invisible for the LHC detectors, just as neutrinos. This particle is very interesting: It connects particle physics and cosmology, as it could constitute dark matter (↓). Measuring its properties could allow us to predict the cosmological development of the Universe much more accurately.

According to the Feynman diagram, two jets would be visible in the detector, generated by the quarks in the final state. In addition, a lot of missing transverse momentum is carried away by the uncharged and undetected neutralinos. The LHC is ideally suited to search for this process, as the scalar quarks interact with the strong interaction. They should be produced abundantly if they are kinematically accessible, and could be rather heavy—just right for the most powerful accelerator with strongly interacting particles in the initial state.

[1] ATLAS Collaboration 2021, *Search for squarks and gluinos in final states with jets and missing transverse momentum [...]*, Journal of High Energy Physics, 143, doi:10.1007/JHEP02(2021)143

What Is a "Blind Analysis"? → p. 166
Fine tuning: Did Our Life Emerge by Chance? → p. 292
Supersymmetry → p. 278

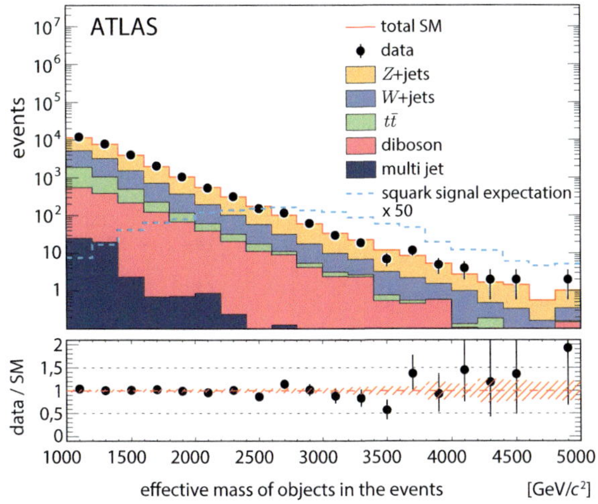

Observed event rates in the search for supersymmetry

The search for this signature is illustrated here using examples from the ATLAS experiment. The observed distribution of the *effective mass* considers a measure of the *invariant mass* (↓) calculated from the momenta of the observed jets and the *missing transverse momentum*. The idea of using this quantity is based on the fact that supersymmetry preserves fundamental properties of the Standard Model, such as energy and momentum conservation (Noether's Theorem ↓). The particles in the initial state (the protons in the accelerator) collide exactly in line with each other. Thus their momentum in the initial state adds up to exactly zero to the right and left or above and below. In particular the entire system in the initial state has no momentum *transverse* to the axis of the proton beams. Due to momentum conservation, the final state must also carry the total momentum of zero in the direction up/down or right/left. If the two observed jets are directed to one side of the detector, then something else (the missing transverse momentum of invisible particles) must balance this momentum in the opposite direction.

Again, the results show a familiar picture for particle physicists: The data agree well with the expectation of the Standard Model, and everything can be understood without new supersymmetric particles. Many possible properties of the scalar quarks and neutralinos could already be excluded in this way (↓). They would have been visible in the data if they existed at light masses. The physicists continue to search. Guided by the idea of supersymmetry or other good ideas, they repeatedly direct the spotlight of the experiments and analyses into the unknown, illuminating a piece of Nature that has never been investigated by a human before.

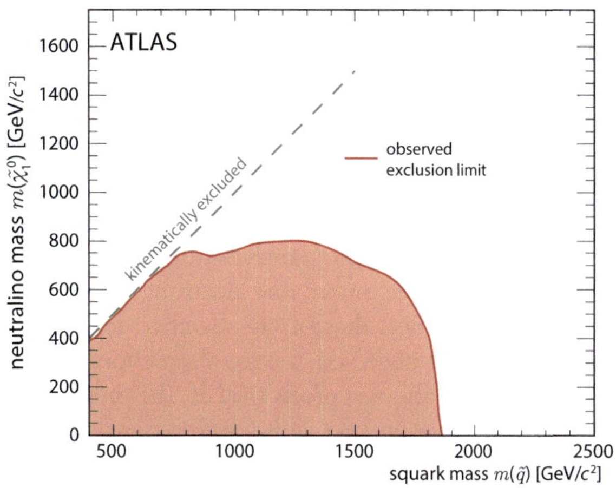

In the range of the masses of the SUSY particles below the red line, supersymmetry should have been discovered at the LHC if the particles with these parameters were present in Nature.

Dark Matter → p. 282
Identifying the Particle Type via the Mass → p. 156
Noether's Theorem → p. 100
Publishing a Non-Observation → p. 162

Searching for New Particles at the LHC II

Searching for the Unknown

In the previous article, the *spotlight theory* of the search for *New Physics* was explained. However there is an obvious criticism: If there is a really clear deviation between measured values and the prediction of the Standard Model that no one has thought of yet, then it is quite possible that no one ever looked for it and no one has ever noticed. This might even occur if the deviation was actually easily measurable. This tragic miss would have happened simply because no one has directed the spotlight there.

To avoid this problem, there are also *model-independent* searches for new physics—jokingly dubbed "MUSiC" ("Model UnSpeciC search") by the CMS collaboration.[1] Even there, not everything is considered. Since many well-known Standard Model processes have large reaction rates, small contributions of unknown origin can easily hide here.

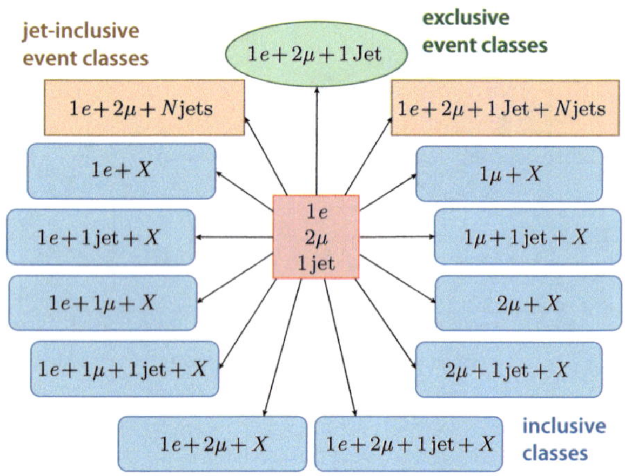

Classification of events

Therefore, it is necessary to make *minimal* assumptions about the unknown signatures before conducting a search. In the example shown here, events that contain at least one muon, one electron, one jet, or a certain combination thereof are searched for. If only these objects are observed, the reconstruction is called *exclusive*, e.g., the top green oval in the graphics. If they occur in connection with missing transverse momentum (Searching for new Particles at LHC ↓) or are accompanied by any other object X, the reconstruction is called *inclusive*, as shown in the blue boxes. The graphic shows some of the categories examined. Of course, this also involves a preselection in terms of detectable phenomena. However, this time the preselection does not come from a prejudice about promising theories but follows the "Here is light, here I can search under the lamppost" principle: We can measure electrons, muons, and jets well, so let's look there no matter whether a theory tells us to looks there or not.

In each of the categories, the number of events expected from the simulation of the Standard Model—or from data driven background estimates—and the number of events observed can be counted. The second graphic shows good agreement within the measurement accuracy, i.e., the measurements can be explained without new physics. From a multitude of event classes, only those with the most significant deviations of the data

[1] CMS Collaboration et al. 2021, *MUSiC: a model-unspecific search for new physics in proton–proton collisions at $\sqrt{s}$=13 TeV*, The European Physical Journal C, 81, 629, doi:10.1140/epjc/s10052-021-09236-z

Searching for New Particles at the LHC I → p. 302

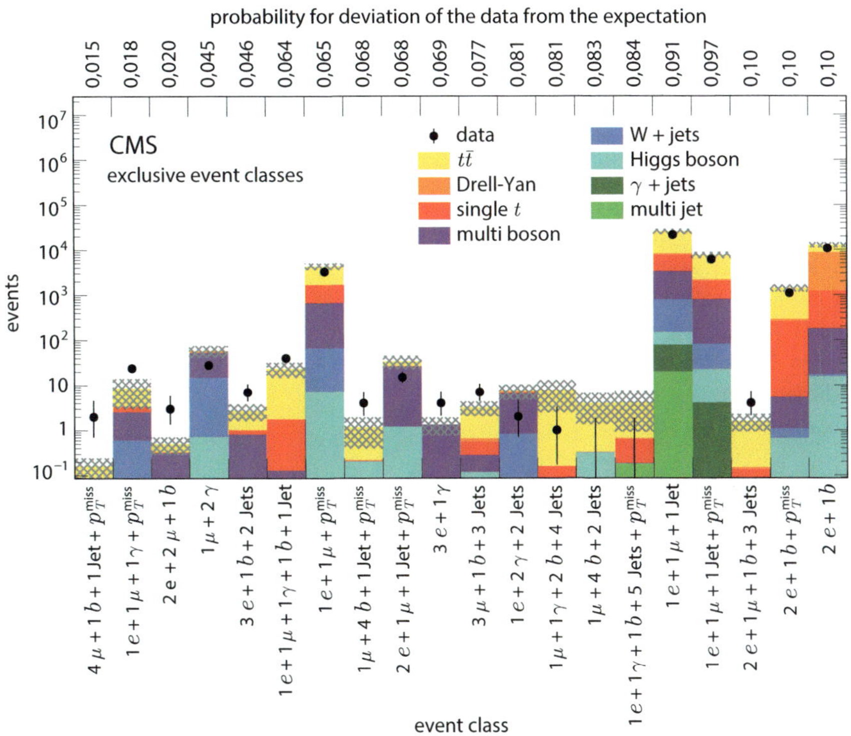

Measured (dots) and expected number of events from the Standard Model (colored bars) in the different event classes. Only the classes with the largest observed deviations are shown.

from the expectation are shown, sorted in descending order of significance. The strongest deviation, far left in the graphic, has a probability of 1.5 % that such data are observed if only the Standard Model is true and nothing else. With hundreds of event classes considered, however, it is not unexpected that at least one event class shows such a discrepancy.

Within each category, the kinematic distribution with the strongest deviation between data and background estimate is also searched for, for example, in a distribution of the combined mass (↓) of different particles. However, then the problem described in the article "What is a 'Blind Analysis'?" (↓) inevitably arises: Somewhere within a wide distribution, there is always a seemingly significant random variation if we only look often enough at different places. Fortunately, if you look at a *known* number of different places, it can be simulated how likely a random deviation between data and a hypothesis is at any point. The distribution of the significance of different regions should then again match the expectation if there is no new physics to observe. In our example, this is again the case; no unexpected new physics has been found so far.

Identifying the Particle Type via the Mass → p. 156
What Is a "Blind Analysis"? → p. 166

Additional Spatial Dimensions at the LHC

Are There Invisible Parts of Gravity?

There are many good reasons why our world may not consist of only three spatial dimensions. If there were more independent spatial dimensions than the three known ones—up/down, forward/backward, and left/right—they would have to be hidden from experiments to this date. For example, they could be rolled up and thus very small in extent (String Theory ↓) or it could be that only the weakest known force, gravity, can spread into these additional dimensions (↓) and the forces of the Standard Model cannot.

In both cases, gravity would not be weak. It would only appear weak because it largely disappears into a part of the multidimensional spacetime that is unreachable for us humans, as we are organisms that are subject to the Standard Model forces. This could mean that gravity, through its exchange particle, the graviton G, is not so weak in particle physics experiments that it can be completely neglected. At very high energies (gravity interacts more strongly the higher the energy), gravitons could interact as often as neutrinos. That is, a gravity-induced interaction could occur in our detectors at very high energies in a similar way as the weak interaction occurs. The graviton then escapes the detector just in the same way as the weakly interacting neutrinos do.

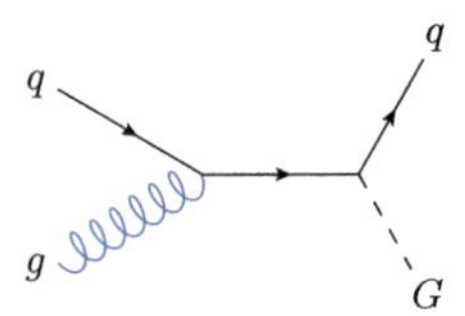

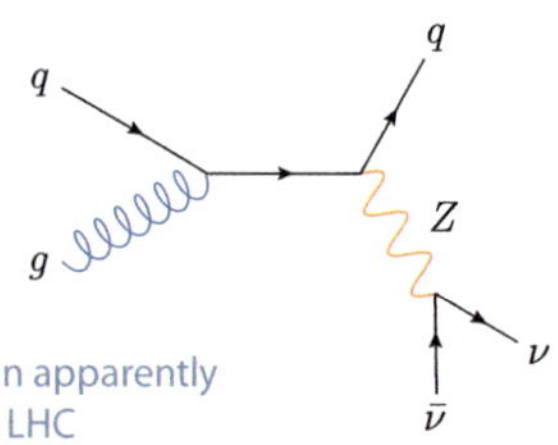

Graviton production (left) and an apparently similar SM process (right) at the LHC

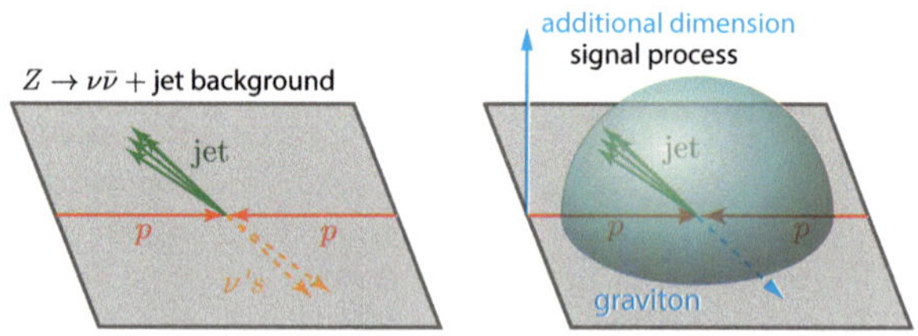

The Standard Model exists in a three-dimensional world—shown on the left as a plane. All Standard Model processes occur figuratively in this plane. A graviton (on the right) could couple much more strongly to known matter in a model with additional dimensions than in the Standard Model, thus interacting at high energies about as often as neutrinos.

This would result in spectacular signatures,[1] in which a single very high-energy jet (*monojet*) is visible in the detector and nothing else—just like with neutrinos, but relatively more frequent at high energies than at low ones.

To search for gravitons, events with high missing transverse momentum and a high-energy monojet are examined (Data Analysis ↓). A good measure is the momentum of the jet projected onto the transverse plane p_T—perpendicular to the beam direction. In the histogram of count rates, one can recognize the background processes of the Standard Model, led by the production of a quark (which is visible as a monojet) and a Z boson, which "repels" against it. The latter then decays into two neutrinos (↓) invisible to the detector. A single event of this kind cannot be unequivocally attributed to the Standard Model or a graviton. What clearly differs, however, is the shape of the distribution of events

String Theory → p. 288
Extra Dimensions → p. 286
Data Analysis → p. 152
Neutrinos → p. 196

A recorded monojet event in the ATLAS detector

in the observable p_T. The magenta dashed curve of the signal drops off more slowly than the Standard Model processes because gravity couples to energy, i.e., the coupling becomes stronger as the energy of the event increases. If the data follow the shape of the Standard Model, then no graviton from an extra-dimension model has contributed to the recorded events.

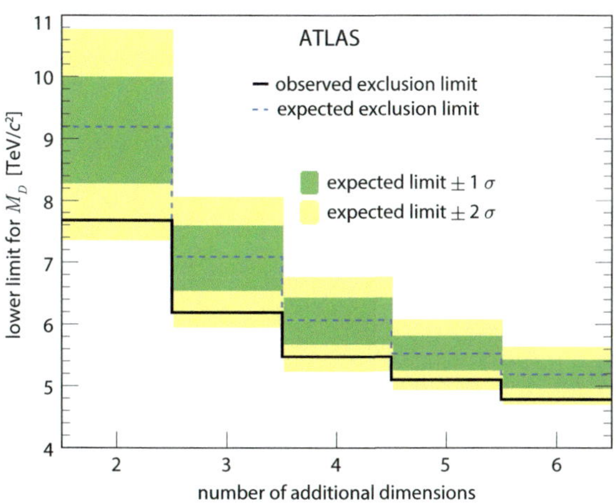

Exclusion limits on the effective mass scale M_D where gravity becomes as strong as the other forces. With two additional dimensions, M_D must be larger than 7.6 TeV.

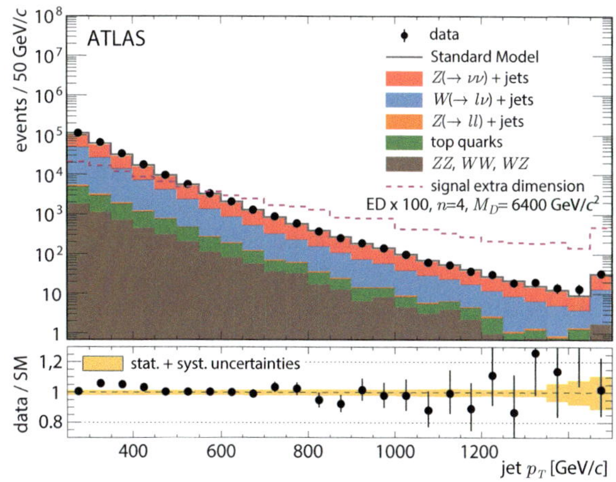

The distribution of the momenta of the monojets agrees with the Standard Model and not with the signal of additional dimensions (magenta dots).

The measured data points agree—once again—very well with the Standard Model expectation. There is no sign of a slower drop-off of the data points than expected by the Standard Model. Therefore, it can be ruled out that additional dimensions exist, unless they meet certain conditions. The resulting exclusion limits, for example, prohibit an effective mass scale M_D, where gravity becomes as strong as the other forces, of about $7.6 \cdot 10^3$ GeV or less, under the condition that there are two additional dimensions (Open Questions ↓). If there were three additional dimensions, the plot says that this energy scale where gravity becomes strong must be higher than about $6.2 \cdot 10^3$ GeV. In a simple model with rolled-up additional dimensions, this corresponds to a diameter of the rolled-up dimension of only about one hundredth of a millimeter or less! With two additional dimensions, gravity would then be a maximum of about $10^{19}/10^3 = 10^{16}$ times stronger than in a three-dimensional world!

[1] ATLAS Collaboration et al. 2018, *Search for dark matter and other new phenomena in events with an energetic jet and large missing transverse momentum using the ATLAS detector,* Journal of High Energy Physics, 126, doi:10.1007/JHEP01(2018)126

Open Questions → p. 270

The Mystery of Dark Matter

How Do We Know What to Look For?

One of the greatest challenges facing modern particle physics is the unresolved question of what most of the matter in the Universe consists of, which we refer to as *dark matter* (↓). As successful as the Standard Model is in explaining all particle reactions observed at accelerators to date, in order to understand dark matter we need to extend the model and introduce at least one new particle. One of the most popular particle candidate types among physicists are the so-called *WIMPs—Weakly Interacting Massive Particles*.

Dark matter does not participate in the electromagnetic and strong interaction, otherwise it would have been observed long ago. If WIMPs nevertheless were to rarely react with normal matter, then at most via the weak interaction.

Dark matter permeates our immediate surroundings on Earth. Calculations predict an average density of about 300 GeV/c^2 in every liter in our vicinity. This means that if WIMPS have a mass of $m = 1$ GeV/c^2, i.e., are as heavy as protons, there would be 300 WIMPs per liter. With the expected velocities of the particles, up to several million particles per second should flow through each cm^2 of us or of a detector we build. This large particle flow gives us hope that, despite the extremely weak interaction of the dark matter particles, we will be able to directly observe a reaction of WIMPs with normal matter in rare cases. The most probable of the very rare reactions of a WIMP is a scattering off an atomic nucleus. In this reaction, as in billiards, one ball (WIMP) transfers kinetic energy to the second ball (atomic nucleus), which recoils. This nuclear recoil is the signal which we have to make visible in the detector material. The big problem is the very low reaction rate and the low energy transfer. We know from previous experiments attempting to detect WIMPs that the rate must be less than one signal event per year in one kilogram of detector material. In comparison, the rate of signals caused by natural radioactivity of the mate-

Experiments for the direct search for dark matter are set up in underground laboratories to shield them from cosmic radiation. The Laboratori Nazionali del Gran Sasso—LNGS is located to the side of a highway tunnel through Abruzzo mountains. In three large halls there are numerous experiments for the investigation of neutrinos and the search for dark matter.

Dark Matter → p. 282

rial in and surrounding the detector is many orders of magnitude higher. Even in our bodies, there are more than 100 decays per kilogram per second and therefore several billion per year. If you want to detect WIMPs, the first important step is therefore to use high-purity materials. You have to be able to control the remaining radioactivity very well, carefully select the materials at great expense and, if possible, constantly cleanse them of radioactive isotopes.

However, even the purest materials will not be successful if the detectors are not additionally protected against cosmic radiation. Cosmic muons reach the Earth's surface with a flux of around 100 muons per second per m^2. In a kilogram detector, cosmic rays easily generate several tens of thousands of signal events per day. Unfortunately, muons are very penetrating, which is why shielding at least one kilometer thick is needed to stop them. To protect against cosmic radiation, all experiments for the direct search for dark matter particles are therefore placed in underground laboratories. The largest in Europe is the Gran Sasso underground laboratory in Italy, the Laboratori Nazionali del Gran Sasso LNGS (↓). Three large halls for such experiments have been built parallel to the ten-kilometer-long highway tunnel under the Gran Sasso mountain in Abruzzo, which connects Rome with the Adriatic Sea. They are protected against cosmic radiation in all directions by at least 1.4 km of rock and the flux of muons is reduced to around one per hour and m^2. In addition to experiments to measure solar neutrinos and to search for double beta decay, the LNGS is home to a series of experiments for the direct search for dark matter particles.

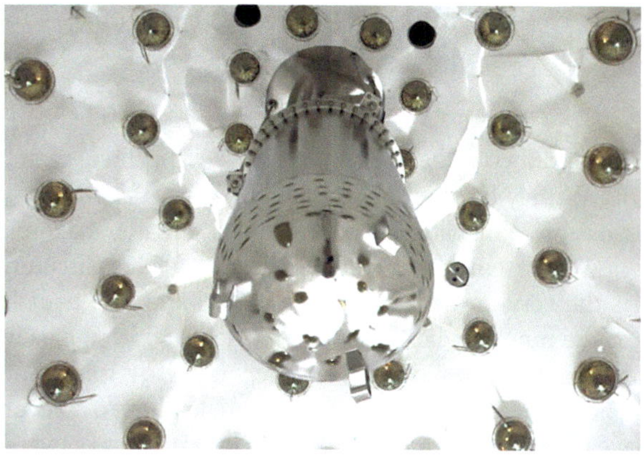

View of the liquid argon detector of the dark-side experiment to search for dark matter particles. The detector in the center is surrounded by photomultipliers. This is used to detect cosmic rays that reach the experiment despite being shielded by the mountain.

View into one of the halls of the LNGS. You can see a cylindrical tank containing the XENON experiment, next to it a building for supplying and controlling the experiment.

However, despite an intensive search, no WIMPs have yet been found. Nevertheless, with every year of measurement, the limits for a possible coupling to normal matter can be narrowed down further and further.

Is the Neutrino Its Own Antiparticle? → p. 300

Direct Search for Dark Matter

Hunting for Invisible Particles

As described in the previous article (↓), the direct search for Weakly Interacting Massive Particles (WIMPs) as a possible explanation of dark matter is a major challenge due to extremely low reaction rates. A further difficulty is the very low recoil energies to be expected, usually at the limit of what just leads to measurable signals in detectors. Fewer charges are generated by ionization than is the case with γ or β radiation, leading to an even weaker signal from nuclear recoil (Interactions between Particles and Matter ↓). Here we discuss how various experiments have approached these problems.

In the DAMA/LIBRA experiment, 25 Na-I scintillation crystals (↓) weighing 10 kg each are used. These were produced with great care and high purity to avoid radioactive contaminations which would cause disturbing signals. In order to see the small signals of WIMPs, each crystal is read out by two photomultipliers. The results of the DAMA/LIBRA measurement show an interesting seasonal variation that could be interpreted as a dark matter signal. The speed and direction of the dark matter particles arriving on Earth is given by the direction of movement of the solar system in the Milky Way.[1] Added to this is the movement of the Earth around the Sun. This leads to a seasonal fluctuation in the speed at which the WIMPs reach us. It should be slightly higher in July and slightly lower in January. This speed modulation leads to slightly higher recoil energies and therefore a higher signal rate above the energy threshold of the detector in July and correspondingly lower signal rate in January. This is consistent with the DAMA/LIBRA measurement. The researchers involved therefore assume that they have discovered the particles of dark matter. Unfortunately, this interpretation is not compatible with the results of many other at least equally sensitive experiments that should have seen the signal, but have not yet observed anything matching it.

Detectors of the DAMA/LIBRA experiment. To avoid radioactive contamination, work on the experiment is carried out under clean room conditions. ↵

Detector crystals of the CRESST experiment are installed in the cryostat, which generates the necessary low temperatures. ↓

[1] It is assumed for good reasons that not all dark matter particles orbit the center of the Milky Way in the same direction and at the same speed as the solar system.

The Mystery of Dark Matter → p. 308
Interactions Between Particles and Matter → p. 68
Scintillators and Photomultipliers → p. 74

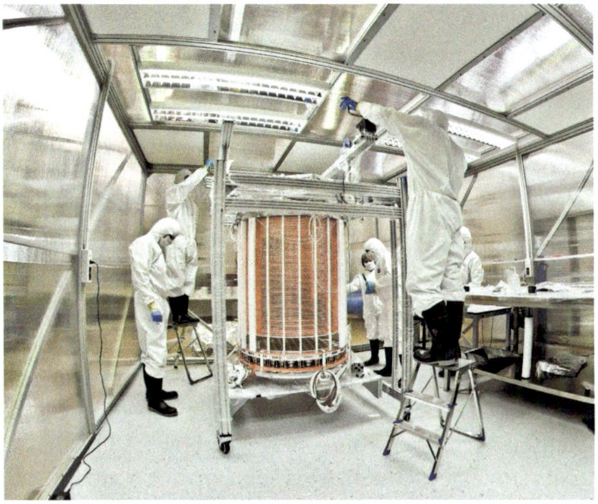

Work on the detector of the XENON experiment. The container will later be filled with liquid xenon. At the lower end you can see the photomultipliers for detecting the scintillation light.

Nuclear recoils generate fewer measurable charges in a detector than radiation generated by radioactivity. This can be used to distinguish nuclear recoil signals from signals caused by radioactivity. The CRESST experiment follows this approach with cryogenic detectors (↓). These, of all detectors, allow the lowest energies to be detected, which is particularly important when searching for WIMPs with a mass below a few GeV. By combining cryogenic and scintillation detectors, it is possible to distinguish from the ratio of temperature to light signal strength whether we are dealing with a nuclear recoil potentially caused by a WIMP or with the well-known α, β or γ radiation. CRESST uses scintillating crystals, such as $CaWO_4$, Al_2O_3 or $LiAlO_2$.

The ability to distinguish between the signals has greatly reduced the risk of producing background interference due to radioactivity and increased the sensitivity in the search for dark matter by many orders of magnitude. This is why most experiments today use a suitable combination of several types of detectors.

In the XENON experiment, light and charge signals are measured simultaneously in liquid xenon (Noble Gas Detectors ↓). The ratio of the rates of the two are used to distinguish nuclear recoils from dark matter from radioactivity. The DarkSide experiment is a very similar project in which liquid argon is used as the detection medium. The detection volume of a liquid such as xenon or argon can be increased much more easily than with the crystals of cryogenic detectors. In addition, the liquid can be circulated to clean it of impurities that may have come loose from the container walls during the measurement. With a larger volume and lower background, one is sensitive to events at a much lower rate. XENON has thus achieved the world's best sensitivity in the search for WIMPs.

All these experiments work at the limits of the measurability of rare events. Time and again, there have been inexplicable signals that have caused excitement and were interpreted as evidence of dark matter. However, it has usually turned out to be some kind of background event that was previously unknown or underestimated.

Even though none of these experiments has yet observed a signal clearly caused by WIMPs, the motivation to push ahead with the projects is great. The results have disproved many particle physics models and are helping to shape theories beyond the Standard Model of particle physics. Physicists are therefore continuing to work on designing larger and therefore ever more sensitive experiments.

Cryogenic Detectors → p. 136
Liquified Noble Gases as Detectors → p. 138

FASER

Using a Collider to Create a Portal Into the Unknown

The particle physics experiments presented in this book are often large behemoths, built after decades of planning. However, sometimes small detectors—assembled from spare components of other experiments—can make an impact far greater than their size suggests. One such example is the Forward Search Experiment, or FASER: a compact particle detector located at the Large Hadron Collider (LHC ↓), the world's most powerful particle accelerator near Geneva, Switzerland.

While most LHC experiments focus on the heavy particles (Discovery of Higgs, Top Quark ↓) produced in violent head-on collisions of protons, FASER looks in a different direction—literally. When protons smash together at nearly the speed of light, the interactions lead to the production of showers of new particles. Many of these particles fly out perpendicular to the beam, where enormous detectors like ATLAS (↓) and CMS (Typical Detectors at Accelerators ↓) capture them. But a large number of particles race straight ahead, tangentially to the LHC ring, continuing along the original beam line. They fly through a hole in the detector (which is for the beam line) and remain completely unnoticed. These forward-traveling particles from the ATLAS collision point are exactly what FASER is built to detect.

A sketch of the FASER location, 480 meter downstream from the ATLAS collision point. The powerful LHC magnets sweep away charged particles travelling in its direction. Its final location is shielded by about 100 meter of rock and concrete. Note that the LHC ring bending away is drawn not to proportion.

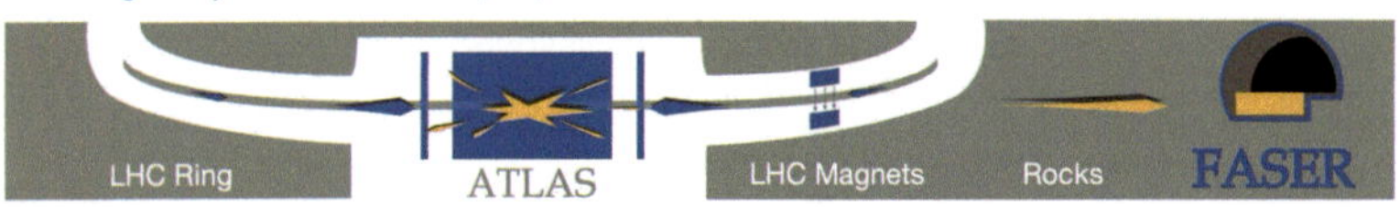

FASER is placed 480 meters downstream from one of the LHC's collision points, tucked away in a side tunnel that once served as part of the accelerator's service infrastructure. From this quiet outpost, it watches for signs of light, long-lived, and weakly interacting particles—exactly the kind of particles that might be responsible for dark matter, or that could help explain why the Universe is made of matter rather than antimatter. These particles would easily slip through the giant detectors but could be spotted by FASER thanks to its strategic location. One such type of matter would be ***dark photons***.

Dark photons are hypothetical particles that are similar to ordinary photons—the particles of light—but belong to a hidden "dark sector" of physics. Just as photons carry the electromagnetic force, dark photons could carry a new force that acts on dark matter, the mysterious substance that makes up most of the matter in the Universe but doesn't emit light. In some theories, dark photons can very weakly interact with normal matter, allowing them to be produced in particle collisions and possibly detected by sensitive experiments like FASER. Discovering dark photons would provide a crucial link between the visible Universe and the invisible world of dark matter. This connection is often referred to as a "portal"—a hypothetical link between the visible world of ordinary matter and the hidden realm of dark matter.

The LHC Accelerator Complex at CERN → p. 116
The Discovery of the Higgs Boson → p. 246, Discovery and Properties of the Top Quark → p. 242
The ATLAS Experiment at the LHC → p. 130
Typical Detectors at Accelerators → p. 128

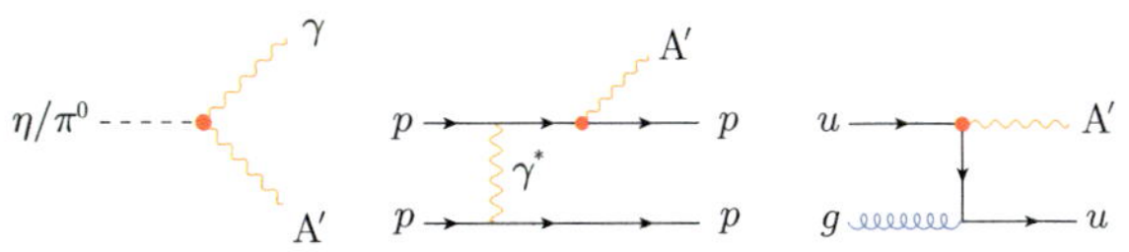

Some example processes how dark photons (A') can be produced in head-on collisions of protons or from the decay of light mesons, which are also produced in such collisions.

In addition to searching for new physics, FASER has already made history by achieving another major goal: detecting collider neutrinos. Neutrinos are among the most elusive particles known to science. They pass through matter almost entirely unnoticed, making them very difficult to study. While neutrinos have been observed from the Sun, supernovae, accelerator beams shot on a fixed target, and nuclear reactors, none had ever been detected coming from particle collisions at a collider—until FASER did just that in 2023. This breakthrough opens a brand-new way to study neutrinos at energies far beyond what's possible in other experiments.

The figure below illustrates how FASER detected its first neutrino candidates. Neutrinos are produced in the decay of mesons created near the ATLAS interaction point and then travel 480 meters through rock to reach the detector. Upon arrival, a neutrino passes unnoticed through the FASERν veto scintillator, which is designed to reject incoming charged particles—such as muons—that leave a visible signal. The neutrino eventually interacts via a charged-current interaction (↓) in a tungsten target, producing a muon and a spray of secondary particles. The resulting muon is tracked and its momentum measured using three precision tracking stations. As it traverses a magnetic field within the decay volume, the charge of the muon can be determined, allowing FASER to distinguish neutrinos from antineutrinos.

The figure also shows how a long-lived particle like a dark photon A', if produced, would enter the detector and decay into a pair of electrons. These electrons would be reconstructed in the tracking system and would initiate an electromagnetic shower in the downstream calorimeter, providing a distinctive signature. Again no charged particle would be noticed entering the detector at the front. Despite its modest size and small budget, FASER is a powerful example of how clever thinking and creative reuse of existing technology can push the boundaries of our understanding. In the vast world of particle physics, sometimes the smallest eyes see the biggest surprises.

Setup of the FASER detector and two possible signatures. In the upper part, a muon neutrino enters the detector from the left and through a charged-current reaction transforms into a muon in the tungsten/emulsion detector (orange). The muon is then observed in the tracking detectors and the calorimeter. In the lower part of the sketch, a signature of a hypothetical dark photon A' is shown. If produced at the LHC, it by chance could decay into an electron pair in the middle of the FASER detector.

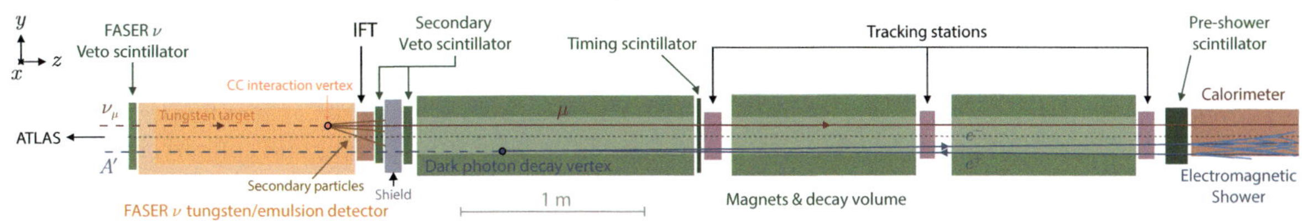

The Weak Interaction → p. 178

Matter and Antimatter

Why Do We Exist at All?

According to our current understanding, a fraction of a second after the Big Bang matter and antimatter (↓) were in equilibrium. There were just as many quarks as antiquarks, just as many electrons as positrons, etc.

Today, however, we are almost exclusively surrounded by matter particles. Antiparticles can only be found in a small proportion in the cosmic radiation, in certain nuclear decays or they have to be produced in accelerators with great effort. So what happened to the antimatter?

Basically, there are two approaches to explain this. Either the antimatter has disappeared, so that today only what we call matter is left. Or the antimatter is hiding in areas of the Universe that are not yet accessible to us. Both possibilities are considered and corresponding experiments are carried out, which will be discussed in separate articles (↓). Here we pursue the first approach a bit further.

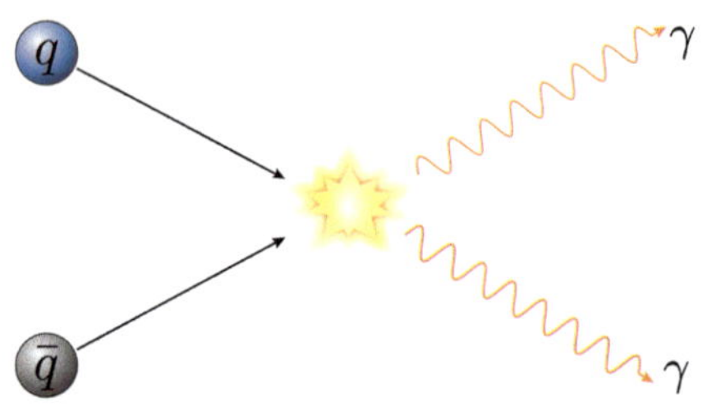

Conversion of a quark and an antiquark into photons

As shown in the figure, photons are created when matter and antimatter collide. From two very different observables, on the one hand the abundance of different nuclei in the Universe (↓) and on the other hand the cosmic background radiation (↓), one can determine the ratio of the density of baryons n_B (recall: protons and neutrons are baryons) to the density of photons n_γ in the Universe. With the additional information that no antibaryons are observed ($n_{\bar{B}} \approx 0$), we obtain the ratio known as the baryon asymmetry:

$$\frac{n_B - n_{\bar{B}}}{n_\gamma} \approx 6 \cdot 10^{-10} \ .$$

This is a rather small number. For every baryon in the Universe visible to us, there are about 10^9 photons from the radiation of matter and antimatter shortly after the Big Bang. What does this ratio tell us? Here are a few considerations. If Nature made no distinction between matter and antimatter, the baryon asymmetry would be zero. If we started with the same number of baryons and antibaryons, each baryon would have annihilated with an antibaryon. This would have mainly produced photons. In the end, there would have been neither baryons (matter) nor antibaryons (antimatter) left. The book you are currently holding in your hands would not exist, nor would there be any hands to hold it. Fortunately, this value is above zero, albeit barely!

The small ratio of about 10^{-9} means that before the mutual annihilation of baryons and antibaryons, for every 1 000 000 001 baryons there were "only" 1 000 000 000

What Is Antimatter? → p. 22
The Search for Electric Dipole Moments → p. 318, Particle Physics in Space → p. 316
Abundance of nuclei: Baryogenesis → p. 328, Nucleosynthesis After the Big Bang → p. 330
Cosmic Microwave Background → p. 332

antibaryons. After mutual annihilation, only one baryon remained from a billion pairs. Our current Universe is built from this tiny difference!

But if matter and antimatter were in equilibrium right after the Big Bang, how could the surplus of matter so important to us arise? As so often, symmetry considerations (↓) play a big role. If the laws of physics made no distinction between matter and antimatter, i.e., the corresponding symmetry were exact, matter and antimatter particles would have annihilated each other. We would not exist today if we followed the first explanation. The Soviet physicist and Nobel Peace Prize laureate Andrei Sakharov pointed out as early as 1967 that the combined violation of mirror symmetry (P) and the symmetry with respect to the exchange of particles and antiparticles (C), which in particle physics is known as ***CP* violation** (↓), is necessary to transition from an initial equilibrium to a dominance of what we call matter today. Although CP violation is a firm integral of the Standard Model, it turns out that this is much too small to explain the imbalance between matter and antimatter. Therefore, there must be additional CP violating processes. How do we track these down? We cannot turn back the wheel of time and observe how matter gained the upper hand over antimatter in the first microseconds of the Universe. However, we can search for symmetry violations beyond the Standard Model with experiments. One example will be discussed in the article on electric dipole moments (↓). A detection of such dipole moments would give a possible explanation for the disappearance of antimatter.

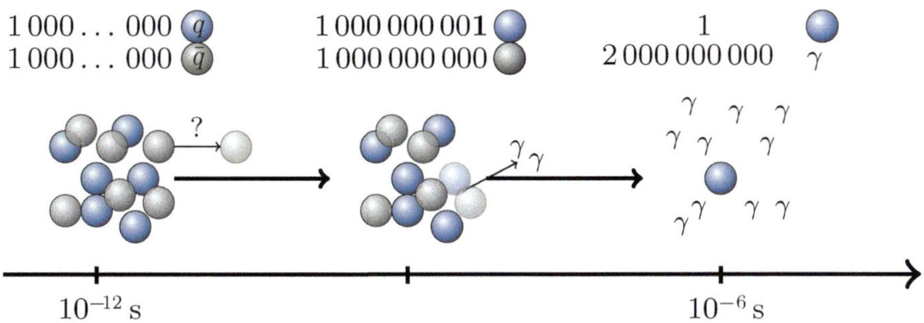

Initially, matter and antimatter were in equilibrium. Through still unknown processes, there was a small surplus of matter. Matter and antimatter annihilate each other. Today we are only surrounded by matter and photons.

Symmetries → p. 60
Discrete Symmetries: Violating P and CP → p. 190 and CP Violation in Beauty → p. 258

Particle Physics in Space

Aiming High

Complex accelerator facilities are operated on Earth to generate high-energy particles. However in cosmic rays (↓) particles of far higher energy are found, which no accelerator on Earth can generate. Depending on the type of particle and energy, these reach the Earth's surface or react beforehand in the Earth's atmosphere with other particles. To detect particles before a possible interaction in the atmosphere, one must therefore operate a detector in space.

The first attempts date back to Victor Hess, who ascended in a balloon at the beginning of the 20th century. The surprising result at the time was that the intensity of the measured radiation increased with increasing altitude. This is an indication that radiation from space hits the Earth and does not, as previously assumed, originate from radioactivity in the Earth's crust.

To investigate this radiation more precisely, a 7.5-ton detector was transported to the International Space Station (ISS) with the last mission of the Space Shuttle Endeavor in 2011. The detector, known as the Alpha Magnetic Spectrometer (AMS), contains numerous components that allow it to detect tracks of various particle types over a wide energy range. The challenge was to design the detector in such a way that it withstands the vibrations during the launch of the Space Shuttle unscathed and then operates for years without major maintenance work. The team of about 500 researchers succeeded in this, as the detector has now been producing data consistently for over 10 years.

The AMS detector on the International Space Station ISS

Cosmic Rays → p. 336

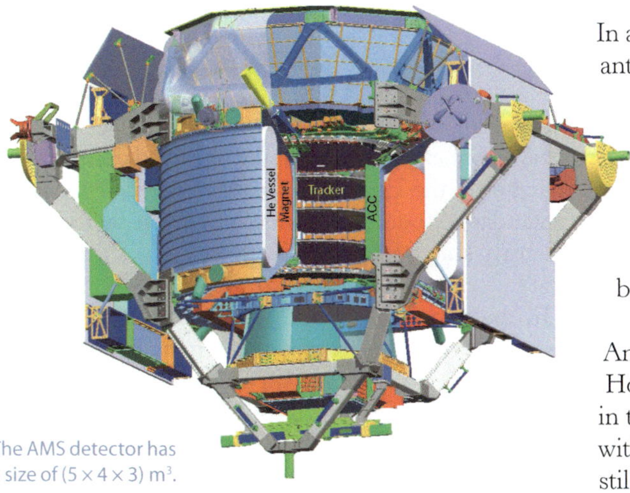

The AMS detector has a size of $(5 \times 4 \times 3)\ m^3$.

In addition to the detection of positrons, the search for antinuclei, such as antihelium or even anticarbon nuclei is of great interest. A few minutes after the Big Bang, conditions in principle prevailed to form antihelium nuclei (↓). As described in the article on the birthplace of nuclei (↓), heavy nuclei like carbon can only be formed in stars. The detection of an anticarbon nucleus would therefore be proof of the formation of a star from antimatter.

An anticarbon nucleus has not yet been detected. However, about one track per year has been found in the AMS detector so far, which could be consistent with an antihelium nucleus. Detailed investigations are still ongoing. Stay tuned!

Of particular interest is the detection of antiparticles. This is related to one of the great mysteries of particle physics and cosmology, the question of the fate of antimatter (↓). There are two possibilities: Either the antimatter has disappeared over time or it is mainly located in areas of the Universe far away from us. Depending on where they are located, every now and then, an antimatter particle should hit the AMS detector and be detected.

The figure at the right shows the measured flux of positrons, i.e., the antiparticles of electrons as a function of their energy. While the gray shaded part of the measurement can be explained by the cosmic radiation we are familiar with, an unknown source must be assumed for the pink shaded part of the spectrum at higher energies.

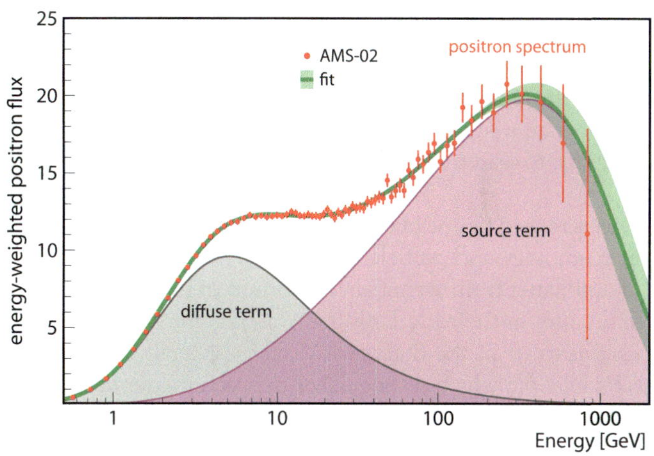

The measured flux of positrons as a function of energy. While the left part is understood, the high-energy part of the spectrum is puzzling.

Matter and Antimatter → p. 314, Baryogenesis → p. 328, *CP* Violation in Beauty → p. 258
Nucleosynthesis After the Big Bang → p. 330
The Birthplace of Nuclei → p. 342

The Search for Electric Dipole Moments

A Hidden Source of Antimatter Generation?

Electric dipole moments, or EDMs for short, occur when positive and negative charges in a system have different centers of gravity (see information box). A regular battery therefore has an electric dipole moment, as the positive and negative charges are separated. This is also the case in every water molecule and is responsible for the attraction between the molecules. The EDM is also a fundamental property of a subatomic particle, like mass, charge, spin or magnetic moment. However, so far, no non-zero EDM has been detected in any of the subatomic particles known to us. There are good reasons for this. Unlike more complex objects like molecules or batteries, EDMs can only occur in subatomic particles like electrons, protons and neutrons, under very specific conditions, namely if there is a combined violation of charge conjugation (*C*) and parity (*P*), called *CP* violation (↓). The *CP* violation known from the Standard Model is simply too small to generate a measurable EDM. Why should we still continue to search for it?

This is related to one of the biggest unsolved mysteries in cosmology and particle physics, the disappearance of antimatter from an initial equilibrium of matter and antimatter in the early Universe (↓). The interactions known to us in the Standard Model, especially their *CP* violation, which distinguishes between matter and antimatter, is far from sufficient to solve this mystery.

Extensions of the Standard Model, such as supersymmetry (↓), however, predict further *CP*-violating interactions that could also lead to detectable EDMs (Higgs boson ↓). This means, if we find an EDM, we would have both explanations for the observed dominance of matter in the Universe as well as physics beyond the Standard Model! No wonder then, that there is an intensive search for it.

To measure EDMs, different methods are used depending on the type of particle: particle traps (↓) for electrically neutral particles like neutrons and atoms, storage rings for charged particles like protons.

Apparatus for measuring the electric dipole moment of the neutron at the Paul Scherrer Institute in Switzerland. The picture shows in the center the two electrodes for generating the electric field surrounded by the magnetic field shielding.

Discrete Symmetries: Violating *P* and *CP* → p. 190 and *CP* Violation in Beauty → p. 258
Matter and Antimatter → p. 314
Is It Really the Higgs Boson? → p. 248
Particle Traps → p. 140

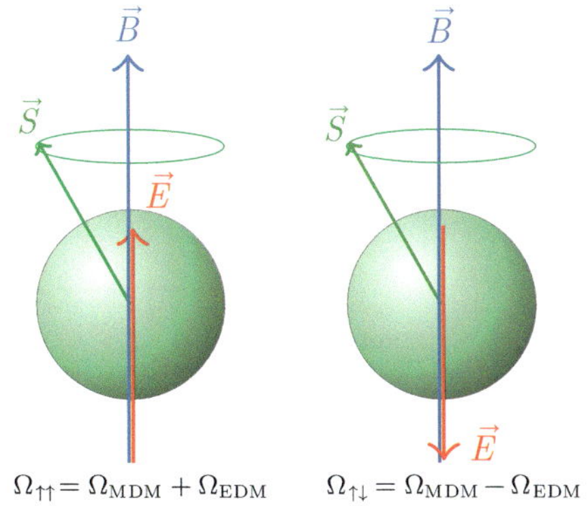

Measurement principle of an EDM experiment. The EDM results from the difference of the spin rotations in a configuration with parallel and anti-parallel positions of an electric ($\vec{E}$) and magnetic field ($\vec{B}$).

Electrical Dipole Moments

If two oppositely charged point charges q are at a distance of r, they generate a dipole moment of size $d = q \cdot r$. The electric dipole moment is usually given in the unit elementary charge times centimeter, $e \cdot \mathrm{cm}$. The dipole is a vector and points from the negative to the positive charge.

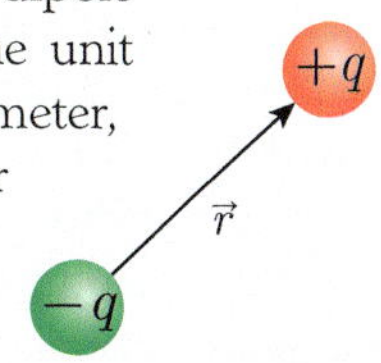

The measurement principle is based on the fact that just as a magnetic dipole moment leads to a rotation of the spin in a magnetic field, an electric dipole moment leads to a rotation of the spin in an electric field. What makes the measurement of the EDM so difficult is the fact that the expected effect is so small. Due to its magnetic moment, a neutron performs about 1500 rotations per second in the Earth's magnetic field, while due to an EDM, with current experimental sensitivities, one rotation every month could be detected! In practice, therefore, the apparatus is shielded against the Earth's magnetic field. Then a very well controllable magnetic field is generated in the laboratory. The precession frequency of the neutrons is measured in a combination of this magnetic field and an applied electric field. If the electric field is reversed, the EDM can be determined from the difference of the two frequencies.

Despite all efforts, no value deviating from zero has been measured for an EDM so far. But even from this, important conclusions can be drawn. From the zero result for the neutron, for example, values greater than $1.8 \cdot 10^{-26}\, e\,\mathrm{cm}$ can be excluded with a confidence (↓) of 90 %. As can be seen from the figure below, this also excludes some areas of the predictions of supersymmetric extensions of the Standard Model, called SUSY (↓).

It is therefore worthwhile to refine the measurement accuracy further and further in order to either find a value different from zero or to move the exclusion area further and further to the left.

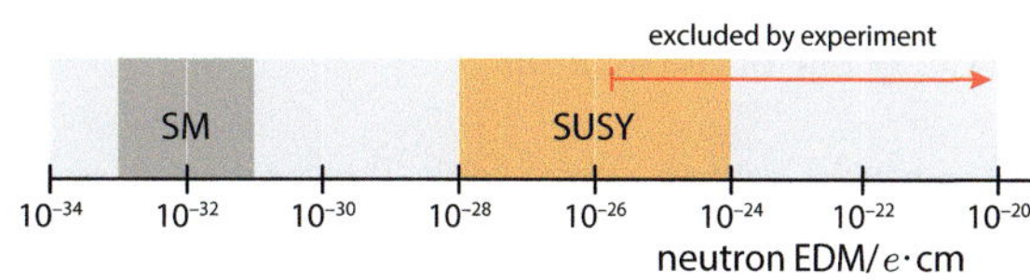

Standard model (SM) and predictions of supersymmetry (SUSY) for the neutron. The area covered by the red arrow is excluded by experiment.

Drawing Conclusions From Statistics → p. 160, Publishing a Non-Observation → p. 162
Supersymmetry → p. 278, Searching for New Particles at the LHC I → p. 302

Proton Decay

Long Live the Proton!

The proton is composed of three quarks (*uud*, ↓) and is therefore classified as a baryon with a baryon number of +1. It is the lightest baryon, lighter than for example the neutron (*udd*) or the lambda (Λ, *uds*). This means that a free proton cannot decay into another baryon. If the proton were to decay, the baryon number would necessarily change, violating the principle of baryon number conservation. In the Standard Model, baryon number is conserved, meaning the number of baryons remains the same before and after any reaction. The electromagnetic, weak, and strong forces all preserve baryon number, as well as lepton number. As a result, the proton is stable within the framework of the Standard Model.

Assuming the mechanism of baryogenesis (↓), we can infer that baryon-number-violating interactions must have been relevant in the very early Universe, at much higher energies. Such interactions typically arise in Grand Unified Theories (GUTs, ↓). We would expect these baryon-number-violating interactions to still exist today at much lower energies. However they should be strongly suppressed, by a factor of approximately 10^{-57}, according to quantum field theory, compared to the decay rate of the neutron.

What would proton decay look like, and how could it be detected? In the simplest Grand Unified Theory (GUT), the most probable mode involves the proton decaying into a neutral pion and a positron, mediated by a new gauge boson (↓), X, as shown in the Feynman diagram[1].

Decay of a free proton into a positron and a neutral pion

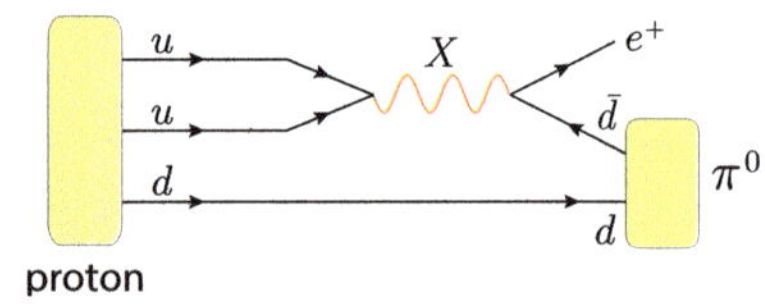

These gauge bosons are also known as ***leptoquarks***, as they can couple to both quarks and leptons simultaneously.

The total energy of the decay positron is approximately 460 MeV, which is about half the rest energy of the proton and is roughly 1000 times greater than the positron's rest energy ($m_e c^2 = 511$ keV), making it highly relativistic (↓). As a result, special relativistic effects will be observed in the detector (see below). The neutral pion decays almost immediately into two high-energy photons, each with an energy around 240 MeV, corresponding to gamma radiation. Some of these decay products must be detected to confirm that at least one proton has decayed.

The proton lifetime should be significantly longer than the age of the Universe, which is around 10^{10} years; otherwise, we would no longer be alive. As we will see, the proton lifetime is even longer than 10^{30} years! Proton decay is a quantum mechanical process that follows the radioactive decay law. One thousand kilograms of pure water contain about $3 \cdot 10^{26}$ protons. With a lifetime of 10^{27} years, one proton would decay roughly every three years. The largest detector currently used to search for proton decay is Super-Kamiokande[2] in Japan, which

[1] The X bosons are often also denoted as $X^{\pm 4/3}$, indicating their electric charge (see GUT → p. 276).

The Structure of the Proton → p. 212
Baryogenesis → p. 328
GUT → p. 276
Gauge Symmetries → p. 98

contains 50 million kilograms of water. Protected from cosmic radiation, it is located deep beneath a mountain in Japan. The massive water tank is surrounded by light-sensitive detectors known as photomultipliers (↓), which convert light into electric pulses. With a proton lifetime of 10^{30} years, there should be around 15 000 decays per year. But how can these decays be observed?

The decaying proton is surrounded by a large amount of water in the detector. The energetic photons from the pion decay cannot penetrate this water to reach the photomultipliers at the wall, so they remain unobserved. However, as the relativistic positron moves through the water it produces Cherenkov radiation (↓) in the visible spectrum. The Cherenkov photons have much lower energy than the photons from pion decay and can travel through the water to reach the detectors. This Cherenkov radiation forms a ring-shaped cone of light, which can be observed. For the decay mode $e^+ + \pi^0$, Super-Kamiokande has determined a lower bound for the proton lifetime of approximately 10^{34} years. To put this in perspective, if we take five people (each weighing 75 kg) who each live to be 100 years old, there would be fewer than one proton decay from the water in their bodies for all five of them over their entire lifetimes. So, we can all rest assured.

The Hyper-Kamiokande detector in Japan is scheduled to begin operations in 2027. It has the shape of an upright cylinder, with a diameter of 68 m and a height of 71 m. Like Super-Kamiokande, it will be located deep underground and equipped with numerous photomultipliers along the inside of the walls. Hyper-Kamiokande will have a volume 20 times larger than Super-Kamiokande, making it correspondingly more sensitive.

The lower limit on the proton lifetime, approximately 10^{34} years, is so stringent that it already rules out the simplest GUT theories, particularly the one *without* supersymmetry. In supersymmetric theories, the decay of a proton into a kaon and a neutrino is preferred: $p \rightarrow K^+ + \bar{\nu}$. This decay is more difficult to detect, and the lower bound on the lifetime is about a factor of 10 weaker.

The Kamiokande experiment was built to search for proton decays. It then became very successful as a neutrino detector. The Super-Kamiokande and Hyper-Kamiokande extensions focused from the outset on both physics aspects. In 2026, the JUNO detector in China will start recording data. This detector was developed for neutrino physics, but can also search for proton decays. JUNO is significantly smaller than the Kamiokande detectors, but made of a much denser material. It is based on a different principle, which makes it easier to detect kaons. It should be able to measure lifetimes in kaon decay by a factor of around 10 longer than Super-Kamiokande. The planned DUNE experiment in the USA is similar to JUNO in this respect.

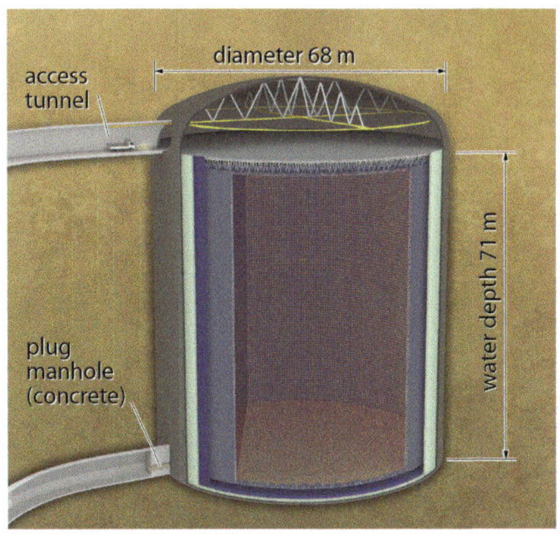

Depiction of the Hyper-Kamiokande detector in Japan, located deep underground

[2] Photos of the Super-Kamiokande detector can be found in the articles Large Detectors Deep in a Mountain → p. 132 and Neutrinos → p. 196.

Special Relativity → p. 38
Scintillators and Photomultipliers → p. 74
Cherenkov Radiation and Particle Identification → p. 82

The Search for Axions

Two Birds With One Stone

Sometimes new particles are invented to solve problems of the Standard Model. Axions are an example of this. They are a by-product of a proposed solution to the strong *CP* problem (↓). The mathematical description of the strong interaction in the Standard Model allows contributions that violate *CP* symmetry (↓). So far, no evidence of these have been found. Thus the *CP*-violating contributions of the strong interaction must vanish or be at least 10 orders of magnitude smaller than expected.

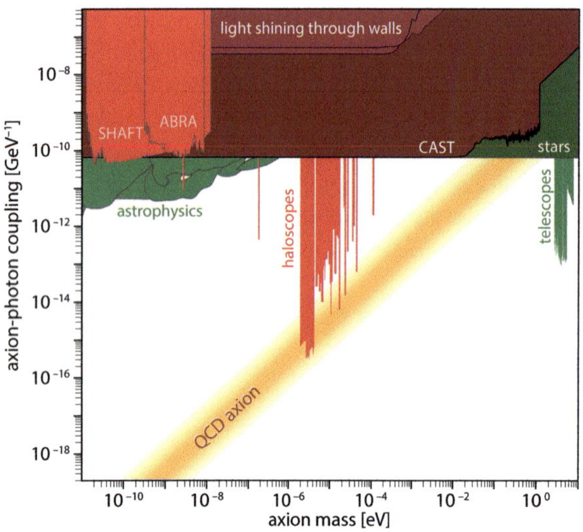

Coupling strength of axions $g_{a\gamma\gamma}$ plotted against their mass m_a. The yellow area shows assumed axions from the solution of the strong *CP* problem (QCD axions). With masses m_a from 10^{-6} eV to 10^{-4} eV they could also explain dark matter. Outside the yellow band axion-like particles (ALPs) could exist. The red and green colored regions are excluded by experiments and astronomical observations.

This can be explained by an additional symmetry, the so-called *Peccei-Quinn symmetry*, which must be spontaneously broken (↓). As a result of the symmetry-breaking, the *CP*-violating term is zero. However, this predicts the existence of a new particle—called *axion*. The properties of the axion also turn the initially annoying by-product into a particle for a possible explanation of what dark matter (↓) in the Universe consists of. The total matter in the Universe is dominated by an invisible dark matter for which there is no explanation within the Standard Model. Axions could be a solution, with the nice property that they could in principle be detected in corresponding experiments: According to theory, axions couple to photons of external fields, e.g., in magnetic fields, producing a real photon. Axions and photons can thus transform into each other in magnetic fields. If axions with properties suitable for solving the strong *CP* problem exist, then there is a direct correlation between their mass m_a and their coupling to photons $g_{a\gamma\gamma}$ (yellow band in the left figure).

If the axions resulting as a solution of the strong *CP* problem, the so-called QCD axions, were the dark matter, they would exist all around us. In the experiment, the axion mass m_a is converted into photon energy, which—in the appropriate mass range—corresponds to microwave photons, in the gigahertz frequency range. The ADMX experiment has been searching for the appearance of such photons for many years. In order to increase the probability of converting axions into photons, this search is carried out at very low tem-

The Strong *CP* Problem → p. 290
Discrete Symmetries: Violating *P* and *CP* → p. 190
Spontaneous Symmetry Breaking → p. 62
Dark Matter → p. 282

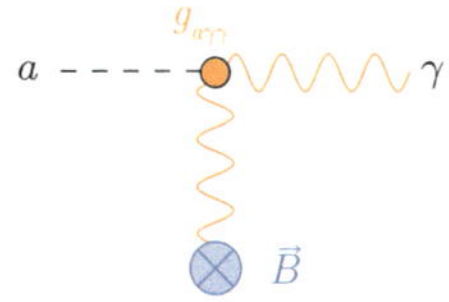

Feynman graph of the coupling of axions to photons. The axion couples to a photon of an external magnetic field and transforms into a real photon.

peratures in a cavity resonator designed for a specific photon frequency. By tuning to a specific frequency, it is only possible to search for axions in a very narrow mass range. New ideas, such as the MADMAX project, are therefore trying to make the search more effective. In a magnetic field, dark matter axions can be converted into potentially detectable photons on the surfaces of dielectric plates. By changing the distance between the plates, different photon frequencies can be amplified.

Axion-like particles that do not follow the QCD relationship between their mass m_a and their coupling to photons $g_{a\gamma\gamma}$ (outside the yellow band in the left figure) are called *ALPs* (*axion-like particles*). ALPs would have nothing to do with the *CP* problem, but could explain dark matter. The search for ALPs is developing spectacularly. The aim of an experiment at DESY in Hamburg, also called ALPS, is to shine light through a wall. A strong laser shines through a magnetic field and hits a wall. If a photon is converted into an ALP before reaching the wall, it can penetrate the latter. Behind the wall is another magnetic field in which the ALPs could turn back into photons. If it were possible to measure photons behind the wall, i.e., to shine through the wall, this would be a very strong indication of the existence of ALPs.

The Sun provides the perfect conditions for generating axions or ALPs. This is why so-called helioscopes such as IAXO or the previous experiment CAST are used to search for axions or ALPs from the Sun. To do this, a telescope is used to look at the Sun through a magnetic field. In the magnetic field, the particles are converted into photons, which can then be observed.

So far, no axions or ALPs have been found. As it is hoped that their existence may solve two problems of the Standard Model of particle physics, the growing efforts to continue searching for them are worthwhile.

Magnet of the ALPS experiment to shine 'light through the wall'

CAST experiment. The Sun is observed through the magnet (blue tube) and photons are searched for, which are produced by the conversion of solar axions in the magnetic field.

Supernovae as Particle Physics Laboratories

Dramatic Events With Unexpected Talents

Supernovae are among the most spectacular events in the Universe—and can be "used" in many ways to test exciting physical ideas. How this works and what it has to do with particle physics will be explained here.

In the article on the Birthplace of Nuclei (↓) we described the phases a star undergoes in its life: when a gas cloud is compressed by gravity, the nuclear fusion of protons begins at a sufficiently high density and temperature. This process produces predominantly photons and neutrinos. The photons build up an outward radiation pressure that stabilizes the star against the inward gravitational pressure. The neutrinos, on the other hand, can leave the star almost unhindered, since they interact so weakly with matter.

As the star matures, it burns faster in the innermost core, where pressure and temperature are highest, than in the outside regions. Therefore, it is in the inside where all protons are used up first and the fusion to heavier nuclei begins, while further out protons are still fusing, leading to a shell-like structure, as shown in figure A. Thus, more and more heavier nuclei accumulate in the interior. These reactions come to a halt with iron. The growing iron block in the interior (see the blue core in image A of the figure) is stabilized because the electrons present there cannot be arbitrarily compressed. This is due to the Pauli exclusion principle, which prevents two electrons from occupying the same quantum state (Fermions and Bosons ↓). However, if the iron block becomes heavier than about 1.4 solar masses and thus reaches the so-called *Chandrasekhar limit* (image B), the gravitational pressure is so large that it overcomes the aforementioned electron pressure: the electrons, which have very high energies as a result of the high compression, are pushed into the nuclei. There they are captured by the protons, which thus become neutrons, releasing high-energy neutrinos. This eliminates the stabilizing electron pressure, causing the iron core of the star to collapse (image C). This continues, until the Pauli principle kicks in for the neutrons, which are also fermions. This is quite a dramatic collapse, since an electron cloud is about 100 000 times larger in diameter than compressed neutron matter. The entire energy of the collapse is hurled back outwards in a shock wave which disintegrates the material of the shell outside the iron core (image D). This is the dynamics of a type II supernova. Depending on the size of the original star, either a neutron star, essentially a giant atomic nucleus composed mostly of neutrons, only a few kilometers in

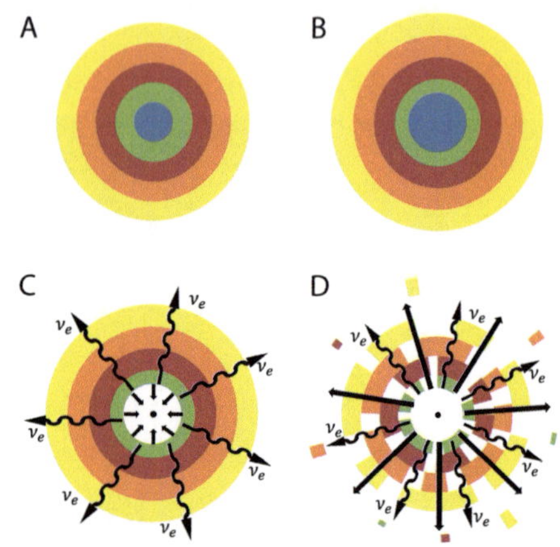

Phases of a star's life

The Birthplace of Nuclei → p. 342
Fermions and Bosons → p. 94

Remnant of supernova SN1987A

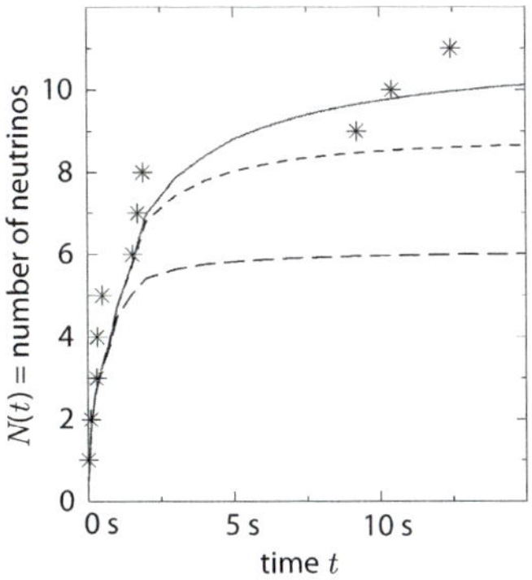

Neutrinos from SN1987A

diameter but several times more massive than the Sun, or even a black hole may remain.

The exciting issue for our discussion is that the total energy of the shock wave is very well known: it corresponds to the energy that was stored in the iron core with 1.4 solar masses. Only one percent of this energy goes into visible light, since the photons are largely trapped inside the dense material. The rest is carried away by the much weaker interacting neutrinos.

One can even calculate rather accurately over what period of time high-energy neutrinos can be expected after the collapse, as can be seen from the comparison of the result with data of a real supernova, whose neutrinos arrived on Earth in 1987. For example, the Kamiokande neutrino detector measured a total of 11 high-energy neutrinos over a time span of 13 seconds; after that, the resulting neutron star had cooled down so much that no more high-energy neutrinos were emitted. In the figure above, the counted neutrinos are marked with asterisks. The diagram is to be read in such a way that, for example, the 9th neutrino arrived after 10 seconds. The mentioned 13 seconds correspond to the timescale that was also calculated: The solid line in the figure is the result of a corresponding numerical simulation.

Now one might ask: "All this is exciting, but what new insights can we gain for particle physics?" The following consideration makes the connection clear: If there were other mechanisms besides neutrinos that couple to the Standard Model particles about as strongly as neutrinos, they could also contribute to the cooling and in this way accelerate the cooling process. This allows us to limit the properties of possible processes beyond the Standard Model.

As an example the two additional curves in the figure show what temporal course of the neutrino signal would have been expected if there had been two additional dimensions for gravity (↓) with a radius of $4 \cdot 10^{-7}$ m (short dashed) or 10^{-6} m (long dashed). In the former case, no more neutrinos would have been expected after 10 seconds (curve becomes flat) and in the latter after 5 seconds. Obviously, the latter curve in particular is not compatible with the data. In other words: The observed neutrino signal excludes that there are two extra dimensions with a radius larger than 10^{-6} m. Thus, the neutrino signal of a supernova has allowed us to learn something about a possible extension of the Standard Model. By the way, the idea also works analogously for other particles that have been proposed as a possible extension of the Standard Model, such as neutralinos (Supersymmetry ↓) or axions (Strong *CP* Problem ↓).

Extra Dimensions → p. 286, Additional Spatial Dimensions at the LHC → p. 306
Supersymmetry → p. 278
The Strong *CP* Problem → p. 290

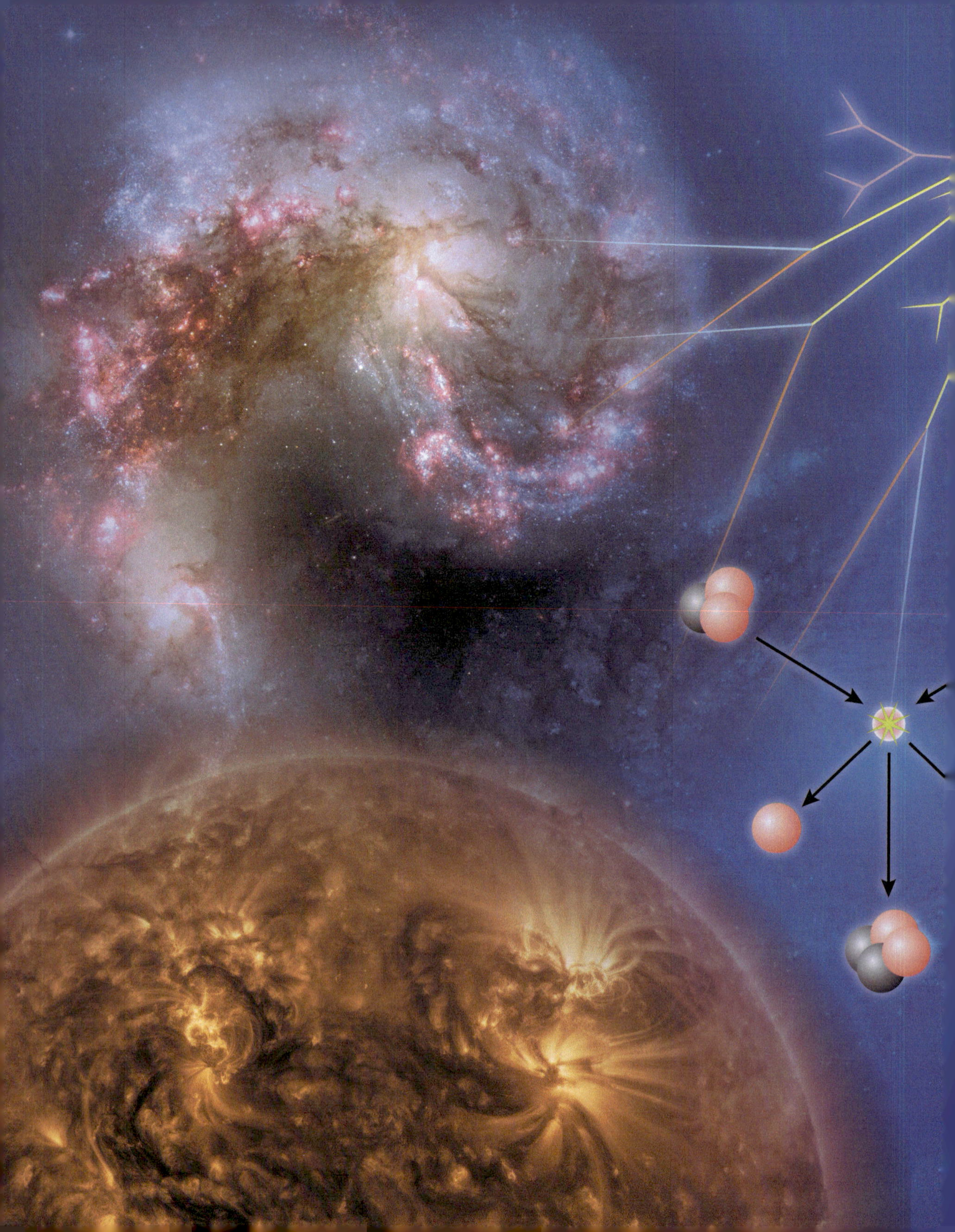

12 Connecting the Smallest to the Largest Scales in the Universe

It is truly remarkable how closely the largest structures of the Universe are interwoven with the tiniest, the elementary particles. In this book, we have explored the smallest and, as far as we know, most fundamental building blocks of Nature. From the very first chapter, we outlined how elementary particles gradually combine into more complex structures, forming nuclei, atoms and molecules on ever larger scales. The appearance and governing laws of the resulting structures at first often have no apparent connection to the laws of the underlying elementary particles.

The journey from the smallest to the largest scales is not merely one of increasing distance, but rather a return to a common origin. In this final chapter of the book, we complete this circle by examining some examples of the deep connection between cosmological observations and discoveries in particle physics.

Baryogenesis

Where Did All the Antimatter Go?

Immediately after the Big Bang our Universe was very hot and contained equal numbers of baryons and antibaryons. Thus the total baryon number was zero and the Universe was symmetrical.[1] Today, however, we observe that all matter in planets, stars and galaxies consists of baryons and there are *no* stable structures made of antibaryons. The Universe has a baryon *asymmetry*. The dynamic theory of baryogenesis describes how today's asymmetric Universe was able to develop from the originally symmetric one. In this article we want to address two questions:

(1) Is the Universe indeed asymmetric today? and

(2) What conditions are required for the Universe to be able to develop an asymmetry?

(1) If we bring matter and antimatter (↓) close together, they annihilate into radiation. When for example an electron meets a positron, they annihilate and two high-energy (511 keV) photons, gamma rays, are emitted (↓). The annihilation of protons and antiprotons leads via pions and muons to electrons and positrons, and thus also to photons.

When the first astronauts landed on the Moon, no such annihilation took place. Sattelites sent to Venus and Mars also did not disappear in a shower of gamma rays. So the Moon, Venus and Mars are made of matter, just like us. The solar wind transports matter from the Sun throughout our solar system. Nowhere do we observe the characteristic gamma rays of annihilation. Our entire solar system therefore consists only of matter, not antimatter.

On larger scales in the Universe, we only have indirect evidence. In our galaxy there are gas clouds in contact with several stars. Here too we do not observe any annihilation. Our entire Milky Way is therefore made up of matter. Collisions of galaxy clusters take place on the largest known scales in the Universe. There also no matter-antimatter annihilation was observed. Our current understanding is that there is *no* accumulation of antimatter in the entire Universe visible to us.

Annihilation of a being and an anti-being

[1] For example protons have baryon number +1, antiprotons have baryon number −1. For equal numbers this adds up to zero.
[2] By the way, Andrej Sacharow received the Nobel Peace Prize in 1975.

What Is Antimatter? → p. 22
Annihilation: Feynman Diagrams II → p. 102, Positron Emission Tomography → p. 144
Symmetries → p. 60

It is therefore asymmetric with regard to the number of baryons. Should aliens ever visit us, we can probably shake their hand with confidence, if they have one. More cautiously you could toss them a tennis ball first.

(2) How can a symmetrical Universe dynamically transform into an asymmetrical one? According to Andrei Sakharov[2] (1967), a model to explain this must have three properties

(i) baryon number violating interactions
(ii) the Universe must temporarily be outside thermal equilibrium, and
(iii) *CP* symmetry (↓) must be violated.

The first condition is necessary, as we want to change the baryon number of the Universe. To illustrate the second condition we consider a saline solution. The solution is in thermal equilibrium when the temperature and the number of dissolved molecules no longer change. For each reaction (salt molecule dissolves) the counter-reaction (salt molecule precipitates) occurs just as often. The same principle applies if we have a reaction $a + b \rightarrow c + d$ that violates baryon number. In thermal equilibrium the reverse reaction $c + d \rightarrow a + b$ occurs at the same frequency, leaving the net baryon number unchanged. However, if the boundary conditions change, e.g., the temperature, the equilibrium can be violated, and the number of dissolved salt molecules or baryons can change.

Because baryon number violation is highly suppressed within the Standard Model, and the *CP* violation is very weak, we consider an extension. For this purpose, we assume that there are new spin-1 particles $X^{4/3}$ that have the electric charge $+4/3\,e$ and can decay in two ways:

$$\text{(a)}\ X^{4/3} \rightarrow u + u\,,$$
$$\text{(b)}\ X^{4/3} \rightarrow \bar{d} + e^{+}.$$

Such particles occur in theories of grand unification (GUT ↓). The two possible final states differ in their baryon number: $+2/3$ (a), and $-1/3$ (b), respectively, and X itself therefore has no specific baryon number. Baryon number is violated in these reactions.

In thermal equilibrium, the reverse reactions take place just as quickly: $u + u \rightarrow X^{4/3}$; $\bar{d} + e^{+} \rightarrow X^{4/3}$. As the Universe expands, the volume increases and the number density of all particles decreases, e.g., at some point, the number of up quarks is so diluted they no longer find each other to react. Then the inverse reactions stop. The $X^{4/3}$ particles decay, but are no longer produced. The thermal equilibrium is disturbed. Since the baryon number is violated in the decays, it is changed!

The antiparticles $\bar{X}^{-4/3}$ also necessarily exist, and in thermal equilibrium have the same abundance as the $X^{4/3}$. The $\bar{X}^{-4/3}$ decay as

$$(\bar{\text{a}})\ \bar{X}^{-4/3} \rightarrow \bar{u} + \bar{u}\,,$$
$$(\bar{\text{b}})\ \bar{X}^{-4/3} \rightarrow d + e^{-}.$$

In total, we still have baryon number zero if, e.g., the decay (a) is just as frequent as (ā), because the baryon number exactly cancels.

If the theory also violates the *CP* symmetry (↓), then the decay rates (a) and (ā) number is violated. This then leads to a baryon asymmetry in the expanding Universe, and could explain current observations.

GUT → p. 276
Discrete Symmetries: Violating *P* and *CP* → p. 190, *CP* Violation in Beauty → p. 258 and The Path to Six Quarks in Three Families → p. 194

Nucleosynthesis After the Big Bang

Helium, Neutrinos and the Matter-Antimatter Asymmetry

In the beginning, the Universe was very hot. In the first few minutes after the Big Bang all kinds of elementary particles, such as quarks, neutrinos and leptons, existed in large numbers. Two minutes after the Big Bang, when the Universe had cooled down sufficiently, the first atomic nuclei were formed. Let's take a look at how this came about.

As described in the article on Matter and Antimatter (↓), almost all particles created in the Big Bang have annihilated with their respective antiparticles to form photons. Due to a tiny asymmetry between matter and antimatter only a small surplus of particles of one in a billion remained (↓). In particular, there were quarks, but no antiquarks, as well as a billion times more photons. Since confinement (↓) does not allow for free quarks, these quarks had to come together to form hadrons. Mesons are made up of quarks and antiquarks but the antiquarks had already disappeared. Therefore, only the lightest baryons were formed, namely protons and neutrons—initially in equal numbers. This meant that the lightest atomic nucleus was already present: the nucleus of hydrogen, which consists of only one proton. Together with the neutrons, all the building blocks for heavier atomic nuclei were available.

As later in stars, the first step in building further nuclei is the fusion of a neutron and a proton to form deuteron, the nucleus of deuterium: $n + p \rightarrow d + \gamma$. The binding energy of 2.2 MeV is emitted in the form of a photon γ. The deuteron can be split into a neutron and a proton by a photon with high enough energy in the reverse reaction. It can therefore only be formed sustainably if the density of high-energy photons has fallen far enough due to the cooling of the Universe. This was only the case around two minutes after the Big Bang. Two minutes doesn't sound like much, but compared to the lifetime of neutrons of fifteen minutes it is already noticeably long. Since neutrons are slightly heavier than protons (Quark Mass Effect ↓) they are less frequently produced in thermal reactions compared to protons, as the Universe cools down. Together with the decay of a fraction of the neutrons during the two minutes, this meant that there were seven times more protons than neutrons before deuterium and then other elements could be formed. Most protons therefore did not find a neutron as a partner for fusion, which is why hydrogen remained the most common element formed in the Big Bang.

From the resulting deuterium nuclei, first tritium and then helium nuclei are produced by fusion with a neutron or a proton. A helium nucleus consists of four nucleons: two protons and two neutrons. As you can see

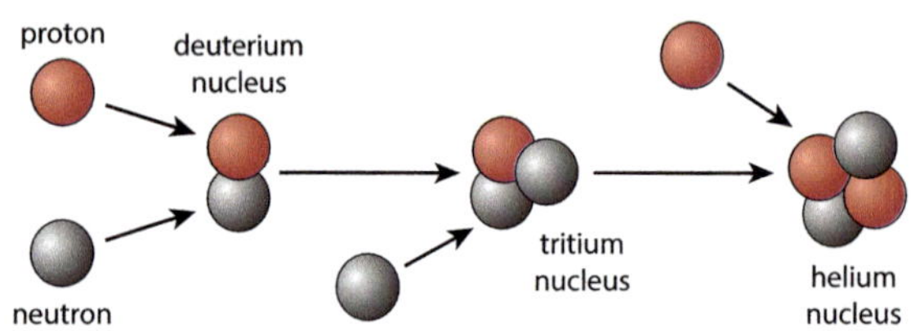

When the first elements are formed, a proton and a neutron first fuse to form deuterium. Further addition of a neutron and a proton creates helium.

Matter and Antimatter → p. 314
Baryogenesis → p. 328
Confinement → p. 206
The Quark Mass Effect → p. 224

Central region of the antennae galaxies. Young stars shine blue and are surrounded by pink-red clouds of ionised hydrogen. Using the abundance of hydrogen and helium in the Universe, it is possible to draw conclusions about the matter-antimatter symmetry.

in the section of the Karlsruhe nuclide map (↓) in the right figure, there is no stable element with five nucleons, element synthesis did not progress any further in the Big Bang. Only small traces of lithium with seven nucleons were created from the fusion of helium and tritium. All elements heavier than helium were essentially only formed later in stars (↓).

The abundances of the light elements in the Universe—and here in particular of hydrogen and helium—reflect the ratio of the abundances of neutrons and protons at the time of the formation of the first elements in the Big Bang. The ratio of protons to neutrons was determined, as described above, by the *baryon-to-photon ratio*, the abundance of protons and neutrons compared to the abundance of photons. The latter corresponds to the asymmetry between matter and antimatter, which therefore can in reverse be determined from the elemental abundances.

As neutrons and protons are converted into each other through beta decay, i.e., reactions with neutrinos, the neutrino density in the early Universe also plays an important role. If there were more than the three known neutrino flavors (electron, muon, tau neutrino), the neutron-proton ratio in element synthesis would have been different. Interestingly, the measurements of element abundances in the Universe can be used to rule out the possibility that there are more than the three known neutrino flavors. (see also Width of Z boson ↓).

The abundance of hydrogen and helium can be determined from spectral lines in the light of gas clouds in the Universe. This makes it possible to conduct particle physics using optical telescopes and astronomical ob-

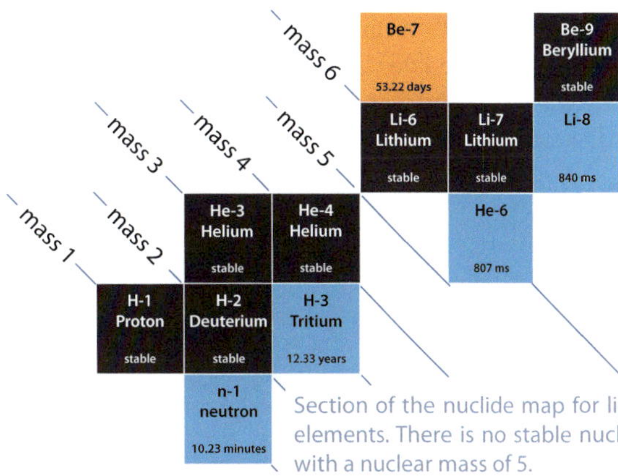

Section of the nuclide map for light elements. There is no stable nuclide with a nuclear mass of 5.

The Karlsruhe Nuclide Chart → p. 228
The Birthplace of Nuclei → p. 342
The Invisible Width of the Z Boson → p. 182

Cosmic Microwave Background

A Look Back Into the Early Universe

In the sixties, Arno Penzias and Robert Wilson worked on a microwave antenna for satellite communication. They had to deal with constant noise in the signal. All attempts to get rid of this noise, e.g., by cleaning away pigeon droppings, were unsuccessful—until they realized what they were up against.

At the same time, astrophysicists Robert Dicke, Jim Peebles and David Wilkinson realized that the Big Bang should have produced microwave radiation that is still measurable today. When Penzias and Wilson saw their unpublished manuscript, they realized what their antenna had measured with the noise: the cosmic microwave background (CMB), a reverberation of the Big Bang. This discovery was the first experimental confirmation of the Big Bang theory. This earned Penzias and Wilson the Nobel Prize in 1978.

The detailed study of the CMB has provided cosmology with a wealth of new insights into the structure of the Universe. We live in an expanding Universe, i.e., it continues to expand and cool down in the process. At earlier times, the Universe was denser and hotter. The rate at which the Universe expands depends on its composition. Which particles make up the Universe? What role does radiation play?

Arno Penzias and Robert Wilson in front of the antenna that led to their discovery of the cosmic microwave background.

According to the general theory of relativity, the energy density of the Universe also influences the geometry of space. If the energy density in the Universe deviates from a certain critical density, space is curved (Evolution of the Universe ↓). How large is the density of the Universe compared to the critical density? This question could be answered for the first time with the precise investigation of the CMB with the American satellites COBE (start 1989) and WMAP (start 2001) and the European Planck satellite (start 2009).

Until about 400 000 years after the Big Bang, all known matter consisted of hydrogen, some helium and small traces of lithium. Up to this time, the temperature was still so high that all atoms were ionized by the photons, i.e., there were negatively charged electrons separated from positively charged atomic nuclei. Radiation was very strongly scattered by these charged particles, so that the photons could only fly straight for short distances before being deflected again by scattering.

It was not until 400 000 years after the Big Bang that the temperature had dropped to such an extent that electrons and atomic nuclei were able to combine and only electrically neutral atoms were left. Without charged particles, the scattering of radiation was eliminated and the photons became free. We call this the *decoupling of radiation*. Since then, the radiation has traveled al-

Evolution of the Universe → p. 346

most undisturbed in the Universe. It forms the CMB as we observe it today.

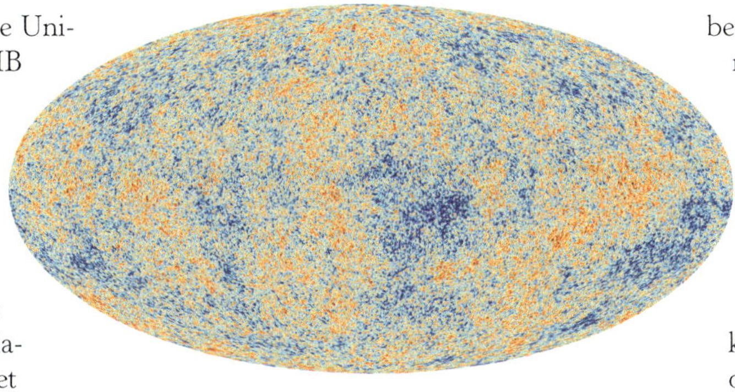

Sky map of the temperature of the cosmic microwave background. Red areas are slightly hotter, blue areas slightly colder. The temperature fluctuations correspond to density fluctuations in the early phase of the Universe.

The pattern in the above picture can be understood as follows: The observed density fluctuations are a consequence of oscillations. The matter was set into oscillation, driven on the one hand by the attraction of the gravitational force and on the other hand by the pressure generated by the radiation due to the scattering of matter. Precise studies of the CMB still show us today a snapshot of the pattern of these fluctuations in the local density of electrons and atomic nuclei at that time. The intensity and size of the oscillations depend very much on the overall energy density in the Universe and the densities of radiation and matter at that time, which is why these can be determined from the study of the CMB.

According to this, the Universe has almost exactly the critical density, i.e., space is not curved on large scales. While the radiation density is very small today and no longer significant for the overall energy budget of the Universe, the matter density is about 30 % of the critical density. Interestingly, the oscillation patterns also show that the particles known to us only contribute about 4 % to this, the remaining 26 % consists of previously unknown, uncharged particles, the *dark matter*. The big question remains: What does the remaining 70 % of the energy budget of the Universe consist of? It must be a different, previously mysterious form of energy, which we call *dark energy* (↓).

70 % dark energy and 26 % dark matter—we have to realize that we do not know 96 % of the content of the Universe and do not know whether and if so which particles are behind it. But the mysteries continue. Particle physics also has no conclusive answer to the question of why there is matter but no antimatter in the Universe (Baryogenesis ↓). And thus we also do not know why the remaining 4 % of known matter exists at all. The extraordinary successes of the investigation of the CMB have thus shown us with high precision that our knowledge of particle physics to explain the Universe essentially consists of ignorance.

The spectrum of density fluctuations of the CMB shows how often fluctuations of a certain extent occur.

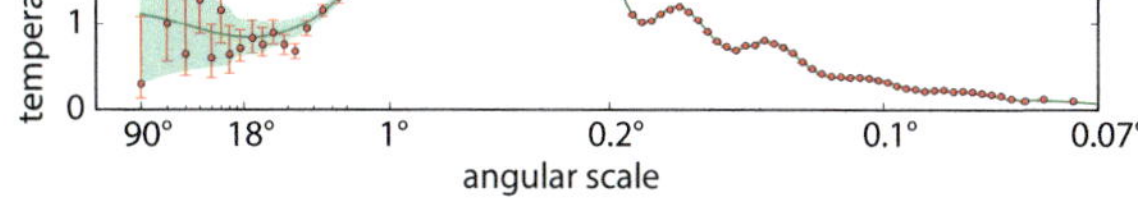

Dark Energy → p. 284
Baryogenesis → p. 328

The Largest and the Smallest

The Neutrino Mass and the Large-Scale Structure of the Universe

There is an astonishing connection between the largest observable objects in the Universe with dimensions of 100 million light years, and neutrinos, the lightest and most elusive of all elementary particles. With a density of around 330 per cm^3, neutrinos are the most common particles in the Universe alongside photons. If the three neutrino masses together weighed only 50 eV/c^2, they would dominate the total mass in the Universe. In comparison, the otherwise lightest known elementary particle of matter, the electron, is four orders of magnitude heavier with a mass of 510 999 eV/c^2. Because they are so numerous, neutrinos could therefore play an important role in the evolution of the Universe and, in particular, significantly influence the structure of the distribution of galaxies.

In the early Universe, particles and antiparticles annihilated or reacted with each other when they collided. The weaker the particles interact, the less frequently this happens. In addition, as the Universe expands and the density of particles decreases, encounters become increasingly less likely. As neutrinos only interact weakly, shortly after the Big Bang (about one second), the average time span to the next reaction between a neutrino and its antiparticle is expected to be longer than the current age of the Universe.

In contrast to the charged particles, the annihilation of neutrinos came to a standstill very early, which is why they survive to this day with a very high density of 330/cm^3. The density of protons and electrons is about a billion times lower. As these cosmic neutrinos today have an extremely low energy, their detection has not yet been successful, but their existence is consistent with all other cosmological observations, such as the abundance of hydrogen and helium (Baryogenesis ↓).

Today we observe a rich structure with clusters of galaxies and empty areas in between. In the early Universe, however, matter was almost evenly distributed. This can be observed with the cosmic microwave background (↓), which shows a constant density of matter

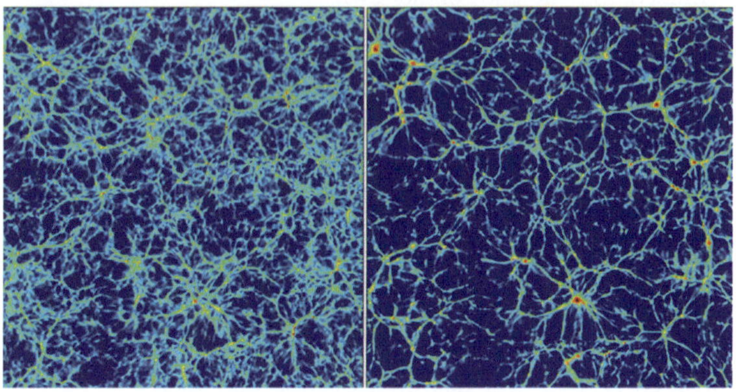

Simulations of the distribution of galaxies in the Universe with large (left) and small neutrino masses (right). The colors indicate the matter density visible in the Universe. A neutrino mass of zero was assumed on the right and 1.9 eV on the left. With a higher mass (left), the influence of neutrinos on the formation of structures increases, which are therefore more blurred.

Baryogenesis → p. 328
Nucleosynthesis After the Big Bang → p. 330
Cosmic Microwave Background → p. 332

about 400 000 years after the Big Bang with fluctuations of only 1/100 000 of the average density. These small fluctuations have since grown into the structures visible today due to the gravitational attraction of matter.

It takes a certain amount of time for a cloud of particles to collapse due to the force of gravity, and thus increase the local density. The collapse competes with the escape of the particles due to their sometimes very high speed. If too many particles fly away before the cloud can collapse, it will disappear before the density of particles and matter can sufficiently increase. Structure formation is then not possible. The lighter the particles are, the longer it takes until they are slow enough to allow structure formation at all due to the cooling of the Universe. Therefore, the greater the contribution of neutrinos as very light particles, the longer the beginning of structure formation would have been delayed in the development of the Universe, with visible consequences for the appearance of today's Universe.

In order to measure the large-scale structure of the Universe, complex sky surveys are carried out with measurements of the positions of several million galaxies. To do this, the light from the many millions of galaxies must be examined spectroscopically. A task that could only be accomplished with the advent of modern computers and digital cameras. The measurements show that the galaxies in the Universe are not distributed evenly, but rather like a sponge, with "walls" of high galaxy density and empty bubbles in between. As we know from model calculations, the size of these bubbles of around 100 million light years in diameter would be different today if neutrinos had played a major role in the formation of the structure. From these measurements, we can now restrict the sum of the three neutrino masses to a maximum of 0.5 eV/c^2. Through the development of new methods in a few years it will be possible to measure the structures in the Universe so precisely that the influence of neutrino masses of only 0.03 eV/c^2 can be seen. From the oscillation experiments (↓) we know that one of the neutrino species must have a mass of at least 0.05 eV/c^2. Therefore, we should soon not only be able to narrow down the influence of neutrinos on the structure of the Universe, we should be able to see it. In fact, the path via the large-scale structure of the Universe is currently our best chance to measure the neutrino mass.[1]

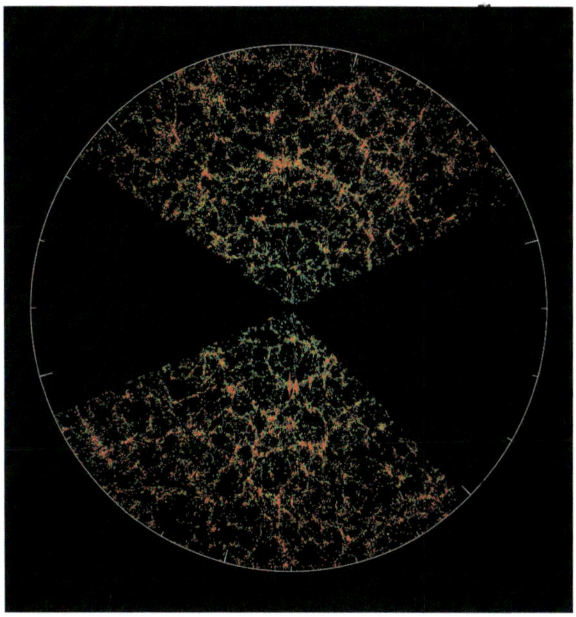

Sections of a measured three-dimensional distribution of galaxies. Each point represents a galaxy with typically 100 billion stars. The outer circle is two billion light years away from the Earth in the center of the image.

[1] To be precise, it is the sum of the three neutrino masses which is measured with the large-scale structure. Direct measurements (↓) via beta endpoint spectroscopy for example, measure a weighted sum of the three masses and could thus provide a cross check of the concept of the cosmological model.

Neutrino Oscillations → p. 198
Measurement of the Neutrino Masses → p. 256

Cosmic Rays

Particles from Outer Space

After the discovery of radioactivity by Henri Becquerel in 1896, it was recognized that some atomic nuclei (↓) are unstable: they decay and emit radiation. In 1912, Victor Hess investigated how the intensity of radiation changes with increasing height above the Earth's surface. If the radiation was emitted by radioactive elements in the ground, he expected the intensity to decrease as he ascended in a balloon with his measuring instruments. Surprisingly, however, from a certain height he found an increase in radiation intensity and concluded this radiation was reaching the Earth from outer space. Today, this cosmic radiation has been studied in great detail. 85 % of it consists of protons and 12 % of helium and other heavier nuclei. Another 2 % are electrons. The energies of these particles detected so far span an enormous range from 10^6 eV up to 10^{20} eV.

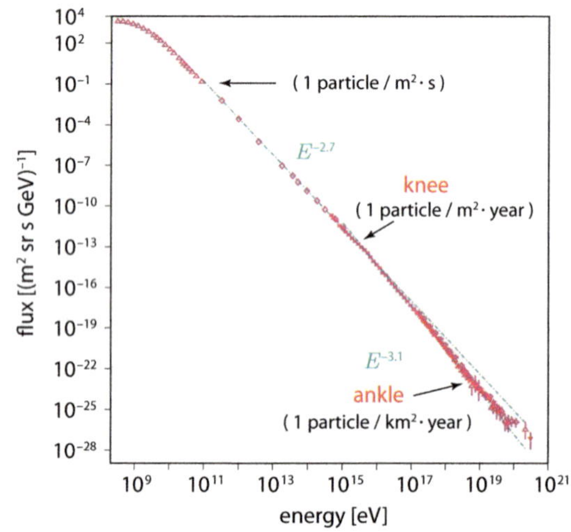

Energy spectrum of cosmic rays above 100 MeV. The numbers on the curve indicate how many particles with energies above the marker hit the Earth.

The intensity of the radiation is relatively constant below 10 GeV and then falls over the entire energy range extending over many orders of magnitude almost uniformly proportional to the 2.7th power of the energy. This means that the intensity of the radiation at $E = 10^{11}$ eV, for example, is a factor of $10^{2.7} = 500$ less than at the ten times lower energy $E = 10^{10}$ eV.

At lower energies, approximately two hundred particles per second hit every square meter of the Earth's atmosphere. At the highest energies, it is only one particle per km² per year. To study the particles with the highest energies, huge areas must therefore be monitored with detectors (↓).

A plausible explanation of where the particles come from and how they are accelerated was provided by Enrico Fermi in 1949. If a particle moves towards a shock wave, it can be thrown back at a higher speed, similar to a tennis ball from a tennis racket. Shock fronts of very hot and ionized matter occur, for example, in supernova explosions (↓) or in the vicinity of black holes. The movement of the ionized matter generates electric currents and thus magnetic fields. Magnetic fields direct the particles on circular paths and thus repeatedly through the shock front. The higher the energy of the particles, the larger their orbital radius (see Accelerators ↓). As long as the radius is smaller than the extent of the

From Nucleus to Nuclei → p. 226
Ultra-High-Energy Cosmic Rays → p. 338
Supernovae as Particle Physics Laboratories → p. 324
How Does a Particle Accelerator Work? → p. 66

region in which the acceleration takes place, the particles will continue to gain energy. Eventually their radius will extend beyond the confining object and they will escape and their energy will no longer increase. Accordingly, the maximum achievable energy depends on the extent and the strength of the magnetic fields of the accelerator region. In the Milky Way, protons can obtain a maximum energy of about 10^{15} eV before their radius reaches the size of the galaxy. In the spectrum of cosmic rays, this is indicated by a kink and a steeper drop in intensity above this energy. It is assumed that extragalactic cosmic rays are dominant at energies above 10^{18} eV as an upward kink is observed in the spectrum of intensities, which then becomes somewhat flatter, again in proportion to $1/E^{2.7}$. The image of this spectrum resembles a human leg, which is why the first bend is referred to as the *knee* and the second as the *ankle*.

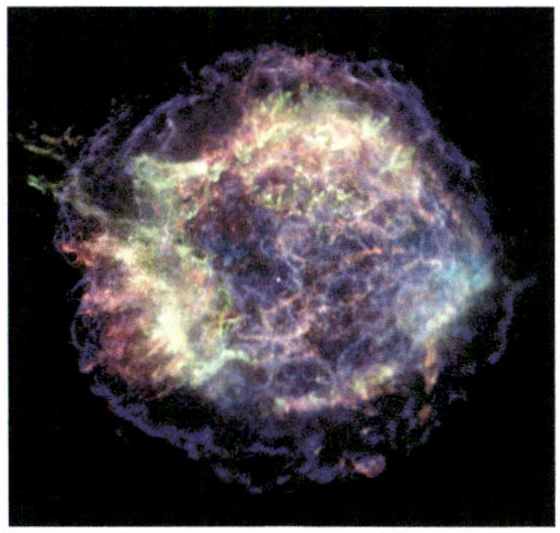

The supernova remnant Cassiopeia A. Supernova explosions are a possible source of cosmic ray particles.

When the cosmic particles reach the Earth's atmosphere, they interact with the air molecules. At low energies, the particles lose all their energy very high up in the atmosphere and don't reach the Earth's surface. The only way to study them is to detect the primary cosmic particles directly high up in balloons or on satellites in space (↓). Since the flux of the particles is sufficiently large at low energies, detector areas of $\sim 1\ \text{m}^2$ are sufficient. Almost the entire variety of detection methods (Experimental Basics ↓) is used to detect particle tracks. Similar to the detectors at CERN (Typical Detectors ↓), the particles can be identified based on the curvature of their path in magnetic fields and the composition of cosmic rays can thus be determined (Position and Momentum Measurements ↓). Examples include the PAMELA satellite or the AMS detector on the International Space Station. Supplemented by measurements of cosmic rays on Earth with detectors such as Auger or the smaller predecessor KASCADE of the Karlsruhe Institute of Technology the composition of cosmic rays at low energies and especially in the region of the "knee" above 10^{15} eV could be studied. This revealed that the end of acceleration in the Milky Way is reached at these energies.

The exact origin of cosmic rays is not yet known. This would require measuring the direction from which the particles come. Why this is easier said than done will be explained in the next article.

AMS experiment on the International Space Station ISS. The AMS spectrometer measures the type and energy of cosmic ray particles. The detector can be seen here piggybacked on a side arm of the ISS.

Particle Physics in Space → p. 316
Chapter 3 Experimental Basics → p. 65
Typical Detectors at Accelerators → p. 128
Position and Momentum Measurement → p. 76

Ultra-High-Energy Cosmic Rays

The Search for High-Powered Cosmic Accelerators

The energies of cosmic particles (↓) hitting the Earth's atmosphere span an enormous range: from 10^6 eV up to 10^{20} eV. The highest measured energy of a particle to date was $3 \cdot 10^{20}$ eV, which is 20 million times what we have been able to achieve with our largest accelerator, the LHC.[1] $3.3 \cdot 10^{20}$ eV is the equivalent of 50 joules. This corresponds to the kinetic energy of a tennis ball traveling at 110 km/h, well beyond what we can create on Earth in a laboratory.

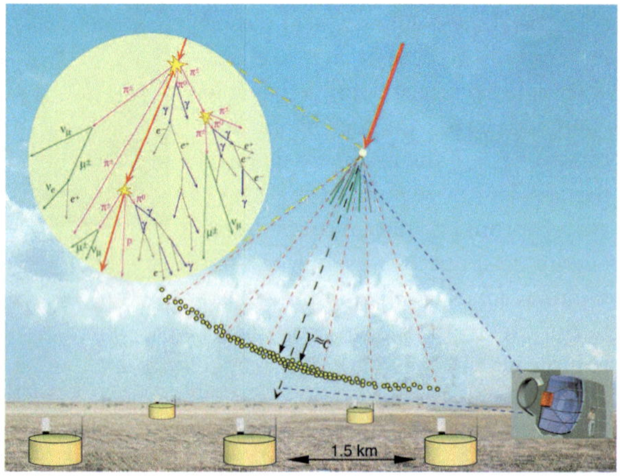

Particle shower after reaction of a high-energy cosmic particle in the Earth's atmosphere. At high energies, the shower reaches the Earth's surface. Particles arriving simultaneously can be detected over distances of several kilometers.

The exciting question is: What are the sources of these particles? Which objects in the Universe have shock fronts and magnetic fields sufficiently large to accelerate particles to such unimaginably high energies? In (optical) astronomy it is possible to observe and localize objects such as stars and galaxies because telescopes can be used to measure the direction from which the light reaches the Earth. In order to find the sources of high-energy cosmic rays, it should be possible to measure the direction from which they originate. Unfortunately, this is anything but easy. Since the cosmic ray particles are electrically charged, their trajectory is deflected by magnetic fields as they travel from the source of acceleration to Earth. Thus the observed arrival direction of the particles has little or nothing to do with the direction towards their source.

Nevertheless, the higher the energies of the particles, the less the trajectories are deflected and the straighter they travel. So in order to recognize where they come from and to create sky maps of the sources, particles with the highest energies must be observed. Unfortunately, as shown in the intensity distribution in the previous article, the higher the energy of the particles, the rarer they are. Furthermore, at the most interesting energies the cosmic microwave background (↓) also gets in the way. At these high energies, for example the proton energy is sufficient for a reaction between cosmic rays and microwave photons to generate particles such as pions or neutrons through a resonance reaction. The original particle is either absorbed or only continues its journey with greatly reduced energy. This *GKZ cutoff*, named after the physicists Kenneth Greisen, Vadim Kusmin, and Georgi Zatsepin, leads to a steep drop in intensity in the spectrum of cosmic rays above 10^{19} eV.

[1] For particle reactions, the energy available in the center of mass system is relevant. At the LHC, this corresponds to the total energy of the two colliding particles, whereas in the collision onto resting particles in the Earth's atmosphere only a fraction of the energy is available for reactions, which is nevertheless more than a hundred times the center-of-mass energy of the LHC.

Cosmic Rays → p. 336

As only very few cosmic particles with the highest energies of interest reach the Earth, huge areas have to be monitored in order to be able to observe them in sufficient numbers. The international Pierre Auger Observatory, built in the Argentinean Pampa, consists of 1600 detectors spread over 3000 km^2. The high-energy cosmic rays can be measured at the Earth's surface, because the particles generate particle showers when they hit the atoms in the atmosphere. This produces pions, for example, which in turn decay into muons, electrons and neutrinos. At high energies, the particles produced can in turn generate more particles, which is why there is an increasing shower of particles. High-energy muons in particular can reach the ground (see Special Relativity ↓). The shower front of simultaneously arriving muons can extend over many square kilometers, the larger the area, the higher the energy of the primary cosmic particle. Such extended showers were first discovered in 1938 by Pierre Auger with widely spaced detectors on the Jungfraujoch in Switzerland. Cosmic ray showers account for around half of our natural radiation exposure and are also the reason why radiation exposure is greater at high altitudes, especially in airplanes, than at sea level.

In a further effect, in the atmosphere, the particle showers excite the nitrogen molecules in the air to glow. This phenomenon, known as fluorescence, makes it possible to also observe the showers with telescopes on moonless nights, even if the showers do not reach the ground. As the structure of the shower can be better measured with fluorescence telescopes and the energy can be better measured on the ground via the arriving particles, the Pierre Auger Observatory uses a combination of both techniques.

The Pierre Auger Observatory observes a steep drop in the spectrum as expected from the GKZ cut-off. However, it is still unclear whether this is really the GKZ cut-off or whether the end of the accelerator energy of extragalactic sources has been reached. This question, as well as the question of which objects can be considered as accelerators at all, is being studied in a collaboration of many projects with *multimessenger astronomy* (↓). Various conceivable scenarios for accelerating cosmic particles simultaneously generate different energy spectra of gamma rays and neutrinos. Many different detectors are therefore used to search for simultaneous signals generated by neutrinos, photons and high-energy cosmic rays in order to track down the sources of the extremely high energy particles.

One of the 1600 detectors of the Pierre Auger Observatory. Such a detector measures incoming muons in a water tank via the Cherenkov effect. The detectors are spaced about 1.5 km apart over an area of 3000 km^2.

Cosmic Microwave Background → p. 332
Special Theory of Relativity in Particle Physics → p. 92
Multimessenger Astronomy → p. 340

Multimessenger Astronomy

New Windows Into the Particle Universe

Are charged particles, high-energy cosmic rays (↓), the only messengers from galaxies far, far away? When the high-energy cosmic ray particles, mostly protons, collide with matter or photons in their regions of origin, pions are created. The pions in turn decay among other things into photons and neutrinos, which can reach us undisturbed, in a direct line and at almost the speed of light. They are therefore perfect messenger particles from the sources of cosmic rays. Depending on the type of suspected source, different correlations are expected between the energy spectra and relative rates of protons, photons and neutrinos. The observation of the different types of particles can therefore provide information about the nature of the sources and the underlying acceleration processes. As large masses are usually moved very quickly in the sources, gravitational waves are often generated. These are distortions of spacetime predicted by Albert Einstein that propagate at the speed of light. They also reach us in a direct line and undisturbed and can complete the picture of the sources. The aim of so-called *multimessenger astronomy* is to observe the four messengers—cosmic particles such as protons, neutrinos, photons and gravitational waves—together.

Separate detectors are used for each of the messengers. High-energy cosmic particles generate particle showers in the Earth's atmosphere, which can be detected on the ground with extensive detector arrays such as Auger. The largest detector for high-energy neutrinos (↓) is IceCube (↓). At the South Pole, an ice volume of about one km^3 has been instrumented with photomultipliers to detect the Cherenkov radiation of the charged electrons, muons, or tauons created by high-energy neutrinos. A very similar project is KM3NeT in the water of the Mediterranean Sea.

Cherenkov telescopes such as H.E.S.S. or MAGIC are used in astronomy to detect *gamma radiation*, which refers to photons at the highest energies. They generate fast electrons in the Earth's atmosphere that emit Che-

The H.E.S.S. telescope in Namibia for measuring high-energy photons. The 28 meter mirror reflects Cherenkov light onto the photomultiplier camera located at the top.

Cosmic Rays → p. 336, Ultra-High-Energy Cosmic Rays → p. 338
Neutrinos → p. 196
The IceCube Detector → p. 134
Cherenkov Radiation and Particle Identification → p. 82
Scintillators and Photomultipliers → p. 74

renkov light (↓). This Cherenkov light is imaged onto a camera made of photomultipliers (↓) and this way the gamma radiation is detected.

The first detection of gravitational waves in 2015 was a major scientific breakthrough. Since then, many other signals, e.g., from the collision of black holes, have been detected with the large laser interferometers LIGO and VIRGO. With the detection of photons via the electromagnetic interaction, protons via the strong interaction, neutrinos via the weak interaction and gravitational waves, the Universe can be studied with all four fundamental forces of Nature (↓).

Previous measurements have revealed protons as well as neutrinos and photons with such high energies that they must be of extragalactic origin, as the extent of the Milky Way and its magnetic fields are not sufficient to accelerate particles to such high energies. If neutrinos, photons and cosmic rays were frequently observed from the same direction, the sources could be identified. So far, this has not been possible due to too few observations. The observation of a short burst that would appear in all messenger particles together with a gravitational wave signal would be spectacular.

One example where this was at least partially successful is the supernova 1987 A (↓), from which neutrinos and, two hours later, optical photons were measured. Another example is the gravitational wave signals measured by LIGO and VIRGO with the almost simultaneous detection of high-energy photons with the Fermi satellite. They presumably originate from the merger of two neutron stars or a neutron star with a black hole. There is also evidence for the simultaneous observation of high-energy neutrinos and gamma radiation.

In order to increase the probability of such events, the instruments for all messengers are being further improved and all projects are connected by a global network, with which signals from all detectors are brought together. With an alarm system, all detectors can be alerted if a signal occurs in one detector.

Multimessenger astronomy has undergone a promising development in recent decades and is on the threshold to expand our view of the particle physics processes in the Universe far beyond classical astronomy.

The VIRGO gravitational wave detector near Pisa in Italy. By superimposing the laser beams bouncing forth and back in the two 3 km long arms, the smallest distortions of spacetime can be measured.

The Four Fundamental Forces of Nature → p. 16
Supernovae as Particle Physics Laboratories → p. 324

The Birthplace of Nuclei

We Are All Star Dust

In the article on nuclear structure (From Nucleus to Nuclei ↓), we saw that there is a huge variety of stable and unstable nuclei, which have all been found on Earth. But where did they come from? After all, we assume that the Universe originated from radiation energy about 14 billion years ago in the Big Bang. While a certain amount of light nuclei (deuteron, helium, and lithium, together with hydrogen) could form in the first three minutes after the Big Bang (↓), this is not the case for the heavier ones. But where do the other light and especially the heavier nuclei come from?

To make progress with this question, one must take a closer look at stars. These emerge from huge gas clouds, which consist mainly of hydrogen. Although the attraction due to gravity between the individual gas particles is very weak, it is still collectively strong enough for a cloud to form and then to contract. In doing so, it becomes denser and especially hotter. This is because the corresponding energy that would be required to disassemble the star into its components is released during its formation and is transferred into radiation and heat energy which is mostly trapped inside the star. At some point, the density and temperature, i.e., the kinetic energy, are so high that two protons can overcome their electromagnetic repulsion. They then come close enough for the strong interaction to kick in and they fuse into deuterium, radiating off a positron and an electron neutrino in the process (see figure). The neutrinos (↓) can then be detected on Earth. In this fusion process, the binding energy of the deuteron is released, minus the electron mass and the proton-neutron mass difference, which is still quite a lot, especially since with 98 % almost all the energy is released as radiation energy—the star begins to shine! This process requires very large densities and temperatures, otherwise the repulsion between the like-charged protons cannot be overcome. This is exactly why it is so difficult to make nuclear fusion usable as an energy source on Earth.

In the next step in the star, another proton can be bound to the deuteron and form ^{3}He (see figure below). Once enough of these helium-3 nuclei are formed, two of them can fuse, radiating two protons, to form ^{4}He, the alpha particle. One might wonder why two deuteron nuclei do not directly fuse into the alpha. The reason is that the latter reaction is much less likely than the production of ^{3}He, as there are many more protons than deuteron nuclei available in the Sun and because the direct transition from two ^{2}H to the alpha is only possible by radiating two photons. Energy conserva-

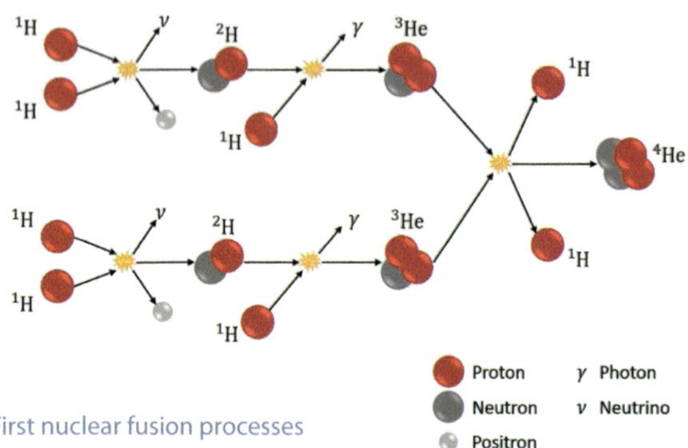

First nuclear fusion processes in a star

From Nucleus to Nuclei → p. 226
Nucleosynthesis After the Big Bang → p. 330
Neutrinos From the Sun → p. 344

tion allows the transition from two stable particles into a single stable one only by radiating at least one additional particle, as the binding energy must be dissipated. The fact that two photons are needed here is due to the quantum mechanical selection rules. Such an electromagnetic transition happens much less frequently than transitions that are made possible by the strong interaction—like the fusion of the ^{3}He nuclei described above.

Our Sun is still in this phase. In older or heavier stars, further fusion reactions take place, in which the very stable alpha particle plays an important role as it acts as a crucial building block: First, two alpha particles merge to form Beryllium-8, which then fuses with another to form Carbon, ^{12}C. But there is a problem: Beryllium-8 is unstable and it decays back into two alpha particles after a short time (it has a lifetime of 10^{-16} s). Thus, carbon can only be produced in large quantities if the rate for the fusion of an alpha particle with ^{8}Be is very high, such that it happens before the ^{8}Be can decay. This is only possible if there is a resonance in the carbon spectrum that precisely matches the energy of the Beryllium-8 nucleus together with an alpha particle (Resonances ↓). Based on these considerations, the British physicist Fred Hoyle predicted this resonance in carbon as early as 1954, which was experimentally found three years later. This fine-tuning of the resonance energy in the Hoyle state and the energy of ^{8}Be + alpha

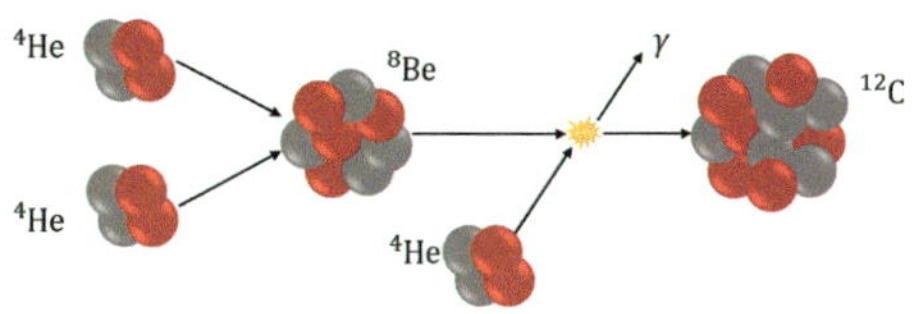

Second stage of nuclear fusion processes in a star

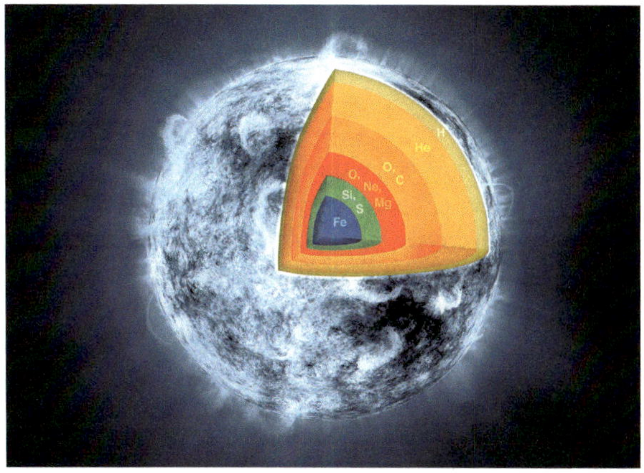

Radial structure of a star

raises the exciting question of how much fine-tuning must exist in the Standard Model for carbon-based life as we know it to have been able to arise (↓).

Further fusion processes lead to the production of all elements up to iron. Then the fusion comes to a halt, as iron has the lowest mass per nucleon, i.e., the highest binding energy per nucleon. Thus, iron gradually accumulates in the interior of stars, while lighter elements can still be found in the outer shells. The structure is as shown in the picture above.

So where do nuclei heavier than iron come from? Some can be produced incidentally by capturing neutrons. The rest, and especially the very heavy nuclei, can only be produced if an atomic nucleus simultaneously absorbs many neutrons. However, this requires a neutron density that can only be found in supernova explosions (↓) or the fusion of neutron stars. Both processes also release huge amounts of energy, thereby distributing the nuclei in space.

Resonances → p. 56
Did Our Life Emerge by Chance? → p. 292
Supernovae as Particle Physics Laboratories → p. 324

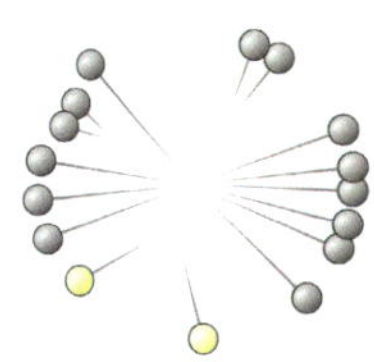

Neutrinos From the Sun

A Deep Look Inside Our Star

Is it possible to observe the Sun from a mine 2000 meters deep underground? You can with neutrinos! Neutrinos are the only elementary fermions that react exclusively via the weak interaction (and via gravity, which is even weaker). On the one hand, this makes the measurement of neutrinos very difficult, on the other hand, it is possible for them to escape directly and unhindered from otherwise inaccessible areas of extremely high density, which means that they can be used to look directly into the center of the Sun.

The Sun's energy is generated by the fusion of hydrogen to helium (↓). This produces electron neutrinos, of which 65 billion should arrive on Earth per second for every square centimeter. The American physicist Ray Davis and his team were the first to attempt to measure solar neutrinos. To do this, they stored over 600 tons of the cleaning fluid C_2Cl_4 in a tank deep in a mine. On average about every two days, an electron neutrino from the Sun should convert a chlorine atom into an argon atom (so-called inverse beta decay). The trick was to fish the few argon atoms out of the tank and detect them via their radioactive decay. What he found was astonishing: only about a third of the expected argon atoms and therefore neutrinos were found. Although the result proves that nuclear fusion takes place in the Sun, it was unclear why so few neutrinos were produced.

The Sun is an active star—visible here in ultraviolet light in the structures of its outer layers. The Sun's energy is generated by nuclear fusion. This produces electron neutrinos, which can be observed on Earth and used to study the processes in the Sun's interior.

Ray Davis at his experiment in the American Homestake Mine. The tank contains 600 tons of C_2Cl_4.

The Birthplace of Nuclei → p. 342

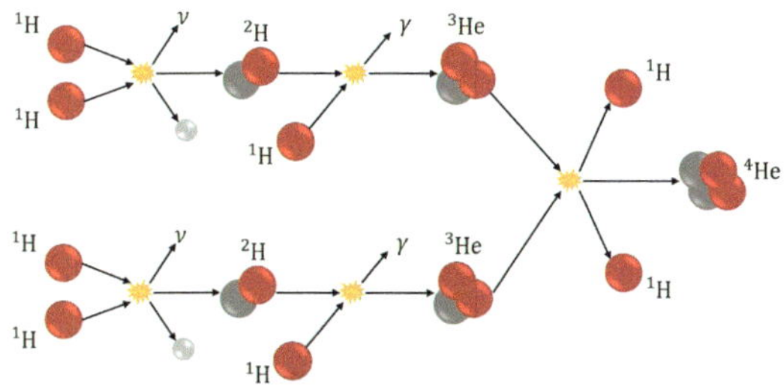

Two hydrogen nuclei fuse in the interior of the Sun to form a deuterium nucleus (^{2}H), which also produces an electron neutrino. Hydrogen and deuterium fuse to helium-3 (^{3}He) and two ^{3}He fuse to helium-4 (^{4}He). The formation of a helium-4 nucleus therefore produces two electron neutrinos.

For a long time, scientists were skeptical about the results and doubted that the experiment and the theoretical calculations were free of errors. However, the alternative explanation that the deficit is caused by neutrino oscillations (↓) is much more exciting. Since only electron neutrinos are produced in the Sun and the detector can only measure electron neutrinos, an oscillation to muon neutrinos on the way to Earth would explain the deficit.

Image of the Sun measured with neutrinos in the Super-Kamiokande experiment. The red to blue colors indicate the direction of high to low neutrino flux.

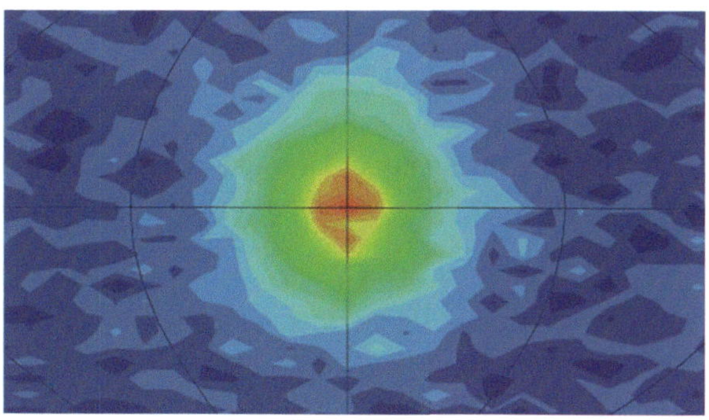

The SNO experiment consists of an acrylic sphere filled with 1000 tons of heavy water (D_2O). The picture shows the approximately 10 000 photomultipliers used to measure Cherenkov light generated by particles inside.

A number of other experiments have been set up to solve the mystery of solar neutrinos: GALLEX and Borexino in Italy, SAGE in Russia, Super-Kamiokande and KamLAND in Japan, and SNO in Canada. Super-Kamiokande was also able to measure the direction of the neutrinos at high energies (> 5 MeV) and thus create an image of the Sun in neutrino "light". SNO, which could also measure muon and tau neutrinos, conclusively proved that neutrino oscillations are responsible for the observed deficit of electron neutrinos and that all neutrinos expected from nuclear fusion in the Sun are indeed emitted. For these findings, the experiments by Davis in 2002 and SNO in 2015 were awarded Nobel Prizes.

In the meantime, many details of solar nuclear fusion processes have been studied, the exact temperature inside the Sun has been determined and the presence of atomic nuclei other than hydrogen and helium in the center of the Sun has been detected. With our increased knowledge of the properties of neutrinos, they now serve as messenger particles from the interior of the Sun.

Neutrino Oscillations → p. 198
Verifiying Neutrino Oscillations → p. 254

Evolution of the Universe

A Glimpse Into the Very Distant Future

Gravity plays no role at the short distances of particle physics. However, the situation is completely different when we look at larger objects, such as stars, where gravity and the effects of other forces influence each other. Gravity is the all-dominant force when we look at the processes in a planetary system, at the formation of stars and galaxies or the evolution of the Universe as a whole, which we will look at in this article.

The best description of gravity is Albert Einstein's general theory of relativity. This describes the effect of gravity through the distortion of spacetime. The mass of an object and the curvature of spacetime in its surroundings are directly related. What we perceive as the deflection of a beam of light in a gravitational field, for example, is therefore a beam of light that propagates in a straight line in a curved spacetime.

You can try to get an idea of curved spaces by looking at curved surfaces. On a curved surface, such as a globe, the sum of the three angles in a triangle is greater than 180°, unlike on a flat surface. This sum is also smaller or greater than 180° in curved spaces. In fact, measuring the curvature of the Universe with the anisotropy of the cosmic microwave background (↓) corresponds to determining the angular sum of a very large triangle.

According to the theory of relativity (↓), there is an equivalence between energy and mass with $E = mc^2$. This is why energy in any form, whether as mass, as radiation or as kinetic energy, is closely linked to the distortion of the geometry of spacetime. As the name suggests, time also plays an important role in spacetime. How the Universe as a whole develops over time is therefore identical to the question of what the energy content of the Universe is. In their simplest form, the distortions can be described by two simple parameters: a *curvature of space* $k = 0, \pm 1$ (angular sum equal, greater, or less than 180°) and a time-varying *scale factor* $R(t)$. When we talk about the Universe expanding, we are referring to the growth of the scale factor $R(t)$. If R increases, all length scales increase. The fact that R is actually growing in our Universe can be observed by the change in the wavelength of light. Light from distant galaxies is shifted into the red, as the Universe and thus also the wavelength of the light has expanded since it was emitted. Since R is getting bigger, it must have been smaller in the past and we assume that everything began with $R = 0$ in the Big Bang.

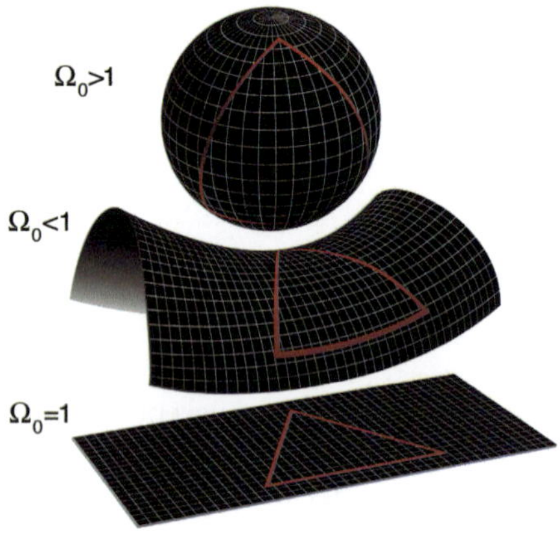

The sum of the angles of a triangle is greater or less than 180 degrees depending on the curvature of the surface.

Cosmic Microwave Background → p. 332
Special Relativity → p. 38

The curvature k and the development of R are linked to the energy content of the Universe. We measure the energy density Ω_0 in units of a critical energy density, which is about 5 GeV/m^3 for the whole Universe. If the energy density is equal to the critical density ($\Omega_0 = 1$), the Universe is flat ($k = 0$). If the energy density is greater or less (Ω_0 greater or less than 1, $k = \pm 1$), it is curved. How R develops over time depends on the curvature and the nature of the energy density. If the flat Universe were dominated by matter, i.e., predominantly filled with massive particles, the energy density would decrease in proportion to $1/R^3$. If the Universe were filled only with radiation, i.e., with massless particles such as photons, then with increasing R the energy density would decrease more strongly in proportion to $1/R^4$ and the expansion rate would decrease more rapidly.

In fact, both contributions play a role, and for the first 10 000 years after the Big Bang, the Universe was dominated by radiation, after that by matter. If the energy density were higher than the critical density, the slowing expansion could even reverse itself at some point and lead to a shrinking of the Universe and thus ultimately to a reversal of the Big Bang, the so-called *Big Crunch*. However, as we know today, the Universe is flat, i.e., the critical energy density prevails within the measurement accuracy. For a long time, it therefore looked as if the future of our Universe would be an eternal, ever slower expansion. However, precise measurements of the redshifts of supernovae show that, on the contrary, the expansion of our Universe is accelerating. This can be explained if the Universe contains a constant (independent of space *and* time) energy density, a so-called *dark energy* (↓). When this dominates, the rate at which R grows is proportional to R itself, which leads to exponential growth.

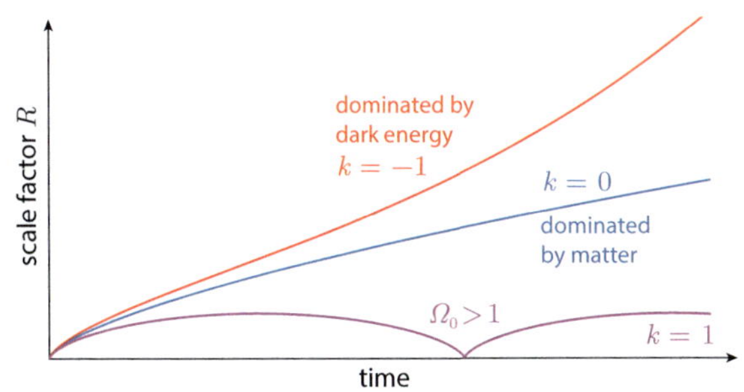

Temporal development of the scale factor R and thus the expansion of the Universe for different forms of energy density.

It stands to reason that all matter, radiation and ultimately also dark energy should be explainable in terms of particle physics. How can we relate the composition of the Universe to the Standard Model of particle physics? Around 70 % of the energy density comes from dark energy, for which there is no explanation in the Standard Model. The remaining 30 % behave for the most part like matter. Of these 30 %, however, only around 4 % consist of the known particles of the Standard Model, which should not even exist due to the symmetry between matter and antimatter (↓). We call the other 26 % whose particle nature is still unknown dark matter (↓).

The Universe does not stay the same, but changes dynamically. Presumably it will continue to expand faster and faster forever. When trying to understand the evolution of the Universe, it becomes clear that despite all the successes in particle physics, we have not yet fully understood many things—especially the big picture of the Universe—and therefore many exciting discoveries still lie ahead.

Dark Energy → p. 284
Matter and Antimatter → p. 314
Dark Matter → p. 282

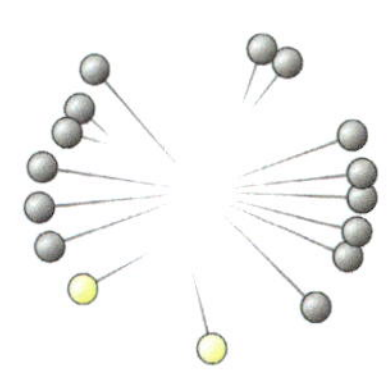

Is it all worth it?

The question is often asked: Why do we do all this? Basic research—especially in particle physics—is far from free of charge. It requires not only the funding of highly skilled personnel but also the construction and operation of complex and costly experimental facilities. At first glance, it might seem reasonable to argue that such resources could be spent more "sensibly". But what does sensible truly mean? Basic research in particle physics is not a luxury—it is a vital expression of the values of a free and open society. It serves a cultural purpose in its own right. And yet, it offers more than that. Much more. We would like to conclude this book with a heartfelt plea: a plea for the continued support of basic research in particle physics.

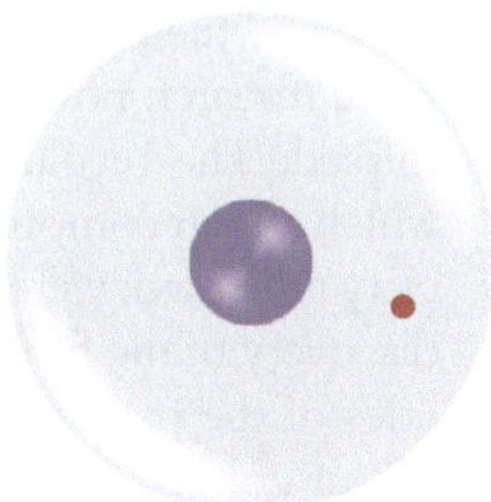

Direct benefits to society are clearly visible in spin-offs like magnetic resonance imaging (MRI) and positron emission tomography (PET), which would be unthinkable without basic research. Moreover, cutting-edge research often demands the development of new technologies. This includes advancements in ultra-fast electronics and novel communication techniques. It is no coincidence that the World Wide Web, which revolutionized global communication and brought the world closer together in a virtual space, was invented at CERN. Its original purpose was to facilitate more efficient communication among researchers working on international particle physics experiments. Another rapidly growing field is tumor therapy using light atomic nuclei for irradiation, which offers several advantages over the more established photon-based methods. This development, too, owes its existence to the work of physicists exploring fundamental questions of nature with particle accelerators.

If we had always only focused on research where the application could be anticipated from the outset, all these mentioned applications would never have been found. The benefit of a development often only becomes apparent when it is available. But not only that: Students can make an essential contribution to addressing current issues in particle physics very early in their studies, as early as their bachelor's thesis—but of course even more so during their master's and doctoral work. Without these contributions, our research would not function (!)—and the students also gain exciting insights into an internationally active research landscape. The latter naturally contributes very positively to the personal development and cosmopolitanism of the students. As they are also trained on the abstract questions of particle physics, they learn, for example, to handle modern methods of machine learning and even to develop such methods. This prepares them ideally to make an important contribution in other sciences as well as industry. Many companies have long recognized this, with the consequence that graduates of our institutes who do not want to pursue a career in science typically find attractive professional fields outside of science in a very short time.

Basic research and nuclear and particle physics have further significance. Naturally, the following discussion is strongly influenced by our personal bias and without claim to completeness. But it would be a pity if researchers working in the field of basic research would stop

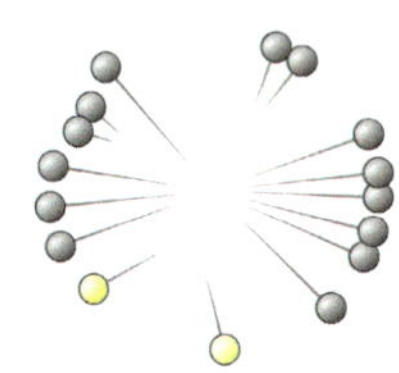

arguing for the necessity of basic research at this point. We are firmly convinced that basic research also has a value or a purpose in itself. This lies primarily in the primal human need to *know*. We want to understand why the world around us looks the way it does. This includes the question of what the fundamental theory of particle physics might look like—maybe there is a formula that unites everything? What are the fundamental principles upon which this theory is built? How much chance contributed to the Universe being structured the way it is?

Insights in this direction could change our view of the world—perhaps even as much as astronomical observations have led to the abandonment of the anthropocentric worldview, when moving away from a stationary Earth at the center of the Universe completely revised our understanding of the role of humans in the world. What is interesting here is looking back: In the history of the great achievements of the cultural history of civilization, scientists play a central role. Almost everyone knows Galileo Galilei and Johannes Kepler, and they played at least as central a role for the cultural leap into the Renaissance as Michelangelo and Leonardo Da Vinci; Erwin Schrödinger and Albert Einstein probably contributed more to the emergence of the cultural upheavals of modernity through their contributions to quantum mechanics and relativity theory than most other people.

We do not know what cultural upheavals particle physics has in store for us—what if it were confirmed that the fundamental parameters of the Standard Model are space- and time-dependent such that the Universe in far-distant regions looks rather different? This is pure speculation as of today—but it could happen. And that is exactly the point: If we stop conducting this kind of basic research now, we will never know what we have missed. The revolutions of quantum mechanics, on which all our computers and our entire information society are based, were just as unpredictable as any other discovery of the future. We should not take the discoveries of the past, which we could only make because of our culture of basic research, for granted, just like possible future discoveries: Both are valuable cultural assets.

Acknowledgements

This book would not have been possible without the initiative and dedication of Lisa Edelhäuser, the responsible editor at Springer Publishing, who was involved from the earliest stages of its conceptualization and writing. She played a key role in helping us develop a structural framework capable of organizing a complex and interrelated subject area within the modular format of this book. We are especially grateful for her thoughtful work on the style of the articles, which still reflect her approach of balancing accessible, straightforward explanations with more abstract and in-depth insights. We also sincerely thank Bianca Alton and Caroline Strunz from Springer for their valuable support and guidance throughout the process.

A multitude of friends and colleagues encouraged us to write this book and critically supported us in its creation and improvement. We especially thank Gunnar Bali, Gabriela Barenboim, Markus Bernhardt, Uta Bilow, Thomas Burwick, Jon Butterworth, Marcela Carena, Sebastian Carena-Wagner, Doug Cowen, Klaus Desch, Jordy de Vries, Abdelhak Djouadi, Rudolf Dreiner, Monica Dunford, Zoltan Fodor, Stefan Förste, Anna Franckowiak, Henning Gast, Howard Haber, Jana Heysel, Paul Hill, Bernhard Holzer, Susanna Jochum, Clifford Johnson, Hermann Karcher, Alex Karim, Kristoffer Kerkhof, Christian Klein-Bösing, Stefan Kowalewski, Michael Krämer, Stefan Krieg, Eric Laenen, Rachel Leuthold, Rainer Mählmann, Victor Martin Lozano, Anne Michel, Cornelia Monzel, Thomas Müller, Johannes Müllers, David Newman, Johann Ostmeyer, Ewald Paul, Timo Poller, Anke Pretz, Pascal Pretz, Jana Puschra, Andrea Raccanelli, Pierre Ramond, Diego Redigolo, Richard Roberts, Graham Ross, Subir Sarkar, Steffen Schaepe, Markus Schumacher, Christoph Schürmann, Martin Schürmann, Christoph Sieg, Anne Stockhausen, Insa Thiele-Eich, Lorenzo Ubaldi, Paula Weber, Detlef Weßling, Birgit Westernstroer, Liv Wiik-Fuchs, Andreas Wirzba, Tiziana von Witzleben, Werner Vogelsang, Fabio Zwirner.

Many of the graphical representations and images in this book were either created by the authors or developed in close collaboration with graphically talented colleagues, friends, and family members. We are deeply grateful to Ella Dreiner, Joseline Heuer, Saurabh Nangia, Anke Pretz, Deborah Rönchen, and Martin Schürmann for their direct contributions of several figures, and valuable support to the visual design and implementation. Our special thanks go to Malin Heuer for the creative idea of including a flip book in this volume.

A book such as this—featuring numerous illustrations, some of them of historical significance—would not have been possible without generous support in organizing image rights for depictions of laboratories and experiments. We gratefully acknowledge the contributions of Eric Adelberger, Dagmar Baroke, Siegfried Bethke, Markus Büscher, Roberto Giacomelli, Klaus Kirch, Kurt Riesselmann, Qiang Zhao, and Thomas Zoufal. We also extend our thanks to Jacob Bourjaily, Angel Ferran Pousa, Eckhard Hofmann, and Stefan Schael for kindly providing additional illustrations. Furthermore, we are grateful to the following institutions for making image material available: CERN, DESY, Fermilab, Jülich Research Center, the Gran Sasso underground laboratory, the Karlsruhe Institute of Technology (KIT), the Paul Scherrer Institute (PSI), the Stanford Linear Accelerator Center (SLAC), and the University of Bonn.

For this book we have translated, updated and extended the original edition, which we wrote in our native language German. Throughout the translation process, Christa Börsch provided invaluable support in helping us achieve a consistent and suitably elegant English style. We thank her from the bottom of our hearts for her dedication and care.

We would like to thank our friend and colleague Michael Kortmann whose support was indispensable in setting up and photographically documenting the demonstration experiments featured in this book. We are also deeply grateful to the entire team of the Bonn Physics Show for many years of inspiring collaboration in the educational and entertaining public communication of physics—an effort that has enriched this book in many ways.

Last but most certainly not least, we would like to express our heartfelt thanks to Rolf-Dieter Heuer for his inspiring foreword to this book.

Properties of Elementary Particles

Overview and Tables

Type	1. Family		2. Family		3. Family		Electric Charge	Strong Charge	Weak Charge
	Name	Mass $[1/c^2]$	Name	Mass $[1/c^2]$	Name	Mass $[1/c^2]$			
Quark	up (u)	$2.16^{+0.49}_{-0.26}$ MeV	charm (c)	$1.27^{+0.02}_{-0.02}$ GeV	top (t)	$172.5^{+0.7}_{-0.7}$ GeV	$+(2/3)\,e$	r, g, b	✓
	down (d)	$4.67^{+0.48}_{-0.17}$ MeV	strange (s)	$93.4^{+8.6}_{-3.4}$ MeV	bottom (b)	$4.18^{+0.03}_{-0.02}$ GeV	$-(1/3)\,e$	r, g, b	✓
Lepton	electron neutrino (ν_e)	< 0.8 eV	muon neutrino (ν_μ)	< 0.19 MeV	tau neutrino (ν_τ)	< 18.2 MeV	0	—	✓
	electron (e^-)	0.511 MeV	muon (μ^-)	106 MeV	tau (τ^-)	1.777 GeV	$-e$	—	✓
Anti-quark	anti-up ($\bar u$)	$2.16^{+0.49}_{-0.26}$ MeV	anti-charm ($\bar c$)	$1.27^{+0.02}_{-0.02}$ GeV	anti-top ($\bar t$)	$172.5^{+0.7}_{-0.7}$ GeV	$-(2/3)\,e$	$\bar r, \bar g, \bar b$	✓
	anti-down ($\bar d$)	$4.67^{+0.48}_{-0.17}$ MeV	anti-strange ($\bar s$)	$93.4^{+8.6}_{-3.4}$ MeV	anti-bottom ($\bar b$)	$4.18^{+0.03}_{-0.02}$ GeV	$+(1/3)\,e$	$\bar r, \bar g, \bar b$	✓
Anti-lepton	electron anti-neutrino ($\bar\nu_e$)	< 0.8 eV	muon anti-neutrino ($\bar\nu_\mu$)	< 0.19 MeV	tau anti-neutrino ($\bar\nu_\tau$)	< 18.2 MeV	0	—	✓
	positron (e^+)	0.511 MeV	anti-muon (μ^+)	106 MeV	anti-tau (τ^+)	1.777 GeV	$+e$	—	✓

Properties of the elementary particles of the three families of fermions (spin 1/2) among the elementary particles. In addition, the six leptons all have lepton number +1 and the antileptons −1. The six quarks have baryon number +(1/3) and the antiquarks −(1/3). Because of the confinement, the quark masses depend on the computation scale.

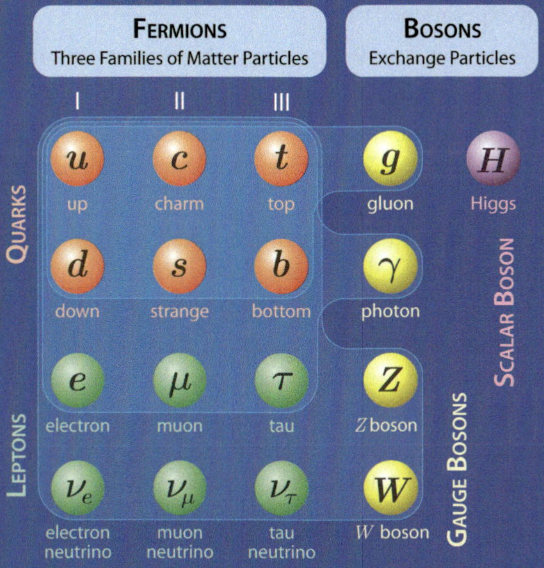

Name	Mass $[1/c^2]$	Spin	Electric Charge	Strong Charge	Weak Charge
photon (γ)	0	1	0	—	—
Z	91.2 GeV	1	0	—	✓
W^+	80.4 GeV	1	$+e$	—	✓
W^-	80.4 GeV	1	$-e$	—	✓
gluon (g)	0	1	0	$r\bar g, r\bar b,$ $g\bar r, g\bar b,$ $b\bar r, b\bar g,$ $r\bar r - g\bar g,$ $r r + g\bar g$ $- 2\,b\,\bar b$	—
graviton	0	2	—	—	—
Higgs	125.1 GeV	0	0	—	—

Comparison of the properties of bosons (integer spin, field particle and Higgs boson)

Glossary

α or α_{QED}	see fine structure constant
α_{QCD}	Strength parameter of the strong interaction or QCD.
accelerator	A facility in which particles are accelerated to the desired energy using electric fields. The resulting beams are focused and usually kept on a circular path using magnetic fields.
ADMX	Experiment searching for axions as a component of dark matter.
ALICE	An experiment at the Large Hadron Collider with a focus on the exploration of the quark-gluon plasma and questions of hadron and nuclear physics.
alpha decay, α-decay	Decay of an atomic nucleus, in which a helium nucleus (alpha particle) is emitted.
alpha particle (α)	The nucleus of the helium atom consisting of two protons and two neutrons.
ALPS	Experiment to search for axions. It investigates whether one can shine an intense laser beam through a wall. If axions exist, photons converted into axions could pass through a wall unhindered.
AMS	An experiment on the International Space Station, ISS, in space. The particle detector has the special feature of measuring the electric charge of cosmic radiation particles in addition to energy and particle type.
analysis (experiment)	The interpretation and evaluation of experimental data.
angular momentum	Kinematic quantity that plays a major role in describing rotational movements in mechanics and especially in quantum mechanics.
anomalous magnetic moment	relative deviation of the so-called gyromagnetic factor g, which relates the magnetic moment of a particle to its spin, from the value 2, which is predicted by the Dirac equation: $a = (g-2)/2$.
antimatter	The partner particle of a matter particle, with quantum numbers that have opposite signs compared to those of the matter particle, including the electric charge.
${}^{40}_{18}\text{Ar}$, ${}^{39}_{18}\text{Ar}$	Two isotopes of argon (atomic number 18), with masses 40 and 39. Liquid argon is commonly used as a scintillator. Natural argon mainly consists of the stable isotope ${}^{40}_{18}\text{Ar}$, but it also contains trace amounts of the radioactive isotope ${}^{39}_{18}\text{Ar}$, which is for example generated by cosmic radiation.
asymptotic freedom	Describes the phenomenon that the strong interaction becomes weaker at high energies, so that quarks and gluons then behave like (almost) free particles.
ATLAS	A multipurpose experiment at the Large Hadron Collider.

atomic spectrum	Each type of atom emits light with characteristic, discrete frequencies when excited. This phenomenon is explained quantum mechanically by the fact that, during a transition from one energy level to another, the energy difference is released in the form of photons.
Auger	see Pierre Auger experiment
axion	A hypothetical particle that arises as a byproduct of an explanation for the vanishing *CP* violation in the strong interaction. It is also considered a promising candidate for dark matter.
B meson	Bound state consisting of a bottom quark and a light antiquark ($\bar{u}$, $\bar{d}$, $\bar{s}$) (or a bottom antiquark and a light quark).
baryon	Hadrons with half-integer spin. The simplest possible baryons, like proton and neutron, contain three light quarks.
baryon asymmetry	The fact that much more matter than antimatter is observed in the Universe.
Belle / Belle II	An experiment to measure *CP* violation and other properties of *B* mesons at the e^+e^- collider KEKB in Japan.
beta decay, β-decay	Transformations between neutrons and protons with the emission of a lepton pair ($e^-\bar{\nu}_e$ in neutron decay, $e^+\nu_e$ in proton decay). In a beta decay, the atomic number of a nucleus changes by one unit. Protons can only decay in this way if they are bound in atomic nuclei and the resulting nucleus is more stable than the initial nucleus.
Big Bang	Hot and dense initial state of the Universe.
black hole	When nuclear fusion in the interior of a star ceases, the star's inner core collapses under gravity, forming a compact object. If the original star's mass is greater than approximately 25 times that of the Sun, this collapse results in the formation of a black hole. In this scenario, the core contracts to such a density that the gravitational force becomes so strong that even photons cannot escape its pull.
BNL	Brookhaven National Laboratory, research center on Long Island, New York, USA.
BOREXINO	Experiment in the Gran Sasso underground laboratory in Italy to study neutrinos from the Sun.
boson	Particles with integer spin. The exchange particles of the interactions (photons, *W*, *Z* and gluons) and the Higgs particle are bosons.
branching ratio	Fraction of decays of a particle into a specific final state, in relation to all possible decays.
c	see speed of light.

calibration	The process in which a certain value of a physical quantity is assigned to an electronic signal of a certain strength.
calorimeter	The part of an experiment that measures the energy of particles.
CAST	Experiment to search for axions from the Sun.
cavity resonator	A (hollow) device made of electrically conductive material, which due to its shape is tuned to a certain frequency for the oscillations of electromagnetic fields inside it. Typically, the frequencies are in the microwave range.
$CaWO_4$	Calcium tungstate.
CEPC	The Chinese Electron Positron Collider. A concept for a possible future e^+e^- Higgs Factory (see Higgs factory) in China
CERN	European Organization for Nuclear Research. International particle physics laboratory, located in Geneva, Switzerland, founded in 1954. The LEP accelerator was once operational here, and the LHC accelerator is still running today.
charged current	A weak interaction mediated by a virtual charged $W^{\pm}$ boson.
Cherenkov effect	When a charged particle moves through a medium (such as water or air) at a speed greater than the speed of light in that medium, it generates light known as Cherenkov radiation. In the medium, the speed of light is slower than in a vacuum, allowing the particle to exceed the medium's light speed, though it can never surpass the speed of light in a vacuum.
Cherenkov telescope	When a particle of cosmic radiation enters the Earth's atmosphere, it and especially the resulting particle shower emit Cherenkov light. This can be observed with special telescopes, from which the direction and energy of the original particle can be determined.
chirality and handedness	Refers to a particle property. Left-handed particles interact with the W and Z boson, right-handed particles only with the Z boson. Left- and right-handed particles both interact with the photon and the gluon, provided they carry electric charge or color charge.
CKM matrix	Mathematically describes the relationship named after the scientists Cabibbo, Kobayashi and Maskawa, according to which the Higgs boson couples to a different combination of quarks than the $W^{\pm}$ bosons. Responsible for CP violation in the Standard Model..
CMS	A multipurpose experiment at the Large Hadron Collider (LHC).
COBE	Satellite-based experiment for measuring the cosmic microwave background.
color charge	Collective name for the three charges of the strong interaction or QCD.
color confinement	see confinement
COMPASS	An experiment at CERN, where experiments on the spin structure of the proton and hadron spectroscopy were conducted.

confidence	see significance
confinement	Describes the empirically observed phenomenon that particles carrying a strong charge (also called color charge) cannot be observed as free particles—they are only found as components of composite objects, whose color charges neutralize each other.
cosmic radiation	High-energy particles, mainly protons, that hit the Earth from space.
COSY	Stands for COoler SYnchrotron and is a particle accelerator for protons and deuterons at the Jülich Research Center.
CP violation	The violation of the combined CP symmetry and thus, according to the CPT theorem, also the T symmetry. In the Standard Model, only observed in the weak interaction through the phase of the CKM matrix.
CPT theorem	Fundamental property of quantum field theory. States that all physical laws remain unchanged under the combined symmetry transformation of charge conjugation (C), parity (P), and time reversal (T).
CsI	Cesium iodide, a transparent, scintillating crystalline material.
CUORE	Experiment with low-temperature detectors in the Gran Sasso underground laboratory to clarify whether the neutrino is its own antiparticle.
D meson	Bound state particle consisting of a charm quark and a light antiquark (u,d,s), or a charm antiquark and a light quark.
D^* meson	Excited state of the D meson.
dark energy	A form of energy motivated by cosmological measurements, which causes an accelerated expansion of the Universe. The nature of this energy is so far unknown.
dark matter	A form of matter required by a multitude of astrophysical and cosmological observations, contributing at least 25 % to the energy density of the Universe. It does not interact electromagnetically or strongly. There are strong indications that dark matter must consist of particles. No particle of the Standard Model fulfills the properties of dark matter. Therefore, dark matter is one of the clearest indications for physics beyond the Standard Model.
DarkSide	Experiment employing liquid argon based in the Gran Sasso underground laboratory searching for the particles of dark matter.
Davis experiment	First experiment to detect neutrinos from the Sun. Realized by Ray Davis in the Homestake Mine in the USA.
DEAP	Experiment employing liquid argon based in SNOLAB in Canada searching for the particles of dark matter.

decay width	see width
DESY	German Electron Synchrotron. A laboratory in Hamburg for research with accelerators in the field of particle physics and with synchrotron radiation. The gluon and the oscillation of B mesons were discovered here. The strength of the strong interaction as well as the structure of the proton were measured very precisely.
detector	In particle physics, this refers to an instrument for observing and measuring the properties of a particle.
deuterium	Hydrogen isotope with mass number two. The atomic nucleus, the deuteron, consists of a proton and a neutron.
Dirac equation	Relativistic equation of motion for spin-1/2 fermions.
Dirac mass	Mass term in the Dirac equation. Distinct from a Majorana mass.
DUNE	Experiment investigating neutrino oscillations to determine the CP violation in the lepton sector.
electric dipole moment	In classical physics, an electric dipole moment arises when opposite charges are spatially separated from each other. For example, an electric dipole moment causes water molecules to attract each other. Point-like particles, such as the electron, can also possess an electric dipole moment, which, however, is of a different nature and can only exist if parity and time reversal invariance are violated.
electromagnetic interaction	Alongside gravity, the weak and the strong interaction, one of the four elementary fundamental forces.
electron	A lepton and a fundamental building block of the Standard Model, as well as all atoms. Experiences the electromagnetic and the weak interaction.
electron volt (eV)	Unit of energy, corresponding to the kinetic energy an electron has acquired when it has been accelerated in a voltage difference of one volt. It is $1\ \text{eV} = 1.60218 \cdot 10^{-19}$ Joule.
ELSA	Electron-Stretcher-Facility, an electron accelerator operated at the University of Bonn, Germany, with a maximal energy of 3.5 GeV.
energy	Important quantity describing the state of a particle/system. It can be composed of different forms, e.g., kinetic energy and potential energy. The total energy of a closed system is always conserved, but energy can be exchanged between particles.
exchange particles	In quantum field theory, forces are transmitted by exchange particles. In electromagnetism, this is the photon, in the weak interaction the $W^{\pm}$ and the Z^0 bosons and in the strong interaction the gluons.
EXO	Experiment employing liquid xenon to clarify the question of whether the neutrino is its own antiparticle.
family of particles	group of elementary particles consisting of a charged lepton, an associated neutrino, and two associated quarks. For example the first family is: (e^-, ν_e); (u, d).

FCCee The Future Circular Collider. A concept for a possible future e^+e^- Higgs factory (see Higgs factory) at CERN.

fermi or femtometer [fm] Length unit used in nuclear and particle physics. 1 fm = 10^{-15} m. One is the typical extension of a hadron.

Fermilab Particle physics laboratory near Chicago, Illinois, USA. Here the bottom quark, the top quark, and the tau neutrino ν_τ were discovered.

fermion Particle with half-integer spin. Follows the Pauli principle. The elementary building blocks of matter are fermions.

fine structure constant Characteristic strength of the electromagnetic interaction, symbol: α_{QED}, often abbreviated as α. Determines, e.g., the binding energies of atoms.

GALLEX Experiment based in the Gran Sasso underground laboratory for measuring neutrinos from the Sun.

gamma radiation, gamma photon, γ-radiation Photon with high energy (> a few 100 keV).

gas-based detector A volume typically filled mainly with a noble gas, in which the ionization of the gas by charged, high-energy particles is measured. Often used as a tracking detector.

gauge symmetry Construction principle of the interactions of the Standard Model: From the requirement of a local, i.e., gauge, symmetry, the existence of gauge fields can be derived, which can be identified with the exchange particles of the electromagnetic, weak, and strong interaction.

$^{76}_{32}$Ge Germanium (atomic number 32), isotope with mass 76. Germanium is a semiconductor and is used in many applications as a detector material. Natural germanium contains 7.4 % of the isotope $^{76}_{32}$Ge, which is one of the few isotopes that decay through double beta decay.

GeV Unit of energy, 1 000 000 000 electron volts = 1000 MeV = 1 GeV

GHz Gigahertz. Hertz is the unit for frequency. Giga stands for one billion (10^9). 1 GHZ denotes one billion cycles per second.

gluon Exchange particle of the strong interaction (QCD), analogous to the photon in electromagnetism (QED). Corresponding to the three strong charges, there are eight gluons.

Gran Sasso National Laboratory Laboratory inside the Gran Sasso massif in the Abruzzo, Italy. The rock cover keeps cosmic radiation away from the experiments, which is why almost no background from, e.g., muons of cosmic origin arrive.

Grand Unified Theory (GUT)	Hypothetical theory that unifies the three forces of the Standard Model at very high energy (10^{16} GeV) into one force.
h and $\hbar$	see Planck's constant
hadron	Bound state of quarks and gluons that does not carry net color charge. Prominent examples of hadrons include the proton and neutron, which are the building blocks of atomic nuclei, as well as the pion.
heavy ion collisions	Heavy ions are electrically charged atomic nuclei of heavy atoms, such as lead or gold nuclei. Heavy ions are collided in accelerators like the LHC at CERN or RHIC at Brookhaven for the study of nuclear matter or of the quark gluon plasma.
helium	An element whose atomic nucleus consists of two protons and typically two neutrons. Helium is a noble gas and is one of the light elements.
HERA	Accelerator at the DESY laboratory in Hamburg, where positrons were accelerated in one direction and protons in the other direction and brought to collision.
Higgs boson	The particle responsible for electroweak symmetry breaking and the mass of elementary particles in the Standard Model. It was discovered in 2012 at the Large Hadron Collider (LHC).
Higgs factory	A concept for a future accelerator complex designed to measure the properties of the Higgs boson with much greater precision than was possible at the LHC. It would also generate significantly more Z and W bosons, as well as top quark pairs, than ever before. Current studies are exploring possible implementations, typically involving an e^+e^- linear or circular collider with a length of about 30 km or a corresponding diameter.
hydrogen	Element whose atomic nucleus in most cases consists of only one proton. It is the lightest element.
Hyper-Kamiokande	Experiment based in the Kamioka Mine in Japan to search for the decay of the proton, as well as the detection of astrophysical neutrinos.
IAXO	Experiment to search for axions from the Sun.
IceCube	Experiment at the South Pole to study high-energy cosmic neutrinos.
ILC	The International Linear Collider. A concept for a possible future e+e− Higgs factory (see Higgs factory)
interaction point	Point where the counter rotating beams in a collider experiment are brought to a collision. The large detectors ATLAS and CMS at the LHC are each centered on an interaction point, respectively.
invariant mass	In a multi-particle system, the invariant mass refers to the square root of the sum of the squares of all energies of the particles minus the square of the sum of all momenta. It is the same in all reference systems, i.e., invariant.

ionization	The process of removing electrons from atoms. The remaining atom is no longer electrically neutral and is referred to as an ion.
isotope	Atomic nuclei of the same element (same number of protons) but different number of neutrons.
JUNO	Experiment in China to determine the order of neutrino masses by precise examination of neutrino oscillations.
KamLAND	Experiment based in the Kamioka Mine in Japan to detect electron antineutrinos from surrounding nuclear reactors. It also observed geological antineutrinos from radioactive decays in the Earth's crust.
KASCADE	Experiment at the Karlsruhe Research Center, Germany, to study cosmic radiation.
KATRIN	Experiment at the Karlsruhe Research Center, Germany, to determine the neutrino mass by precise measurement of the beta decay of tritium.
KM3NeT	Experiment in the waters of the Mediterranean Sea to measure high-energy cosmic neutrinos.
Λ_{QCD}	Characteristic energy scale of QCD of about 250 MeV. For energies/masses much larger than Λ_{QCD}, the QCD coupling becomes small. For energies smaller than Λ_{QCD}, the coupling becomes large and other methods are needed for systematic calculations (e.g. lattice QCD).
Large Hadron Collider (LHC)	The largest and most powerful particle accelerator at the time of this book's publication. Operated by the CERN laboratory in Geneva, Switzerland.
lattice QCD	Method in which space-time is discretized in order to numerically solve the dynamical equations of QCD using complex simulations on supercomputers.
law of gravity	Mathematical description of the dependence of the gravitational force on the distance between the gravitating masses.
LEGEND	Experiment with germanium detectors to clarify whether the neutrino is its own antiparticle.
LEP accelerator	Large Electron-Positron Collider. Electron-positron collider at CERN in operation from 1989 to 2000.
lepton	Spin-1/2 elementary particle that does not feel the strong interaction. In the Standard Model, these are the electron, the muon, the tau, as well as the associated neutrinos.
leptoquark	Hypothetical particle that appears for example in grand unified theories (GUTs) and is associated with the unification of the strong and electroweak force.

LHCb	Experiment at the Large Hadron Collider, specialized in the study of *B* mesons and other hadrons.
$LiAlO_2$	Lithium aluminate. In crystal form, a transparent non-conductive material.
lifetime	The average time in the rest system of the unstable particle until it decays into other particles.
light year	Distance that light travels in one year (9460 billion km).
low-temperature calorimeter	Device for the measurement of (particle) energies based on temperature increases. The method is usually more sensitive the lower the temperature of the calorimeter.
$LuAlO_2$	Lutetium aluminate. In crystal form, a transparent non-conductive material. Used as a scintillator, e.g., in PET.
MACHO	Massive Compact Halo Object. Compact celestial bodies made of baryons. Candidate for dark matter.
MADMAX	Experiment to search for axions as a possible component of dark matter in the Universe.
magnetic moment	Fundamental property of a particle. Measure for how it reacts to an external magnetic field.
Majorana mass	Mass term for a Majorana fermion, a particle which is its own antiparticle. In the Standard Model only possible for neutrinos.
Majorana neutrino	Neutrino, which is its own antiparticle. Perhaps the observed neutrinos are Majorana neutrinos.
measurement uncertainty	see uncertainty
meson	Hadron with integer spin and baryon number 0, thus made up of equal numbers of quarks and antiquarks. Example: Pion.
MINOS	Experiment in the USA for measuring neutrino oscillations over a distance of about 750 km, the distance from the production point at Fermilab to the MINOS detector.
momentum	Quantity to describe the state of a particle/system. The momentum is a vector, so it is characterized by a direction and a magnitude. Usually, the magnitude of the momentum, $\|\vec{p}\|$, of a particle is linked with its energy, *E*, via $\|\vec{p}\|^2c^2 = E^2 - m^2c^4$ (exception: virtual particles). This relation also holds when the particle is massless. It then has a momentum given by $\|\vec{p}\| = E/c$. At low energies, the momentum of a massive particle is equal to the velocity of the particle multiplied by its mass. The sum of the momenta of all particles in a particle reaction is always conserved, but momenta can be exchanged between particles.
MRI	Magnetic Resonance Imaging. Imaging method in medicine. Uses the magnetic moment of protons.
muon μ	Heavy sister particle of the electron. The second heaviest known lepton.

NaI	Sodium iodide, transparent, scintillating crystalline material.
natural radioactivity	Radioactive radiation, emitted by naturally occurring unstable isotopes.
$^{150}_{60}$Nd	Neodymium (atomic number 60), isotope with mass 150. Natural neodymium contains 5.6 % of the isotope $^{150}_{60}$Nd, which is one of the few isotopes that decay through double beta decay.
neural network	A sequence of nonlinear mathematical operations in a computer code, with which various measured quantities can be combined into one quantity. A decision based on the latter can be made as sensitively as possible, or a size can be predicted as accurately as possible. It can also be used for the simulation of events or completely different questions outside of the natural sciences.
neutral current	A weak interaction mediated by a (neutral!) virtual Z^0 boson.
neutrino	Electrically neutral lepton. Fundamental building block of the Standard Model. There are three neutrinos: each family has its own neutrino.
neutrino mass	The masses of the neutrinos are very small compared to all the other fermions in the Standard Model. Their absolute values are still unknown, only the differences are well measured through oscillation experiments. Since there are three neutrino flavors (electron, muon, and tau neutrino), there can in principle be three different neutrino masses. We do not know if the neutrino's have a Dirac or a Majorana mass.
neutron	Uncharged baryon, consisting of three quarks (down-down-up). Along with the proton, one of the two building blocks of atomic nuclei.
neutron star	When the energy generation processes by nuclear fusion in the interior of a star come to an end, the inner core of the star collapses due to gravity into a compact object. If the star has a mass between 8 and 25 times the mass of the Sun, a neutron star is formed. In it, the matter is compressed to a density higher than in atomic nuclei and an excess of neutrons is created. A typical neutron star has a mass of about 1.5 solar masses, but measures only 25 km in diameter.
nEXO	Experiment with liquid xenon to clarify the question of whether the neutrino is its own antiparticle.
nuclear fusion	Fusion of light atomic nuclei into heavier atomic nuclei. For nuclei lighter than iron nuclei, energy is released. Stars produce their energy through nuclear fusion.
nucleon	Collective term for the nuclear building blocks proton and neutron.
OPERA	Experiment in the Gran Sasso underground laboratory for measuring neutrino oscillations over a distance of about 750 km. The neutrinos were generated at CERN in Geneva.

oscillation (particles)	Here the periodic transformation of a particle into another (e.g. neutrinos) or between particle and antiparticle (e.g. neutral mesons).
PAMELA	Satellite experiment to investigate cosmic radiation.
parity violation	The violation of mirror symmetry P in the weak interaction.
Pauli principle	Identical fermions (e.g. two electrons) cannot match in all quantum numbers, such as energy, spin orientation, orbital angular momentum and its orientation, in a quantum system (e.g. an atom).
$PbWO_4$	Lead tungstate. In crystal form, a transparent, non-conductive and scintillating material.
PET	Positron Emission Tomography. Imaging method in medicine. The annihilation radiation from the annihilation of positrons with electrons is used.
photographic plates	Light sensitive plates, as used in early cameras. They also react to the passage of charged particles and can thus be used as a detector.
photomultiplier	Very sensitive measuring device for photons (light quanta). The electrical signal is generated by multiplying electrons produced by the photoelectric effect.
photon	Exchange particle of QED and the quanta of the electromagnetic field, also known as the "light" particle.
physics case	The expected measurements of a future experiment, set in relation to what can be learned from them: which theories could be refuted or confirmed, and which new phenomena could potentially be discovered
Pierre Auger experiment	Experiment in the Argentine Pampa with many smaller detectors widely distributed over an area of 3000 km^2 for the investigation of ultra-high-energy cosmic radiation.
pion	The lightest particle built from quarks. Pions belong to the mesons, which consist of a quark and an antiquark. Building blocks of the pions are up and down (anti)quarks.
Planck satellite	Satellite experiment for measuring the cosmic microwave background.
Planck's constant (h or $\hbar$)	Natural constant. It is $h = 6.62607015 \cdot 10^{-34}$ Js. Often $\hbar = h/(2\pi)$ is used. The energy of a photon is $E = h\ f$, with f the frequency. Also known as Planck's quantum of action.
Planck's quantum of action	see Planck's constant
potential	In non-relativistic systems, the potential associated with an interaction can be assigned to each location: If a particle is brought to a certain place in this potential, a potential energy can be assigned to it. In QFT the interaction energy between the particles becomes in the non-relativistic limit the potential energy.
potential energy	The part of a particle's energy that is not in motion or kinetic energy.

proton Baryon with positive electric charge, consisting of three quarks (up-up-down). Alongside the neutron, one of the two building blocks of atomic nuclei. The number of protons determines the nuclear charge and thus the element.

QCD Quantum Chromodynamics. The quantum field theory of the strong interaction.

QED Quantum Electrodynamics. The quantum field theory of the electromagnetic interaction.

QFT Quantum Field Theory. The general, relativistic quantum theory for fields. In quantum field theory, particles can be created and annihilated.

quantum number The state of a particle is characterized by a complete set of quantum numbers. For an electron in a hydrogen atom, these are, for example, the energy value (principal quantum number), the total angular momentum (including spin) and its projection on the quantization axis, and the orbital angular momentum.

quark Fundamental building block of the Standard Model. Experiences all interactions, especially the strong interaction (QCD). There are six quarks, three light (up, down, strange) and three heavy (charm, bottom, top), where the energy Λ_{QCD} serves as a comparison scale.

quark-gluon plasma (QGP) A state of matter that is so dense and hot that quarks do not anymore bind in individual, separable hadrons or nuclei, but from one continuous "soup" of quarks and gluons. It can be produced in heavy ion collisions at the accelerators LHC and RHIC. It is expected that the quark gluon plasma also played an important role in the early phases of the Universe.

recoil energy If a nucleus emits a particle, part of the released energy is transferred as recoil to the remaining nucleus, in order to conserve momentum. Even when another particle scatters off a nucleus and transfers part of its energy to the nucleus, this is referred to as recoil energy.

reconstruction The mathematical process by which the experimental measurement data (e.g. the strength of individual electronic signals in different components of the detector) are converted into physical quantities such as the track or the energy of a particle.

relativity theory The special theory of relativity allows observables to be converted between different inertial systems (systems moving relative to each other at constant speed). A basic tennet is that the speed of light is the same in all systems indepedent of the speed of the emitter. The general theory of relativity extends to accelerated reference systems and is the theory of gravitation.

resonance	When a system is excited in such a way that the excitation frequency exactly matches the system's characteristic frequency, energy transfer is maximal. In particle physics, a resonance is typically characterized by a significant increase in transition probability, resulting in a noticeable rise in count rates.
SAGE	Soviet-American experiment in the Caucasus Mountains to measure neutrinos originating from the Sun.
scattering	Reaction between two particles, in which momentum and possibly energy is transferred between the particles. Also particles can be annihilated or new particles can be produced in a scattering reaction.
Schrödinger equation	Fundamental equation of non- relativistic quantum mechanics which the wave function must satisfy
scintillator	A detector made from transparent plastic, crystal, or liquid, where the energy of passing particles is converted into visible light. This light can then be detected by photodetectors, such as photomultipliers, allowing the energy of the particles to be measured.
shells	Closely spaced energy levels, e.g., in atoms or nuclei, are grouped into shells.
significance	The statistical certainty of a statement, derived from the measure for uncertainty. Within $\pm 1\,\sigma$ of the central measured value, 68 % of all results of an experiment are expected, when repeated. (Here σ is the unit of significance and uncertainty. σ is the Greek letter "s" and is pronounced "sigma".) Within $\pm 2\,\sigma$, it is about 95 %. Outside the range corresponding to the significance of $5\,\sigma$, a result is only expected in $2.7 \cdot 10^{-7}$ of all cases.
silicon detector	A detector made from the semiconductor material silicon, typically (but not always) divided into thin strips or tiny rectangles, which measures ionization when charged particles pass through it.
SLAC Laboratory	Particle physics laboratory in California, USA.
SNO	Experiment in SNOLAB in Canada to measure neutrinos originating from the Sun.
solar neutrinos	Neutrinos generated inside the Sun. In the nuclear fusion reaction of hydrogen to helium, there is a net conversion of two protons into two neutrons by inverse beta decays, where among other things two electron neutrinos are emitted.
spectral line	Excited atoms emit light at characteristic frequencies. This can be used to identify the atom and even the isotope. These frequencies correspond to the energy differences of the atom's levels. They appear as lines on the screen due to the shape of the aperture.
spectroscopy	Study of the frequencies emitted by a radiation source, e.g., star as a light source.
speed	This is the absolute value of the velocity and has no direction.
speed of light (c)	Natural constant. It is $c = 299\,792\,458$ m/s.

spin	Characteristic quantum mechanical property of a particle, also referred to as intrinsic angular momentum, which contributes to the total angular momentum. The spin of the electron is 1/2.
statistical uncertainty	The natural fluctuation of a measurement result (e.g., a count rate). Can be reduced by more frequent measurements.
statistics	The mathematical description of randomness.
string theory	A hypothetical theory, which assumes that particles are not point-like, but instead extended objects, strings, which can vibrate. The strings would be about 10^{-35} meters small. In string theory the three forces of the Standard Model are naturally unified with gravity.
strong *CP* problem	Refers to the fact that the strong interaction preserves the *CP* symmetry to high experimental accuracy, although it should actually violate it for formal theoretical reasons.
strong interaction	Alongside gravity, electromagnetic and weak interaction, one of the four fundamental forces.
$SU(3)_C \times SU(2)_L \times U(1)_Y$	The gauge symmetry of the Standard Model. Here, $SU(n)$ denote certain symmetry transformations in a space of $n^2 - 1$ dimensions. $U(1)$ are symmetry transformations in a one-dimensional space. The "×" (tensor product) describes that the operations in the different spaces can be performed independently of each other.
Super-Kamiokande	Experiment in the Kamioka Mine in Japan for measuring atmospheric, solar, and supernova neutrinos and for searching for proton decay.
superconductivity	Some materials lose all electrical resistance below a critical temperature. This phase is referred to as superconducting.
supersymmetry (SUSY)	A symmetry transformation between fermions and bosons. Hypothetical extension of the Standard Model.
symmetry (transformation)	A symmetry exists when an object remains unchanged under the application of a symmetry transformation. For example, one cannot distinguish a rotated circle from an unrotated one. It has a rotational symmetry.
symmetry breaking	This occurs when the symmetry in a system is not (exactly) realized. A symmetry can be explicitly or spontaneously broken.
synchrotron	A circular type of accelerator, in which the magnetic field is adjusted synchronously with the increasing particle energy in order to keep it on the same orbit radius.

systematic uncertainty, systematics	An uncertainty that corresponds to the range of possible mismodeling of the measuring apparatus, the measurement process, or the numerical inaccuracy of a calculation. It often has to be determined by elaborate comparisons of independent methods.
tau lepton, tauon, τ	The heaviest known lepton, sister particle of electron and muon.
$^{130}_{52}Te$	Tellurium (atomic number 52), isotope with mass 130. Natural tellurium contains 33.8 % of the isotope $^{130}_{52}Te$, which is one of the few isotopes that decay through double beta decay.
TeO_2	Tellurium oxide. In crystal form, a transparent non-conductive material.
Tevatron	An accelerator at Fermilab near Chicago, USA.
theory	A mathematical description of Nature well supported by measurement results. Typically a complex combination of fundamental assumptions and methods for explaining and predicting measurement results.
threshold energy	Corresponds to the minimal energy needed in a reaction to produce a specific final state.
time dilation	According to the special theory of relativity, in two distinct reference frames the measurement of time gives different results. The difference is the time dilation. It is particularly large if the relative speed between the two reference frames is close to the speed of light.
top quark	The heaviest known quark. Discovered in 1995 at the Tevatron accelerator.
tracking detector	A part of an experiment, usually used to measure the path of a particle flying through the detector in a magnetic field. The curvature of the track can be used to determine the momentum of the particle. Often implemented as a gas-based or silicon detector.
trigger	The part of the readout electronics of an experiment, with which it is decided whether a certain event should be recorded.
tritium	^3H, an isotope of hydrogen; decays through beta decay to ^3He. The energy released corresponds to the very precisely known mass difference between tritium and ^3H and is 18.574 keV.
uncertainty	An interval within which a future measurement result of the same quantity is expected due to statistical fluctuations or systematic short comings of the measuring apparatus.
vector	Mathematical object characterized by its direction and length. Typical vectors are velocity and momentum.
velocity	A vector, $\vec{v} = (v_x, v_y, v_z)$, describing the motion of an object, which has an absolute value, the speed $v = \sqrt{v_x^2 + v_y^2 + v_z^2}$ and a direction.
vertex detector	Detector component, which is typically made up of thin silicon plates and is located near the interaction point in accelerator experiments. By extrapolation, the origin of observed charged particles and thus the location where the reaction took place can be inferred.

virtual particle	A particle whose energy, mass and momentum do not correspond to the relation $E^2 = m^2c^4 + p^2c^2$, and therefore cannot exist as a real free particle. However, virtual particles play an important role as intermediate states in quantum mechanical systems and reactions. Their existence is allowed for a short time due to the Heisenberg uncertainty principle.
wave function	A function which describes a quantum mechanical particle or collection of particles. For a single particle the square of the magnitude of the wave function corresponds to the probability distribution of the particle's location.
weak interaction	Alongside gravity, the electromagnetic, and strong interaction, one of the four fundamental forces in Nature.
width	The energy range over which a resonance appears as an enhancement in the data. The width of a resonance curve is inversely proportional to the lifetime of the resonance.
WMAP	Satellite experiment for measuring the cosmic microwave background.
$W^\pm$ boson	The exchange particle of the weak interaction that causes the charged current.
${}^{A}_{Z}\mathrm{X}$	Notation for atomic nuclei. X stands for the chemical symbol, Z is the charge number, i.e., the number of protons. A is the mass number. The number of neutrons is given by $N = A - Z$. Since the chemical symbol determines the number of protons, one often also finds the notation ${}^{A}\mathrm{X}$.
${}^{136}_{54}\mathrm{Xe}$	Xenon (atomic number 54), isotope with mass 136. Liquid xenon is used as a scintillator. Natural xenon contains 8.9 % of the isotope ${}^{136}_{54}\mathrm{Xe}$, which is one of the few isotopes that decay through double beta decay.
XENON	Experiment employing liquid xenon based in the Gran Sasso underground laboratory and used for the search for dark matter particles.
Z^0 boson	The exchange particle of the weak interaction that causes the neutral current.

Image Credits

Following abbreviations are using for the image credits:

PB	Philip Bechtle
FB	Florian Bernlochner
HD	Herbi Dreiner
CH	Christoph Hanhart
JJ	Josef Jochum
JP	Jörg Pretz
KR	Kristin Riebe

Preface

S. VII: top (spiral galaxy NGC 6956): NASA, ESA, and D. Jones (University of California – Santa Cruz); Processing: Gladys Kober (NASA/Catholic University of America); bottom (bubble chamber image): CERN

S. VIII: CERN, ATLAS experiment

Preface to the First English Edition

S. IX: CERN, ATLAS experiment

About This Book

S. X: KR

S. XI: left: CERN, CMS experiment

S. XI: right: CERN, ALICE experiment

Content

S. XII: top: Marcus Breig/KIT

S. XII: bottom: Wikimedia Commons / Parent Géry

S. XIII: KR, background image from Joshua Woroniecki at Pixabay

S. XIV: top: ASSOCIATED PRESS / Uncredited / picture alliance

S. XIV: bottom: CERN

S. XV: Jan Pavelka / Wikimedia Commons, CC BY 4.0

S. XVI: CERN, ATLAS experiment

S. XVII: Roy Kaltschmidt, Lawrence Berkeley National Laboratory

S. XVIII: ESA & NASA/Solar Orbiter/EUI team; Data processing: E. Kraaikamp (ROB)

S. XIX: top: NASA, ESA and the Hubble Heritage Team (STScI/AURA)-ESA/Hubble Collaboration

S. XIX: flip book images: KR, after drafts from JP and PB

1 What is Particle Physics?

Chapter introduction page

S. 1: proton with quarks (top left) and hydrogen atom (bottom right): Joseline Heuer; Bohr atom: KR; detector in the center (KATRIN): Marcus Breig, KIT; bottom left (Daya-Bay): Roy Kaltschmidt, Lawrence Berkeley National Laboratory; bottom right (BaBar/SLAC): Peter Ginter; composition by KR

Particle Physics as a Global Endeavor

S. 2: FB

S. 3: from top to bottom, left to right: BaBar/SLAC: Peter Ginter; Fermilab; Pierre Auger Observatory/Miguel Salvadores; IceCube Collaboration/NSF; H.E.S.S.: Christian Föhr, LHC: CERN; KATRIN: Marcus Breig/KIT; Daya-Bay: Roy Kaltschmidt, Lawrence Berkeley National Laboratory; Super-Kamiokande: Kamioka Observatory/ICRR (Institute for Cosmic Ray Research)/The University of Tokyo; SABRE experiments hall: SABRE Collaboration

The Crystal

S. 4: left: P. & M. Bechtle

S. 4: right: Wikimedia Commons/Parent Géry

S. 5: top: P. & M. Bechtle

S. 5: bottom: PB

Molecules and Atoms

S. 6: PB

S. 7: PB, KR

What is Particle Physics?

S. 8: Netzwerk Teilchenwelt

S. 9: Joseline Heuer, inset images left and bottom: CERN

Origins of Particle Physics

S. 10: top: Rolf Kickuth/Wikimedia Commons

S. 10: bottom: Wikipedia

S. 11: CERN

Length Scales and Units

S. 12: Jülich Research Center

The Elementary Particles

S. 14: Jülich Research Center

S. 15: CH

The Four Fundamental Forces of Nature

S. 16: Joseline Heuer

What Makes a Particle Elementary?

S. 18: HD

S. 19: HD

The Quark Model: A First Look

S. 20: Anke Pretz

S. 21: JP, KR

What is Antimatter?

S. 22: HD

S. 23: top: HD

S. 23: middle: KR

Detectors in Everyday Life

S. 24: PB, KR

S. 25: PB

How Does Physics Make Progress?

S. 26: CH, KR

S. 27: PB

Experimental Research in Particle Physics

S. 28: CERN

S. 29: CERN

A Deep Dive Into the Scientific Method

S. 31: PB

Imagining Reality

S. 32: PB

S. 33: PB, KR

2 Fundamental Principles

Chapter introduction page

S. 35: snowflake: Hayri Er/Getty images/iStock; light bulb: macrovector at freepik; magnets in the background: Joseline Heuer; Stern-Gerlach experiment: B. Bahr, J. Resag, K. Riebe, Faszinierende Physik, Springer-Verlag 2019; remaining: KR; composition by KR

Energy and Power

S. 36: H. Blossey/blickwinkel/picture alliance

S. 37: JP

Special Relativity
S. 38: HD, KR

Kinematics
S. 40: HD
S. 41: HD, Ella Dreiner

Acceleration and Radiation
S. 42: Saurabh Nangia
S. 43: Saurabh Nangia

Spin
S. 44: JP
S. 45: from: B. Bahr, J. Resag, K. Riebe, Faszinierende Physik, Springer-Verlag 2019; small grey image: from W. Gerlach, O. Stern 1922, Zeitschrift der Physik, doi:10.1007/BF01326983

Scattering Experiments
S. 46: JP
S. 47: JP

Feynman Diagrams
S. 48: HD
S. 49: HD

Concepts of Quantum Mechanics
S. 50: Joseline Heuer

Chance, Interference and Predictions
S. 52: Joseline Heuer
S. 53: CH, KR

Particles' Births and Deaths
S. 54: Joseline Heuer
S. 55: P. A. Zyla et al. (Particle Data Group), Prog. Theor. Exp. Phys., 2020, 083C01 (2020) and 2021 update

Resonances
S. 56: top: CH, KR
S. 56: bottom: CH
S. 57: Deborah Rönchen

On the Ranges of Interactions
S. 59: CH

Symmetries
S. 60: left and top: JP
S. 60: bottom right: U.S. Postal Service
S. 61: left: snowflake: Hayri Er/Getty images/iStock, dome: Peter Schickert/picture alliance, synagogue: Dimitrios Karamitros/Getty Images/iStock, mosque: RudyBalasko/Getty Images/iStock
S. 61: right: JP

Spontaneous Symmetry Breaking
S. 62: Joseline Heuer
S. 63: Andreas Wirzba, KR

3 Experimental Basics

Chapter introduction page
S. 65: top left (AGATA): S. Akkoyun et al., NIMA 668(2012)26–58 and https://www.agata.org/; bottom left (detector): CERN; center background (particle traces): CERN; center front: CH; atom with electrons: PB; graph at the right: JP; composition by KR

How Does a Particle Accelerator Work?
S. 66: JP
S. 67: top: JP
S. 67: bottom right: Jülich Research Center

Interactions Between Particles and Matter
S. 69: top: CERN
S. 69: bottom: PB, KR

Semiconductor Detectors
S. 70: left: Product Image courtesy of Mirion Technologies, Inc.
S. 70: right: Yury Suvorov INFN Naples
S. 71: left: S. Akkoyun et al., NIMA 668(2012)26–58 and https://www.agata.org/
S. 71: right: CERN

Gaseous Detectors
S. 72: PB
S. 73: top: PB
S. 73: bottom left and right: Physical Insitute, University of Bonn (left: edited and simplified by KR)

Scintillators and Photomultipliers
S. 74: top: CERN
S. 74: center: https://www.gammaspectacular.com.au
S. 75: left: JJ
S. 75: right top: Hamamatsu Photonics K.K.
S. 75: right bottom: Kamioka Observatory, ICRR (Institute for Cosmic Ray Research), The University of Tokyo

Position and Momentum Measurement
S. 76: center: PB
S. 76: bottom: CERN
S. 77: left: IOP Publishing Ltd and Sissa Medialab, CC BY 4.0
S. 77: right: CERN

Energy Measurement
S. 78: top: PB
S. 78: bottom left: CERN
S. 78: bottom right: Saint-Gobain Crystals
S. 79: CERN

Signatures of Particles
S. 80: KR
S. 81: CERN

Cherenkov Radiation and Particle Identification
S. 82: JP
S. 83: top left:CERN
S. 83: bottom left: JP
S. 83: bottom right: ASSOCIATED PRESS/Uncredited/picture alliance

Statistics
S. 84: PB
S. 85: PB

Measurement Uncertainties
S. 86: JP
S. 87: JP

Simulations in Experimental Physics
S. 89: top left: PB
S. 89: top right: CERN
S. 89: bottom: from J. Allison et al. 2016, Nuclear Instruments and Methods in Physics Research Section A, doi:10.1016/j.nima.2016.06.125, CC BY 4.0

4 Basic Theoretical Concepts

Chapter introduction page
S. 91: models of atoms: JP; water waves in the background: Michael Kortmann (Physical Institute, University of Bonn); supercomputer: Jülich Research Center; curves: HD, gluons (top) and Feynman diagrams: HD, CH and KR; composition by KR

Special Theory of Relativity in Particle Physics
S. 92: HD
S. 93: top: HD
S. 93: bottom: KR, background image from Joshua Woroniecki at Pixabay

Fermions and Bosons
S. 94: Joseline Heuer
S. 95: JP

External and Internal Symmetries
S. 96: Michael Kortmann
S. 97: Michael Kortmann

Gauge Symmetries
S. 98: HD
S. 99: HD

Noether's Theorem
S. 100: Chris Yarzab, Warner Bros Studios, flickr.com, CC BY 2.0
S. 101: top: Michael Kortmann
S. 101: bottom: Ella Dreiner

Feynman Diagrams II
S. 102: HD
S. 103: HD

Quantum Field Theory
S. 104: HD
S. 105: HD

Loops
S. 106: CH, KR
S. 107: CH, KR

Renormalization
S. 108: CH, KR
S. 109: CH

Accurate Calculation of the Unsolvable
S. 110: CH, KR
S. 111: CH, KR

Simulations in Theoretical Physics
S. 112: Jülich Research Center

5 Accelerators and Detectors

Chapter introduction page
S. 115: particle traces top left: CERN, ALICE experiment; sphere top right: DEAP Collaboration; bottom left: CERN, CMS experiment; bottom right: CERN; center: JP; composition by KR

The LHC Accelerator Complex at CERN
S. 116: left: CERN
S. 116: right: JP
S. 117: top: CERN, ALICE experiment
S. 117: bottom left: CERN
S. 117: bottom right: CERN, JP

Beam Optics
S. 118: JP
S. 119: top and bottom left: JP
S. 119: bottom right: JP, Jülich Research Center

Plasma Acceleration
S. 121: top: Anke Pretz
S. 121: bottom: Ferran Pousa, DESY/University of Hamburg, doi.org:10.1140/epjst/e2020-
S. 121: 000127-8

Accelerators in Medicine
S. 122: JP
S. 123: top: JP
S. 123: bottom left: Siemens Healthineers
S. 123: bottom right: Scanderbeg Sauer Photography

X-Rays
S. 124: JP
S. 125: top: image by W. Röntgen
S. 125: bottom right: Eckhard Hofmann, Ruhr-University Bochum

Synchrotron Radiation
S. 126: JP
S. 127: left: mmac72/Getty Images/iStock
S. 127: right: Reprinted with permission from Joris Dik, Koen Janssens, Geert Van Der Snickt, et al, Analytical Chemistry, American Chemical Society. Copyright 2008 American Chemical Society

Typical Detectors at Accelerators
S. 128: CERN, CMS experiment
S. 129: BGO-OD

The ATLAS Experiment at the LHC
S. 130: CERN, ATLAS experiment
S. 131: CERN, ATLAS experiment

Large Detectors Deep in a Mountain
S. 132: Borexino Collaboration
S. 133: left: Reidar Hahn, Fermilab
S. 133: right: KM3NeT

The IceCube Detector
S. 134: top (IceCube lab): Sven Lidstrom, IceCube/NSF
S. 134: left (DOM single): Robert Schwarz, NSF
S. 134: right (string in ice): Mark Krasberg, IceCube/NSF
S. 135: IceCube/NSF

Cryogenic Detectors
S. 136: SRON Netherlands Institute for Space Research
S. 137: left: Yury Suvorov INFN Naples
S. 137: right: JJ

Liquified Noble Gases as Detectors
S. 138: XENON Collaboration/LNGS-INFN
S. 139: DEAP Collaboration

Particle Traps
S. 140: JP
S. 141: left: JP
S. 141: right: Jan Pavelka/Wikimedia Commons, CC BY 4.0

A Penning Trap at CERN
S. 142: Ulmer, S., Smorra, C., Mooser, A. et al. High-precision comparison of the antiproton-to-proton charge-to-mass ratio. Nature 524, 196–199 (2015). https://doi.org/10.1038/nature14861

Positron Emission Tomography
S. 144: Fabian Schmidt, University Hospital Tübingen & Jens Maus
S. 145: left: Fabian Schmidt, University Hospital Tübingen
S. 145: right: University Hospital Tübingen

Magnetic Resonance Imaging
S. 146: top: Dominik Blum, University Hospital Tübingen
S. 146: bottom: JJ
S. 147: top left: University Hospital Tübingen
S. 147: bottom right: Dominik Blum, University Hospital Tübingen

6 Recording and Analyzing Data

Chapter introduction page
S. 149: supercomputer: CERN; particle traces on the left: CERN, ATLAS experiment; Earth: CERN, WLCG; curve: PB, KR; event display (bottom right): CMS, CERN; composition by KR

Trigger and Data Processing
S. 150: PB
S. 151: PB

Data Analysis
S. 152: PB
S. 153: CERN, ATLAS experiment

From Event to Result
S. 154: CERN, ATLAS experiment
S. 155: top: PB
S. 155: bottom: CERN, ATLAS experiment

Identifying the Particle Type via the Mass
S. 156: up: PB
S. 156: bottom: CERN, ATLAS experiment
S. 157: top: PB
S. 157: bottom: G. Aad et al. 2012, Physical Review D, doi:10.1103/PhysRevD.85.012001

Determining Measurement Uncertainties
S. 158: PB
S. 159: PB

Drawing Conclusions From Statistics
S. 161: PB

Publishing a Non-Observation
S. 162: PB
S. 163: PB

The Discovery of a Particle
S. 164: CERN, ATLAS experiment
S. 165: CERN, ATLAS experiment

What Is a "Blind Analysis"?
S. 167: PB

How to Fix False Discoveries
S. 168: B. McKinnon et al. (CLAS Collaboration) 2006, Physical Review Letters, doi:10.1103/PhysRevLett.96.212001
S. 169: CERN, ATLAS experiment

Neural Networks and Machine Learning
S. 170: PB, FB
S. 171: PB, FB

The Invention of the World Wide Web
S. 172: CERN
S. 173: top: Image provided by arXiv and used with permission
S. 173: bottom: CERN, WLCG

7 The Electroweak Standard Model of Particle Physics

Chapter introduction page
S. 175: lightnings: HD and Michael Kortmann; background top left: HD, KR; background top right (supernova remnant): X-ray: NASA/CXC/Caltech/S. Kulkarni et al.; optical: NASA/STScI/UIUC/Y. H. Chu & R. Williams et al.; IR: NASA/JPL-Caltech/R. Gehrz et al.; bottom left (potential): Martin Schürmann; particle overview: KR, after a figure from DESY; composition by KR

Electromagnetism
S. 176: HD and Michael Kortmann
S. 177: HD

The Weak Interaction
S. 178: HD
S. 179: HD

Up, Down, Left, Right
S. 180: HD
S. 181: Ella Dreiner

The Invisible Width of the *Z* Boson
S. 182: bottom: HD
S. 182: top and next page: The ALEPH Collaboration, the DELPHI Collaboration, the L3 Collaboration, the OPAL Collaboration, the SLD Collaboration, the LEP Electroweak Working Group, the SLD electroweak, heavy flavour groups 2005, Physics Reports, doi:10.48550/arXiv.hep-ex/0509008; adjusted by KR;

Spontaneous Symmetry Breaking I
S. 184: Martin Schürmann
S. 185: HD, KR

Spontaneous Symmetry Breaking II
S. 186: HD
S. 187: HD

Spontaneous Symmetry Breaking III
S. 188: HD
S. 189: HD

Discrete Symmetries: Violating *P* and *CP*
S. 190: left: HD
S. 190: right: HD and Ella Dreiner
S. 191: HD

The Weak Decay of Pions
S. 192: PB
S. 193: PB, FB

The Path to Six Quarks in Three Families
S. 194: PB, FB
S. 195: PB, FB

Neutrinos
S. 196: Peter Grabmayr, KR
S. 197: top: FB
S. 197: bottom: Kamioka Observatory, ICRR (Institute for Cosmic Ray Research), The University of Tokyo

Neutrino Oscillations
S. 198: KR
S. 199: Anke Pretz

8 The Peculiar Strong Interaction

Chapter introduction page
S. 201: grid in the background: CH; atomic nucleus: KR; atomic nucleus right: Joseline Heuer; big proton with quarks and gluons: JP; diagram bottom right: Gunnar Bali; tetra quark: Sören Lange, KR; composition by KR

The Quark Model Revisited
S. 202: CH, KR
S. 203: CH, KR

Quantum Chromodynamics
S. 204: from B. Bahr, J. Resag, K. Riebe: Faszinierende Physik, Springer Verlag 2019, translated
S. 205: CH

Confinement
S. 206: CH
S. 207: Gunnar Bali

The Search for Free Quarks
S. 208: Mc Cusker et al. 1969, Physical Review Letters, doi:10.1103/PhysRevLett.23.658, American Physical Society
S. 209: JP

The Running Coupling
S. 210: CH, KR
S. 211: P. A. Zyla et al. (Particle Data Group), Prog. Theor. Exp. Phys., 2020, 083C01 (2020) and 2021 update, supplemented by CH

The Structure of the Proton
S. 212: JP
S. 213: JP

Quark Distributions Inside the Proton
S. 214: JP
S. 215: left: JP
S. 215: right: German Electron Synchrotron DESY, Hamburg

Seeing Quarks and Gluons?
S. 216: CH
S. 217: top: JADE experiment, DESY
S. 217: bottom: KR

Quark-Gluon Plasma
S. 218: left: CERN
S. 218: right: JP
S. 219: top: CERN, CMS experiment
S. 219: bottom: Anke Pretz

Where Does Our Mass Come From?
S. 220: Joseline Heuer

Lattice QCD
S. 222: CH
S. 223: Christian Hoelbling, for data see S. Durr et al. 2008, Science, doi:10.1126/science.1163233

The Quark Mass Effect
S. 224: CH, KR
S. 225: CH, KR

From Nucleus to Nuclei
S. 226: CH
S. 227: CH

The Karlsruhe Nuclide Chart
S. 228: spread: Z. Sóti, J. Magill, R. Dreher 2019, EPJ Nuclear Sci. Technol., doi:10.1051/epjn/2019004

Exotic States
S. 230: Sören Lange, KR
S. 231: R. Aaij et al. (LHCb Collaboration) 2019, Physical Review Letters, doi:10.1103/PhysRevLett.122.222001

Radiocarbon Dating With ^{14}C
S. 232: Anke Pretz
S. 233: left: JP
S. 233: right: LUHUANFENG/Getty Images/iStock

9 Incredible Success and Yet Possible Failure of the Standard Model

Chapter introduction page
S. 235: plot and event display to the left: CERN, ATLAS experiment; bubble chamber (bottom): CERN; cone in the center: PB, particle paths below: CERN; composition by KR

The Discovery of Neutral Current Interactions
S. 236: PB
S. 237: CERN

The Accidental Discovery of the *J/ψ*
S. 239: top: PB
S. 239: left: J. J. Aubert et al. 1974, Physical Review Letters, doi:10.1103/PhysRevLett.33.1404
S. 239: right: J.-E. Augustin et al. 1974, Physical Review Letters, doi:10.1103/PhysRevLett.33.1406

The Discovery of the *W* and *Z* Bosons
S. 240: PB
S. 241: CERN

Discovery and Properties of the Top Quark
S. 242: PB
S. 243: top: T. Aaltonen et al. 2012 (CDF Collaboration), Physical Review Letters, doi: 10.1103/PhysRevLett.109.152003
S. 243: right: CERN

Why the Standard Model?
S. 244: KR, after: G. Giacomelli & R. Giacomelli 2005, doi:10.1142/9789812701893_0016
S. 245: top: PB
S. 245: bottom right: KR, after ALEPH Collaboration et al. 2005, Physics Report, doi:10.1016/j.physrep.2005.12.006

The Discovery of the Higgs Boson
S. 247: top left and right: PB
S. 247: bottom left: ATLAS Collaboration 2012, Phys. Lett. B, doi:10.48550/arXiv.1207.7214
S. 247: bottom right: CMS Collaboration 2013, JHEP, doi:10.48550/arXiv.1303.4571

Is It Really the Higgs Boson?
S. 248: CERN, ATLAS experiment
S. 249: top: PB
S. 249: bottom: CERN, CMS experiment

Precision Experiments
S. 250: top: Jacob Bourjaily/Wikimedia Commons
S. 250: bottom: JP
S. 251: JP

The Muon
S. 252: Reider Hahn, Fermilab
S. 253: top: JP
S. 253: bottom left: Physics Journal June2021
S. 253: bottom right:JP

Verifiying Neutrino Oscillations
S. 254: RCNS (Research Center for Neutrino Science), Tohoku University; text adjusted by JJ
S. 255: top: CERN, adapted by JJ, KR
S. 255: bottom: Tom Caroll, adjusted by KR

Measurement of the Neutrino Masses
S. 256: top: KR
S. 256: bottom: JJ, KR, after an image by KATRIN Collaboration
S. 257: top: JJ, map: NASA, Visible Earth, Blue Marble Collection
S. 257: bottom: Markus Breig/KIT

***CP* Violation in Beauty**
S. 258: PB, FB
S. 259: top: Y. Chao et al. (Belle Collaboration) 2004, Phys. Rev. Lett., doi:10.48550/arXiv.hep-ex/0408100
S. 259: bottom: PB, FB

Rare Decays of Beauty
S. 260: PB
S. 261: top: BaBar experiment
S. 261: bottom: PB

Shapeshifters Among Themselves
S. 262: FB
S. 263: top: DESY, ARGUS experiment
S. 263: bottom: CERN, LHCb experiment

Measuring the CKM Triangle
S. 264: FB
S. 265: left: Belle Collaboration, KEK
S. 265: right: CKMFitter Group

The Hurdles of Success
S. 266: top: CERN, ATLAS experiment
S. 266: bottom: CERN, CMS experiment
S. 267: CERN, ATLAS experiment

10 Limitations of the Standard Model

Chapter introduction page
S. 269: left: KR; balloons: HD and Michael Kortmann; bottom center: Joseline Heuer; center right: KR; supernova (bottom right): ESO, M. Kornmesser; composition by KR

Open Questions
S. 271: HD

Neutrino Masses
S. 272: HD
S. 273: Ella Dreiner

Coupling Unification
S. 274: HD
S. 275: Stephen Martin

GUT
S. 276: HERA experiment, adjusted by KR
S. 277: HD

Supersymmetry
S. 279: HD

Big Bang
S. 280: KR, after an image by CERN; spiral galaxy: NASA, ESA, and D. Jones (University of California—Santa Cruz)

Dark Matter
S. 282: JJ, galaxy: NASA, ESA and B.W.Holwerda
S. 283: KR, after an image by NASA/WMAP Science Team

Dark Energy
S. 284: HD and Michael Kortmann
S. 285: bottom left: KR
S. 285: top right: ESO, M. Kornmesser

Extra Dimensions
S. 286: Joseline Heuer
S. 287: Eric Adelberger

String Theory
S. 288: KR
S. 289: Ella Dreiner

The Strong *CP* Problem
S. 290: Andreas Wirzba, KR
S. 291: Anke Pretz

Did Our Life Emerge by Chance?
S. 292: Dean Lee, KR, landscape images by MenielDM, ThankYouFantasyPictures and JCK.H at Pixabay
S. 293: ESA/Hubble & NASA, V. Antoniou; Acknowledgement: Judy Schmidt

11 Searches for Physics Beyond the Standard Model

Chapter introduction page
S. 295: crab nebula (top left): NASA, ESA, NRAO/AUI/NSF and G. Dubner (University of Buenos Aires); mountains at bottom left: Laboratori Nazionali del Gran Sasso; center: nEDM collaboration at PSI, picture: M. Fischer/PSI; particle stream: ATLAS collaboration, CERN; accelerator at bottom right: DESY, Heiner Müller-Elsner; composition by KR

Searching for a New Experiment
S. 296: DUNE Experiment
S. 297: PB, KR

The Case for a Future Higgs Factory
S. 298: PB, KR
S. 299: right: PB, KR
S. 299: left: upper: CERN; lower: Jie Gao, Yuhui Li, Chenghui Yu, "China's designs for a future circular collider", CERN Courier, March 2024, p. 39

Is the Neutrino Its Own Antiparticle?
S. 300: GERDA Collaboration
S. 301: left: CUORE Collaboration, Yury Suvorov INFN Naples
S. 301: right: Kai Freund

Searching for New Particles at the LHC I
S. 302: CERN, ATLAS experiment
S. 303: CERN, ATLAS experiment

Searching for New Particles at the LHC II
S. 304: CERN, CMS experiment
S. 305: CERN, CMS experiment

Additional Spatial Dimensions at the LHC
S. 306: PB
S. 307: CERN, ATLAS experiment

The Mystery of Dark Matter
S. 308: Laboratori Nazionali del Gran Sasso
S. 309: bottom left: Corrieri-De Perio/LNGS-INFN
S. 309: top right: Dark Side collaboration, Yury Suvorov INFN Naples

Direct Search for Dark Matter
S. 310: top: DAMA/LIBRA collaboration
S. 310: bottom: CRESST collaboration
S. 311: Luigi Di Carlo, XENON collaboration

FASER
S. 312: Tobias Blesgen
S. 313: FASER Collaboration

Matter and Antimatter
S. 314: JP
S. 315: top: Anke Pretz
S. 315: bottom: JP

Particle Physics in Space
S. 316: NASA
S. 317: left: S. Schael/Wikimedia Commons
S. 317: right: M. Aguilar et al. (AMS Collaboration) 2019, Phys. Rev. Lett., doi:10.1103/PhysRevLett.122.041102

The Search for Electric Dipole Moments
S. 318: nEDM collaboration at PSI, picture: M. Fischer/PSI
S. 319: JP

Proton Decay
S. 320: HD
S. 321: Kamioka Observatory, ICRR (Institute for Cosmic Ray Research), The University of Tokyo

The Search for Axions
S. 322: Ciaran O'Hare, doi:10.5281/zenodo.3932430, https://cajohare.github.io/AxionLimits
S. 323: top: FB, KR
S. 323: bottom left: DESY, Heiner Müller-Elsner
S. 323: bottom right: CERN

Supernovae as Particle Physics Laboratories
S. 324: Joseline Heuer
S. 325: left: ESA/Hubble, NASA
S. 325: right: Hanhart et al. 2001, Phys.Lett.B, doi:10.1016/S0370-2693(01)00544-5

12 Connecting the Smallest to the Largest Scales in the Universe

Chapter introduction page
S. 327: antenna galaxies (top left): NASA, ESA and the Hubble Heritage Team (STScI/AURA)-ESA/Hubble Collaboration; spiral galaxy: ESA/Hubble & NASA, V. Antoniou; Acknowledgement: Judy Schmidt; Sun: ESA & NASA/Solar Orbiter/EUI team; Data processing: E. Kraaikamp (ROB) ; Telescope: H.E.S.S. Collaboration, Christian Föhr; bottom right: US Department of Energy; cosmic rays and nuclear fusion (center): KR; composition by KR

Baryogenesis
S. 328: Ella Dreiner

Nucleosynthesis After the Big Bang

S. 330: FB, KR
S. 331: left: JJ
S. 331: right: NASA, ESA and the Hubble Heritage Team (STScI/AURA)-ESA/Hubble Collaboration

Cosmic Microwave Background

S. 332: Nokia Corporation and AT&T Archives
S. 333: ESA and the Planck Collaboration

The Largest and the Smallest

S. 334: S.Agarwal, H.A.Feldman 2011, Monthly Notices of the Royal Astronomical Society, doi:10.1111/j.1365-2966.2010.17546.x
S. 335: Sloan Digital Sky Survey

Cosmic Rays

S. 336: H. Blümer, Auger Collaboration; adjusted by KR
S. 337: top: NASA/CXC/MIT/UMass Amherst/M.D. Stage et al.
S. 337: bottom: NASA

Ultra-High-Energy Cosmic Rays

S. 338: Auger Collaboration
S. 339: Auger Collaboration, adjusted by KR

Multimessenger Astronomy

S. 340: EGO/Virgo/Nicola Baldocchi
S. 341: H.E.S.S. Kollaboration, Christian Föhr

The Birthplace of Nuclei

S. 342: Joseline Heuer
S. 343: top right: NASA/CXC/M.Weiss
S. 343: bottom left: Joseline Heuer

Neutrinos From the Sun

S. 344: top right: ESA & NASA/Solar Orbiter/EUI team; Data processing: E. Kraaikamp (ROB)
S. 344: bottom left: US Department of Energy
S. 345: top left: Joseline Heuer
S. 345: right: Lawrence Berkeley Lab, SNO Collaboration
S. 345: bottom left: RCNS (Research Center for Neutrino Science), Tohoku University

Evolution of the Universe

S. 346: NASA/WMAP Science Team
S. 347: JJ
S. 348: top: CERN, WLCG
S. 348: center: CERN, CMS experiment
S. 348: bottom: ESA/Hubble & NASA, V. Antoniou; Acknowledgement: Judy Schmidt

Is it all worth it?

S. 349: Joseline Heuer
S. 350: NASA, ESA, CSA, and STScI

Acknowledgements

S. 352: KR, with images from Ella Dreiner, CERN, CMS experiment, HD and Michael Kortmann, Joseline Heuer, NASA/CXC/M.Weiss, Anke Pretz

Properties of Elementary Particles

S. 353: data from: https://pdg.lbl.gov/2023/html/authors_2023.html; image: KR

The Team of Authors

S. 384: Philip Bechtle: Philipp König
S. 384: Florian Bernlochner: FB
S. 384: Herbi Dreiner: Beatrice Mack—foto-box.org
S. 385: Christoph Hanhart: Bastian Kubis
S. 385: Josef Jochum: C. Strandhagen, Uni Tübingen
S. 385: Jörg Pretz: JP
S. 385: Kristin Riebe: AIP

Index

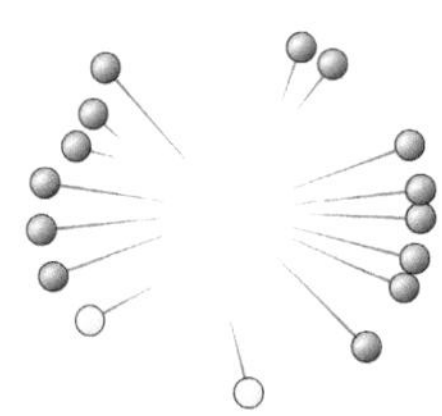

The Team of Authors

Philip Bechtle

Philip Bechtle is an experimental particle physicist and is teaching at the University of Bonn in Germany. His work comprises the search for new phenomena and precision measurements of the Standard Model, including the Higgs boson, at collider experiments at SLAC at Stanford, USA, CERN in Geneva, Switzerland, and DESY in Hamburg, Germany. Since 2007 he is a member of the ATLAS collaboration at the Large Hadron Collider. He regularly forages into other areas of science—be it renewable energies or epidemiology—in an attempt to transfer techniques learned in basic science into other fields.

Florian Bernlochner is a Professor of Experimental Particle Physics at the University of Bonn, Germany. He studies the properties of heavy quarks, neutrinos, and the Higgs boson to search for new unknown particles and phenomena. He does this with the FASER and ATLAS experiments at the Large Hadron Collider and has been elected as spokesperson of the Belle II experiment starting from mid 2025.

Florian Bernlochner

Herbi Dreiner studied physics at the University of Bonn, Germany, and the University of Wisconsin-Madison, USA, where he completed his Ph. D. He has worked at DESY in Hamburg, Germany, the University of Oxford, UK, ETH Zurich, Switzerland, and the Rutherford Appleton Laboratory, UK. Since 2000, he has been a professor at the University of Bonn, where he conducts research in theoretical physics, focusing mainly on physics beyond the Standard Model, as well as on collider physics, astrophysics, and cosmology. In 2001, he co-founded the Bonn Physics Show with Michael Kortmann, for which they jointly received the 2009 High Energy Physics Outreach Prize from the European Physical Society (EPS). Herbert Dreiner is a member of the renowned Particle Data Group (PDG).

Herbi Dreiner

Christoph Hanhart

Christoph Hanhart studied physics in Bonn, Germany, and Sydney, Australia. Today as a theoretical physicist he works on phenomena of the strong interaction at the boundary between nuclear and particle physics at the Jülich Research Centre, Germany, and at the University of Bonn, where he also holds a professorship. He is a member of the prestigious Particle Data Group (PDG) and received the Teaching Award of the University of Bonn in 2018.

Jörg Pretz

Jörg Pretz studied physics in Mainz, Germany, and Marseille, France. After receiving his doctorate in Mainz, he worked at Yale University, USA, at CERN in Geneva, Switzerland, and at the University of Bonn, Germany. Since 2012, he has been a professor of experimental physics at RWTH Aachen University, Germany and senior scientist at Jülich Research Center, Germany. His research areas are the structure of hadrons and precision experiments at storage rings.

Josef Jochum studied physics in Tübingen, Germany and Munich, Germany. After completing his doctorate at the Technical University of Munich, Germany, he worked in Berkeley, USA, on the development of superconducting detectors for the search for dark matter particles. He has been Professor of Experimental Particle Physics at the Eberhard Karls University of Tübingen since 2004. In addition to investigating the nature of dark matter, he also conducts experiments to determine the properties of neutrinos.

Josef Jochum

Kristin Riebe studied physics in Heidelberg, Germany, and completed her doctorate at the University of Potsdam, Germany. She has been working at the Leibniz Institute for Astrophysics Potsdam since her dissertation in the field of cosmology. After working on data management and publishing astronomical data in the Virtual Observatory, she now devotes herself to the development of websites and public outreach activities. She has made several contributions to successful book projects as a graphic designer and co-author.

Kristin Riebe

9783662720431